Handbook of Gemmology

Handbook of Gemmology

Edited by Elisha Price

SYRAWOOD
PUBLISHING HOUSE

New York

Published by Syrawood Publishing House,
750 Third Avenue, 9th Floor,
New York, NY 10017, USA
www.syrawoodpublishinghouse.com

Handbook of Gemmology
Edited by Elisha Price

International Standard Book Number: 978-1-64740-405-5 (Hardback)

Trademark Notice: Registered trademark of products or corporate names are used only for explanation and identification without intent to infringe.

Cataloging-in-publication Data

Handbook of gemmology / edited by Elisha Price.
    p. cm.
Includes bibliographical references and index.
ISBN 978-1-64740-405-5
1. Gemology. 2. Precious stones. 3. Mineralogy. I. Price, Elisha.
TS725 .H36 2023
739.22--dc23

TABLE OF CONTENTS

Permissions

List of Contributors

Index

PREFACE

Every book is initially just a concept; it takes months of research and hard work to give it the final shape in which the readers receive it. In its early stages, this book also went through rigorous reviewing. The notable contributions made by experts from across the globe were first molded into patterned chapters and then arranged in a sensibly sequential manner to bring out the best results.

A gemstone is a precious or semi-precious stone, which is used for making jewelry and is used as an ornament due to its antiquity. Gemmology is the branch of mineralogy, which deals with the study of natural and artificial gemstones. The Earth's crust is composed of three types of rocks, including metamorphic, igneous and sedimentary. Each type of rock originates under various conditions resulting in a diverse range of gemstones. They can be categorized as precious gemstones and semi-precious gemstones. Precious gemstones include sapphires, diamonds, emeralds and rubies. All other gemstones are classified as semi-precious gemstones. Geological conditions like redox conditions, temperature and chemistry determine the amount of trace chemical components. The difference between a gemstone and mineral specimens can be determined through such trace chemical components. The color of the gemstone is determined by these components and can provide a basis for identifying the gemstone's origin. This book provides comprehensive insights on gemmology. Researchers and students in this field will be assisted by it.

It has been my immense pleasure to be a part of this project and to contribute my years of learning in such a meaningful form. I would like to take this opportunity to thank all the people who have been associated with the completion of this book at any step.

Editor

An X-Ray Absorption Near-Edge Structure (XANES) Study on the Oxidation State of Chromophores in Natural Kunzite Samples from Nuristan, Afghanistan

Habib Ur Rehman [1,2], Gerhard Martens [3], Ying Lai Tsai [4], Chawalit Chankhantha [1], Pinit Kidkhunthod [5] and Andy H. Shen [1,*]

[1] Gemmological Institute, China University of Geosciences, Wuhan 430074, China; habib.rehman@nwfpuet.edu.pk (H.U.R.); charles.valydth@gmail.com (C.C.)

[2] Gems & Jewellery Centre of Excellence, University of Engineering & Technology Peshawar, Peshawar 25120, Pakistan

[3] Private Scientific Consultant, D-24558 Henstedt-Ulzburg, Germany; GerhardMartens@t-online.de

[4] Department of Jewelry Technology, Dahan Institute of Technology, Hualian, Taiwan; laser@ms01.dahan.edu.tw

[5] Synchrotron Light Research Institute, 111 University Avenue, Muang, Nakhon Ratchasima 30000, Thailand; pinit@slri.or.th

[*] Correspondence: shenxt@cug.edu.cn

Abstract: Kunzite, the pink variety of spodumene is famous and desirable among gemstone lovers. Due to its tenebrescent properties, kunzite always remains a hot research candidate among physicists and mineralogists. The present work is continuing the effort towards value addition to kunzite by enhancing its color using different treatments. Before color enhancement, it is essential to identify the chromophores and their oxidation states. In this paper, the authors investigated the main impurities in natural kunzite from the Nuristan area in Afghanistan and their valence states. Some impurities in the $LiAlSi_2O_6$ spodumene structure were identified and quantified by using sensitive techniques, including Laser Ablation Inductively Coupled Plasma Mass Spectrometry (LA-ICP-MS), UV−VIS and X-ray absorption near-edge structure (XANES). LA-ICP-MS indicated many trace elements as impurities in kunzite, among which Fe and Mn are the main elements responsible for coloration. The oxidation states of these two transition elements were determined by the XANES technique. The study reveals that Mn is present in both Mn^{2+} and Mn^{3+} oxidation states, while Fe is present only in Fe^{3+} oxidation state.

Keywords: kunzite; oxidation state; XANES; LA-ICP-MS; UV−VIS

1. Introduction

The silicate materials are of special interest for science and technology due to their wide applications in optical and semi-conductor devices [1]. Spodumene ($LiAlSi_2O_6$) is widely utilized in ceramics and glass manufacturing and is an important source of Li. It is a member of the pyroxene mineral group and normally the stable low-temperature type of spodumene (α-spodumene) crystallizes in a monoclinic crystal system with space group $C2/c$. According to Figure 1 it has two inequivalent metal cation sites $M1$ and $M2$ in the same crystallographic plane [2]. Aluminum occupies the slightly distorted $M1$ site, which is octahedrally coordinated with an average metal oxygen distance of 1.92Å; while the $M2$ site is occupied by Li, which is also six-fold coordinated with an average metal-oxygen distance of 2.23Å [3,4]. There is only one type of tetrahedral site, which is completely occupied by silicon in pyroxene and is not involved in producing colors. $O_{(1)}$, $O_{(2)}$ and $O_{(3)}$ are three crystallographically non-equivalent oxygen atoms in the spodumene crystal structure. The oxygen atom "$O_{(3)}$" is referred to as a bridging oxygen and is bonded to two silicon atoms, whereas $O_{(1)}$ and $O_{(2)}$ are both non-bridging oxygen atoms [5].

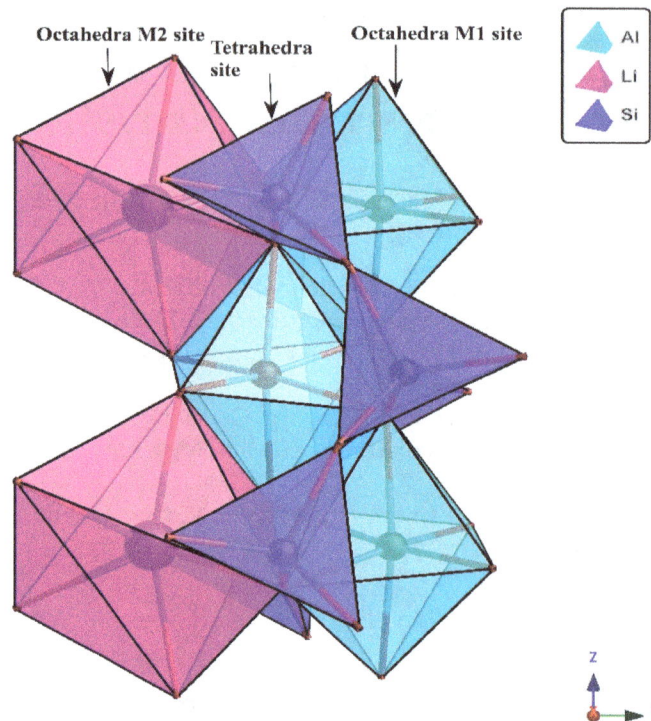

Figure 1. This polyhedral model of a spodumene crystal, using the crystal refinement data of [5] was drawn through CrystalMaker X software, with the following unit cell parameters (a: 9.46, b: 8.412 and c: 5.224 Å, α: 90, β: 110 and γ: 90 degree). There is a total of 40 atoms in one complete unit cell. The basic building blocks: AlO_6 ($M1$) and LiO_6 ($M2$) are at octahedral, while SiO_4 are at tetrahedral positions as labeled in the picture. The crystallographic orientation is shown at the bottom right corner, projected down the x-axis.

In spodumene, both atoms (Li and Al) can be replaced by transition metal elements, such as Mn, Fe, Cr and others, as it belongs to the group of allochromatic minerals. According to the gemmological point of view, the minor and trace elements are mostly responsible for producing color in allochromatic gemstones. Furthermore, the gemstones and common minerals can be differentiated through minor and trace elements [6], and in particular, when V^{3+}, Cr^{3+}, Mn^{3+}, Fe^{2+}, Fe^{3+} and Mn^{2+} exist in high concentration, they can produce strong coloration in gemstones. The d-orbital electrons of these transition elements are responsible for coloration [7]. The distribution of trace elements in the gemstones also works as a "fingerprint" in finding the locality of their mining origin [7]. In a gemstone, the existence of trace and light lithophile elements can also be used to distinguish between natural and synthetic samples, as well as to give evidence about its synthesis methodology [8–10].

Spodumene possesses three varieties (kunzite, triphane and hiddenite) with brilliant colors and they are used as semi-precious gemstones. The two well-known varieties kunzite and hiddenite were named after the esteemed American geologists and collectors George F. Kunz and William E. Hidden, respectively [11]. Kunzite is a major source of Li, which is used in different materials, such as mobile phones and automotive batteries, in lithium carbonate production, ceramics and as flux agent [12,13].

Due to its pink to lilac hue kunzite is more acceptable for use in jewelry. The lilac color shows up when the Mn/Fe concentration ratio is larger than 1 [14]. Spodumene is a comparatively new mineral in scientific studies, being discovered in the last two centuries while its gem quality has only been known for about 100 years [15]. Spodumene in the green and dark pink color (kunzite) is acceptable to people for use in jewelry, but if the color is light pink then people do not like it, just like pink topaz. In fact, the color of gemstones is very important in the jewelry industry and one important factor is that gemstones are valued by their color. The light pink colored kunzite is abundantly available on the Pakistan gemstone market and needs color enhancement for value addition and to enable it for use in jewelry. People perform color treatment locally without knowing any standard parameters and ways, resulting in a lot of kunzite going to waste every year, although different quantitative analyses and EPR were already used and the results were published regarding the oxidation state of kunzite. However, there has not yet been reported an X-ray absorption near edge structure (XANES) study for checking the chromophore oxidation states in kunzite. It is the purpose of this paper to evaluate the oxidation states of the color-causing chromophore in kunzite by using the XANES technique.

X-ray absorption spectroscopy using intense synchrotron radiation looks to be the method of choice for materials that are generally low in concentration and/or small in total sample size [16]. This technique is atomic specific and is even sensitive to low concentration (10 ppm or lower) as well as low sample mass (~µg with third-generation synchrotron sources). Chemical information such as valence, coordination geometry, bond distances, and ligand coordination numbers can be obtained with high accuracy [17]. As gemstones are precious, non-destructive techniques are usually preferred for their analysis. XANES is a non-destructive and element-selective technique and the samples need virtually no chemical parameters. Furthermore, this technique applies to both crystalline materials and also to amorphous phases [18]. Here we only focus on the energy position of the absorption edges from which the oxidation numbers can be derived. Evaluation of the extended X-ray absorption fine structure (EXAFS) showing up in a photon energy regime about 50 eV above the edge energy cannot be evaluated because of noise issues. Thus, the bond lengths and coordination numbers are not derived here.

The aim of this study is to develop methods and parameters in order to enhance the kunzite color for value addition but before doing any treatment it is necessary to confirm the chemistry and the oxidation states of the coloring elements in the samples used for the current research.

2. Samples Site, Materials and Analytical Methods

2.1. Samples Site

The sample material for this study was kunzite from the Nuristan District, Afghanistan. Literally hundreds of thousands of carats of fine kunzite have emerged from the Nuristan region, northwest of Kabul since active mining began there in the early 1970s [11]. The spodumene crystals from the Nuristan region are among the best variety of this mineral ever found [19].

The kunzite samples for the current research were collected from a mine owner (named Khalid Habib, main dealer of Afghanistan kunzite and Swat emeralds) in Namak Mandi Peshawar, Khyber Pakhtunkhwa (KPK), Pakistan.

2.2. Sample Preparation

Figure 2A,B and Figure 3 show kunzite samples before cutting and after cutting, the samples for the current research work were selected on color variation basis. The kunzite crystals were cut into 21 cut stones with different cutting styles (Figure 2B) and 6 wafers with dimensions 2 cm × 2 cm × 0.5 cm (length, width and thickness, respectively) and both surfaces were polished with diamond powder as shown in Figure 3.

Figure 2. (A) Photograph of the kunzite crystals before cutting, (B) photograph of the 21 cut kunzite crystals.

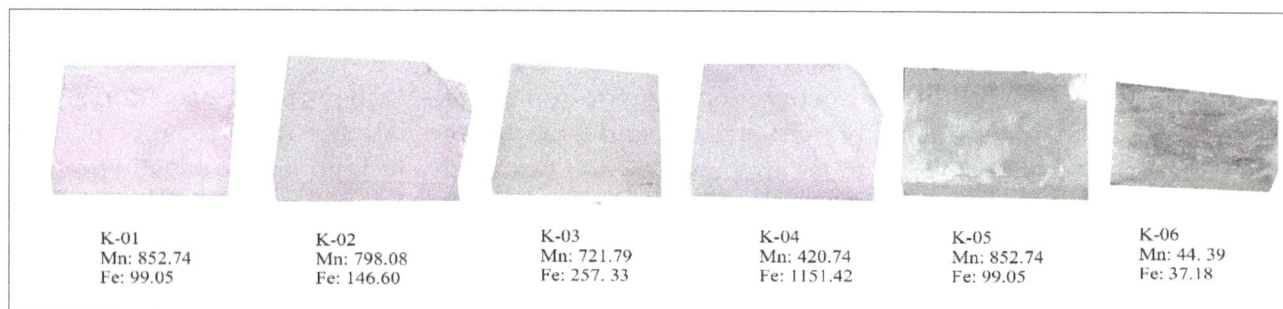

K-01	K-02	K-03	K-04	K-05	K-06
Mn: 852.74	Mn: 798.08	Mn: 721.79	Mn: 420.74	Mn: 852.74	Mn: 44. 39
Fe: 99.05	Fe: 146.60	Fe: 257. 33	Fe: 1151.42	Fe: 99.05	Fe: 37.18

Figure 3. Six kunzite samples which were cut into wafers with dimensions 2 cm × 2 cm × 0.5 cm, (length, width, and thickness, respectively). These samples were used for X-ray absorption near-edge structure (XANES) and UV–VIS spectra. These samples were selected according to the Mn and Fe concentration and each sample's value is written below the sample in ppm (average).

2.3. Analytical Methods

2.3.1. LA-ICP-MS

Quantitative chemical analyses of major and trace elements were performed with LA-ICP-MS at Wuhan Sample Solution Analytical Technology Co., Ltd., Wuhan, China. The GeolasPro laser ablation system(Coherent, Inc., Santa Clara, CA, USA) used in the tests was developed by COMPexPro102 (Coherent, Inc., Santa Clara, CA, USA) and the ArF 193nm Laser by MicroLas optical system (Coherent, Inc., Santa Clara, CA, USA) The ICP-MS system used in the 7900 ICP-MS was from Agilent Technologies, Inc., Santa Clara, CA, USA. The carrier gas was helium and argon was used as the makeup gas to adjust the sensitivity. The laser beam spot diameter is 44 μm having an intensity of 5.5 J/cm^2. Glass reference materials BHVO-2G, BCR-2G, BIR-1G and NIST 610 were used. Every two samples, analyses were followed by calibration of the instrument by SRM610, in order to correct the time-dependent drift of sensitivity and the mass discrimination.

2.3.2. X-ray Absorption Near-Edge Structure (XANES)

Pre-edge and XANES spectra at the Fe K-edge and Mn K-edge were collected at the Beam Line BL5.2 XAS, Synchrotron Light Research Institute (SLRI), Thailand. The storage ring operation conditions were 1.2 GeV electron energy and 100 mA. The beam size at the sample position was 20 mm (width) × 1 mm (height). A Ge (220) double-crystal monochromator with 2d spacing of 4.001 Å and a silicon drift detector were used for these experiments. The XANES data were collected in the fluorescence mode with the sample positioned 45° with respect to the beam. Moreover, the fluorescence spectra were collected by fixed energy-range windows which did not include both the elastic and the inelastic/Compton signals. The measured fluorescence intensity I_F is proportional to the X-ray absorption coefficient $\mu(E)$ and the incident intensity I_0 ($I_F \sim \mu(E)*I_0$), from which $\mu(E) \sim I_F/I_0$ is derived.

2.3.3. UV−Visible Spectroscopy

In order to correlate the observed color to the specific chromophore, precise measurements of the absorption feature through a known optical path in the UV−visible spectrometer were required. For UV−visible spectroscopy the same kunzite samples as for XANES were used. In addition, to check the absorption spectrum along the c-axis, a wafer from each sample was cut perpendicular to the c-axis and polished. UV−visible spectra were collected with a PerkinElmer Lambda650 UV/Vis spectrometer equipped with mercury and tungsten light sources and photomultiplier tube/PbS detector that were built into an integrating sphere. In order to do the precise and accurate quantitative analyses, a custom-made sample holder was used to conform to the precise positioning of the sample area in a 3 mm diameter window. The spectra were collected at the Gemmological Institute, China University of Geosciences, Wuhan.

3. Results and Discussion

3.1. Gemmological Observations

Spodumene belongs to the monoclinic crystal system and is usually found in prismatic form, highly flattened, and strongly striated along the c-axis {100} direction. Spodumene shows good {110} cleavage with the angle between {110} and {110} at about 87 degrees and has a parting on {100} and {010}. The fracture is from uneven to sub-conchoidal and is brittle having a directional hardness from 6.5 to 7.5 [3,20]. The refractive index was checked by using a standard refractometer and the values were noted, $n\alpha= 1.661$, $n\beta= 1.663$ and $n\gamma= 1.676$. Similarly, the pleochroism was checked in different directions by using a calcite dichroscope and showed colors from pink-green and pink to colorless (trichroic).

3.2. LA-ICP-MS

The quantitative chemical analyses of 27 kunzite samples were performed and the concentration average of major and trace elements are shown in Tables 1 and 2, respectively. The main chromophores for producing color are Mn and Fe. The lilac color in kunzite occurs when the Mn/Fe ratio is larger than 1 [14], while the Mn/Fe concentration ratios in the investigated samples were 1.96.

Table 1. Elemental concentration of major elements in wt.% in the kunzite samples.

Oxides	Minimum	Maximum	Average
SiO_2	64.026	64.960	64.051
Al_2O_3	27.513	28.216	27.974
Li_2O	7.528	7.739	7.621
TiO_2	0	5.667	0.0003
FeO	0	0.149	0.0362
MnO	0.003	0.117	0.0694
Na_2O	0.064	0.155	0.133
K_2O	0	0.035	0.0015
P_2O_5	0.035	0.061	0.055
CaO	0	4.331	0.3795

3.3. X-ray Absorption Spectroscopic Analysis

X-ray absorption spectroscopy (XAS) is the study of X-ray beams absorbed by an electron near and above the core-level binding energies of an atom. XAS is associated with the chemical and physical states of the absorbing atoms. Therefore, it is sensitive to the valence state, bond lengths and the coordination numbers of the neighboring atoms [21,22]. The study of a sample based on XAS provides significant evidence of the above properties on the atomic scale. Fundamentally, each XAS spectrum from an absorbing atom is divided into three characteristics regions named the pre-edge, X-ray absorption near edge structure (XANES) and the extended X-ray absorption fine structure

(EXAFS). The pre-edge region in the spectrum is associated with the transitions from core electrons to bound states, such as from 1s to nd, $(n + 1)$ s or $(n + 1)$ p orbitals of the K edge. The second part is the XANES region which gives information about the valence states of the absorbing atom, while the EXAFS analysis gives us information about the geometric structure, such as bond distances and coordination numbers of the neighboring atoms. The fine structure in the EXAFS region, is due to the interference of outgoing and incoming electron waves, and is thus dependent on the immediate environment surrounding the absorbing atom [23]. In the current study the XAS experiments were performed in order to evaluate the oxidation states of Mn and Fe in kunzite samples. The XANES spectra of Mn and Fe are shown below in Figures 4 and 5, while the EXAFS spectra were very weak, and indistinguishable from background noise due to the low concentration of Mn and Fe in the samples.

Table 2. Elemental concentration of the trace elements in ppm in the kunzite samples.

Elements (ppm)	Minimum	Maximum	Average
Sn	19.74	300.424	126.1673
Ga	50.542	86.637	73.6148
Ge	7.794	68.007	28.4366
B	8.035	84.441	23.2575
Zn	0	25.176	6.5206
Cd	0	2.590	0.7602
Be	0	2.050	0.6742
Pb	0	5.923	0.63
Cu	0	1.167	0.3265
Sc	0	0.859	0.2615
Rb	0	0.585	0.2319
Mo	0	0.656	0.155
Hf	0	0.463	0.15
Cr	0	1.498	0.1249
Cs	0	0.750	0.1

The elements detected during the analyses having a concentration less than 0.1 ppm are, V, Co, Sr, Y, Zr, Nb, Ag, Ba, La, Ce, Pr, Nd, Sm, Eu, Gd, Tb, Dy, Ho, Er, Tm, Yb, Lu, Ta, W, Hg, Tl, Bi< 0.1 ppm.

The above mentioned prepared samples (shown in Figure 3) were used for XANES analyses, the XANES spectra and the EXAFS interference functions were extracted from the measured absorption spectra using the Athena and Artemis software programs [24].

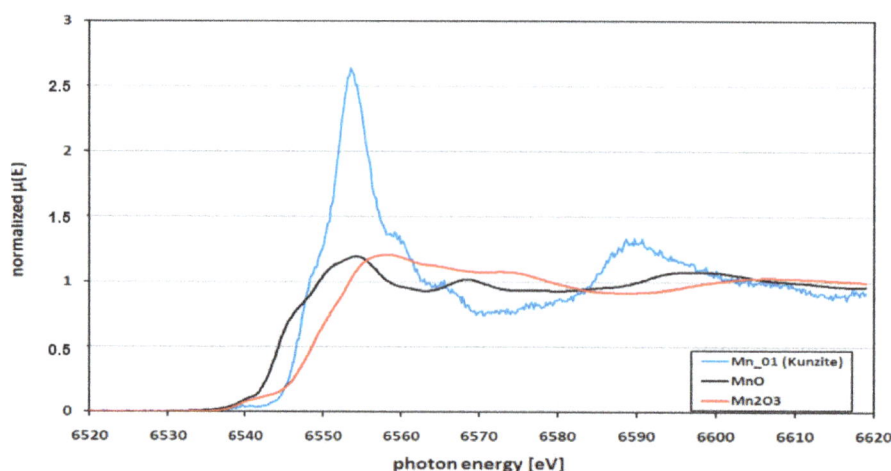

Figure 4. Mn K-edge XANES spectrum of kunzite from Nuristan, Afghanistan, (Mn-01, represents kunzite sample K-01) versus MnO and Mn_2O_3 standards. The edge position of the kunzite is just between the MnO and Mn_2O_3.

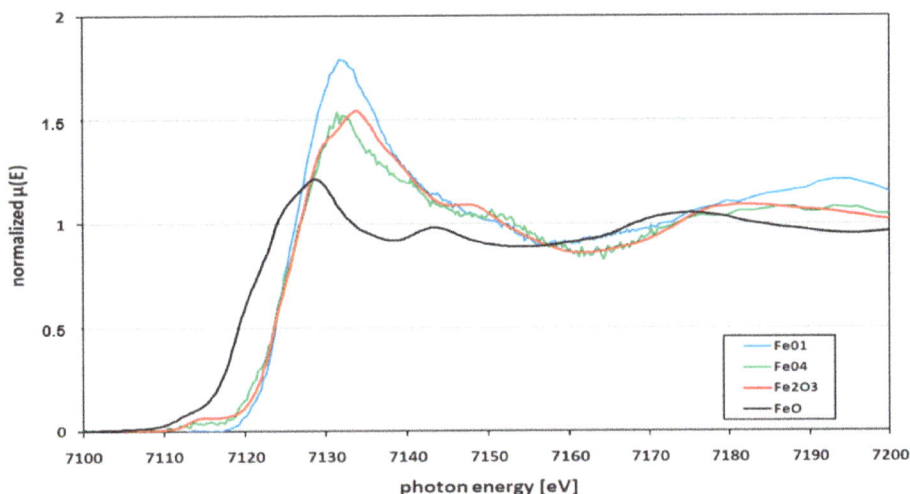

Figure 5. Fe K-edge XANES spectra of kunzite from Nuristan, Afghanistan, (Fe-01, Fe-04 represent kunzite samples) versus FeO and Fe_2O_3 standards.

3.3.1. Distribution of Mn

Figure 4 shows the Mn K-edge XANES spectra of the kunzite sample K-01 (spectrum: Mn-01) together with those of standard oxides MnO and Mn_2O_3. In total six kunzite samples were checked; all showed the same shape, very similar to the one in Figure 4. The edge position was defined as the first maximum of 1st derivative of the normalized XANES spectra. The edge of the present kunzite samples was just between those of the standard oxides, suggesting a mixed valence state of Mn (i.e., Mn^{2+} and Mn^{3+}). This showed that the investigated samples of kunzite contain Mn in Mn^{2+} and Mn^{3+} oxidation states. A second definition of the position of the K-edge was the use of the inflection point at the edge jump. Thus, the positions were alternatively evaluated by normalizing the obtained data for Mn K-edge of kunzite samples then taking the first derivative and by using the linear combination process to compare the obtained spectrum with a mixture of the standard MnO and Mn_2O_3 spectra.

3.3.2. Distribution of Fe

Figure 5 shows the Fe K-edge XANES spectra of kunzite sample K-01 and K-04 (spectra Fe-01 and Fe-04) as an example together with those of standard oxides of FeO and Fe_2O_3. In this case the same six samples were tested for Fe oxidation state. The edge positions of the analyzed kunzite samples were exactly matching with standard oxide Fe_2O_3, which clearly shows the Fe present solely in Fe^{3+} oxidation state.

3.4. UV—Visible Spectroscopy

Kunzite belongs to the monoclinic crystal system and is therefore trichroic, i.e., shows three colors when checked through the dichroscope. The pleochroism was checked in all optical directions by using a gemmological handheld calcite dichroscope. The pleochroism varied from green-pink and pink to colorless, the UV—visible spectra of all the samples were checked in these optical directions and also along the direction parallel to the *c*-axis. All the kunzite samples, except sample K-06 exhibited the absorbance peaks in the same wavelength regions, with little variation in peak intensity due to the variation in chromophore concentration.

The sample K-06 (white spodumene) due to the very low concentration of chromophores, did not show any significant absorbance in the visible region as depicted in Figure 6; there was only a little absorption at 376 nm and 480 nm. All the samples (K-01 to K-05) showed very weak absorption bands in the ultraviolet region at 376 nm and in the visible region between 400 to 500 nm (410, 440, 450 and 480 nm), the absorption is due to Fe^{3+} [25,26], in Figure 7 only one spectrum is represented for explanation. There was a strong absorption band at 530 nm and the absorbance was intense along c-axis as compared to another optical axes. The absorption in this region was due to Mn^{3+} ions, which was correlated to the original lilac color of kunzite [1,27].

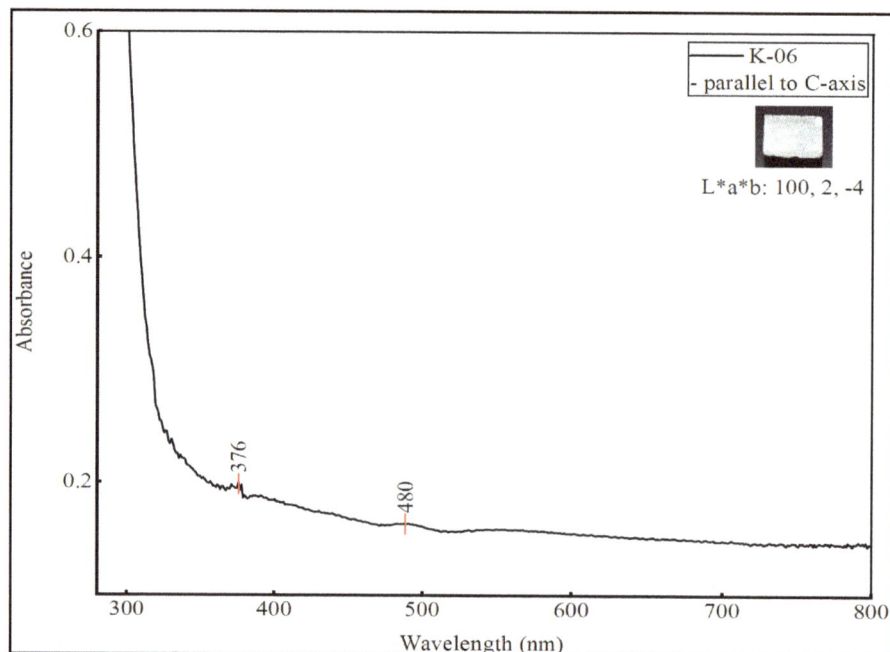

Figure 6. The graph shows the UV−visible spectrum of sample K-06, whose color is totally white and only shows a little absorbance due to Fe when measured along the C-axis. Color coordinates (CIE L*a*b) were taken by using uvwinlab\meth900 and D65 illuminant in transmission mode.

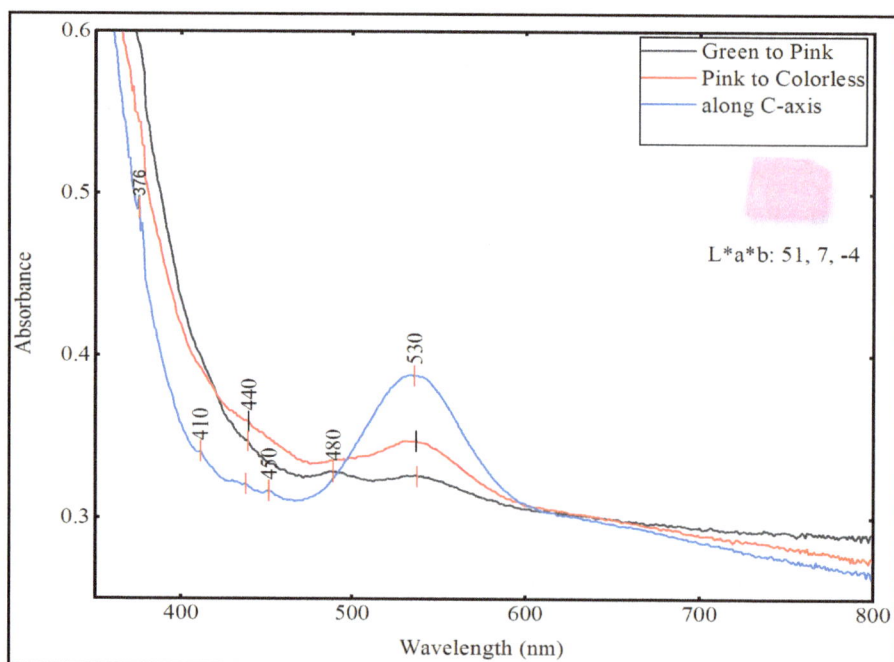

Figure 7. All kunzite samples (K-01 to K-05) show the absorbance peaks at the same wavelengths, with little variation in peak intensity, therefore only one spectrum is taken here for explanation. The spectra were taken in three different optical directions by using gemmological calcite dichroscope and they are marked as: 'green to pink', 'pink to colorless' and 'along C-axis'. Color coordinates (CIE L*a*b) were taken by using uvwinlab\meth900 and D65 illuminant in transmission mode.

3.5. Ionic Structure Consideration

The valence state and ionic size play an important part in their incorporation as best possible chromophores. Mn^{3+} and Fe^{3+} have the same valence states, i.e., isovalent and charge-balanced. Their ionic radii are very close to those of Al^{3+} in six-folded coordinated octahedral sites, and thus facilitate substitution into the octahedral site of Al^{3+}, i.e., the M1 site in kunzite [10,28]. Therefore, considering the ionic radius, isovalent Mn^{3+} is a preferred chromophore due to the close matching of both the ionic charge and the ionic size with Al^{3+}. On the other hand, Mn^{2+} has an appreciably greater ionic size than Al^{3+} and requires additional charge compensation to fit into the Al^{3+} octahedral site. The ionic radius of Mn^{2+} is close to Li and thus facilitates substitution into the M2 site in kunzite, while the tetrahedral site is occupied by Si [13,20]. However, in the case of Mn^{2+} d-d electronic transition is spin-forbidden and also no Jahn–Teller distortion is possible for this ion [29]. Due to the absence of Jahn–Teller distortion and spin-forbidden transition, Mn^{2+} produces near colorless or weak coloration as compared to Mn^{3+} [7,30]. Therefore, Mn^{2+} is not the best choice for pink color in kunzite.

Similarly, trivalent ferric ion Fe^{3+}, having the same electronic structure as Mn^{2+}, is also a weak chromophore and produces noticeable color only at higher concentrations. For example, to get an intense yellow color in sapphire, it requires at least 1000 ppm Fe^{3+}. This is an important consideration to understand the comparatively weak contribution of Fe^{3+} to coloration [28,30].

Both Mn^{2+} and Mn^{3+} can occur in six-folded octahedral coordination, and are the known cause of red or pink colors in a variety of minerals e.g., rhodonite, andalusite, grossular, morganite, tourmaline (pink and red) and kunzite [31]. Mn^{2+}, when coordinated to oxygen ligands (Mn^{2+}-O^{2-}) produces relatively weak absorption bands as compared to Mn^{3+}, due to weak oscillator strength or a low cross-section involving spin-forbidden transitions [30].

4. Conclusions

In the current study the quantitative analyses of chromophores responsible for coloration was carried out by LA-ICP-MS. The results show that the concentration of Mn is higher in kunzite as compared to Fe. The oxidation states of Mn and Fe were checked through XANES, which confirm the presence of Mn in both the Mn^{2+} and the Mn^{3+} oxidation states in kunzite while Fe is present in the Fe^{3+} state. The UV–visible spectra confirm that lilac-pink colored kunzite is due to a broad peak given by Mn^{3+} at 530 nm. It is also concluded that the higher the concentration of Mn^{3+} in kunzite, the deeper the pink color will be and it will be more valuable. The locality used in this study has both Mn^{2+} and Mn^{3+} in kunzite and has a light pink color. It is suggested that its color can be enhanced by changing the Mn from Mn^{2+} to Mn^{3+} by using some treatments. This change will make its color become deeper and will help in value addition to the kunzite.

Author Contributions: Formal analysis, H.U.R., C.C., P.K.; conceptualization, methodology, software, writing original draft, H.U.R; writing-review and editing, G.M., Y.L.T., C.C., A.H.S.; funding acquisition, resource, Y.L.T., A.H.S.; conceptualization, project administration, supervision, A.H.S. All authors have read and agreed to the published version of the manuscript.

Acknowledgments: This paper is GIC contribution CIGTWZ-2020008. We are also grateful to Siam Photon Laboratory, Synchrotron Light Research Institute (SLRI), Nakhon Ratchasima 30000, Thailand, for providing lab facilities free of cost. We are also thankful to the editorial staff of *Minerals* and the anonymous reviewers for their positive and constructive reviews which polished the article.

References

1. Souza, S.; Watanabe, S.; Lima, A.; Lalic, M. Thermoluminescent mechanism in lilac spodumene. *Acta Phys. Pol. Ser. A.* **2007**, *112*, 1001. [CrossRef]
2. d'Amorim, R.A.P.O.; de Vasconcelos, D.A.A.; de Barros, V.S.M.; Khoury, H.J.; Souza, S.O. Characterization of α-spodumene to OSL dosimetry. *Radiat. Phys. Chem.* **2014**, *95*, 141–144. [CrossRef]

3. Walker, G.; Jaer, A.E.; Sherlock, R.; Glynn, T.J.; Czaja, M.; Mazurak, Z. Luminescence spectroscopy of Cr [3+] and Mn [2+] in spodumene (LiAlSi$_2$O$_6$). *J. Lumin.* **1997**, *72–74*, 278–280. [CrossRef]

4. Isotani, S.; Watari, K.; Mizukami, A.; Bonventi, W.; Ito, A.S. UV optical absorption spectra analysis of spodumene crystals from Brazil. *Phys. B Condens. Matter* **2007**, *391*, 322–330. [CrossRef]

5. Cameron, M.; Sueno, S.; Prewitt, C.T.; Papike, J.J. High-temperature crystal chemistry of acmite, diopside, hedenbergite jadeite, spodumene and ureyite. *Am. Mineral. J. Earth Planet. Mater.* **1973**, *58*, 594–618.

6. Mattson, S.M.; Rossman, G.R. Identifying characteristics of charge transfer transitions in minerals. *Phys. Chem. Miner.* **1987**, *14*, 94–99. [CrossRef]

7. Rossman, G.R. The geochemistry of gems and its relevance to gemology: Different traces, different prices. *Elements* **2009**, *5*, 159–162. [CrossRef]

8. Schmetzer, K. Surface treatment of gemstones, especially topaz—An update of recent patent literature. *Gemmology* **2004**, *7*. [CrossRef]

9. Shigley, J.E.; McClure, S.F. Laboratory-treated gemstones. *Elements* **2009**, *5*, 175–178. [CrossRef]

10. Rossi, M.; Dell'Aglio, M.; De Giacomo, A.; Gaudiuso, R.; Senesi, G.S.; De Pascale, O.; Capitelli, F.; Nestola, F.; Ghiara, M.R. Multi-methodological investigation of kunzite, hiddenite, alexandrite, elbaite and topaz, based on laser-induced breakdown spectroscopy and conventional analytical techniques for supporting mineralogical characterization. *Phys. Chem. Miner.* **2014**, *41*, 127–140. [CrossRef]

11. Bowersox, G.W. A status report on gemstones from Afghanistan. *Gems Gemol.* **1985**, *21*, 192–204. [CrossRef]

12. Lagache, M.; Sebastian, A. Experimental study of Li-rich granitic pegmatites: Part II. Spodumene + albite + quartz equilibrium. *Am. Miner.* **1991**, *76*, 611–616.

13. Ogundare, F.; Alatishe, M.; Chithambo, M.; Costin, G. Thermoluminescence of kunzite: A study of kinetic processes and dosimetry characteristics. *Nucl. Instrum. Methods Phys. Res. Sect. B Beam Interact. Mater. At.* **2016**, *373*, 44–51. [CrossRef]

14. Ito, A.S.; Isotani, S. Heating effects on the optical absorption spectra of irradiated, natural spodumene. *Radiat. Eff.* **1991**, *116*, 307–314. [CrossRef]

15. Yonghua, D.; Lishi, M.; Ping, L.; Yong, C. First-principles calculations of electronic structures and optical, phononic, and thermodynamic properties of monoclinic α-spodumene. *Ceram. Int.* **2017**, *43*, 6312–6321. [CrossRef]

16. Koningsberger, D.C. *X-Ray Absorption: Principles, Applications, Techniques of EXAFS, SEXAFS, and XANES*; John Wiley and Sons: Hoboken, NJ, USA, 1988.

17. Ressler, T.; Wong, J.; Roos, J.; Smith, J.L. Quantitative speciation of Mn-bearing particulates emitted from autos burning MMT-added gasolines using XANES spectroscopy. *Environ. Sci. Technol.* **2000**, *34*, 950–958. [CrossRef]

18. Hayashi, H.; Abe, H. Gel-state dependencies of brown patterns of Mn–Fe-based prussian blue analogues studied by combined X-ray spectroscopies. *Bull. Chem. Soc. Jpn.* **2017**, *90*, 807–819. [CrossRef]

19. Bariand, P.; Poullen, J.F. Famous mineral localities: The pegmatites of Laghman, Nuristan, Afghanistan. *Miner. Rec.* **1978**, *9*, 301–308.

20. Cook, R.B. Connoisseur's choice: Spodumene var. Kunzite, Nuristan, Afghanistan. *Rocks Miner.* **1997**, *72*, 340–343. [CrossRef]

21. Teo, B.K. *EXAFS: Basic Principles and Data Analysis*; Springer Science & Business Media: Berlin, Germany, 2012; Volume 9.

22. Phan, T.L.; Zhang, P.; Yang, D.S.; Nghia, N.X.; Yu, S.C. Local structure and paramagnetic properties of Zn1-xMnxO. *J. Appl. Phys.* **2011**, *110*, 063912. [CrossRef]

23. Haeger, T. Study of impurity in blue spinel from the Luc Yen mining area, Yen Bai province, Vietnam. *Vietnam J. Earth Sci.* **2018**, *40*, 47–55.

24. Ravel, B.; Newville, M. ATHENA, ARTEMIS, HEPHAESTUS: Data analysis for X-ray absorption spectroscopy using IFEFFIT. *J. Synchrotron Radiat.* **2010**, *12*, 537–541. [CrossRef] [PubMed]

25. Oliveira, R.A.P.; Mello, A.C.S.; Lima, H.R.B.R.; Campos, S.S.; Souza, S.O. Radiation detection using the color changes of lilac spodumene. In Proceedings of the INAC 2009: International Nuclear Atlantic Conference Innovations in Nuclear Technology for a Sustainable Future, Rio de Janeiro, RJ, Brazil, 27 September–2 October 2009.

26. Farges, F.; Panczer, G.; Benbalagh, N.; Riondet, G. The grand sapphire of Louis Xiv and the ruspoli sapphire: Historical and gemological discoveries. *Gems Gemol.* **2015**, *51*. [CrossRef]

27. Souza, S.; Ferraz, G.; Watanabe, S. Effects of Mn and Fe impurities on the TL and EPR properties of artificial spodumene polycrystals under irradiation. *Nucl. Instrum. Methods Phys. Res. Sect. B Beam Interact. Mater. At.* **2004**, *218*, 259–263. [CrossRef]

28. Lu, R. Color origin of lavender jadeite: An alternative approach. *Gems Gemol.* **2012**, *48*, 273–283. [CrossRef]

29. Fridrichová, J.; Bačík, P.; Ertl, A.; Wildner, M.; Dekan, J.; Miglierini, M. Jahn-Teller distortion of Mn3+-occupied octahedra in red beryl from Utah indicated by optical spectroscopy. *J. Mol. Struct.* **2018**, *1152*, 79–86. [CrossRef]

30. Burns, R.G. *Mineralogical Applications of Crystal Field Theory*, 2nd ed.; Cambridge University Press: Cambridge, UK, 1993; Volume 5. [CrossRef]

31. Sugiyama, K.; Arima, H.; Konno, H.; Kawamata, T. Distribution of Mn in pink elbaitic tourmaline from Mogok, Myanmar. *J. Miner. Petrol. Sci.* **2016**, *111*, 1–8. [CrossRef]

2

Agate Genesis: A Continuing Enigma

Terry Moxon [1] and Galina Palyanova [2,3,*] ⓘ

[1]   55 Common lane, Auckley, Doncaster DN9 3HX, UK; moxon.t@tiscali.co.uk
[2]   Sobolev Institute of Geology and Mineralogy, Siberian Branch of Russian Academy of Sciences, 630090 Novosibirsk, Russia
[3]   Department of Geology and Geophysics, Novosibirsk State University, Pirogova str., 2, 630090 Novosibirsk, Russia
[*]   Correspondence: palyan@igm.nsc.ru

Abstract: This review covers the last 250 years of major scientific contributions on the genesis of agates found in basic igneous host rocks. From 1770 to 1955, the genesis question was frequently limited to discussions based on observations on host rock and agate thick sections. Over the next 25 years, experimental investigations examined phase transformations when silica glass and various forms of amorphous silica were heated to high temperatures. This work demonstrated that the change from the amorphous state into chalcedony was likely to be a multi-stage process. The last 40 years has seen modern scientific instrumentation play a key role in identifying the physical and chemical properties of agate. The outcome of this work has allowed limited evidence-based comment on the conditions of agate formation. There is a general consensus that agates in these basic igneous hosts form at <100 °C. However, the silica source and the nature of the initial deposit remain to be proven.

Keywords: agate; chalcedony; XRD; genesis; moganite; crystallite growth; age

---

1. Introduction

Agate and chalcedony are the compact varieties of silica that are primarily composed of minute crystals of α-quartz. Chalcedony has a fairly uniform colour but is band free while agate is banded chalcedony. Sectioned agates from igneous host rocks generally show the banding as one of two major types with wall lining (sometimes called fortification) banding being the most common type. These agates show initial bands appearing to replicate the supporting cavity wall. Subsequent bands follow with an approximate repetition that frequently continues towards the agate centre. A second type demonstrates horizontal bands that are apparently gravity controlled. Both types can be found in the same agate.

Basalt and andesite are the most widespread agate host rocks but agates can also be found in some fossil wood and sedimentary host rocks, particularly limestone. However, there is no reason why agates in different environments should follow identical genesis routes. Discussion in the present paper is mainly limited to scientific contributions that have been made over the last 250 years on the complex question of agate genesis in basaltic and andesitic host rocks.

The present review is in three parts. From 1770 to 1955, scientific contributions to the genesis question were generally based on host rock observations and agate thick sections. Experimental investigations played a relatively small part and contributions during this period are considered chronologically. During the next 25 years phase transformations in silica glass and amorphous silica were investigated. Many of these studies were not directly concerned with agate but the work has implications for the agate genesis question. Modern scientific instrumentation has played a key role in identifying the physical and chemical properties of agate. In order to avoid information repetition, some relevant pre-1980 work is considered with 1980–2020 in the final discussion. The review includes some contributions by research workers from Russia and Germany, who only published in Russian and German.

2. Scientific Contributions over the Years 1770–1980

2.1. 1770–1900

Agates have been collected and fashioned since the earliest civilisations and many must have wondered about the formation of these beautiful minerals. Collini [1] states, "Not only is the red colour in agate due to iron but it is also the cause of the green, red and brown in moss agate". According to Liesegang [2], Collini [1] was the first to suggest that these colours had their origins from iron compounds that had been transported in circulating waters from the neighbouring host rocks. The iron rich solutions were then able to penetrate the still soft agate. Hardening of the agate prevented further iron solution penetration. Liesegang [2] also cites Lasius as an early worker who suggested silica enters the gas vesicles in solution. In 1931, Liesegang [3] added more experimentation and comment to his earlier observations.

Germany hosts some of the world's finest agates and the agate industry based around Idar Oberstein has produced high quality agate products since the 14th century [4]. This large commercial background accounts for the fact that agate and its genesis was very much the preserve of the German scientists. In 1849, Noeggerath [5] and Haidinger [6] published a series of open letters in which they stated their opposing views on the method of silica solution entry into the gas vesicle. The conference records of the 20th July in the Freunde der Naturwissenschaften, Vienna stated the areas of agreement and their differences. Both agreed that gaseous bubbles are formed in the erupted lavas and the subsequent parallel vesicular flows were due to the movement of the lava. Misshapen amygdales are the result of vapour bubbles meeting within the high viscosity lava; this leads to twin and triplet bubbles. Noeggerath [5] emphasised the agate variations were a feature of the different local environments. He pointed to amygdales in Westphalen where the calcite had been dissolved in the upper reaches but not at the lower levels and claimed the firm rock protected the calcite from dissolution.

Noeggerath [5] asserted that the silicic acid, necessary for the agates, comes from hot springs and appears after the decomposition of the host rock. He noted that agates had been found in compact bedrock while entirely friable rocks can be free from agates. However, it is the method of silica entry into the vesicle that produces major differences of opinion.

Haidinger [6] took a contrary view arguing that the "mountain sweat" carries the separate silica into the gas vesicle. Noeggerath [5] acknowledged that water could pass through the host rock but would not be able to penetrate the impermeable agate layers. As an alternative he pointed to the infiltration canals as means of silica entry and this would allow the continual development of successive layers. These canals are found in some agates and appear to breach the outer layer.

At the end of his open letter of May 1849, Noeggerath [5] writes, "I beg you to compare the permeable and impermeable layers, which allow the art of colouring banded quartz. How is it possible for the silica entry to happen and form successive precipitations from outside when the earliest formed impermeable layers presents a barrier to all further additions?". Haidinger [6] did not answer the question and the infiltration canal was accepted until challenged by Liesegang in 1915 [2]. This interesting infiltration feature is still proposed by some authors as a point of solution input or water exit. The canals feature regularly in agate genesis and are discussed later in the paper.

2.2. 1900–1945

In 1901 Heddle [7] published an important work: "The Mineralogy of Scotland". He had spent many years collecting Scottish agates and the book included a chapter on the question of agate genesis. He suggested that silica could enter the gas vesicle as a result of osmosis with the early deposits acting as a membrane. As a semipermeable membrane and solute concentrations are not mentioned he is really describing diffusion. The desilicified solution then leaves the agate via tubular openings: Noeggerath's [5] infiltration canals.

The first detailed thin section study of agates that we identified was carried out by Timofeev in 1912 [8]. Three of his micrographs are shown in Figure 1. Using a polarizing microscope he was able to reveal the typical chalcedonic pattern with white bands temporarily halting the growth. His work was based upon the agate amygdales found in the Suisari Island basalts in Lake Onega, Karelia, Russia. He proposed a crystallization sequence of chalcedony → quartzine → quartz. Some of the agates had developed a stalactite type growth. Cross sections of the stalactite are described as alternating zones of chalcedony and quartzine separated by a pigment. Some stalactites had the occasional calcite deposit within the agate.

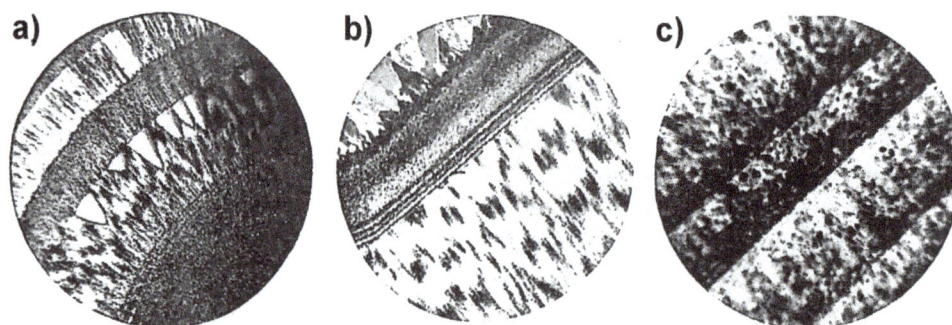

Figure 1. Timofeev micrographs [8]. The original captions describe the (**a**) alternation of zones of quartz (light) and chalcedony (dark); (**b**) the formation of quartzine filaments on the faces of the quartz rhombohedra; and (**c**) zones showing a changing orientation from chalcedony to quartz.

Liesegang [2] had made several publications on rhythmic banding in silica gels and by 1915 he was able to publish his major work Die Achate. Apart from infiltration canals, Liesegang was able to synthesize various agate-like patterns by allowing metal ions to diffuse in silica gels containing suitable anions. The common wall lining agate can be simulated if a single curved line of $Fe^{3+}_{(aq)}$ is drawn across a flat surface of solidified silica gel containing an alkali. A simulated wall lining agate pattern develops due to the ions diffusing across the gel (Figure 2a). Liesegang banding can also be shown when separate partially miscible liquids are poured from a container (Figure 2b). Horizontal bands are readily formed by $OH^-_{(aq)}$ on top of solidified gelatine in a large boiling tube containing $Mg^{2+}_{(aq)}$ or transition metal ions. Liesegang's [2] Die Achate has had a major impact on the discussion of agate genesis and 80 years later it was still being linked with agate genesis.

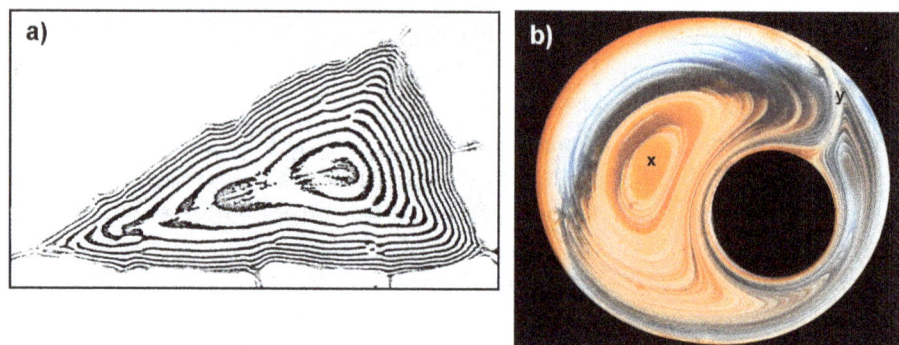

Figure 2. (**a**) Liesegang rings of silver chromate. They are formed when aqueous silver ions slowly diffuse through a gel containing chromate ions producing a precipitate of silver chromate (from Liesegang [3]). (**b**) Liesegang banding that has been produced when samples of blue, orange, and white dyed polyester resin are simultaneously poured round a solid circular disc. The pouring point was at (x) with an apparent exit canal at (y). Disc hole diameter = 2 cm.

Reis [9] published the largest single work on agate and later he produced a further long treatise in which he gave his conclusions on agate genesis: Reis [10]. He was not the first worker to recognise that the wall contact layer was different from the rest of the agate but he was the first to suggest that it was of a different generation. Reis [9,10] opposed the silica entry osmosis mechanism as it depended upon a semipermeable membrane being complete. He believed that the temperature gradient within the lava flow created a sucking effect that drew the silica solutions into the cavity. Iron oxides played a dual role by precipitating the silica and adding colour to the agate. The work of Reiss was largely ignored until Fischer in 1954 [11] added key support for the Reiss hypotheses.

Jessop in 1930 [12] questioned the view that chalcedony was a mixture of quartz and opal. This was based on the fact that the refractive index would not be the approximately constant value found in samples of all geological ages. Jessop [12] proposed that the silica was deposited as a gel containing gas and solid impurities in a solution. As crystallisation advanced from the outside edge, the dissolved substances move with the crystallisation front until a point of saturation is reached. The gaseous or solid impurities are entombed in the crystalline silica and the thickness of the band depends upon the concentration of impurity and the rate of crystallisation. As the solid and gaseous solutes do not necessarily reach saturation at the same point, they allow bands to be partially or totally distinct.

The white bands cannot be stained and this was demonstrated using thin sections. These bands reveal an acicular structure and were parallel to the fibres. The bands that can be dyed are uniform. He proposed that the white bands contain the solid inclusions and the clear bands contain the gaseous bubbles. Such a hypothesis allows dyeing to be a question of relative retentivity and permeability.

In 1931, Liesegang [3] admitted that a pre-existing gel is not always necessary but objected to the length of time taken to form agate via any rhythmic deposition. In this paper, he proposed an alternative mechanism for the formation of the horizontal layering in agate. Layers of varying concentrations of hot sugar solutions can reflect light differently. The salt lakes in Siebenbürgen where the bottom of the lake is 50 °F hotter than the more dilute upper layers show a similar effect. Such variations can also be produced in water glass and he proposed that the horizontal layers are caused by a similar salt-like effect that was eventually "frozen".

Pilipenko in 1934 [13] supported the Noeggerath hypothesis and opposed the Liesegang propositions. Pilipenko [13] was one of a few to base his argument on the results of a detailed study of thin sections, polished sections, and the outer surface of agates. He put forward a hypothesis for the deposition of chalcedony from "aqueous silica solutions" that penetrated the empty chambers in effusive rocks through pores, hair cracks, and supply channels in the walls of voids. He proposed that chalcedony deposition was in a liquid and gaseous medium. He discounted the Liesegang hypothesis on a number of points:

(1)   Some samples of agates are found with a hollow space. This demonstrates that the vesicle is not always filled with a gel;

(2)   The vesicular infill occurs by the layering of the first generation and can be accompanied by metasomatic substitution of crystals, which have been previously precipitated;

(3)   The hypothesis does not explain why, in the same sample, colourless banded chalcedony varies with the pigmented chalcedony;

(4)   The role of pigmentation is not clear with (a) well-formed amethyst crystal centres and (b) brown pigments found in various zones with the disappearance of strong pigment zones;

(5)   In artificial preparations, the deposits are torn and split while in nature this phenomenon is rare.

As an alternative Pilipenko [13] proposed that, "a study of the outer surface of agate reveals a number of pores or openings in the walls of sections which surround the agate". It is presumably these openings that allow the silica to enter the vesicle.

A study of the outer surface of agates in contact with effusive rocks showed the various types of silica deposition in the chambers is grouped into seven types. The first six types of agates are formed in chambers that are not completely filled by silica solutions. In this case, silica deposition occurs horizontally starting at the bottom of the chambers and builds layer upon layer. Silica stalactites and various sagging formations descend from free walls. The seventh type occurs when the chamber is completely filled with a silica solution. Here, deposition of chalcedony goes along the walls of the chamber layer by layer. The number of incoming solutions into the chamber travel through conducting canals and pores of various sizes. Solution inflow equals outflow.

Pilipenko [13] investigated the structure of the infiltration channels and gave great importance to their formation. Infiltration canals develop only within the agate body. In longitudinal sections through the canal, it is clearly shown that traces of agate in the form of a sleeve are pulled out. The sleeves themselves are folded from chalcedony tubes, concentrically embedded into each other.

2.3. 1945–1955

Kuzmin in 1947 [14] included a chapter on agate in a PhD thesis that was based on "Periodic-rhythmical phenomena in Mineralogy and Geology". In 2019, the thesis was published in his memory. He investigated the banded structures of agates and rejected Liesegang's hypothesis of agate formation. As agate banding could form as the wall lining or horizontal type, he proposed that the vesicle silica solutions would be colloidal or molecular leading to the respective wall lining and horizontal banding. He credited molecular or highly diluted solutions for the development of sharply defined horizontal layers in the chamber. However, the so-called infiltration canals in agates were interpreted as a point of "output". He proposed that these exit points occur when the chamber is full of a silica solution and osmotic pressure forces the excess solution to exit via the channels. Once diffusion has started, the channels lower the pressure and allow fresh solutions to diffuse into the cavity. The outward pointing structure shows that they exist as a departure point. The structure of these output channels convincingly shows that the agate structure follows a sequential deposition of rhythmic layers.

In a paper that is more tentative in its suggestions than many of its predecessors, Nacken in 1948 [15] suggests that the silica precursor for agate is the existence of an immiscible liquid state within the magma. Perhaps in an attempt to pre-empt the inevitable criticism, Nacken comments on the problem presented by the melting point of silica at 1600 °C existing within a basaltic magma of 1100 °C. He quotes the work of the Geophysical Laboratory at the Carnegie Institute, Washington, where experiments have been performed on "peculiar acid magma systems" containing excess silica in the molten state. Such systems divide into silica rich and silica poor immiscible mixtures". The silica rich component would lead to drops with a density of 2.3 g/cm^3 within a magma density of 2.6 g/cm^3. This density difference would allow the silica drops to rise and eventually form agates. With silica drops suitably in place Nacken [15] returns to his original synthesis of chalcedony from glass that had been subject to a hydrothermal conversion at 400 °C. The conversion was sensitive to pH, impurities, and the temperature gradient within the glass. When the glass had been converted to chalcedony, there is a volume decrease, which allows further solution attacks along capillary cracks. Any impurities are pushed along with the crystallisation front and precipitate in layers.

Fischer in 1954 [11] reviewed the 20th century German literature on agate genesis and commented on the theories of Liesegang [3], Nacken [15], and Reis [9,10] in some detail. He supported the Reis [9,10] contribution and rejected the Nacken [15] hypothesis on the grounds that a magmatic temperature of 1100 °C would not support a silica melt of 1600 °C. Calcite is often found in agate and at these temperatures calcite would be converted into wollastonite. Liesegang's hypothesis [3] required the existence of a pre-existing gel but his explanation failed to correlate with laboratory findings. Fischer [11] accepted the low solubility of silica in water was a problem but the enormous volumes of water generated and the length of time for the demise of thermal springs are sufficient to support the theories of Reis [9,10].

2.4. 1955–1982

Studies during 1955–1982 allowed modern day instrumentation to play a greater part in the investigation of agate. Potential precursors, such as amorphous silica and silica glass, were examined at high temperature and pressure. Operating in the temperature range of 330–450 °C and pressures of 10–40 kPa, Carr and Fyfe [16] established the conversion of near amorphous silica into quartz. Intermediate silica phases were identified by powder X-ray diffraction (XRD). They established the transition as: near amorphous silica → cristobalite → silica–K → quartz. Silica-K or keatite was first synthesised by Keat [17]. It is very rare in the natural state and has never been identified in agate.

White and Corwin [18] investigated the formation of synthetic chalcedony by heating glass in hydrothermal solutions at 400 °C and 34 MPa. Chalcedony was made by the direct transformation of silica glass or cristobalite. They found that no conversion took place in weakly acid solutions but a complete and rapid conversion occurred under slightly alkaline conditions. The transformation proceeded via cristobalite or keatite.

Experimental investigations on the formation of quartz from amorphous silica were carried out in alkaline hydrothermal conditions with the addition of alkali metal hydroxides or carbonates [19]. Experiments were carried out over a temperature range of 100 and 250 °C with transformations occurring in a matter of days. The results show the change of amorphous silica to quartz went through two intermediate phases of silica-X (a new polymorph of silica) and then cristobalite. The transformation rate was increased by a high temperature and pH.

Ernst and Calvert [20] investigated the changes in porcelanite when heated hydrothermally at temperatures of 300, 400, and 500 °C, 200 MPa pressure. Natural porcelanite from the Monterey Formation, CA, USA, was used in the experiments. The porcelanite was 98% cryptocrystalline with finely crystalline cristobalite together with very minor quartz. Their prime aim was to establish the conversion rates with regard to the eventual formation of chert. The change of cristobalite to quartz is relevant to the question of agate genesis and was found to be zero order.

Colloidal silica has been frequently suggested as the starting material for agate. However, Oehler [21] was the first to carry prolonged heating experiments on colloidal silica solutions. He added HCl to neutralise the alkaline silica colloid bringing the pH from 10 to pH 7 in order to produce a solid gel. The gel was sealed in silver capsules and heated between 100 and 300 °C at 300 MPa pressure for 25–5200 h. The solidified gel crystallised as quartz that was predominantly in the form of chalcedonic spherulites. The spherulites were either length slow or length fast or both. The formation of each type depended upon the conditions. It should be noted that the fibrosity was just in quartz microspheres and was not a synthetic version of an agate amygdale.

Previously, detailed chemical analysis in agate literature was rare but in 1982, Flörke and co-workers [22] investigated several Brazilian agates for their trace metal ion content. Additional work examined the two types of water that are found in agate: free molecular water was low while the silanol water (SiOH) was high. They were able to demonstrate that the wall banding and the horizontal banding in the same agate were clearly of different generations.

3. Scientific Contributions over the Years 1980–2020

3.1. Genesis Contributions in the Early Years

The origins of the silica source, means of transportation, and its physical state on deposition have been discussed over the last 250 years. Unfortunately with the unknowns, every suggestion is met with valid queries or objections. The various problems have been reviewed in detail with Godovikov et al. [23], Goncharov et al. [24], Landmesser [25,26], Moxon [27], and Götze [28].

From 1770 to 1930, the agate genesis hypotheses were largely based on the agate appearance and found in basic igneous host rocks. Our comments are made with hindsight but some suitable techniques were available and rarely applied to the agate genesis question. In particular, there was a general failure to examine agate in the thin section. Sorby [29] was the first person to use a polarizing microscope for the examination of rock thin sections. This initial publication was used to differentiate agate from calcite in the calcareous grit in Yorkshire, UK. The collection held in the Technische Universität Bergakademie, Freiberg, Germany shows that rock thin sections were being studied at various times during the late 19th century [30]. Yet agate thin sections rarely featured in these early years of the agate genesis discussion. The chalcedonic fibrosity and its apparent fibrous growth are shown by a thin section examination (Figure 3).

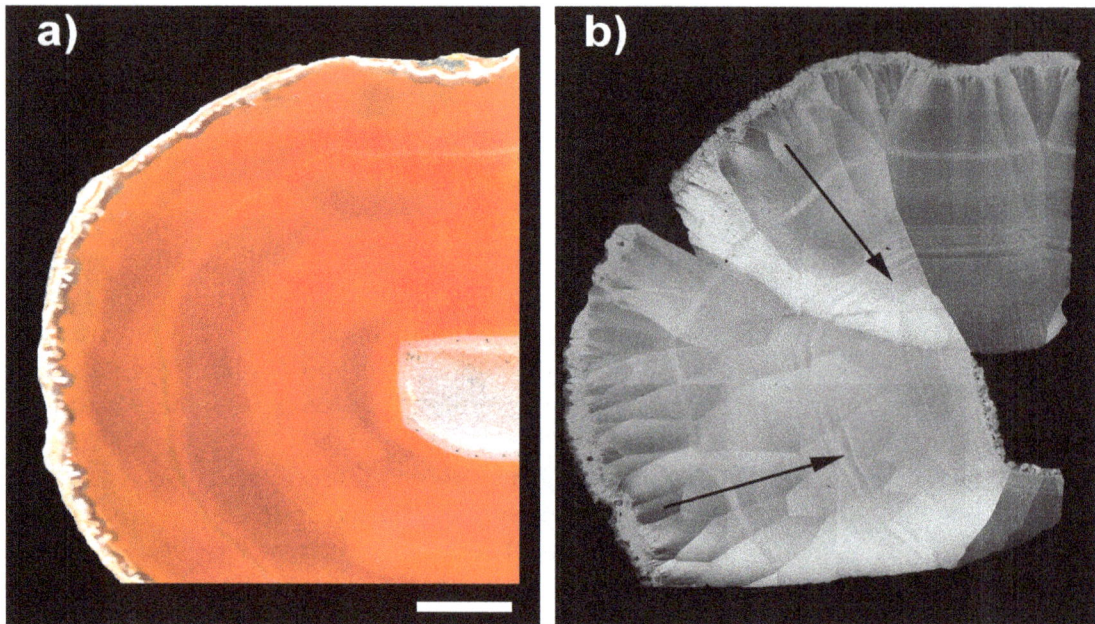

Figure 3. (a) A Brazilian agate that has been dyed to a depth of around 2 mm. (b) The thin section that was made from a slab just below the dyed surface. Note the wall contact layer shows a different texture to the main microstructure. The arrows indicate the direction of growth leading to a final development of macrocrystalline quartz crystals. Polars crossed. Scale bar = 2 cm.

By 1920, a number of inorganic crystal structures had been investigated using XRD. However as far as we are aware, the first agate investigation using XRD was by Heinz (1930) [31]. Undoubtedly, it was the work of Liesegang [2,3] with his experiments on silica gels that had the greatest influence. His ideas dominated the agate genesis question for much of the 20th century. The simulation of the rhythmic agate patterns in silica gels and in natural settings has ensured that the term "Liesegang bands or rings" will be preserved in perpetuity (Figure 4).

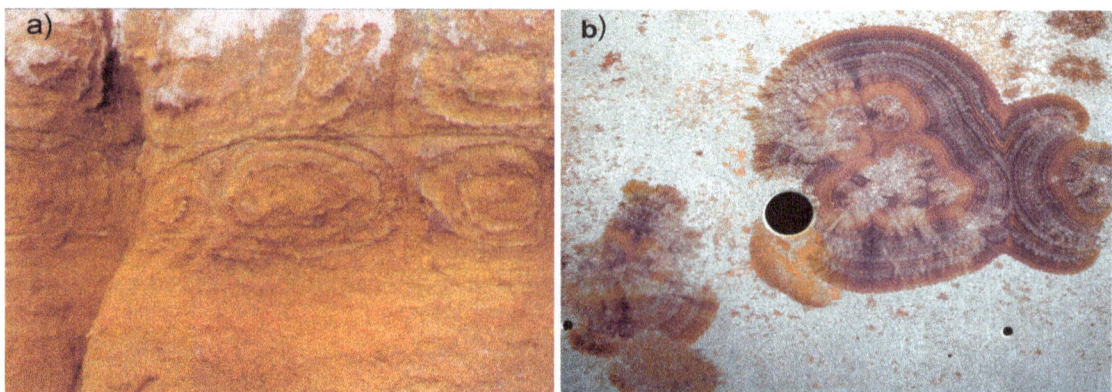

Figure 4. Two natural settings that have developed Liesegang rings. (a) An elliptical development in the sandstone cliffs at Hunstanton, Norfolk, England. (b) Rust rings that have formed on the support plate of a battery driven clock.

The need to have a pre-existing gel and its reliance of a later diffusion by a suitable cation resulted in some objecting to the Liesegang [2] hypothesis. Ion diffusion across silica gel in a Petri dish and vertical diffusion in a boiling tube produces a clear impression of wall lining and horizontal banding respectively. This initial gelatinous deposit is still favoured by some recent authors. Pabian and Zarins in 1994 [32] invoked genesis from a silica gel deposit with banding caused by the Belousov-Zhabotinsky (B-Z) reaction. The B-Z reaction creates banding oscillations between two chemicals and can produce rhythmic banding in gels [33,34]). The Pabian and Zarins [32] hypothesis is based upon an initial vesicular infill of silica gel and contact is made with alkaline ground water. Based on the B-Z reaction, they propose that this creates a continual electrochemical wave front banding.

Suggestions of an initial gel deposit as a starting material for agate genesis are flawed from the start. A comment was never made about the temporary nature of the generated patterns. After several weeks, the gels become cracked and dehydrated leaving only a coloured silica powder. A further comment is made in the discussion.

3.2. Chalcedony from Silica Glass

During the late 1950s, the electronics industry experienced an increased demand for high quality quartz. There were many attempts to synthesise quartz from various forms of silica glass. As chalcedony was often a byproduct, some data might have potential links to agate genesis. Following on from a previous study, White and Corwin [18] produced synthetic chalcedony from a silica glass that had been subjected to a hydrothermal solution at 400 °C and 34 MPa pressure. Conversion only occurred in weakly alkaline conditions with the transformation proceeding from silica glass → cristobalite → keatite → chalcedony → quartz. During long runs, the glass was completely converted to quartz. Synthetic chalcedony had the same anomalous properties as those of natural quartz: a brown colour in transmitted light and a refractive index lower than quartz. The transformation process is repeated in many changes of amorphous silica to quartz. Unfortunately, the silica glass starting material discounts any direct link with agate genesis.

3.3. Agate Formation Temperature

Götze in 2011 [28] carried out an extensive review of the methods used to study this temperature question. Here, we will deal with just three of the techniques. Oxygen and hydrogen isotope analysis was first used on Scottish Devonian and Tertiary agates by Fallick et al. [35]. Their work demonstrated agate formation was 50 °C. Later, Harris [36] examined Namibian agates and assigned a formation temperature of 120 °C. The Harris data were re-examined by Saunders [37] who placed the agate formation temperature between 39 and 85 °C. A second method is based on the concentration of Al^{3+} ion in macrocrystalline quartz that is sometimes found at the centre of the agate [38]. Here, agate formation temperature was estimated at between 50 and 200 °C. A mineralothermometric study of gas-liquid inclusions in quartz of agates located in andesite-basalts from many deposits in the north-east of Russia showed that the homogenization temperatures were <120–170 °C: Lebedinoe (140–170 °C), Kedon (100–135 °C), Yana (130–135 °C), Yakanvaam (100–120 °C), Takhtayama and Ryrkalaut (110–120 °C) [24].

Most recent investigators into the question of agate genesis accept that agate formation occurs at <100 °C. Any question of agate genesis occurring at temperatures higher than 200 °C has to be totally discounted. Agates when heated at higher temperatures produce irreversible property changes.

A thermogravimetric sketch plot of heated agate powder is drawn in Figure 5a. This typical curve shows three basic changes that have all been investigated using infrared (IR) by Yamagishi et al. [39] and they demonstrated:

(1)   Loosely bound water molecules are lost at <200 °C.
(2)   Tightly bound water and various forms of silanol water are lost up to 800 °C.
(3)   The maximum mass loss is reached at 850 °C.

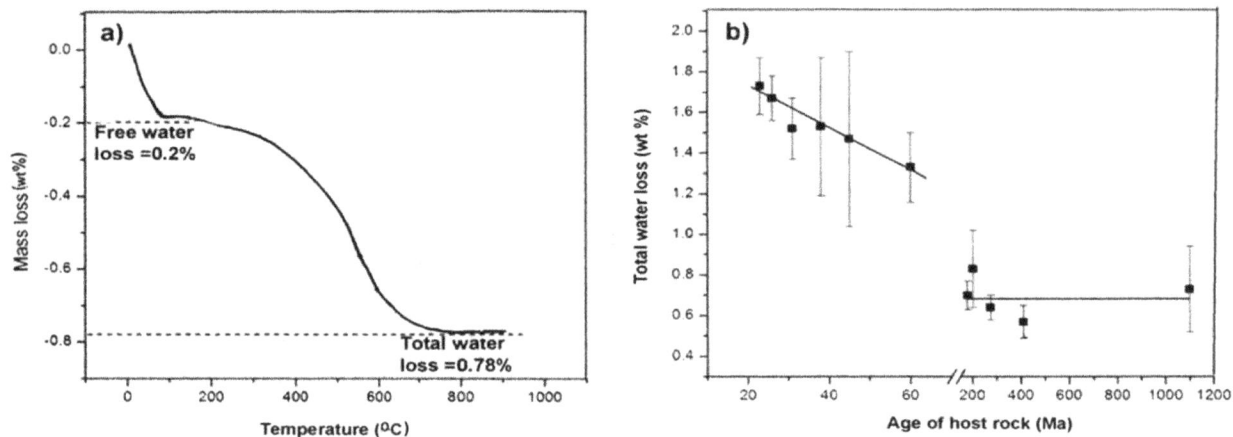

Figure 5. (a) A sketch plot of typical TGA (thermo gravimetric analysis) changes when powdered agate is heated to 1000 °C (after Yamagishi et al. [39]). **(b)** Total water loss from agate with respect to age of the host rock. Agate powders (<50 μm) were heated at 1200 °C (after Moxon [40]).

The variations in the total water in agate with respect to host age are shown in Figure 5b. The determined total water took into account the water loss in preparing <50 μm powder, which was then heated to 1200 °C. The plot shows a linear decrease in total water loss in agates over the first 60 Ma years with an approximate constant total water loss in agates from host rocks aged between 180 and 1100 Ma [40].

3.4. The Discovery of Moganite

The application of oxygen and hydrogen isotope analysis of agate formation temperatures by Fallick et al. [35] was a key contribution to the agate genesis enigma. Of equal importance was the discovery of moganite in the cracks and flows of the Mogan formation, Gran Canaria by Flörke et al. [41].

Since 1976, there have been many investigations into the properties of moganite but it was not until 1999 that it was officially recognised as a separate mineral by the International Commission on New Minerals and Mineral Names. Totally pure moganite has never been synthesised nor found pure in the natural state. The samples from Gran Canaria still provide the highest concentrations of moganite. It was the work of Heaney and Post [42] that effectively opened up the moganite story in agate by examining more than 150 samples of agate, chalcedony, chert, and flint. One of the data plots demonstrated chalcedony and agate differences in moganite ranged from 23 to 2 wt%. In a separate study, XRD was used to analyse 10 samples of Mogan moganite; the only mineral phases present were quartz and moganite with the moganite content greater than 80 wt% in half the samples. However, moganite was as low as 52 wt% in the remainder [43].

Raman spectroscopy is also a common technique for determining the moganite content. Götze et al. [44] used the two techniques to examine agate and showed that Raman can give higher values than was determined by XRD. These differences were attributed to the limited contribution to the Bragg reflection due to the small size and distribution of moganite. There have been a number of trace cation and moganite analyses of chalcedony. One was carried out by Petrovic et al. [45] with quartz and moganite (22 wt%) once again the only mineral phases. Al^{3+} at 0.52 wt% produced the largest concentration of the trace metal ions. Moganite is more soluble than quartz and the release of water during the geological time scale results in moganite slowly dissolving and reforming as microcrystalline quartz. The importance of this change is examined in the discussion.

3.5. Cation and Silica Loss from Basalt Host and Their Potential Role in Agate Formation

Suggested mechanisms have frequently been based upon an initial silica solution entering the gas vesicle with the subsequent water loss leaving a deposit of amorphous silica. Process repetition fills the vesicle with silica and eventually agate forms. Noeggerath [5] and Haidinger [6] first introduced aspects of this in 1849. However, the nature of silica in solution and its solubility were unknown until the 1950s. Previously, it had been assumed that aqueous silica existed in the colloidal state together with higher forms of silicic acid. Krauskopf [46] summarised earlier work that had established silicon exists primarily as monomeric silicic acid (H_4SiO_4). He used the yellow colour formation between silica solutions and ammonium molybdate to a great effect. Colorimetric methods could now be used to identify the concentration of $H_4SiO_{4(aq)}$.

Krauskopf [46] was interested in the conditions for silica solubility and the precipitation of amorphous silica. He added to, and summarised, the solubility changes with temperature. Silica solubility was found to be 100–140 ppm at 25 °C; 300–380 ppm at 90 °C. He also demonstrated that the solubility has shown little change between pH values of 1 and 9. Krauskopf [46] was dealing with silica in all sedimentary environments. However, the data clearly has implications for silica transport and deposition into gas vesicles.

The alteration of the host rock has frequently been cited as a possible source of silica. Patterson and Roberson [47] carried out an analysis on the depth of boreholes from the eastern part of Kauai Island, Hawaii. Seven boreholes of varying depth together with fresh rock were analysed for their chemical composition. Water from the host rock was also included in the analysis. Just four metal ions with the highest listed concentrations have been selected for comment (Table 1). The depth of basalt bore holes ranged from 0.6 to 11 m (hole numbers 1 and 7) and the data were compared to fresh basalt (number 8).

Table 1. Trace Analysis of Kauai Basalt Taken from Boreholes at Various Depths.

Analysis of Rock Samples							Water Analysis from the Rock					
Hole	Depth	SiO_2	Fe_2O_3	Al_2O_3	MgO	Na_2O	H_2O	SiO_2	Fe^{3+}	Al^{3+}	Mg^{2+}	Na^+	pH
No	m			(wt%)						(ppm)			
1	0.6	5.7	40.2	25.6	0.65	0.04	18.6	2.9	0.00	0.11	1.9	11	4.6
6	9.5	21.5	30.7	25.1	1.0	0.06	14.6	2.5	0.00	0.12	1.5	8.5	4.7
8	-	45.0	4.3	11.8	12.0	3.1	3.1	42	0.00	0.06	11	21	7.6

Data after Patterson and Roberson [47]. The depths shown in Table 1 are the mid point of the range given in the paper. Hole 8 is fresh basalt.

Kauai Island basalt has been dated at 5.1 Ma [48]. Table 1 shows the changes with borehole depth from a relatively young host. The analysis of water from the rock shows the total loss of Fe^{3+} and demonstrates that this ion has been transported out of the host rock system at an early stage. Such a young island gives an indication of a possible cation and silica mobility from a basalt rock and to a gas vesicle. We are not aware of any agate or chalcedony forming on Kauai but the Patterson and Roberson [47] data has relevance regarding mineral loss in an agate basalt host. Equally important is a chemical analysis of trace metal ions that have been identified in agate and the agate host.

Götze et al. [49] quantified 27 trace metal ions in 18 worldwide agates and host rocks. We selected host and agate data from two Scottish sites and one each from Mexico and Brazil. Agate host rocks for this selection of sites ranged in age from 38 to 412 Ma. Both the Götze et al. [49] and Patterson and Roberson [47] studies identified a larger number of ions, but we limited our speculative comments to those cations that were likely to have the greatest effect on H_4SiO_4. Iron (III) oxide offers colour to many types of agate and the trivalent ions of Fe^{3+} and Al^{3+} would be expected to have the greatest effect on any silica solutions or sols. Magnesium ions are generally the largest concentration of the divalent ions and Na^+ has by far the largest concentration of the monovalent ions.

Montrose and Ardownie Quarry share the same glassy andesite that formed 412 Ma years ago in the Eastern Midland Valley, Scotland. The distance between the two areas is 38 km. Chihuahua is world famous for the high quality agates and, like Brazil, is one of the few countries that mine and export agates. Brazil is by far the world leader in bulk agate export but the agates tend to be very large but dull and unusually, they can be dyed. Trace element analysis in agate and host rock from Mexico, Brazil, and Scotland are shown in Table 2.

Table 2. Cation Concentrations in Agate and Host in Samples from Mexico, Brazil, and Scotland.

Agate Source	Country	Host Rock Type	Age of Host * (Ma)	Fe^{3+}	Al^{3+} (ppm)	Mg^{2+}	Na^+
Chihuahua	Mexico	Andesite	38	6695	nd	5460	25,000
		Agate	-	215 **	925 **	151 **	320 **
Rio Grande do Sol	Brazil	Basalt	135	14,700	nd	nd	19,000
		Agate	-	137	54	10	140
Montrose	Scotland	Andesite	412	33,400	nd	nd	23,000
		Agate	-	70	456	15	230
Ardownie Quarry	Scotland	Andesite	412	37,300	nd	nd	27,000
		Agate	-	36	232	18	210

Data after Götze et al. [49], nd-not determined, ** two colours tested in the agate and the highest concentration is shown. * Citations for host rock age in Moxon and Carpenter [50]. The physical properties of Brazilian agate show its formation was 26–30 Ma.

The two neighbouring Scottish agate sites had similar concentrations of Mg^{2+} and Na^+ and the concentrations of Fe^{3+} in agate are much lower than in the younger host rocks. This is in spite of the older host rocks presently having Fe^{3+} at a greater concentration. The agates from these two areas are attractive but not typically highly coloured with oxides of iron. The concentration of Na^+ in all four host rocks is high and interestingly only relatively small concentrations survive in the agate. Presumably, much remains in solution and is lost with the departing vesicular water.

Two studies have used the silico-molybdate method to investigate the effect of iron ions/compounds on silica solutions and soils. A pressure membrane extractor was used to obtain H_4SiO_4 solutions from various soils [51]. In addition to the effect of iron (III) oxide, they separately added aluminium oxide to the silica solution at pH values 1–12. The addition of iron (III) oxide on the concentration of monosilicic acid resulted in a silica fall from 120 to 80 ppm at pH 4 and 70 ppm at pH 9. Aluminium oxide was more effective in reducing concentrations over the same two pH values producing a respective fall of 35 and 25 ppm.

Moxon [52] examined the separate effects of Fe^{3+} and Mg^{2+} in removing H_4SiO_4 and colloidal SiO_2 from solutions and sols after a short run of 45 min. The various runs were within the normal groundwater pH range of 4–9. A relatively high $Fe^{3+}:H_4SiO_4$ ratio of 1:3 could only remove 18% while a 1:96 ratio of Fe^{3+}: colloidal SiO_2 removed 90% of the silica. Over a 16 month period, 10, 20, and 40 ppm of $Fe^{3+}_{(aq)}$ and $Mg^{2+}_{(aq)}$ were separate additions to silica sols containing 1600, 3300, and 6800 ppm of silica respectively. These runs included fresh silica controls. In each case the solution pH was adjusted to 8.2. After 16 months, the added cations had been no more effective in precipitating silica than the control.

The solutions were stored in a cupboard at room temperature and reinvestigated after 14 years. All the H_4SiO_4 solutions had reached an equilibrium concentration of 122 ppm. There were a number of gelatinous flocs in suspension, on the side and floor of the containing flask. Apart from one of the 9 solutions tested, the pH fell from 8.2 to 6.1. Samples of the gels were dehydrated by oven heating and examined using XRD that included a new fresh gel control for comparison. In all cases XRD produced one single broad reflection. A full width at half maximum profile intensity (FWHM) was measured. With standard deviation included, all the 14 years old gels showed FWHM values were greater than the new fresh gel control. Unfortunately, FWHM could not differentiate between the Mg^{2+}, Fe^{3+} samples and the original gel control. Nevertheless, over the 14 years some silica reorganisation had taken place in all samples. The new fresh silica gel control showed only very minor unidentifiable peaks in the diffractogram scan from 15 to 36° 2θ. However, a number of the scans did show minor crystalline development peaks 21.5 and 22° 2θ. The XRD signals from the 14-year-old gels (Figure 6b) were compared to those of an agate from New Zealand. The host age was 89 Ma but its agate properties show agate formation was 40 Ma (Figure 6a).

The International Centre for Diffraction Data [53] gives the 100% signal intensity for synthetic tridymite and cristobalite as 21.58 and 21.98° 2θ respectively. There is the possibility of tridymite amongst the signal noise to the left of the cristobalite (Figure 6a) but XRD identification of cristobalite and tridymite together is not straightforward. Additionally, both silica polymorphs show some data variability with samples obtained from natural sources. Agates less than 60 Ma produce a similar collection of peaks and are discussed later. XRD details for Figure 6a: data was obtained from a Bruker D8 diffractometer in the reflection mode; the <10 μm powder was scanned over 17° < 2θ < 25° with a step size of 0.02° 2θ and a scan speed of 20 s/step (after Moxon and Carpenter [50]). Further examples and discussion about this cristobalite signal is described using the samples shown in Figure 13. The data for Figure 6b was obtained from a Philips vertical goniometer with runs over 15° < 2θ < 38°. Full details were given in Moxon [52].

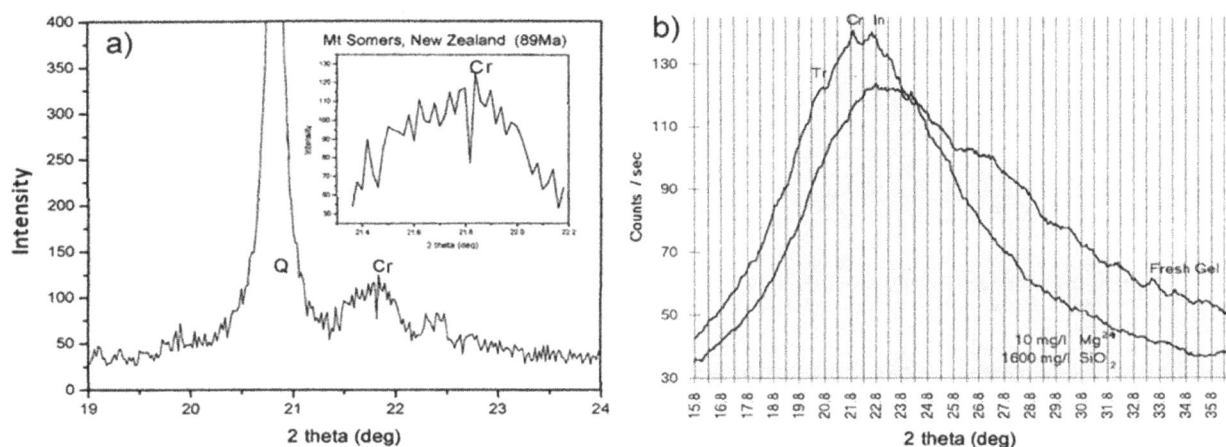

Figure 6. (**a**) An expanded examination of the cristobalite (Cr) and a possible tridymite signal. (**b**) XRD diffractogram of a 14-year-old silica gel and a fresh gel control. The aged gel shows small development peaks that were attributed to tridymite and cristobalite and an unidentified intermediate signal (In) (after Moxon [52]).

3.6. Age of Agate and Host

Unfortunately, there has only been one attempt to date both the age of the agate and volcanic host rock. Yucca Mt., NV, USA was proposed as a nuclear waste storage site and safety concerns have ensured three decades of continual and varied research. Here, chalcedony started development 4 Ma after the formation of the 13 Ma host [54]. Water movement within the mountain was an important part of the investigation. However, the low level of uranium in agates is partially the reason for the lack of any further U-Pb dating.

The availability of a variety of agates from around the world allowed changes of the quartz crystallite size and moganite content to be investigated with respect to host rock age [50]. Agates from 26 host rocks aged between 13 and 3480 Ma were examined using powder XRD to determine the crystallite size and moganite content. A total of 180 and 144 agates were investigated for the respective crystallite size and moganite content determinations. Two of the three agate regions had been involved in later metamorphic events and this had affected their properties [55,56]. The crystallite size (nm) of the remaining 23 regions shows four distinct changes of development (Figure 7a). At (A) the first 60 Ma followed a linear trend; (B) cessation of growth for the next 240 Ma; (C) a small growth spurt over the next 100 Ma; and (D) cessation of growth that continued for the next 700 Ma. The change in moganite content with respect to host rock age is shown in Figure 7b. Here, there was a linear trend over the first 60 Ma with an approximate constancy with agates from older hosts.

Three of the 23 regions were obviously off trend: agates from Brazil, New Zealand and the older agates from Northumbria, England. Brazil and New Zealand agates appeared to have an age of 26 and 30 Ma respectively. Similar comments can be made about the moganite content of the Brazilian and New Zealand agates shown in Figure 7b. The same agates show further evidence of this agate immaturity in the measured various types of water. All show the Brazilian and New Zealand agate regions are not outliers but have clear agate property trends supporting a later agate deposition [40]. Nevertheless for the majority of investigated agates, formation was generally penecontemporaneous with the host rock (Figure 7).

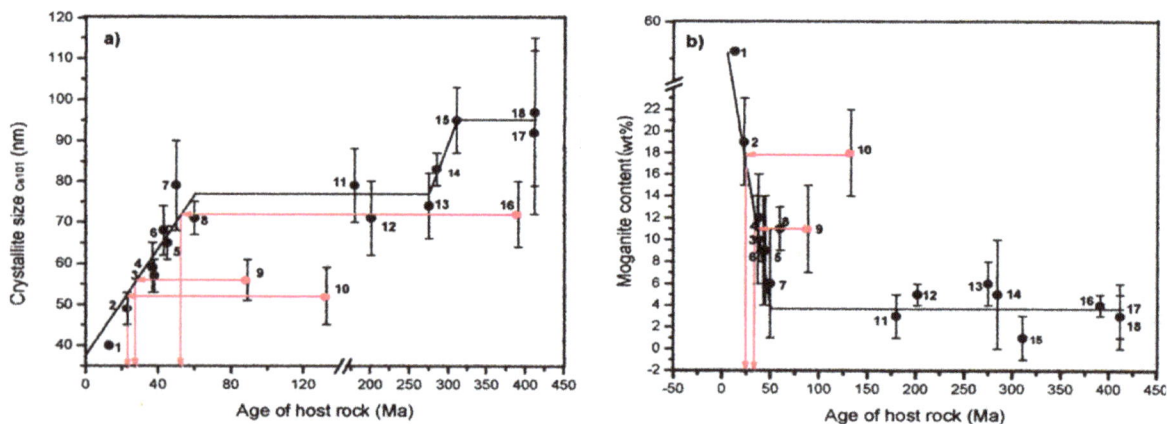

Figure 7. (**a**) Mean crystallite size as a function of age. The plot divides into four age regions: (A) an initial linear growth found in agates from host rocks aged up to 60 Ma: (B) over the next 210 Ma there is minimal growth: (C) growth restarts for 30 Ma: (D) after 300 Ma there is a final cessation of growth up to 1100 Ma. (**b**) Mean moganite content as a function of age. Agates formed from host rocks <60 Ma show a linear decline in the moganite content and little change in the older hosts. The apparent outliers (shown in red) suggest agate formation long after the host age. Agate regions: (1) Yucca Mt., USA; (2) Mt. Warning, Australia; (3) Chihuahua, Mexico; (4) Cottonwood Springs, USA; (5) Washington, USA; (6) Las Choyas, Mexico; (7) Khur, Iran; (8) BTVP, Scotland; (9) Mt Somers, New Zealand; (10) Rio Grande do Sul, Brazil; (11) Semolale, Botswana; (12) Nova Scotia, Canada; (13) Agate Creek, Australia; (14) Thuringia, Germany; (15) Derbyshire, England; (16) East Midland Valley, Scotland; and (17) West Midland Valley, Scotland. A further 8 regions from host rocks aged between 430 and 3750 Ma are not shown as minimal change occurs after 400 Ma. Error bars show one standard deviation. XRD details: Grain size < 10 μm using a Brucker D8 diffractometer in the reflection mode, crystallite size was based on the main (101) quartz reflection recorded on scans $17° < 2θ < 30°$ with step size of $0.01°$ $2θ$ and a scan speed of 10 s/step. The changed details for the moganite determination were $17° < 2θ < 25°$, step size of $0.02°$ $2θ$ and a scan speed of 20 s/step. The moganite and quartz peak areas were determined by fitting unconstrained Lorentzian functions using the Advanced Fitting Tool in "Origin" Moganite content has been taken as the proportion (in %) of moganite peak area/total area with an estimated resolution of ±2%. Full data is given in Moxon and Carpenter [50].

The changes in moganite content and quartz crystallite size with respect to host age is due to the release of structural silanol water:

$$Si\text{-}OH + HO\text{-}Si \rightarrow Si\text{-}O\text{-}Si + H_2O \tag{1}$$

This age-related change shown in Equation (1) has been established using cathodoluminescence. Released water is then able to dissolve the more soluble moganite and increase the size of recrystallizing quartz crystallites. Collectively, this produces a decrease in the moganite and silanol content with an increase in the age-related quartz crystallite size [57].

3.7. The Role of the Infiltration Canal

This peculiar feature has been noted since the 19th century and has been discussed as a means either of silica solution entry or as a water conduit for eventual water loss. It is not difficult to find these canals in some agates (Figure 8). Equally, it is not difficult to slice an agate into many sections and fail to find any canals. Walger et al. [58] examined the structures of these canals and commented that agates can be found with and without the canal. The Ayr to Girvan railway in Scotland was constructed at the beginning of the 20th century and allowed Smith in 1910 [59] to collect many hundreds of agates. He comments on the rarity of the canals in these Scottish agates. It is clear that the canals are a peculiar and interesting feature but cannot play a universal role in agate genesis.

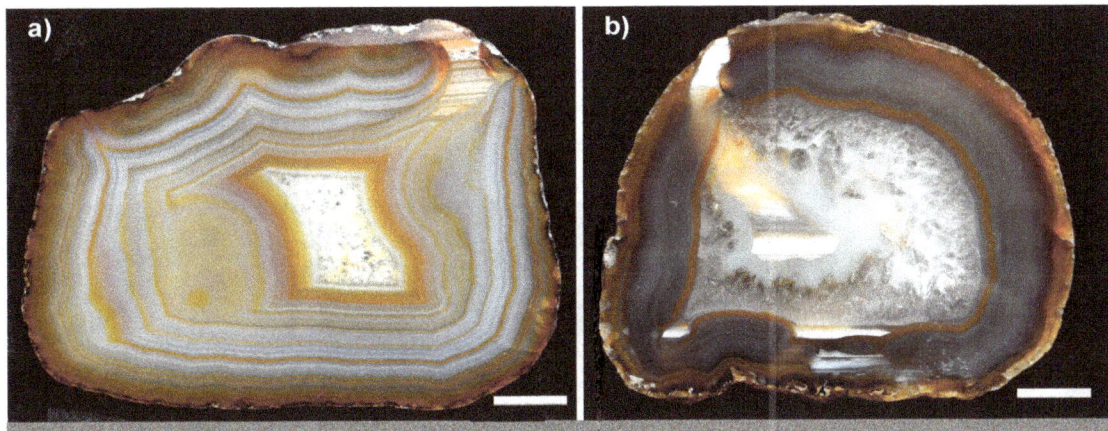

Figure 8. (**a**,**b**) Two Brazilian agates showing similar horizontal banding in the infiltration canals. Scale bar = 1 cm.

3.8. The Wall-Contact Layer

Agates from basic igneous host rocks always have an initial wall-contact layer that is different from the rest of the agate. This layer varies in thickness, generally 1–3 mm, but may appear to be absent. However, a thin section examination with polars crossed will always demonstrate an initial wall contact layer that is different from the bulk of the agate. These agate contact layers evenly coated the vesicle and did not appear to be influenced by gravity. The Botswana agate in Figure 9a was unusual with an apparent 5 mm contact layer. However, a thin section shows that the wall contact lining was the usual 1–2 mm (x in Figure 9b). The brown layer (y) shows a contrasting micro texture with the central region (z) demonstrating that these two areas were different generations.

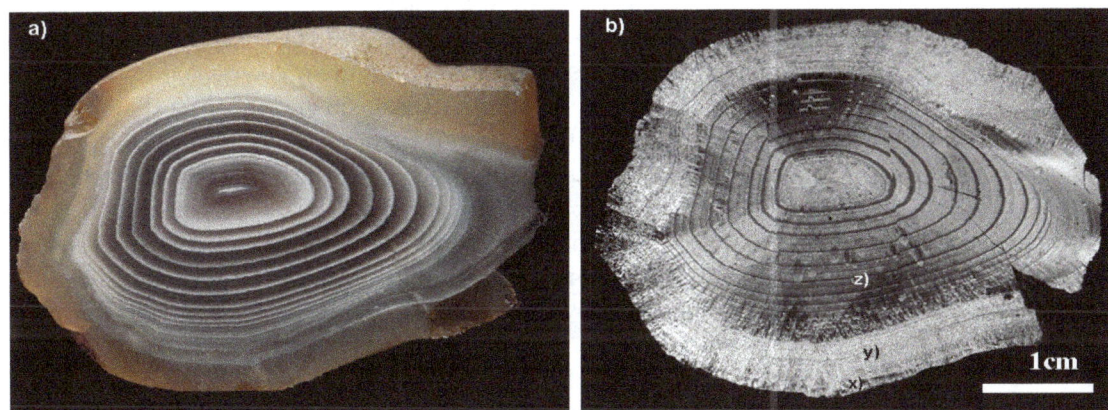

Figure 9. (**a**,**b**) Botswana agate showing an apparent large wall lining layer. However, a thin section (**b**) shows that there are 3 distinct growth formations with (x) a thin actual wall contact layer that is 0.2 mm; (y) the rest of the brown layer and (z) the black and white agate bulk.

Reis [10,11] proposed this wall-lining was a different generation. Comment has been made about Brazilian agates [60]; bituminous agates [61]; and sedimentary agates [62]. An agate wall lining is also found in the Brazilian amethyst geodes. This led to the proposition that aqueous silica in the forming solutions diffuse through agate fractures [63]. Fractures or even hairline cracks are rare in agate and when observed they are sufficiently significant to be assigned to host rock movement. The genesis implications of this agate contact layer are of prime importance. A first deposit of wall lining would always present a potential barrier for bulk entry of silica into the vesicle.

This wall lining deposit is sometimes apparent in a thick section: as shown by the Brazilian agates in Figure 8b. Notice the near uniformity of the wall lining layers in these agates. It would be remarkable if the first silica solution could fill the gas vesicle and evenly coat the vesicle wall, crystallize, and then allow subsequent silica solutions to enter.

One study characterised nine Brazilian agates that had a lower horizontally banded section and an upper chamber with the normal wall lining pattern [64]. The volume division between the wall lining and horizontal banding varied with different agates. The horizontal banding in these agates is between 30 and 70% of the total. Thin sections of four of these Brazilian agates are shown in Figure 10a–d.

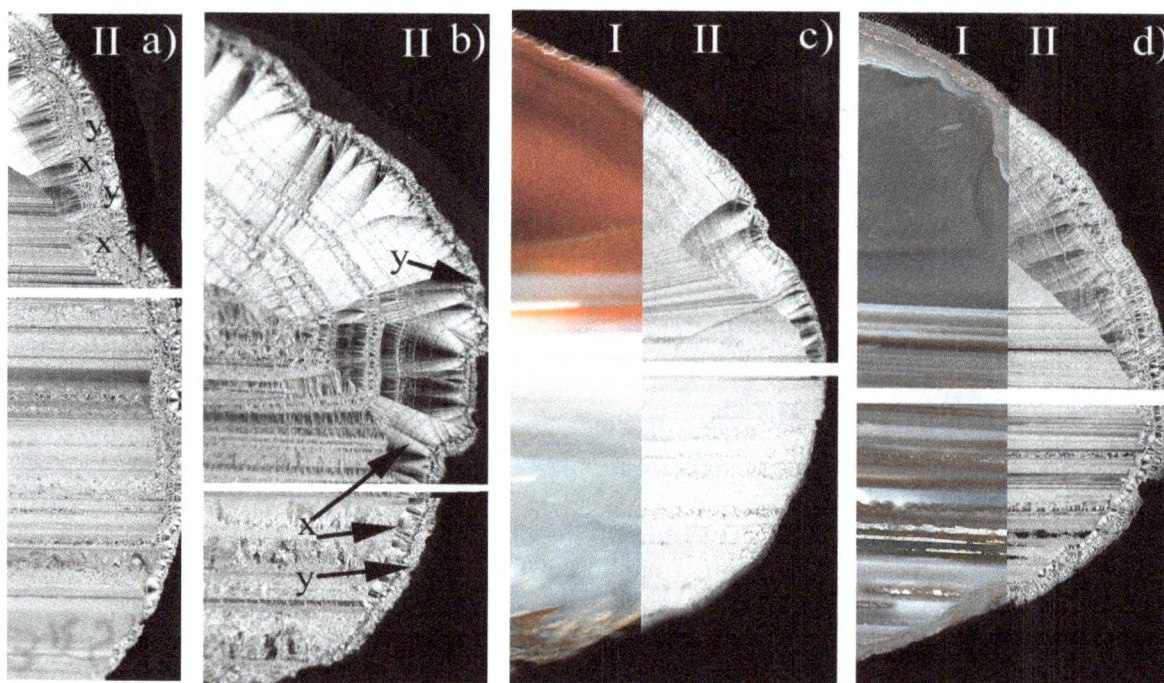

Figure 10. (a–d) The four monochrome micrographs are thin sections of four Brazilian agates with wall lining and horizontal banding, these are labelled as II (a–d). There are two thick sections with the original banding (Ic, Id). The discrete layers of horizontal bands are separated from the wall lining sections by the white dividing space. In Brazil slab II (a), there are two initial wall lining deposits, labelled x and y: these are reduced in the horizontally banded section. The thickness of the initial wall lining layer in Brazil slab II (b) shows tapering in the base of the horizontally banded section. The continuous sheaf-like growth (x) from the wall lining also coats the gap at the edge of the horizontally banded layers. Brazil slab (c) shows the initial wall lining deposit is absent in the horizontally banded layers. Brazil slab (d) demonstrates four separate wall lining deposits line the cavity wall. Scale of thin sections = 2 cm at the maximum width.

The wall contact layer in Figure 10IIa,b,d was present in the upper and lower chambers but absent in the lower chamber of IIc. There are two wall contact layers shown in agate IIa (labelled x and y). These two layers gradually faded into one at the bottom of the horizontal region. In IIc the wall contact layer was absent in the horizontally banded region. Additionally, these layers got thinner in the horizontal banded region in IId.

The observations on these combined horizontal and wall lining agates suggest the following development. Initially, enough silica is deposited on the empty vesicle floor to form a single horizontal band. At some point, there was a cessation of silica. During a later period fresh silica was deposited and followed by another pause of silica deposition. The process is repeated producing several layers. The evidence for these separate depositions is shown by the differing layer textures that range from chalcedony to prismatic quartz. Clearly this is a stop/start activity. Eventually the upper chamber now had a horizontal base but the silica input continued until sufficient silica had filled the remaining space. These agates were examined using XRD and some areas demonstrated trace cristobalite. The presence of cristobalite in agate supports the development mechanism proposed by Landmesser [25,26] and will be discussed later:

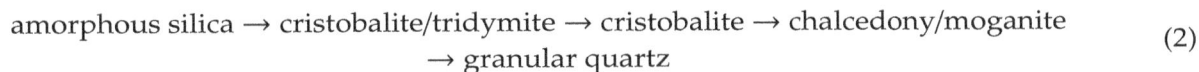

$$\text{amorphous silica} \rightarrow \text{cristobalite/tridymite} \rightarrow \text{cristobalite} \rightarrow \text{chalcedony/moganite} \\ \rightarrow \text{granular quartz} \tag{2}$$

Each of the silica transformations shown in Scheme (2) has a greater density than its predecessor. When the transformation of layers and upper chamber is complete then shrinkage will occur but the larger bulk in the upper chamber will produce the greatest shrinkage. Here, the gap will be wider between the agate and the vesicle wall. Finally, silica deposition fills this wall lining gap. Agate IIb shows a much wider fibrous growth in the upper chamber and with a continuous but thinner fibrosity at the bottom of this chamber where thin horizontal bands have formed. The wall lining falls to around a fifth of the thickness when it fills the contact wall in the lower horizontal bands. Hence, the wall lining layer is the last generation.

3.9. Agate under the Scanning Electron Microscope

Lange et al. [65] and Holzhey [66] used the scanning electron microscope (SEM) to find evidence of the petrographic fibrosity observed in agate thin sections. The fibrosity was not found and instead, a surface composed of globulites was observed. The globulites varied in size between 0.1 and 3 μm [66] or subparticles with sizes ranging from 0.07 to 0.7 μm [66]. Additionally, there have been a number of publications citing the use of the SEM in a support role when investigating agate and chalcedony. Examples include, the effects of heating chalcedony at 500 °C [67]; agates from Permian rocks [68].

Early work had shown age-related trends in the development of quartz crystallite size and the SEM was used to observe differences between agates from young and old host rocks [69]. There are clear trends of increasing globulite size with age (Figure 11). However apart from agates from the youngest and oldest host rocks, there were too many size variables in each micrograph for worthwhile size comparison data. The same study showed host-age related differences between the white bands in agate. Agates from young host rocks will frequently show hints of faint white banding that becomes more intense in agates from host rocks greater 50 Ma. The white bands develop a structural form that is reminiscent of the edges of a stack of plates (Figure 12). Heinz [31] boiled the white and coloured bands in KOH$_{(aq)}$ and found that white bands offered a greater resistance to the KOH$_{(aq)}$: this was credited to an increased opal content. Opal is absent in agate and the SEM shows the observed differences are structural: the "plate" edges offering more resistance to the alkali attack.

3.10. Hydrocarbon Inclusions in Agate

In recent years, various hydrocarbon deposits have been reported in agates. Barsanov and Yakovleva [70] found oil and water in both horizontal layered bands from North Timan in the Russian Urals and in wall lining banded agate from the Kyzal-Tugan deposit, Taldy-Kurgan, Kazakhstan. The identification of hydrocarbons between horizontal agate layers is particularly interesting as the different formations in these horizontal layers is indicative of their long time-related formations.

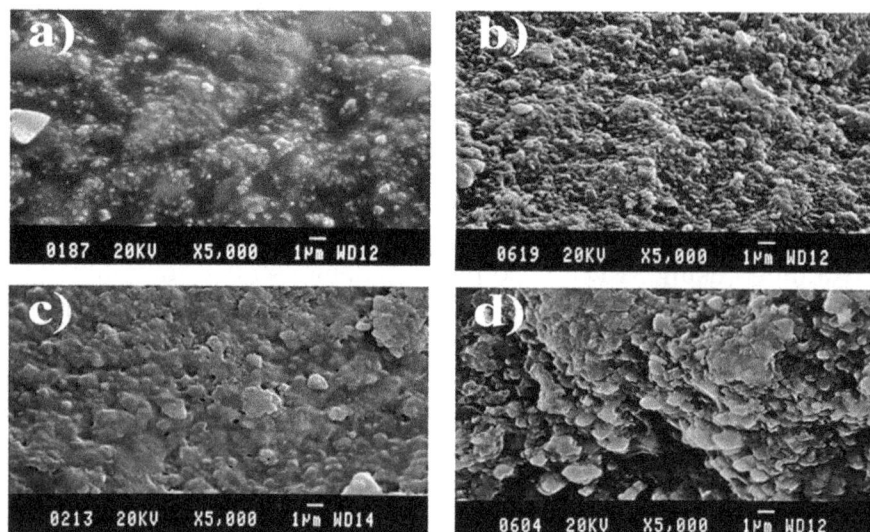

Figure 11. SEM micrographs showing the development in clear bands: (**a**) Rio Grande do Sul, Brazil 135 Ma, (52 nm) 135 Ma; (**b**) Isle of Mull, Scotland, 60 Ma (72 nm); (**c**) Agate Creek, Queensland, Australia 275 Ma (74 nm); and (**d**) Ethiebeaton Quarry, Scotland 412 Ma (96 nm). The stated age is the host age and crystallite size is the mean crystallite size for that particular region. The properties of the Brazilian agate, including crystallite size, indicate a formation age of ~26 Ma. SEM details: small agate slices were coated with gold and examined in a JEOL 820 SEM. The experimental conditions used a 20 kV accelerating voltage with a current of 1 nA.

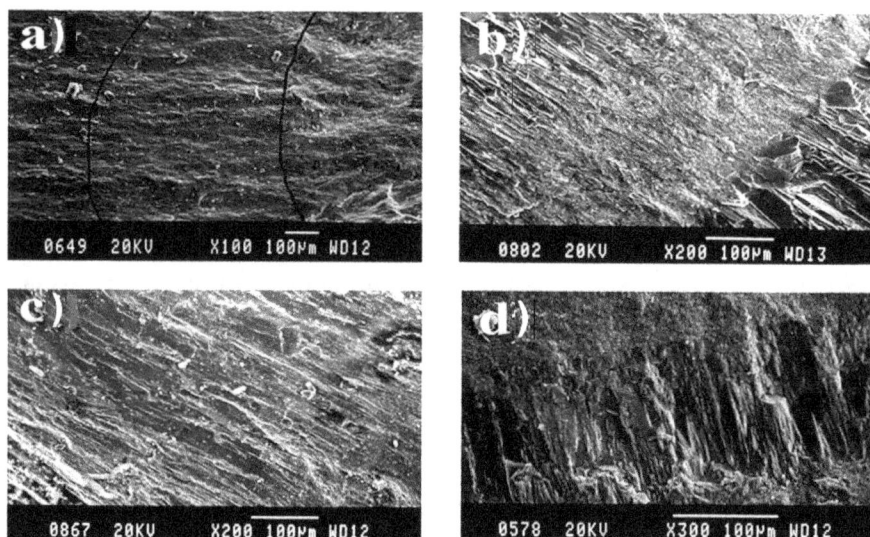

Figure 12. (**a–d**) are the white banded regions from the same agates shown in Figure 11: (**a**) Rio Grande do Sul, Brazil 135 Ma, (52 nm) 135 Ma. The black lines mark a faint white band; (**b**) Isle of Mull, Scotland, 60 Ma (72 nm); (**c**) Agate Creek, Queensland, Australia 275 Ma (74 nm); (**d**) Ethiebeaton Quarry, Scotland 412 Ma (96 nm). The stated age is the host age and crystallite size is the mean crystallite size for that particular region. The properties of the Brazilian agate, including crystallite size, indicate a formation age of ~26 Ma. SEM details as given in Figure 11.

Silaev et al. [71] studied the composition of gases in agates from the epithermal deposits from the Russian Polar Urals. Most of the secretion bodies have a lens shape with intermittent and wedge-shaped protrusions. Using pyrography, they identified the main composition of the inclusions. In order of decreasing concentration these were H_2O, CO_2, CO, CH_4, and N_2. The total content of the gases was in the range of 100–600 µg/g. Hydrocarbons in the fluid inclusions varied from 0.25 to 7 µg/g. The presence of hydrocarbons in agate was confirmed by radioscopic data. As a result of the determinations of the gas–liquid inclusions, they proposed a formation temperature in the range of 100–200 °C.

A third study of hydrocarbons in agates was made using samples from Nowy Kościoł Lower Silesia, Poland by Dumeńska-Slowik et al. [61]. They describe the composition of the bitumen as mainly asphaltenes (56%). The remaining fractions in order of composition are saturated hydrocarbons (18%), resins (16%), and aromatic hydrocarbons (10%). Carbon isotope analysis revealed an algal-humic or algal origin.

4. Discussion

Agates are mainly found in basic igneous host rocks and this discussion is limited to the origins of agates from these host rocks. Any solution to the genesis problem requires answers to the following questions.

(a)   What is the formation temperature?
(b)   What is the source of the silica?
(c)   What is the nature of the silica deposit?

(a) Formation temperature: this is the only question that has been answered and the majority of interested workers accept the oxygen and hydrogen isotope analysis of Fallick et al. [35] and Saunders [37]. The collective present evidence shows that the agate formation temperature in these basaltic host rocks was <200 °C and most likely <100 °C. Any proposal of a higher formation temperature has to explain the conflict of evidence found by the various irreversible silanol water losses that are observed between 200 and 850 °C that were described by Yamagishi et al. [37]. All the observed changes would not be observed if agate had formed at a temperature >250 °C. Furthermore, when agate is heated at higher temperatures there are clear visible and further irreversible physical changes, as the agate turns white and porous. Depending upon the temperature, agate is also likely to become fractured.

(b) Silica source: Host rock leaching would be the most popular choice suggested by interested workers. One of us (T.M.) has examined many of the 412 Ma agate-bearing host rocks of the Midland Valley, Scotland. There is ample evidence of weathering/alteration in these old host rocks. Although fresh feldspars were observed, most did show signs of change. The ferromagnesian content was highly altered with thin sections showing indistinct remnants as a common feature. These rocks had the potential to release the silica for agate formation and Fe^{3+} for oxide colouring. However, the most telling feature for some of the Eastern Midland Valley agates is their clear lack of red/brown iron oxide colouring. Conversely, the Burn Anne agates from the Western Midland Valley provide Scotland's most colourful red and yellow agates. Extensive research on the Yucca Mt., NV, USA shows the downward percolation of water is depositing amorphous silica in these 13 Ma Miocene Tuffs. Chalcedony forms in fissures and as a host coating [72]. However, these tuffs are not the typical agate host rocks.

The link between host age and agate crystallinity demonstrates agate formation is mostly a penecontemporaneous deposit of host and agate [50]. This timing would offer support for late hydrothermal solutions or hot springs carrying higher concentrations of H_4SiO_4. Scientists investigating hot spring silica differentiate between two types of silica sinter: a direct deposition from hot solutions and silica released by steam attacking the host rock. Either directly or indirectly these actions are a potential supply of silica for agate formation.

(c) The nature of the silica deposit: There are two potential alternatives for the initial deposit: silica gel or powder. This would be followed by a later transformation into chalcedony. Harder and Flehmig [73] showed that iron, magnesium, and aluminium hydroxides could absorb silica from low concentration of silica solution (0.5 ppm). The silica enriched precipitates formed quartz crystals within days. Elsewhere, quartz crystals have been grown in the laboratory from a solution containing 4.4 ppm of silicic acid at 20 °C [74]. Both these studies require quartz formation to be from solutions that are saturated with respect to quartz but under saturated with respect to amorphous silica. These conditions are limiting and the developing crystals are prismatic quartz and not agate.

Godovikov et al. [23] presented the results of experiments investigating the diffusion processes in silica gel. They used the silicate (office) glue, which is mainly an aqueous solution of Na_2SiO_3 and iron, nickel, or copper sulphate in the experiments. A sparingly soluble salt in solution was poured onto a silica gel containing the same salt. Maximum supersaturation occurs at the interface between solution and gel. At this interface, they observed the appearance of complex branched dendrites. When the degree of supersaturation was minimal, there was a growth of individual full-faceted crystals. They proposed that spherulites arise under conditions of supersaturation. Additionally the authors proposed that the colour of agates has various causes. Essentially, a) uneven distribution of pores and their different shapes in separate layers of agate; b) a different mineral composition in individual zones; and c) finely dispersed inclusions of brightly coloured minerals.

Heaney [75] developed a theoretical model to explain the formation of chalcedony and prismatic quartz in the same sample. This was based on the differences of structure and solubility between chalcedony and prismatic quartz. A final single deposit of macrocrystalline quartz is shown in Figure 3. When this occurs prismatic quartz is usually found around the centre of a sample with a clean break from the surrounding chalcedony. An immediate impression is distinct separate entries and depositions. This is not possible and the Heaney [75] mechanism is based on a weakly polymerised silica solution allowing the formation of chalcedony and leaving a more dilute solution of monomeric silicic acid. Eventually, the very dilute silicic acid solution is now under saturated with respect to amorphous silica but saturated with respect to prismatic quartz: ideal conditions for a final deposit of prismatic quartz.

Over the last century and prompted by the work of Liesegang [2], there has been support for the direct precipitation of silica gel as an agate precursor. Reported direct formations of silica gel are very rare but Naboko and Silnichenko [76] commented on a silica gel (52% silica) that was forming on the solfataras of the Golovnin volcano in Russia. However, gel cannot diffuse through the host rock. Any gel precursor must form within the gas vesicle. To overcome this objection, a hypothesis was proposed by Harder and Flehmig [73] who suggested that a reaction between solutions containing silica sols and $Fe^{3+}_{(aq)}$ or $Al^{3+}_{(aq)}$ enter the vesicle separately. This mixture produces a gel-like state that could eventual form agate. Present day solutions circulating in host rocks do not contain the high concentrations of these two ions or silica; these would be necessary to generate a silica gel. Dry gels quickly dehydrate and produce a powder of amorphous silica. As a hypothesis, there seems little point in adding one more step to this amorphous silica beginning.

Agate, chalcedony, and chert are all forms of microcrystalline quartz. Although chert commonly shows a granular texture when examined petrographically, patches of chalcedony in chert are not rare. Chalcedony has been identified occurring with amorphous silica in the Yucca Mt, Nevada. Additionally it is amorphous silica (opal-A) that is commonly found around hot springs in New Zealand and Yellowstone Park, USA. Unfortunately, investigators of silica sinter do not tend to differentiate between the various forms of microcrystalline quartz: chalcedony is not often reported. Nevertheless, chalcedony is presently being formed in the Tengchong geothermal area China [77]. Additionally, Marcoux et al. [78] identified black, grey-brown and white chalcedony that formed in the siliceous sinter of the 295 Ma French Massif Central.

Studies on the genesis of chert and sinter have produced a much greater volume of work than agate but similarities and differences between all three provide potential development links for agate investigation. Stages of development for chert [79] and sinter [80] have been shown to follow sequence:

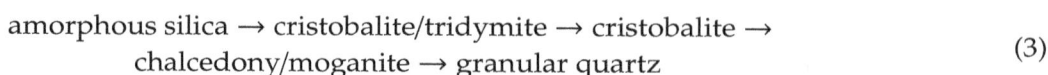

$$\text{amorphous silica} \rightarrow \text{cristobalite/tridymite} \rightarrow \text{cristobalite} \rightarrow$$
$$\text{chalcedony/moganite} \rightarrow \text{granular quartz} \tag{3}$$

Landmesser [25,26] argued the same transformations for agate (2). A comparison of the rate of phase transformations for agate, chert, and sinter demonstrates the major origin differences between these three forms of microcrystalline quartz. Silica sinter discharges from weakly alkaline chloride waters at ≤100 °C. The time scale for completion of the sinter diagenesis is typically 50 ka [81]. However, a faster rate of conversion of sinter into quartz has been achieved when a partial conversion into quartz occurred within weeks [82]. As expected, the completion of each silica sinter transformation step is generally not rapid. For example, a single XRD diffractogram can show evidence of amorphous silica, cristobalite and tridymite.

Moganite is found in sinter, chert, and agate but the survival rates differ greatly. In sinters with ages 20–220 ka the moganite content is at a maximum of 13 wt% whereas Tertiary sinters are generally free of moganite [83]. Chert hosts of the Cretaceous age have been identified with a moganite content of >20 wt% [84]. This contrasts with 56 wt% moganite identified in the Yucca Mt chalcedony and it has been identified in agate from the 1100 Ma Lake Superior host [50]. However, detecting moganite from agate in host rocks >400 Ma using XRD is not always possible.

Chert diagenesis is much more prolonged where surface amorphous silica can survive for 85 Ma. Transformation to cristobalite/tridymite can be 5–10 Ma and is still found in rocks aged 120 Ma but is not observed in rocks >140 Ma. Quartz requires a minimum of 40 Ma at depths of 500 m [85]. The phase transformation changes and moganite differences in sinter, agate, and chert have a link to the particular formation temperatures: silica sinter shows a variable but the most rapid changes with the silica deposits held at 100 °C.

Agate development bears a close link to the age of the host and polymorph precursors have occasionally been identified: cristobalite/tridymite layering in Brazilian agate [22]. Moxon and Carpenter [50] identified cristobalite in agates from Mexico, Iran, Rum, and Scotland: these are all hosts younger than 60 Ma. Cristobalite was also found in the hosts of New Zealand agate (89 Ma) and Brazil (135 Ma): providing further evidence for a formation age much later than their hosts. The basic XRD data identifying the weak cristobalite XRD signals in agates from Brazilian, and Iranian hosts are shown in Figure 13. Cristobalite identification in agate has not been reported in agate from hosts older than 60 Ma elsewhere. As discussed earlier, the properties of Brazilian agate demonstrate a formation at 26–30 Ma.

The prime aim of the Moxon and Carpenter (2009) [50] study was to investigate the moganite to quartz transformation. The quantification of the moganite content was carried out using a Bruker diffractometer with a step size of 0.02° 2θ and a scan rate of 20 s/step. A total of 32 agates were examined from hosts ≤60 Ma (that also included Brazilian and New Zealand agates). Exactly half were free of cristobalite but the remainder gave the cristobalite XRD signal. Three agates from the cristobalite batch were selected for a more intense scan and at least two of the next cristobalite most intense signals were identified at 28.4, 31.5, and 36.1°2θ. The detected cristobalite was noted in the paper but not examined any further.

The same data has been examined in more detail and two of the diffractograms are shown in Figure 13. The Brazilian agate shows the presence of cristobalite and the plot was similar to that produced by the New Zealand agate in Figure 6a. All the agates, apart from those from Brazil and New Zealand, were from hosts younger than 60 Ma.

There was a contrast between the cristobalite peaks shown by the two agates in Figure 13. There were two peaks in the Brazilian agate against the single peak in the Iranian agate. We speculated that the second outer peak could be tridymite and the change from younger to older agate was demonstrating an opal CT → opal C change. The outer peaks of the agates showing this peaked feature were measured and produced a mean value of 21.89(6)° 2θ and a cristobalite mean value of 21.99(4)° 2θ. Data shown in parentheses are ± one standard deviation. The JCPDDS [53] data for synthetic tridymite and synthetic cristobalite were 21.58 and 21.98° 2θ respectively. The identification of tridymite and cristobalite with XRD when found together in the natural state is not straightforward. The regular appearance and consistency of the outer signal in these samples would rule out instrument noise.

Of the 16 agates that showed these two signals, 5 were from Mt Somers New Zealand and 4 in each of the batches from Chihuahua, Mexico, and Khur, Iran. Cristobalite identification in agate has not been found elsewhere in agates older than 60 Ma. Hence, the identification of cristobalite in

Brazilian and New Zealand agates further supports their late agate formation. Dates of the late entry of some secondary minerals into the gas vesicles of the Brazilian Paraná Continental Flood basalts have been summarised by Gilg et al. [86]. Various secondary infills occurred up to 60 Ma after the basalt formation. However, the properties of Brazilian agates (water and moganite content and crystallite size) show a formation age around 100 Ma after host formation.

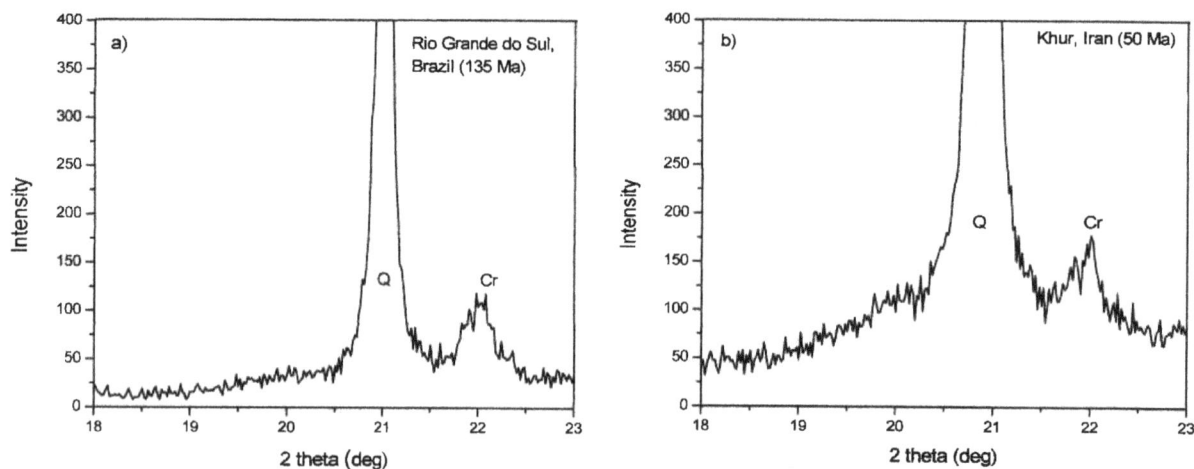

Figure 13. Cristobalite (Cr) identified in agates from Brazil (**a**) and Iran (**b**). The stated ages are the host ages but the properties of Brazilian agate suggest an agate age of 25–30 Ma.

5. Summary

The most prolific source of agate is from basalt and andesite and we concentrated our deliberations on agates from these hosts. Such agates provide by far the greatest scientific interest and involvement. Agates found in acidic igneous and sedimentary rocks are more rare and this is reflected by the smaller number of literature publications.

Until the mid 20th century there were few scientific publications on agate and these were mainly limited to speculative comments based on agate banding and their hosts. Viewed from 2020, clear exceptions would be the early observations and micrographs produced by Timofeev [8]; the detailed thin section examinations of Pilipenko [13] and the gel studies of Liesegang [2,3]. Liesegang's works on gels were still producing links to agate genesis until the end of the last century. However, there is now little support for an explanation of agate formation based upon the concept of metal ion diffusion into a silica gel.

From 1960, industrial requirements for high quality quartz were the driving force for a number of high temperature and pressure investigations on various forms of silica. Unfortunately, none of these experiments had a direct application to agate genesis. Additionally, studies using high temperature/pressure conditions and those under normal earth conditions can proceed by different mechanisms giving conflicting results. Nevertheless, advances in scientific equipment over the last 60 years have allowed a more detailed understanding of the properties and composition of agate.

In these basic igneous host rocks, enough evidence has now been obtained for an agate formation temperature at mostly < 100 °C. The agate silica content was initially assumed to be just fine-grained α-quartz. However, the more recent discoveries of moganite and cristobalite have allowed evidence-based speculation to propose that agate starts as amorphous silica. Over geological time, it progresses to cristobalite before transforming into agate.

It is probable that tridymite forms before cristobalite but the evidence has still to be conclusively found. On initial formation, agate is on a lengthy journey of development. Agates are unusual in ageing over the geological time scale. The changes are due to the release of water and the very slow transformation of moganite into quartz. The result is an increase in quartz crystallite size. SEM examinations demonstrate globular development in the clear bands and further maturity in the white bands. This development in the white bands is often apparent in agate thin sections. White bands in agates older than 180 Ma generally show a more intense white than is found in younger agates (compare the Brazilian and Botswana samples in the respective Figures 8 and 9). A thin section of a 3480 Ma agate from Australia shows that the end of agate development is only reached when fibrous quartz is transformed into microcrystalline granular quartz [55].

Presently, agate genesis research does lack any commercial driving force. The genesis problem has been in the literature since the 18th century and will not be to be solved with a 3-year funding grant. One important incentive for chert and silica sinter research has been the early forms of life that have been trapped in the material. This might change for agate as the Mars Rover units have been investigating silica links to water and microbiological activity and this does include chalcedony.

For the future, agate properties that change w.r.t host age might be a simpler and a cheaper direction of study. In order to make valid comparisons, any such investigation will require sufficient agates from host rocks of various ages. As always, there is the proviso regarding outliers that have formed later than the host. As described in this paper, three of the 26 agate regions demonstrated an earlier but consistent age of formation that was much later than the host.

The most likely area of profitable investigation would involve agate from very young host rocks. We described data from a single sample of Yucca Mt (13 Ma) chalcedony providing a crystallite size and moganite content. While this agate had a crystallite size of 40 nm that is in line with expectations, the 56 wt% moganite did appear to be very high when compared to the 19% moganite in agate from the 23 Ma Mt Warning Australia host. At face value, this would suggest that agates in their very early years of development are exceptionally high in moganite. In which case, extrapolation back to zero time would not be possible. Alternatively, this sample is a freak result and further analyses would produce other data that is more in line with expectations. Where possible a minimum of five samples from the same area should be examined. It would appear that agates from the young host rocks have the most to offer future investigations into agate genesis. In this category, the islands of Hawaii would be a prime source for agate study. These islands have formed as a result of movement across a hot spot in the Earth's mantle. Several islands offer formation dates ranging from 0.43 to 5.1 Ma [48]. It helps that the geology of these islands has been thoroughly investigated by the US Geological Survey and many others.

Author Contributions: Conceptualization, T.M.; methodology, T.M.; writing and editing, T.M. and G.P.; visualization, T.M. and G.P.; supervision, T.M.; translation of works in Russian into English, G.P. All authors have read and agreed to the published version of the manuscript.

Acknowledgments: The agates used in the studies by T.M. have been donated by Glenn Archer (Outback Mining, Australia), Rob Burns, Jeannette Carrillo (Gem Center USA, Inc.), Roger Clark, Nick Crawford, Brad Cross, Robin Field, Gerhard Holzhey, Brian Isfield, Herbert Knuettel (Agate Botswana), Reg Lacon, Brian Leith, Maziar Nazari, Dave Nelson, Leonid Neymark, John Raeburn, John Richmond, Vanessa Tappenden, Bill Wilson, and Johann Zenz. Links between agate properties and host age could not have been investigated without these donations. T.M. is indebted to the Dept. of Earth Sciences, Cambridge University, UK for access to the laboratories. The contributions of co-workers and support staff were essential for the progress made. Thanks are due to Tony Abraham, Michael Carpenter, Ian Marshall, Chris Parish, Chiara Petrone, Stephen Reed, Susana Rios, Paul Taylor, Martin Walker, and Ming Zhang. Gerd Schmid translated the early German papers into English. We thank Evgeniya Svetova for help in finding rare pre-1934 Russian agate articles. We thank three reviewers for their comments after a first submission of the manuscript.

References

1. Collini, C. *Tagebuch Einer Reise, Welches Verschiedene Mineralogische Beobachtungen, Besonders über Die Achate und den Basalt Enthält*; C.F. Schwan: Mannheim, Germany, 1776; p. 582.

2. Liesegang, R.E. *Die Achate*; T. Verlag von Theodor Steinkopff: Dresden/Leipzig, Germany, 1915; p. 122.

3. Liesegang, R.E. Achat Theorien. *Chem Erde* **1931**, *4*, 143–152.

4. Zenz, J. *Fascinating Idar Oberstein in Agates*; Bode: Lauenstein, Germany, 2009; pp. 164–185.

5. Noeggerath, J. Sendschreiben an den K. K. wirklichen Bergrath und Professor Haidinger in Wien, über die Achat-Mandeln in den Melaphyren. *Verh. Nat. Ver. Preuss. Rheinl. Westphalens* **1849**, *6*, 243–260. (In German)

6. Haidinger, W. Versammlungsberichte. In *Berichte über die Mittheilungen von Freunden der Naturwissenschaften in Wien*; Braumüller u. Seidel: Wien, Germany, 1849; pp. 61–69. (In German)

7. Heddle, M.F. *The Mineralogy of Scotland*; David Douglas: Edinburgh, Scotland, 1901; Volume 1, p. 148.

8. Timofeev, V.M. Chalcedony of Sujsar Island. *Proc. Soc. St. Petersburg Nat.* **1912**, *35*, 157–174. (In Russian)

9. Reis, O.M. Einzelheiten über Bau und Entstehung von Enhydros, Kalzitachat und Achat I. *Geognost. Jahresh.* **1917**, *29*, 81–298. (In German)

10. Reis, O.M. Einzelheiten über Bau und Entstehung von Enhydros, Kalzitachat und Achat II. *Geognost. Jahresh.* **1918**, *31*, 1–92. (In German)

11. Fischer, W. Zum Problem der Achatgenese. *N. J. Miner. Abh.* **1954**, *86*, 367–392. (In German)

12. Jessop, R. Agates and cherts of Derbyshire. *Geol. Assoc.* **1930**, *42*, 29–43. [CrossRef]

13. Pilipenko, P.P. Zur Frage der Achat genese. *Bul. Soc. Nat. Mosc.* **1934**, *12*, 279–299. (In German and Russian)

14. Kuzmin, A.M. *Periodic-Rhythmical Phenomena in Mineralogy and Geology*; Scientific & Technical Translations: Tomsk, Russia, 2019; p. 336. (In Russian)

15. Nacken, R. Über die Nachbildung von Chalcedon–Mandeln. *Natur Folk* **1948**, *78*, 2–8.

16. Carr, R.M.; Fyfe, W.S. Some observations on the crystallization of amorphous silica. *Am. Mineral.* **1958**, *43*, 908–916.

17. Keat, P.P. A New Crystalline Silica. *Science* **1954**, *120*, 328–330. [CrossRef] [PubMed]

18. White, J.F.; Corwin, J.F. Synthesis and origin of chalcedony. *Am. Mineral.* **1961**, *46*, 112–119.

19. Heydemann, A. Untersuchungen über die Bildungsbedingungen von Quarz in Temperaturbereich zwischen 100 °C und 250 °C. *Beiträge zur Mineralogie und Petrographie* **1964**, *10*, 242–259. (In German) [CrossRef]

20. Ernst, W.G.; Calvert, S.E. An experimental study of the recrystallization of porcelanite and its bearing on the origin of some bedded cherts. *Am. J. Sci.* **1969**, *267*, 114–133.

21. Oehler, J.H. Hydrothermal crystallization of silica gel. *Geol. Soc. Am. Bul.* **1976**, *87*, 1143–1152. [CrossRef]

22. Flörke, O.W.; Köhler-Herbetz, B.; Langer, K.; Tönges, I. Water in microcrystalline quartz of volcanic origin: Agates. *Contrib. Mineral. Petrol.* **1982**, *80*, 324–333. [CrossRef]

23. Godovikov, A.A.; Ripinen, O.I.; Motorin, S.G. *Agates*; Nedra: Moscow, Russia, 1987; p. 368. (In Russian)

24. Goncharov, V.I.; Gorodinsky, M.E.; Pavlov, G.F.; Savva, N.E.; Fadeev, A.P.; Vartanov, V.V.; Gunchenko, E.V. *Chalcedony of North-East. of the USSR*; Science: Moscow, Russia, 1987; p. 192. (In Russian)

25. Landmesser, M. Mobility by metastability: Silica transport and accumulation at low temperatures. *Chem. Erde* **1995**, *55*, 149–176.

26. Landmesser, M. Mobility by metastability in sedimentary and agate petrology: Applications. *Chem. Erde* **1998**, *58*, 1–22.

27. Moxon, T. *Agate Microstructure and Possible Origin*; Terra Publications: Doncaster, UK, 1996; p. 106.

28. Götze, J. Agate-fascination between legend and science. In *Agates III*; Zenz, J., Ed.; Bode-Verlag: Salzhemmendorf, German, 2011; pp. 19–133.

29. Sorby, H.C. On the microscopical structure of Calcareous Grit of the Yorkshire coast. *Quart. J. Geol. Soc. Lond.* **1851**, *7*, 1–6. [CrossRef]

30. Eberspächer, S.; Lange, J.-M.; Zaun, J.; Kehrer, C.; Heide, G. The Historical Collection of Rock Thin Sections at the Technische Universität Bergakademie Freiberg and Evaluation of Digitization Methods. 2015. Available online: https://www.researchgate.net/publication/273118591e (accessed on 1 September 2020).

31. Heinz, H. Die Entstehung Die Entstehung der Achate, ihre Verwitterung und ihre künstliche Färbung. *Chem. Erde* **1930**, *4*, 501–525. (In German)

32. Pabian, R.K.; Zarins, A. *Banded Agates—Origins and Inclusions. Educational Circular. No. 12*; University of Bebraska: Lincoln, NE, USA, 1994; p. 32.

33. Belousov, B.P. A periodic reaction and its mechanism. In *Sbornik Referatov po Radiatsioanoi Medicine for 1958 Year*; Medgiz: Moscow, Russia, 1959; p. 145. (In Russian)

34. Zhabotinsky, A.M. Periodic liquid phase reactions. *Proc. Acad. Sci. USSR* **1964**, *157*, 392–395. (In Russian)

35. Fallick, A.E.; Jocelyn, J.; Donelly, T.; Guy, M.; Behan, C. Origin of agates in the volcanic rocks of Scotland. *Nature* **1985**, *313*, 672–674. [CrossRef]

36. Harris, C. Oxygen-isotope zonation of agates from Karoo volcanic of the Skeleton Coast, Namibia. *Am. Mineral.* **1989**, *74*, 476–481.

37. Saunders, J.A. Oxygen-isotope zonation of agates from Karoo volcanic of the Skeleton Coast, Namibia: Discussion. *Am. Mineral.* **1990**, *75*, 1205–1206.

38. Götze, J.; Plötze, M.; Tichomirova, M.; Fuchs, H. Aluminium in quartz as an indicator of the temperature of formation of agate. *Miner. Mag.* **2001**, *65*, 407–413. [CrossRef]

39. Yamagishi, H.; Nakashima, S.; Ito, Y. High temperature infrared of hydrous spectra microcrystalline quartz. *Phys. Chem. Miner.* **1997**, *24*, 66–74. [CrossRef]

40. Moxon, T. A re-examination of water in agate and its bearing on the agate genesis enigma. *Miner. Mag.* **2017**, *81*, 1223–1244. [CrossRef]

41. Flörke, O.W.; Jones, J.B.; Schmincke, H.U. A new microcrystalline silica from Gran Canaria. *Zeitschrift für Kristallographie-Crystalline Materials* **1976**, *143*, 156–165. [CrossRef]

42. Heaney, P.J.; Post, J.E. The Widespread Distribution of a Noel Silica Polymorph in Microcrystalline Quartz Varieties. *Science* **1992**, *255*, 441–443. [CrossRef]

43. Zhang, M.; Moxon, T. Infrared absorption spectroscopy of SiO_2-moganite. *Am. Mineral.* **2014**, *99*, 671–680. [CrossRef]

44. Götze, J.; Nasdala, L.; Kleeberg, R.; Wenzel, M. Occurrence and distribution of "moganite" in agate/chalcedony: A combined micro-Raman, Rietveld, and cathodoluminescence study. *Contrib. Mineral. Petrol.* **1998**, *133*, 96–105. [CrossRef]

45. Petrovic, I.; Heaney, P.J.; Navrotsky, A. Thermochemistry of the new silica polymorph moganite. *Phys. Chem. Miner.* **1996**, *23*, 119–126. [CrossRef]

46. Krauskopf, K. Dissolution and precipitation of silica at low temperatures. *Geochim. Cosmochim. Acta* **1956**, *10*, 1–26. [CrossRef]

47. Patterson, S.H.; Roberson, C.E. Weathered basalt in the Eastern part of Kauai, Hawaii. *Prof. Paper 424 C 219 US Geol. Surv.* **1961**, *424*, 195–198.

48. Fleischer, R.C.; McIntosh, C.E.; Tarr, C.L. Evolution on a volcanic conveyor belt: Using phylogeographic reconstruction and K-Ar based ages of the Hawaiian Islands to estimate molecular evolutionary rates. *Mol. Ecol.* **1998**, *7*, 533–545. [CrossRef]

49. Götze, J.; Tichomirow, M.; Fuchs, H.; Pilot, J.; Sharp, Z.D. Chemistry of agates: A trace element and stable isotope study. *Chem. Geol.* **2001**, *175*, 523–541. [CrossRef]

50. Moxon, T.; Carpenter, M.A. Crystallite growth kinetics in nanocrystalline quartz (agate and chalcedony). *Miner. Mag.* **2009**, *73*, 551–568. [CrossRef]

51. Jones, H.P.; Handreck, K.A. Effects of iron and aluminium oxides on silica in solution in soils. *Nature* **1963**, *198*, 852–853. [CrossRef]

52. Moxon, T. The co-precipitation of Fe^{3+} and SiO_2 and its role in agate genesis. *Neues Jahrb. Fur Mineral. Mon.* **1996**, *1*, 21–36.

53. File, P.D. *JCPDS International Centre for Diffraction Data*; ICDD: Newtown Square, PA, USA, 1998.

54. Neymark, L.A.; Amelin, Y.; Paces, J.B.; Peterman, Z.E. U-Pb ages of secondary silica at Yucca Mountain, Nevada: Implications for the paleohydrology of the unsaturated zone. *Appl. Geochem.* **2002**, *17*, 709–734. [CrossRef]

55. Moxon, T.; Nelson, D.R.; Zhang, M. Agate recrystallisation: Evidence found in the Archaen and Proterozoic host rocks, Western Australia. *Aust. J. Earth Sci.* **2006**, *53*, 235–248. [CrossRef]

56. Moxon, T.; Reed, S.J.B.; Zhang, M. Metamorphic effects on agate found near the Shap granite, Cumbria: As demonstrated by petrography, X-ray diffraction spectroscopic methods. *Miner. Mag.* **2007**, *71*, 461–476. [CrossRef]

57. Moxon, T.; Reed, S.J.B. Agate and chalcedony from igneous and sedimentary hosts aged from 13 to 3480 Ma: A cathodoluminescence study. *Miner. Mag.* **2006**, *70*, 485–498. [CrossRef]

58. Walger, E.; Matthess, G.; von Seckendorf, V.; Liebau, F. The formation of agate structure for silica transport, agate layer accretion, and flow atterns and models flow regimes in infiltration canals. *N. Jb. für Geologie und Palāontologie Abh.* **2009**, *186*, 113–152.

59. Smith, J. *Semi-Precious Stones of Carrick*; Kilwining: Spean Bridge, UK, 1910; p. 84.

60. Wang, Y.; Merino, E. Self-organisational origin of agates: Banding fiber twisting, composition and dynamic crystallization model. *Geochim. Cosmochim. Acta* **1990**, *54*, 1627–1638. [CrossRef]

61. Dumeńska-Slowik, M.; Natkaniec-Nowak, L.; Kotarba, M.J.; Sikorska, M.; Rzymełka, J.A.; Łoboda, A.; Gaweł, A. Mineralogical and geochemical characterization of the "bituminous" agates from Nowy Kościol Lower Silesia. *J. Mineral. Geochem.* **2008**, *184*, 255–268.

62. Götze, J.; Möckel, R.; Kempe, U.; Kapitonov, I.; Vennemann, T. Characteristics and origin of agates in sedimentary rocks from the Dryhead area, Montana, USA. *Miner. Mag.* **2009**, *73*, 673–690. [CrossRef]

63. Commin-Fischer, A.; Berger, G.; Polve, M.; Dubois, M.; Sardini, P.; Beaufort, D.; Formoso, M. Petrography and chemistry of SiO$_2$ filling phases in amethyst geodes from Sierra Geral Formation deposit, Rio Grande do Sul, Brazil. *J. Am. Earth Sci.* **2010**, *29*, 751–760. [CrossRef]

64. Moxon, T.; Petrone, C.M.; Reed, S.J.B. Characterisation and genesis of horizontal banding in Brazilian agate: An X-ray diffraction, thermogravimetric and microprobe study. *Miner. Mag.* **2013**, *77*, 227–248. [CrossRef]

65. Lange, P.; Blankenburg, H.-J.; Schron, W. Rasterelectronmikroscopische Untersuchgen an Vulkanachaten. *Z. Geol. Wiss.* **1984**, *12*, 669–683. (In German)

66. Holzhey, G. Mikrokristalline SiO$_2$ Minerallisation in rhyolithischen. Rotliegendvulkaniten des Thüringer Waldes (Deutschland) und ihre Genese. *Chem. Erde* **1999**, *59*, 183–205. (In German)

67. Fukuda, J.; Yokoyama, T.; Kirino, Y. Characterization of the states and diffusivity of intergranular water in a chalcedonic quartz by high temperature situ infrared spectroscopy. *Miner. Mag.* **2009**, *73*, 825–835. [CrossRef]

68. Götze, J.; Möckel, R.; Vennemann, T.; Muller, A. Origin and geochemistry of agates in Permian volcanic rocks of the Sub-Erzgebirge basin, Saxony (Germany). *Chem. Geol.* **2016**, *428*, 77–91. [CrossRef]

69. Moxon, T. Agate: A study of ageing. *Eur. J. Mineral.* **2002**, *14*, 1109–1118. [CrossRef]

70. Barsanov, G.P.; Yakovleva, M.E. Mineralogy, macro- and micromorphological features of agates. *New Data Miner.* **1982**, *30*, 3–26. (In Russian)

71. Silaev, V.I.; Shanina, S.N.; Ivanovskii, V.S. Inclusions of oil gases in agate-type secretions: Implications for forecast of the oil- and gas-bearing potential of the Polar Urals. *Dokl. Earth Sci.* **2002**, *383*, 246–252. (In Russian)

72. Whelan, J.B.; Paces, J.B.; Peterman, Z.E. Physical and stable-isotope evidence for the formation of secondary calcite in the unsaturated zone, Yucca Mountain, Nevada. *Appl. Geochem.* **2002**, *17*, 735–750. [CrossRef]

73. Harder, H.; Flehmig, W. Quarzsynthese bei tiefen temperature. *Geochim. Cosmochim. Acta* **1970**, *34*, 295–305. [CrossRef]

74. Mackenzie, F.T.; Gees, R. Quartz synthesis at earth surface conditions. *Science* **1971**, *173*, 533–535. [CrossRef]

75. Heaney, P.J. A proposed mechanism for the growth of chalcedony. *Contrib. Mineral. Petrol.* **1993**, *115*, 66–74. [CrossRef]

76. Naboko, S.L.; Silnichenko. The formation of silica gel on the solfataras of the Golovin volcano on Kunashir Island. *Geochemistry* **1957**, *3*, 253–256. (In Russian)

77. Meixiang, Z.; Wei, T. Surface hydrothermal minerals and their distribution in the Tengchon geothermal area, China. *Geothermics* **1987**, *16*, 181–195. [CrossRef]

78. Marcoux, E.; Le Berre, P.; Cocherie, A. The Meillers Autunian hydrothermal chalcedony: First evidence of a 295 Ma auriferous epithermal sinter in the French Massif Central. *Ore Geol. Rev.* **2004**, *25*, 69–87. [CrossRef]

79. Knauth, L.P. Petrogenesis of chert. *Rev. Mineral.* **1994**, *29*, 233–258.

80. Rodgers, K.A.; Browne, P.R.L.; Buddle, T.F.; Cook, K.L.; Greatrez, R.A.; Hampton, W.A.; Herdianita, N.R.; Holland, G.R.; Lynne, B.Y.; Martin, R.; et al. Silica phases in sinters and residues from geothermal fields of New Zealand. *Earth Sci. Rev.* **2004**, *66*, 1–61. [CrossRef]

81. Herdianita, N.R.; Browne, P.R.L.; Rodgers, K.A.; Campbell, K.A. Mineralogical and textural changes accompanying ageing of silica sinter. *Miner. Depos.* **2000**, *35*, 48–62. [CrossRef]

82. Lynne, B.Y.; Campbell, K.A.; Perry, R.S.; Browne, P.R.L.; Moore, J.N. Acceleration of sinter diagenesis in an active fumarole, Taupo volcanic zone, New Zealand. *Geology* **2006**, *34*, 749–752. [CrossRef]

83.	Rodgers, K.A.; Cressey, G. The occurrence, detection and significance of moganite (SiO_2) among some silica sinters. *Miner. Mag.* **2001**, *65*, 157–167. [CrossRef]

84.	Heaney, P.J. Moganite as an indicator for vanished evaporites: A testament. *J. Sediment. Res.* **1995**, *65*, 633–638.

85.	Hesse, R. Origin of chert: Diagenesis of biogenic siliceous sediments. *Geosci. Can.* **1988**, *15*, 171–192.

86.	Gilg, H.A.; Morteani, G.; Kostitsyn, Y.; Preinfalk, C.; Gatter, I.; Streider, A.J. Genesis of amethyst geodes in basaltic rocks of the Serra Geral formation (Ametista do Sul, Rio Grande do Sul, Brazil): A fluid inclusion, REE, oxygen, carbon and Sr isotope study on basalt, quartz and calcite. *Miner. Depos.* **2003**, *38*, 1009–1025. [CrossRef]

3

Corundum Anorthosites-Kyshtymites from the South Urals, Russia: A Combined Mineralogical, Geochemical and U-Pb Zircon Geochronological Study

Maria I. Filina [1,*], Elena S. Sorokina [1,2], Roman Botcharnikov [2], Stefanos Karampelas [3], Mikhail A. Rassomakhin [4,5], Natalia N. Kononkova [1], Anatoly G. Nikolaev [6], Jasper Berndt [7] and Wolfgang Hofmeister [2]

[1]   Vernadsky Institute of Geochemistry and Analytical Chemistry Russian Academy of Sciences (GEOKHI RAS), Kosygin str. 19, 119991 Moscow, Russia; elensorokina@mail.ru (E.S.S.); nnzond@geokhi.ru (N.N.K.)

[2]   Institut für Geowissenschaften, Johannes Gutenberg Universität Mainz, J.-J.-Becher-Weg 21, 55128 Mainz, Germany; rbotchar@uni-mainz.de (R.B.); hofmeister@uni-mainz.de (W.H.)

[3]   Bahrain Institute for Pearls & Gemstones (DANAT), WTC East Tower, P.O. Box 17236 Manama, Bahrain; stefanos.karampelas@gmail.com

[4]   Institute of Mineralogy SU FRC MiG UB RAS, 456317 Miass, Chelyabinsk Region, Russia; miha_rassomahin@mail.ru

[5]   Ilmen State Reserve SU FRC MiG UB RAS, 456317 Miass, Chelyabinsk Region, Russia

[6]   Department of mineralogy and lithology, Institute of Geology and Petroleum Technologies, Kazan Federal University, 420008 Kazan, Russia; anatolij-nikolaev@yandex.ru

[7]   Institut für Mineralogie, Westfälische Wilhelms Universität Münster, Corrensstrasse 24, 48149 Münster, Germany; jberndt@uni-muenster.de

*   Correspondence: makimm@mail.ru

Abstract: Kyshtymites are the unique corundum-blue sapphire-bearing variety of anorthosites of debatable geological origin found in the Ilmenogorsky-Vishnevogorsky complex (IVC) in the South Urals, Russia. Their mineral association includes corundum-sapphire, plagioclase (An_{61-93}), muscovite, clinochlore, and clinozoisite. Zircon, churchite-(Y), monazite-(Ce), and apatite group minerals are found as accessory phases. Besides, churchite-(Y) and zircon are also identified as syngenetic solid inclusions within the sapphires. In situ Laser Ablation Inductively Coupled Plasma Mass Spectrometry (LA-ICP-MS) U-Pb zircon geochronology showed the ages at about 290–330 Ma linked to the Hercynian orogeny in IVC. These ages are close to those of the syenitic and carbonatitic magmas of the IVC, pointing to their syngenetic origin, which is in agreement with the trace element geochemistry of the zircons demonstrating clear magmatic signature. However, the trace element composition of sapphires shows mostly metamorphic signature with metasomatic overprints in contrast to the geochemistry of zircons. The reason for this discrepancy can be the fact that the discrimination diagrams for sapphires are not as universal as assumed. Hence, they cannot provide an unambiguous determination of sapphire origin. If it is true and zircons can be used as traces of anorthosite genesis, then it can be suggested that kyshtymites are formed in a magmatic process at 440–420 Ma ago, most probably as plagioclase cumulates in a magma chamber. This cumulate rock was affected by a second magmatic event at 290–330 Ma as recorded in zircon and sapphire zoning. On the other hand, Ti-in-zircon thermometer indicates that processes operated at relatively lower temperature (<900 °C), which is not enough to re-melt the anorthosites. Hence, zircons in kyshtymites can be magmatic but inherited from another rock, which was re-worked during metamorphism. The most probable candidate for the anorthosite protolith is carbonatites assuming that metamorphic fluids could likely leave Al- and Si-rich residue, but removed Ca and CO_2. Further, Si is consumed by the silicification of ultramafic host rocks. However, kyshtymites do not show clear evidence of pronounced metasomatic zonation and evidence for large volume changes due to metamorphic alteration of carbonatites. Thus, the obtained data still do not allow for univocal reconstruction of the kyshtymite origin and further investigations are required.

Keywords: blue sapphire; anorthosites; kyshtymites; sapphire geochemistry; Ilmenogorsky-Vishnevogorsky complex; in situ LA-ICP-MS U-Pb zircon dating

1. Introduction

Corundum α-Al_2O_3 is the common mineral of many magmatic and metamorphic rocks. However, its blue gem-quality variety (i.e., sapphire) colored mainly by iron and titanium is rare and commonly found in secondary placers of debatable origin [1]. Recent studies of blue sapphires from different deposits worldwide demonstrate growing interest in their genesis due to findings of blue sapphires in situ within the primary rocks, for instance, those from alkali basaltic terrains [2–8].

Corundum in anorthosites is seldomly found worldwide and rarely of gem-quality, e.g., there is anorthosite occurrence with gem-quality pink corundum discovered in Fiskenaesset complex of W. Greenland. Pink corundum associated with coarse-grained, radial anthophyllite, green pargasite, green or red spinel, sapphirine, cordierite, and phlogopite [9]. Anorthosites with pink corundum are known from the Sittampundi Layered Complex, Tamil Nadu in India. These anorthosites belong to the rare group of metamorphosed Archean layered complexes, which are the part of oceanic crust formed in back-arc settings [10]. Anorthosites with colorless corundums were also found in Central Fiordland, New Zealand, where anorthositic complex is a part of the Tuyuan orogenic belt undergone a multiphase metamorphism of the amphibolite facies [11]. Another corundum anorthosite from the Chunky Gal Mountain (North Carolina, USA) [12] with pink-colored corundum is located in association with amphibolites and peridotites within alpine-type orogenic belt [13].

Corundum anorthosites (kyshtymites) are known for more than two centuries in South Urals of Russia; however, the genesis of these rocks remains enigmatic with the latest research results performed more than 50 year ago. For better understanding the origin of kyshtymites, their mineralogy, in situ LA-ICP-MS trace-element geochemistry and geochronology of zircon, and UV-Vis-NIR spectroscopy were studied by modern analytical techniques. The obtained data provide new insights into the origin of blue sapphires in anorthosites of Ilmenogorsky-Vishnevogorsky complex (IVC). Mineralogy and geochemistry of sapphires within kyshtymites were also compared to those found in other primary occurrences in IVC of South Urals (corundum-blue sapphire syenite pegmatites, and sapphire-bearing metasomatites within meta-ultramafic host rocks) and to those from secondary placer occurrences with similar geochemical and mineralogical features.

2. Geological Setting

The studied corundum deposit named "the 5th versta" was discovered by Karpinsky in 1883 [14]. Three kyshtymite veins were discovered during exploration of the deposit. Corundum was used mainly as an abrasive material, some of the crystals were gem-quality, however, these rough crystals did not exceed 1 carat. The exploration of the deposit was prosecuted until the 1930s and, currently, the occurrence is almost exhausted.

Blue sapphires in kyshtymites (the 5th versta deposit and the larger occurrence called Borzovsky deposit [15–17]) are located at the western flank of Vishnevogorsky nepheline syenite (miascite)-carbonatite alkaline complex of the South Urals with unique REE-mineralization [18] (Figure 1). These two deposits are accompanied by other primary sapphire occurrences in syenite pegmatites (mines 298 and 349) [2] and sapphires in metasomatites within meta-ultramafic host rocks (mine 418) (Figure 1) [3].

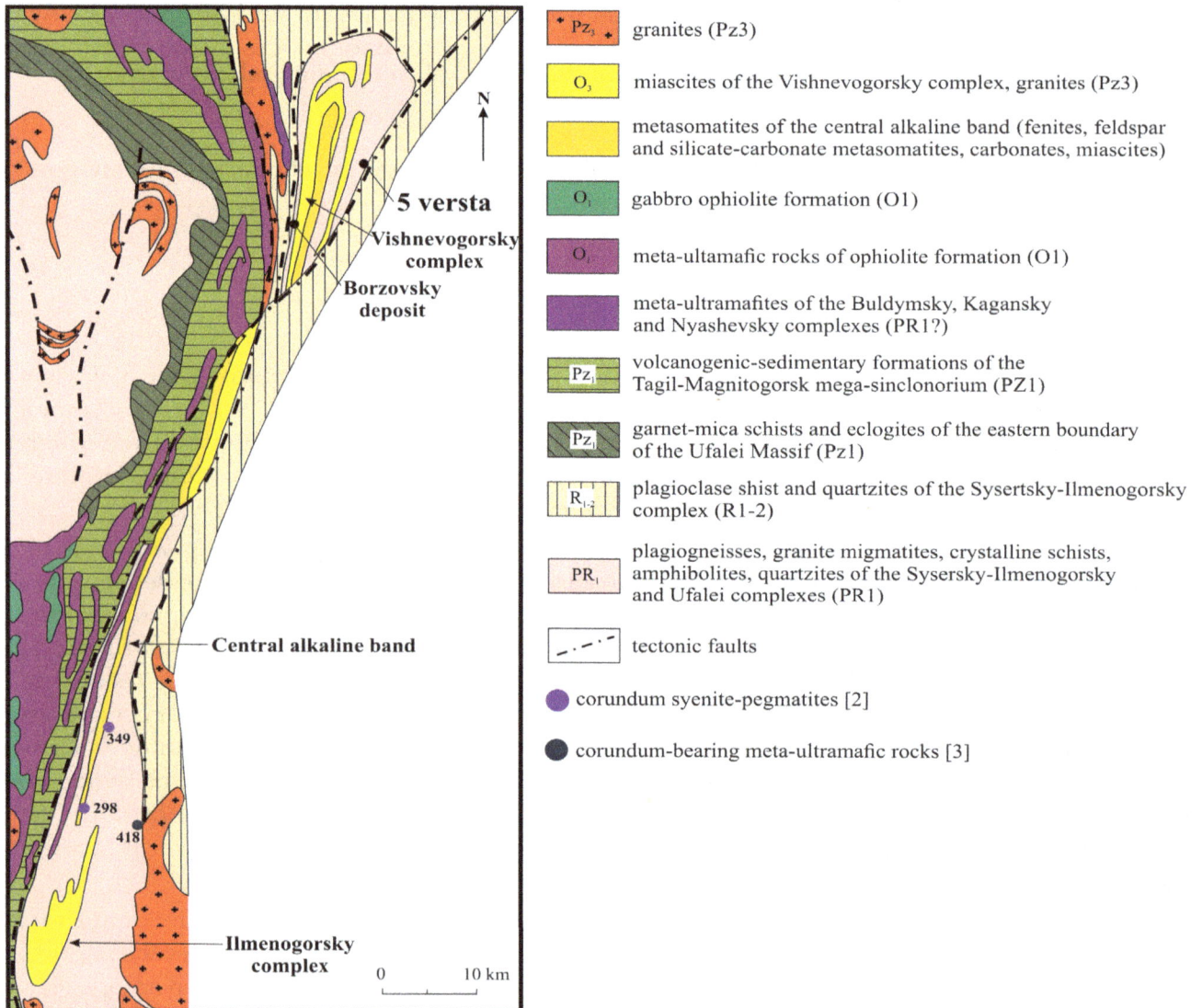

Figure 1. Geology of the Ilmenogorsky-Vishnevogorsky alkaline-carbonatite complex [18]. Coordinates of the 5th versta deposit are 55°54'03"N, 60°41'16"E.

The kyshtymites are found within the meta-ultramafic host rocks located among the quartzite shales of the meta-terrigenous Saitovsky series. Saitovsky series is one of the structural units within the Ilmenogorsky-Vishnevogorsky polymetamorphic zone, which is a deep fragment of the regional post-collisional shear [19]. The series with meta-ultramafites undergone the re-working during several thermal events (SHRIMP U-Pb zircon geochronology). The age of the mantle protolith was dated at ~1.3 Ga [20]. A stage of metamorphic evolution linked to the miascite intrusions was at about 450–420 Ma [21–23]. The stage of metamorphism and granite formation in the Sysertsky-Ilmenogorsky block linked to the Hercynian orogeny was at 360–320 Ma [24], whereas the ages of ~330–270 Ma corresponds to the collision processes [24].

At the beginning of the 20th century, the studied vein of kyshtymites of a lenticular body (Figure 2) was explored from the surface by quarry extended currently to the depth of about 3–4 m. The meta-ultramafic host rocks (initial orthopyroxenites) were composed mainly of enstatite, which undergone metasomatism (serpentinization). The reaction rim with a thickness of 10–25 cm consisting of chrysotile-asbestos was detected at the contact of the meta-ultramafic host rock with the kyshtymites [25].

Figure 2. Vein of corundum-blue sapphire anorthosites-kyshtymites: (**a**) the kyshtymite outcrop, thickness is about 3 m; and (**b**) A kyshtymite vein is at a contact with the meta-ultramafic host rock, their contact showed by a green-colored reaction rim consisting of chrysotile-asbestos.

Kyshtymites with meta-ultramafic host rocks are adjacent from the west by the miascites (nepheline syenites) of the Vishnevogorsky complex consisting of potassium feldspar (20–60 wt. %), nepheline (20–30 wt. %), lepidomelane (5–20% wt. %), amphibole (up to 20 wt. %), and plagioclase (up to 20 wt. %). Additionally, calcite (up to 3 wt. %), cancrinite, and sodalite were identified in miascites [26].

3. Materials and Methods

Fourteen kyshtymite samples with blue corundum-sapphires, one sample from the reaction rim, and one sample of meta-ultramafic host rock were investigated in this study. The list of research methods is provided in Table S1. Minerals of kyshtymites, meta-ultramafic host rock, and the reaction rim between them were identified by the optical microscopy in the petrographic thin-sections. Some of the samples were studied by the Raman spectroscopy at the Renishaw Moscow using the Renishaw inVia Raman spectrometer coupled with Ar^+ green Stellar-REN Modu-Laser (Renishaw plc, Gloucestershire, UK) with $\lambda = 514$ nm and 50× magnification at room temperature. The laser power was 10 mW on the sample with a 60 s acquisition time (three cycles), at a resolution of about 1.5 cm^{-1}. Rayleigh scattering was blocked using a holographic notch filter. Backscattered light was dispersed using a grating with 1800 grooves/mm. The spectrometer was calibrated at 520.7 cm^{-1} using Si as a reference.

The mineral chemistry was studied by the electron micro-probe analyses (EMPA) using Cameca SX 100 electron microprobe (CAMECA, Gennevilliers, France) in the wavelength-dispersive detection mode (WDS) at GEOKHI RAS, Moscow. The accelerating voltage was 15 kV, current was 30 nA, and beam size was from 3 to 5 μm. Both natural and synthetic reference materials were used for the instrument control: andradite for Si, jadeite for Na, orthoclase for K and Al, augite for Ca and Fe, olivine for Mg, rhodonite for Mn, TiO_2 for Ti, vanadinite for V, Cr_2O_3 for Cr, apatite for P, galena for Pb, $Rb_2Nb_4O_{11}$ for Nb, xenotime-(Y) for Y, metallic Gd for Gd, Pr_3PO_4 for Pr, Sm_3PO_4 for Sm, La_3PO_4 for La, and homogeneous glasses for Zr, Ta, Th, and U. The detection limits for almost all elements were less than 0.01 wt. %, the lower limits of the determined values were less about 0.03 wt. %. Correction coefficients are determined by PAP correlation (atomic number, fluorescence, and absorption correction).

Trace element composition of three representative sapphire crystals from samples K-8 and K-12 was determined using Laser Ablation–Inductively Coupled Plasma–Mass Spectrometry (LA-ICP-MS) at the Institute of Geology of Ore Deposits, Petrography, and Mineralogy RAS (IGEM RAS), Moscow. The analyses were conducted using New Wave Research UP-213 Nd:YAG laser (New Wave Research, Inc., Fremont, CA, USA) combined with the XSERIES 2 ICP-Mass Spectrometer (Thermo Scientific, Waltham, MA, USA). Trace-element concentrations were determined by the monitoring of ^6Li, ^9Be, ^{24}Mg, ^{27}Al, ^{44}Ca, ^{47}Ti, ^{51}V, ^{53}Cr, ^{57}Fe, ^{71}Ga, and ^{91}Zr and ablating a material with a spot size of 60 μm at a repetition rate of 10 Hz, and an energy density of about 14–15 J/cm^2. Warm up/background time was 15 s, dwell time was 40 s, and wash out time was 20 s. The NIST SRM 610 and NIST SRM 612 glasses were used as reference materials and the BHVO–2G glass as a quality control material (QCM). The time-resolved signal was processed in Igor (IOLITE) commercial software using ^{27}Al as the internal standard applying the theoretical value of 52.93 wt. % of Al in pure crystalline α–Al_2O_3 for analyzing the corundum unknowns. The measured concentrations of reference material and QCM agree for all elements within 10% and 15%, respectively, of preferred values by Jochum et al. 2011 [27]. This larger discrepancy between measured and preferred values can be attributed to isobaric interferences that cannot be resolved with the instrumentation used [28].

UV-Vis-NIR (Ultraviolet-Visible-Near Infrared) absorption spectroscopy was acquired at Kazan Federal University on one sapphire (sample 12-K). Spectra were recorded at a room temperature using SHIMADZU UV-3600 spectrometer (Shimadzu Corp, Kyoto, Japan) with 2 light sources (one for UV and one for Vis-NIR) and 2 detectors (FEU R928 for UV-Vis and InGaAs for NIR) from 185 to 3300 nm range with a data interval and spectra bandwidth of 1 nm and a scan rate of 300 nm/min.

Chemical composition of six selected kyshtymite rock samples macroscopically representing different textures and containing from 30 to 50 wt. % of sapphires, one sample of meta-ultramafic host rocks, and one sample of reaction rim were used for whole-rock major element analyses at GEOKHI RAS, Moscow using Energy Dispersive X-ray Fluorescence spectrometer (EDXRF) AXIOS Advanced (PANalytical B.V., Almelo, The Netherlands). The equipment provides the determination of quantitative concentrations of elements from oxygen to uranium from about 10^{-4} to 100 wt. %.

Three selected kyshtymite rock samples containing from 50 to 70 wt.% of sapphires with gem-quality zones in crystals, one sample of meta-ultramafic host rocks, and one sample of the reaction rim between kyshtymites and host rocks previously analyzed by EDXRF were used for the whole-rock trace-element analyses at the Institute of Oceanology RAS, Moscow. Inductively Coupled Plasma–Mass Spectrometer (ICP-MS) Agilent 7500 (Santa Clara, Ca, USA) was applied to determine the contents of REE and trace-elements. Calibration of the sensitivity over the entire mass scale was carried out using 68-element solutions references (ICP-MS-68A, HPS, solutions A and B). The STM-2 standard was used as a quality control material. Indium was added to all sample solutions in the concentration of 10 ng/g to control signal stability. The detection limits of the elements were 0.1 ng/g for the heavy and medium elements, and 1 ng/g for the light elements. Analytical uncertainties were less than 1–3%.

Trace-element composition and U-Pb isotopic analysis of 6 zircon grains identified in the thin-section of samples 8-K were analyzed using Element2 ICP-Mass Spectrometer (ThermoFisher Scientific, Waltham, MA, USA) coupled with an Analyte G2 (Photon Machines Inc, Redmond, WA, USA) laser at the Westfälische Wilhelms Universität Münster. Before the analysis, the backscattered electron (BSE) and cathodoluminescence (CL) images, as well as maps in average weighted atomic numbers were taken at GEOKHI RAS, Moscow using the same equipment as for the EMPA WDS to identify the internal structures within the zircon grains. During the LA-ICP-MS measurements, gas flow rates were about 1.1 L/min for He, 0.9 L/min and 1.1 L/min for the Ar-auxiliary and sample gas, respectively. Cooling gas flow rate was set to 16 L/min. Trace-elements concentrations were determined by measuring of ^{29}Si, ^{43}Ca, ^{49}Ti, ^{51}V, ^{53}Cr, and REE with a spot size of 20 μm at a repetition rate of 10 Hz, and an energy density of about ~3–4 J/cm^2. Background time was 15 s, dwell time was 40 s, and wash out time was 20 s. NIST SRM 612 was used as the reference materials, and 91,500 zircon and BIR-1G glass as unknowns for quality control (QCM). The time-resolved signal was processed in GLITTER commercial software using Zr as the internal standard applying the stoichiometric value of Zr in pure crystalline ZrSiO$_4$ for analyzing the zircon unknowns. The measured concentrations of reference material and QCM agree for all elements within 10% and 15%, respectively, of preferred values by [29]. For U-Pb zircon dating ^{204}Pb, ^{206}Pb, ^{207}Pb, and ^{238}U were measured along, with ^{202}Hg to correct the interference of ^{204}Hg on ^{204}Pb, which is important to apply, if necessary, for a common Pb correction. Repetition rate was 10 Hz using an energy of ~3 J/cm^2 and a spot size of 25 μm. Ten unknowns were bracketed with three calibration standards GJ1 [29] to correct for instrumental mass bias. The data reduction was performed using in-house Excel spreadsheet [30]. Along with the unknowns, 91,500 reference zircon [31] was measured to monitor accuracy and precision of the analysis. Long term reproducibility of the reference zircon standard 91,500 [31] yielded a Concordia age of 1074.8 ± 8.8 Ma, which is indistinguishable to the age of 1065.4 ± 0.3 Ma for the 91,500 reference zircon determined by TIMS [31]. The Concordia diagrams and age calculations were made using Isoplot v. 4.13 (Ludwig, 2009) [32]. Weighted mean age calculations are given at 95% confidence level.

4. Results

4.1. Petrology and Mineralogy of Kyshtymites

Mineral composition and chemistry are shown in Tables S2–S4. Kyshtymite consists mainly of idiomorphic blue-colored transparent to translucent corundums–sapphires (up 50 wt. %) and plagioclase with composition varying from labradorite to anorthite An$_{61-93}$ (30–50 wt. %) (Table S3 and Figure 3).

Figure 3. Photography of the samples used in this study with dipyramidal-prismatic corundum-sapphires (Crn) crystals elongated along the *c* axis and fine-grained plagioclase (Pl): (**a**) sample 2-K; and (**b**) sample 7-K.

The rock has a porphyritic structure, where large corundum-sapphire crystals are located among the fine-grained plagioclase, muscovite, clinozoisite, and clinochlore (Figure 4). Other rock-forming minerals—muscovite, clinochlore, and clinozoisite—occupy up to 10 wt. %. Accessory minerals were detected as zircon, churchite-(Y), apatite, and monazite-(Ce) (Figure 5).

Table 1. Rim-to-rim LA-ICP-MS profiles of studied corundum-blue sapphire from anorthosites- kyshtymites (in µg/g).

Sample/No. of Spot	Color	Li	Be	Mg	Ti	V	Cr	Fe*	Ga	Ga/Mg	Fe/Ti	Cr/Ga	Fe/Mg
K-8-1 1	Wt	32.45	1.85	199.85	972.77	5.59	3.18	1166.20	36.40	0.18	1.20	0.09	5.84
2	Bl	8.34	1.67	183.70	700.93	5.96	12.34	1477.19	38.57	0.21	2.11	0.32	8.04
3	Bl	bdl	bdl	334.58	782.98	6.49	2.59	2099.17	36.90	0.11	2.68	0.07	6.27
4	Wt	bdl	bdl	74.20	702.78	6.35	3.31	855.22	36.53	0.49	1.22	0.09	11.53
5	Wt	bdl	2.20	69.22	231.35	7.54	2.59	1399.44	33.72	0.49	6.05	0.08	20.22
6	Wt	bdl	3.12	67.05	115.41	8.52	1.61	1710.43	33.38	0.50	14.82	0.05	25.51
7	Bl	bdl	2.51	66.78	101.64	9.08	2.44	2332.41	33.03	0.49	22.95	0.07	34.92
8	Bl	bdl	bdl	291.17	458.20	5.35	3.86	1865.92	32.00	0.11	4.07	0.12	6.41
9	Wt	bdl	bdl	388.58	297.79	4.39	7.07	1710.43	30.36	0.08	5.74	0.23	4.40
10	Wt	bdl	bdl	741.16	585.52	4.74	6.51	855.22	30.76	0.04	1.46	0.21	1.15
K-8-2 1	Wt	bdl	bdl	131.29	986.27	5.53	6.04	1788.18	32.77	0.25	1.81	0.18	13.62
2	Bl	bdl	bdl	59.29	782.19	6.27	1.89	2021.42	36.16	0.61	2.58	0.05	34.09
3	Bl	bdl	bdl	46.67	706.48	6.38	1.60	2021.42	37.30	0.80	2.86	0.04	43.32
4	Bl	bdl	bdl	296.46	559.58	6.17	2.40	1865.92	37.19	0.13	3.33	0.06	6.29
5	Wt	bdl	3.47	127.59	164.38	8.63	2.52	1166.20	34.76	0.27	7.09	0.07	9.14
6	Wt	bdl	2.67	94.50	122.82	8.55	2.90	1166.20	34.46	0.36	9.50	0.08	12.34
7	Bl	bdl	2.33	92.62	133.67	8.60	1.86	1943.67	35.50	0.38	14.54	0.05	20.99
8	Bl	bdl	bdl	77.21	583.13	6.59	0.92	2410.15	39.81	0.52	4.13	0.02	31.21
9	Wt	6.83	bdl	95.29	656.99	5.98	0.73	1943.67	37.72	0.40	2.96	0.02	20.40
10	Wt	bdl	bdl	91.85	711.78	5.17	0.93	1399.44	36.40	0.40	1.97	0.03	15.24
K-12-1 1	Bl	bdl	bdl	164.38	939.95	3.97	bdl	1010.71	42.99	0.26	1.08	-	6.15
2	Wt	bdl	2.81	167.82	294.08	6.70	bdl	621.97	56.25	0.34	2.11	-	3.71
3	Wt	bdl	1.96	188.47	210.70	6.22	bdl	1321.70	61.33	0.33	6.27	-	7.01
4	Wt	bdl	3.60	147.68	216.79	6.49	bdl	4198.33	57.49	0.39	19.37	-	28.43
5	Wt	bdl	3.81	94.97	125.20	6.11	bdl	855.22	48.65	0.51	6.83	-	9.00
6	Wt	bdl	bdl	222.08	205.14	3.71	bdl	1554.94	50.95	0.23	7.58	-	7.00
7	Wt	bdl	bdl	127.77	156.44	6.14	bdl	1321.70	46.38	0.36	8.45	-	10.34
8	Bl	bdl	1.85	157.23	374.82	8.07	bdl	5209.04	52.41	0.33	13.90	-	33.13
9	Bl	bdl	bdl	190.32	524.11	8.71	0.03	3265.37	54.58	0.29	6.23	-	17.16
10	Bl	bdl	bdl	237.44	555.61	7.91	0.35	3109.87	50.90	0.21	5.60	0.01	13.10

Wt, Bl, white and blue colors of sapphire. * Calculated from WDS EMPA; bdl, below the detection limit.

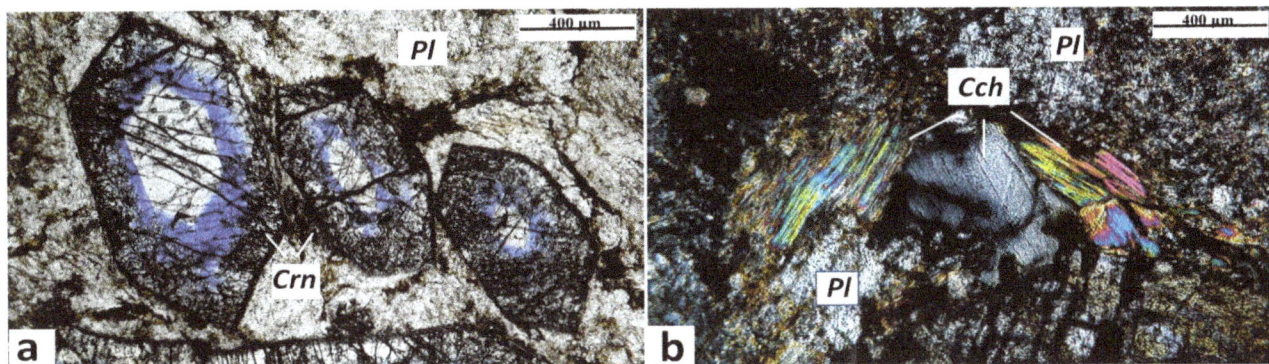

Figure 4. Photomicrographs of corundum-blue sapphire anorthosite-kyshtymite samples 8-K: (**a**) idiomorphic crystals of corundum-sapphire (Crn) with oscillatory white-blue zonation within the matrix of fine-grained plagioclase (Pl), parallel polarized light. The dotted arrows show the profiles measured by LA-ICP-MS (cf. Table 1). (**b**) Xenomorphic crystals of clinochlore (Cch) associated with plagioclase, crossed polarized light.

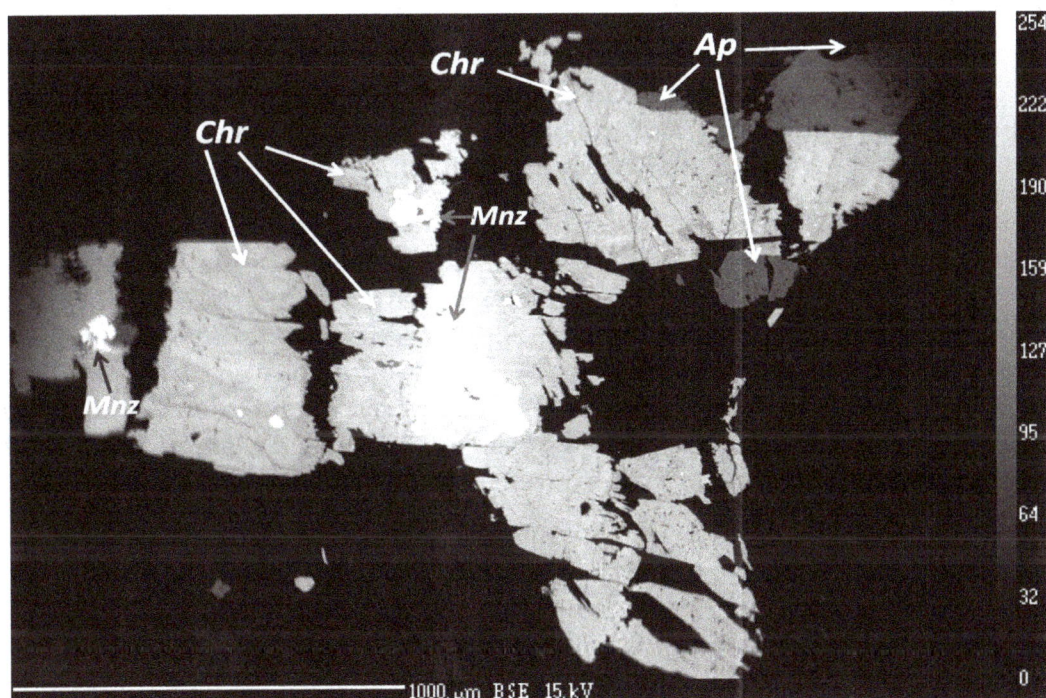

Figure 5. BSE image of churchite-(Y) (Chr), monazite-(Ce) (Mnz), and apatite group minerals (Ap) embedded in the plagioclase–muscovite–clinochlore matrix.

Muscovite occurs as plates of 0.1–0.2 mm in size common around plagioclase grains. The mineral is detected in association with clinochlore and clinozoisite, both epigenetic by nature. High MgO contents in muscovite (up to 1.71 wt. %) and K_2O in clinochlore (up to 7.11 wt. %) are associated with the replacement of muscovite by a clinochlore. Clinozoisite forms small rounded grains up to 0.1 mm in size replacing plagioclase. Clinozoisite contains up to FeO_{tot} of 2.07 wt. %.

Churchite-(Y), forms small syngenetic xenomorphic crystals of 30–70 μm in size commonly found as solid micro-inclusions within corundum-sapphire (Figure 5). Churchite-(Y) is also found in intergrowths with the monazite-(Ce) and likely apatite group minerals (Table S3). Zircon forms syngenetic and epigenetic prismatic or dipyramidal crystals with a size up to 40–100 μm, and contains up to 2.45 wt. % HfO_2 (Table S4).

4.2. Mineralogy, Geochemistry, Solid Inclusions, and UV-Vis-NIR-Spectroscopy of Corundum–Sapphire

Colorless to blue-colored translucent to transparent corundum-sapphires with the fractures passing through the entire crystals were found in kyshtymites. All sapphires show oscillatory zonation in the elongated dipyramidal-prismatic crystal sized up to 4 cm in length (Figure 3b). The most developed crystal faces are hexagonal prism (11$\bar{2}$0), pinacoid (0001), and hexagonal dipyramid (22$\bar{4}$3) (Figure 4a).

Sapphires from kyshtymites are almost inclusion-free except occasional finding of churchite-(Y) and zircon solid inclusions. Churchite-(Y) was found to be syngenetic with sapphires (Figure 5). The mineral contains traces of Gd_2O_3 (1.8–2.19 wt. %), Pr_2O_3 (0.10–0.17 wt. %), La_2O_3 (0.09–0.12 wt. %), Sm_2O_3 (0.91–1.05 wt. %), U_2O_3 (1.77–2.14 wt. %), and ThO_2 (0.55–72 wt. %). Inclusion of churchite-(Y) in sapphire has not been previously described. Therefore, it is likely the first identification and chemical analysis of churchite-(Y) solid inclusions within sapphires to the best of our knowledge.

Zircon solid inclusions in corundum-sapphire showed concentration of HfO_2 from 0.72 to 3.00 wt. % (one measurement showed 5.38 wt. %; see Table S5), which is almost the same as in zircon found in mineral association of kyshtymites (0.97–2.45 wt. %). Besides, the epigenetic zircon, muscovite, and clinochlore filling the sapphire fractures were also detected.

The UV-Vis-NIR spectra of the studied sapphires are comparable to those observed on sapphires of metamorphic or metasomatic origin [33] and of the other two occurrences in the region linked with syenitic pegmatite [2] and metasomatites within meta-ultramafic host rocks [3].

Corundum-sapphire trace element measurements by LA-ICP-MS are shown in Table 1. The Fe content in blue colored zones varies from 1010 to 5209 µg/g. The concentration of Mg (47–335 µg/g) and Ti (101–940 µg/g) in sapphire from kyshtymites are higher than that in blue sapphires from syenite pegmatites (mines 298 and 349) [2,34] and those in sapphires from metasomatites in meta-ultramafic host rocks (mine 418) [3] (Table S6), both located in Ilmen Mountains (see Figure 1). The Ga content remains low (30–61 µg/g) for all studied samples as in the case of sapphires in meta-ultramafites (mine 418), however, lower than in those of syenite pegmatites (Table S6). Cr is from b.d.l. to 12 µg/g and V from 4 to 9 µg/g. Detected Li and Be concentrations in the measured spots are presumably due to micro-inclusions within sapphires.

The 10,000 Ga/Al ratio is above 0.60–0.80, Ga/Mg ratio is 0.11–0.80, Fe/Mg is 6.15–43.32, Cr/Ga is 0.01–0.32, and Fe/Ti is 1.08–22.95. These ratios are common for metamorphic sapphires [35,36]. On the Fe versus Ga/Mg diagram, studied sapphires with blue color fall in the field of metamorphic sapphire, similar (but not overlapping) to those of meta-ultramafic host rocks of Ilmen Mountains. The results are partially overlapping with those of magmatic sapphires in Yogo Gulch in USA, Gortva in Slovakia, Baw Mar in Myanmar (filled symbols in Figure 6), metamorphic sapphires from Ratnapura in Sri-Lanka, and metasomatic sapphires from Kashmir in India (dotted lines in Figure 6).

On the Fe–Mg*100–Ti*10 ternary plot (Figure 7), the studied sapphires from kyshtymites also fall in the field of "metamorphic" sapphires overlapping those within Ilmen meta-ultramafic host rocks (mine 418). Meanwhile, blue sapphires from anorthosites-kyshtymites overlap most of the other known metamorphic (Ratnapura in Sri-Lanka—purple dotted lines in Figure 7), metasomatic (Kashmir in India—pink dotted lines in Figure 7), magmatic (Mogok in Myanmar, Yogo in USA, and Gortva, Slovakia—filled symbols in Figure 7), and placer sapphire occurrences (Balangoda in Sri Lanka, Pailin in Cambodia, and Ilakaka in Madagascar, and Montana in USA).

The sapphires from kyshtymites are also plotted to the "metamorphic" field on Fe/Mg vs. Ga/Mg diagram by Peucat et al. (2007) and Sutherland et al. (2009) [35,36] (Figure 8).

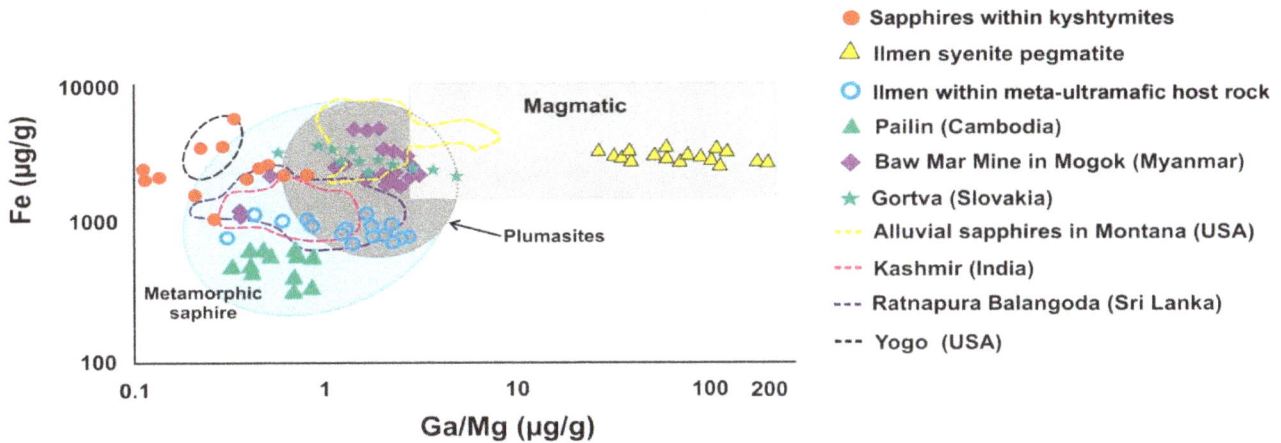

Figure 6. A Fe versus Ga/Mg diagram showing the boundaries for magmatic and metamorphic sapphires, modified after Peucat et al. (2007) [35] and Zwaan et al. (2015) [37] with blue sapphires within kyshtymites, Ilmen sapphires within syenite pegmatites [2], Ilmen sapphires within meta-ultramafic host rocks [3], and sapphire deposits from other regions after [35]: Ratnapura and Balangoda (Sri lanka), Kasmir (India), Yogo Gulch (USA), and Baw Mar Mine in Mogok. The plot of alluvial sapphires from Montana (USA) is after Zwaan et al. (2015) [37]; data on sapphires from the Hajacka, Gortva (Slovakia) are after Uher et al. (2012) [38].

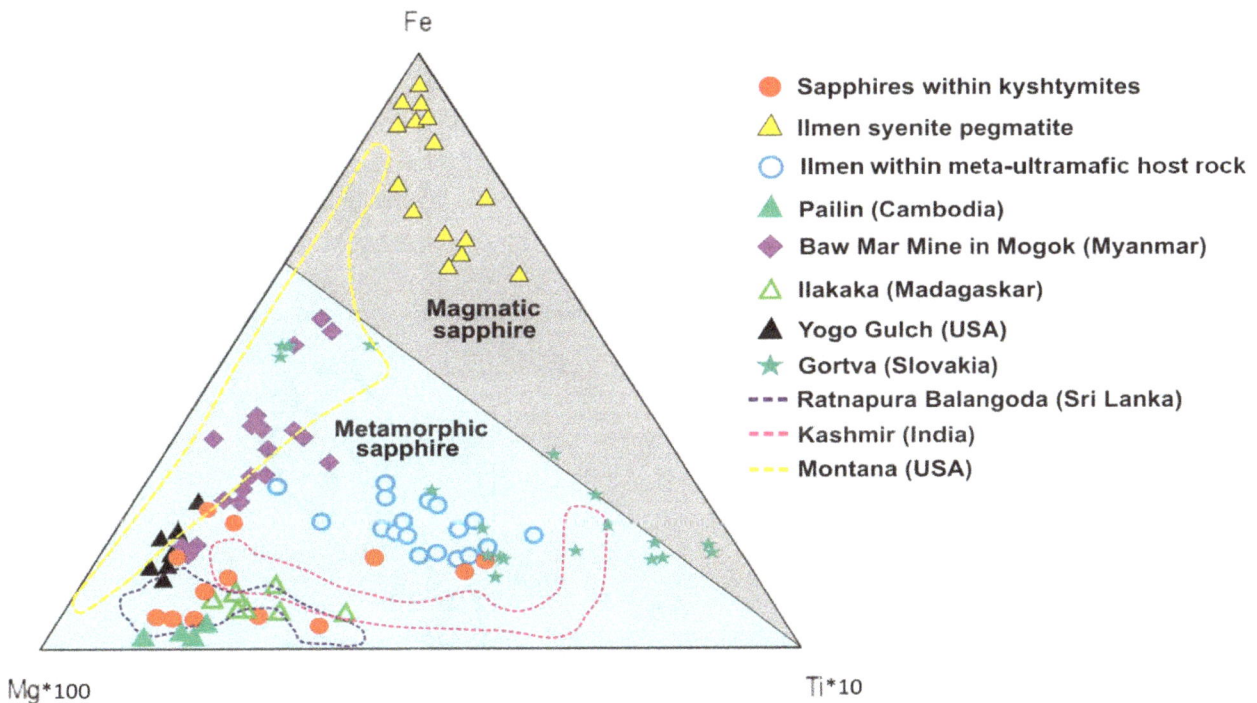

Figure 7. A Fe–Mg*100–Ti*10 ternary plot modified after Peucat et al. (2007) [35] of corundum-sapphire within kyshtymites, Ilmen sapphires within syenite pegmatites [2], Ilmen sapphires within the meta-ultramafic host rocks [3], and sapphire deposits from other regions after [35]: Pailin (Cambodia), Baw Mar Mine in Mogok (Myanmar), Ilakaka (Madaskar), Ratnapura and Balangoda (Sri lanka), Kasmir (India), and Yogo Gulch (USA). The plot of Gortva sapphire (Slovakia) is after Uher et al. (2012) [38] and Montana (USA) sapphires are after Zwaan et al. (2015) [37].

On the Fe–Cr*10–Ga*100 ternary plot (Figure 10), the sapphires from kyshtymites fall in the "magmatic field" similar to those "magmatic" sapphires from Australia [36] likely due to intermediate Fe and Ga contents, and absence of Cr values. However, they were plotted to "metasomatic" and "plumasitic" fields on discriminant factors diagram by Giuliani [40] (Figure 11), as in the case of those from Ilmen metasomatites within meta-ultramafic host rocks [3].

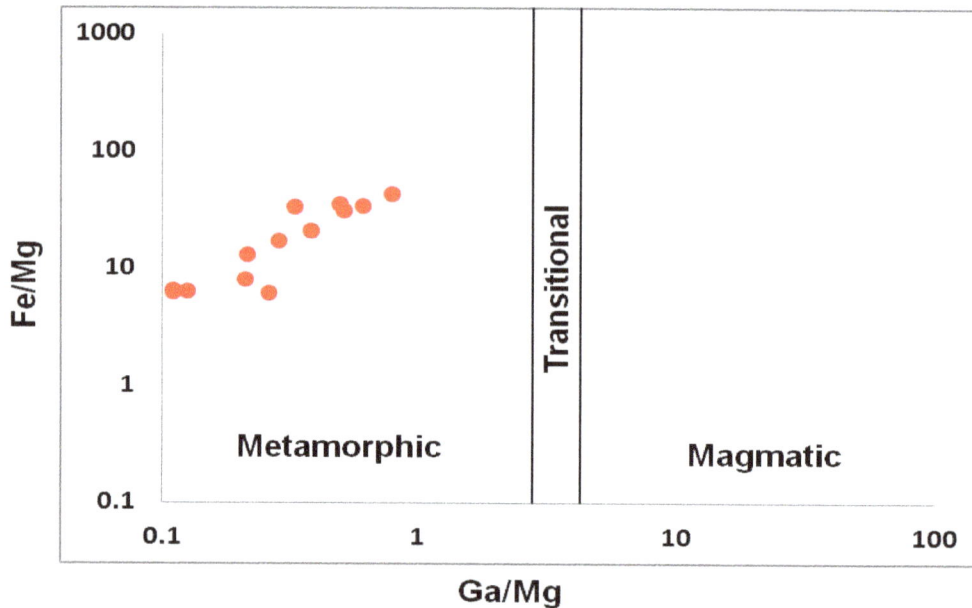

Figure 8. Fe/Mg vs. Ga/Mg discrimination diagram modified after [35,36].

Blue sapphires within kyshtymites overlap the "metasomatic" fields on $FeO - Cr_2O_3 - MgO - V_2O_3$ vs. $FeO + TiO_2 + Ga_2O_3$ discriminant diagram (Figure 9) by [35] as in case of those sapphires within Ilmen syenite pegmatites, Ilmen sapphires within meta-ultramafic rocks, sapphires from Gortva syenite xenoliths within alkali basalts [38], and sapphires from lamphrophiric dyke in Yogo Gulch [39].

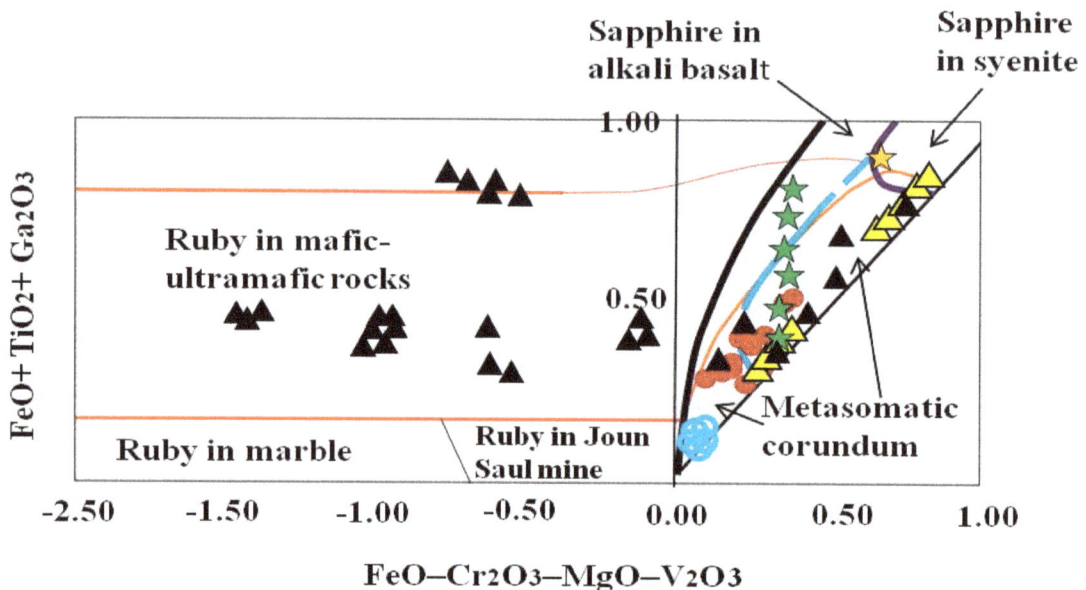

Figure 9. A $FeO - Cr_2O_3 - MgO - V_2O_3$ versus $FeO + TiO_2 + Ga_2O_3$ discriminant diagram (wt. %) is modified after Giuliani et al. (2014) [40] with extended "magmatic/syenitic" field with plotted corundum-sapphire within kyshtymites (red circles), Ilmen sapphires within syenite pegmatites (yellow triangles) [2], and within the meta-ultramafic host rocks (blue circles) [3]. The data on sapphires associated with xenoliths in alkali basalts (Gortva in Slovakia—green star; Loch Roag in Scotland—yellow star) were modified after Uher et al. (2012) [38]. The data on sapphires associated with ultramafic lamprophyre dike in Yogo Gulch (USA) (black triangles) are after [39]. Blue dotted line shows the extension of "magmatic/syenitic" sapphire field.

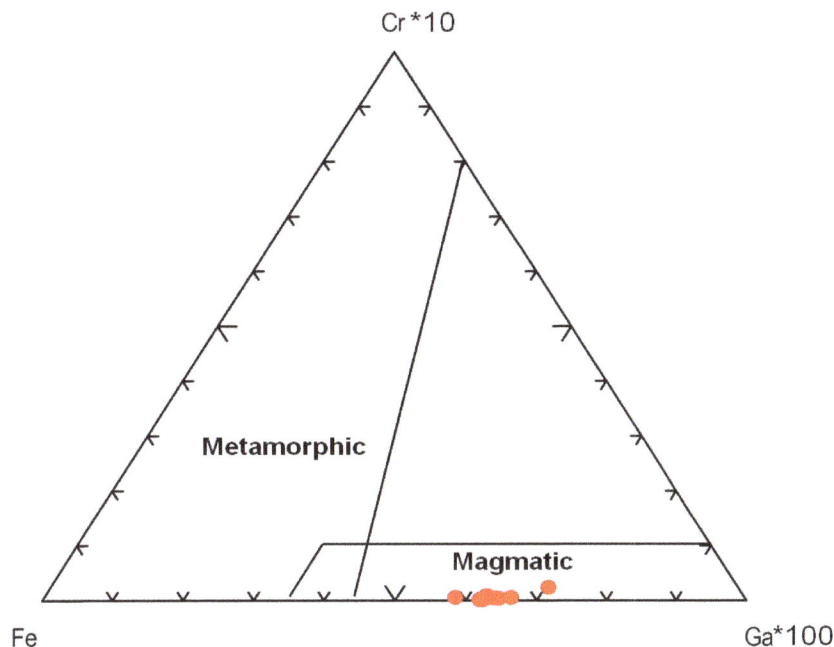

Figure 10. A Fe–Cr*10–Ga*100 ternary plot modified after Sutherland et al. (2009) [36].

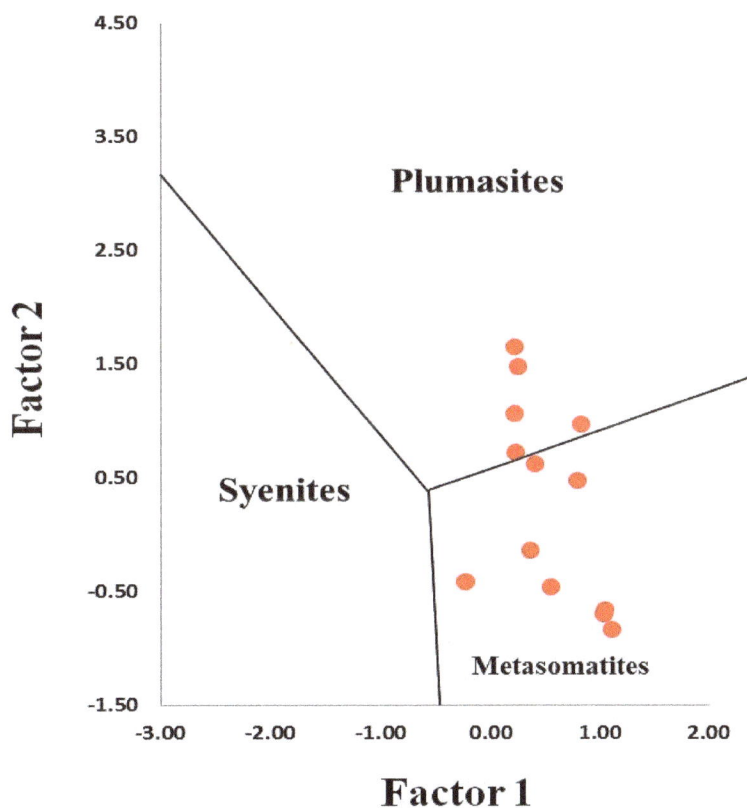

Figure 11. Discriminant factor diagram for identification of sapphire origin modified after Giuliani et al. (2014) [40].

4.3. Whole Rock Geochemistry of Kyshtymites and Elements Mobility

Major elements of kyshtymites measured by EDXRF are reported in Table S7. Kyshtymites have high Al_2O_3 (34.76–42.94 wt. %), Na_2O (0.82–4.01 wt. %), and K_2O (0.50–1.29 wt. %), but low SiO_2 content (40.84–42.72 wt. %) as well as variable CaO (5.89–15.79 wt. %) when compared with those of meta-ultramafic host rocks. Kyshtymites are characterized by low concentration of Fe_2O_3 (0.12–1.32 wt. %), MgO (0.60–2.86 wt. %), and TiO_2 (0.04–0.15 wt. %).

Meta-ultramafic host rocks are enriched in SiO_2 (69.72 wt. %), MgO (17.97 wt. %), and Fe_2O_3 (6.22 wt. %), while depleted in CaO (up to 0.25 wt. %), Al_2O_3 (up to 1.18 wt. %), and alkaline elements ($Na_2O + K_2O$ is 0.24 wt. %). The reaction rim between kyshtymites and meta-ultramafic host rocks also contains more MgO (28.08 wt. %), Fe_2O_3 (6.72 wt. %), and SiO_2 (45.09 wt. %) and less Al_2O_3 (4.29 wt. %), CaO (0.42 wt. %), and alkaline elements ($Na_2O + K_2$ is 0.14 wt. %) compared to those of kyshtymites. The concentration of TiO_2 was below the detection limit (Table S7, Figure 12).

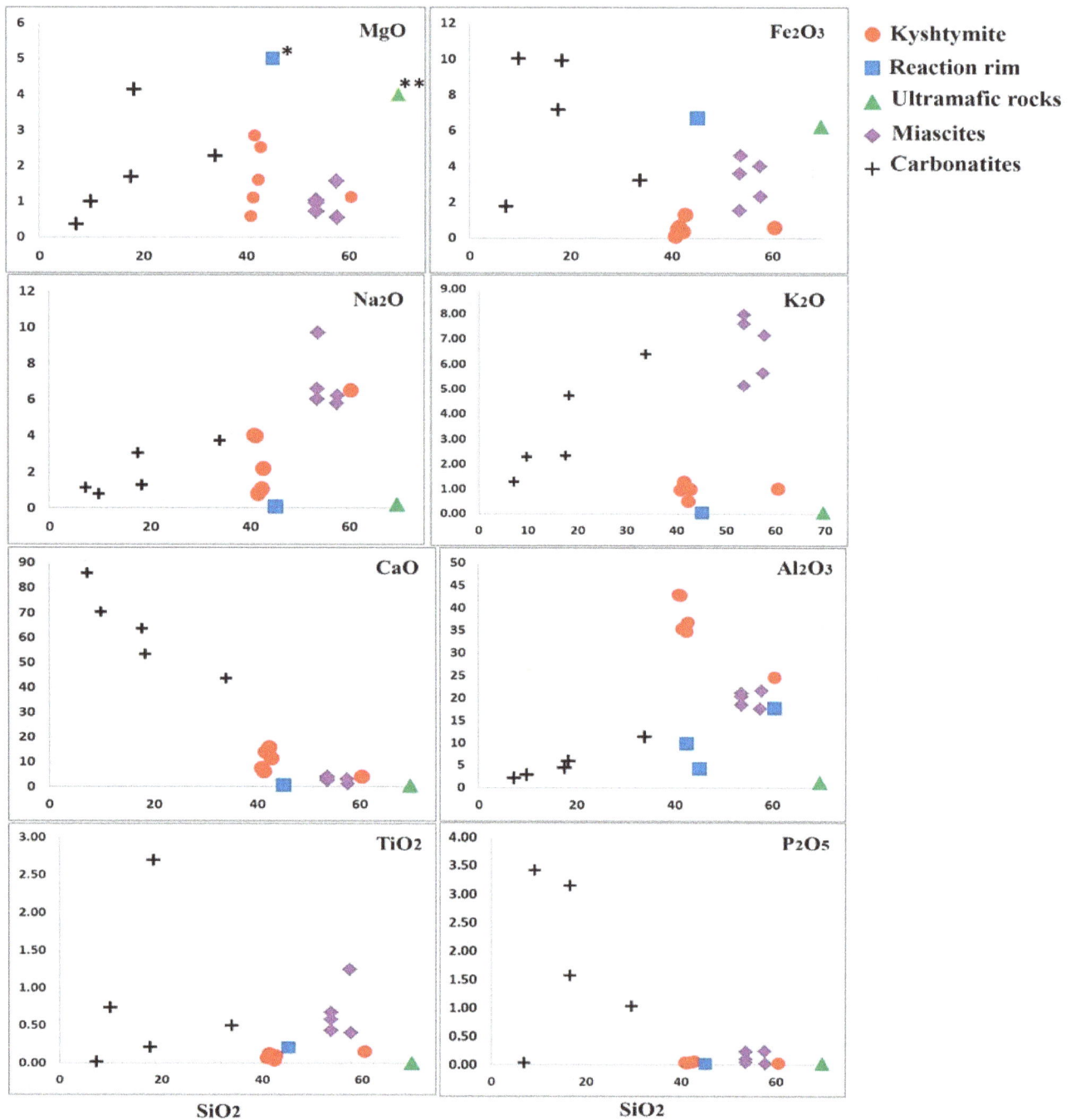

Figure 12. Harker diagrams for analyzed kyshtymites, reaction rim (* MgO, 28.08 wt. %) between kyshymites and meta-ultramafic host rocks, meta-ultramafic host rocks (** MgO, 17.97 wt. %), miascites, and carbonatites of Vishnevogorsky complex (chemistry of miascites and carbonatites is modified after Nedosekova et al. (2009) [18].

On the chondrite-normalized REE spider diagram, the main trend is the enrichment of the LREE comparing to the HREE (Figure 13). This trend is similar to those detected in nepheline syenites (miascites) and carbonatites of Vishnevogorsky complex [18]. However, kyshtymites are more enriched in REE than some miascites, and depleted in REE when compared to carbonatites (Figure 13). Some samples of kyshtymites show positive Eu patterns.

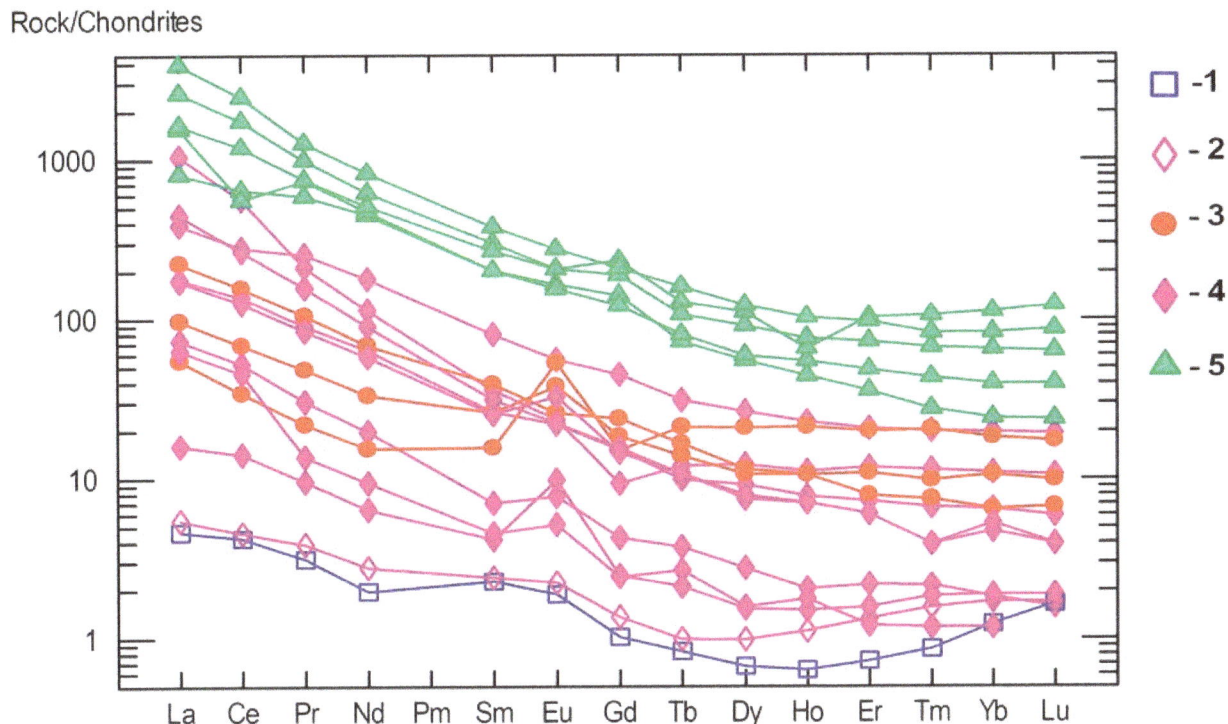

Figure 13. Chondrite-normalized REE spider diagram for the reaction rim between kyshymites and meta-ultramafic host rock (blue squares—1), meta-ultramafic host rocks (purple diamonds—2), kyshtymites (red circles—3), miascites (purple diamonds—4), and carbonatites (green triangles—5) of the Ilmenogorsky-Vishnevogorsky complex modified after [18] and Medvedeva E.V. (unpublished data); REE values in chondrite are after [41].

4.4. Trace Element Chemistry of Zircons and In Situ LA-ICP-MS U-Pb Zircon Geochronology

Nine zircon grains syngenetic with the sapphires from the kyshtymite sample 8-K were chosen for trace-elements measurements (Table S8) and in situ LA-ICP-MS U-Pb geochronological research (Table S9). The zircon grains ranged from about 50 μm × 50 μm to about 100 μm × 200 μm and showed common "magmatic" oscillatory zonation [42] visible in cathodoluminiscence images and maps in average weighted atomic numbers (Figure S1). Yellow and orange colors on the maps in Figure S1 correlate with the higher U and Th contents in zircons (see Table S8).

The REE spider diagram shows enrichment in HREE relative to LREE which is common for magmatic zircons [43] (Figure 14 and Figure S2). However, the zircons from kyshtymites show significantly higher values for most REE except for Tm, Yb and Lu. The U/Th ratio in studied zircons varies from 0.05 to 0.60 as in zircons from nepheline syenites (miascites) and carbonatites [43] with high Hf content of 6845–17482 μg/g (in the range of Hf values detected in zircon inclusions within sapphires, see Table S4), and Y concentration of 380–2370 μg/g. Zircons are also characterized by a positive Ce pattern Ce/Ce* = 1.39–27.36, whereas Eu pattern is absent (Eu/Eu* = 0.35–1.04). Titanium concentrations were below the detection limit except for 4 measurements with contents of 43-182 μg/g.

One spot showed 2523 µg/g of Ti, most likely due to the ablation of inclusions. The kyshtymite crystallization temperature calculated by Ti-in-zircon thermometer $T(°C)_{zircon} = \frac{5080\pm30}{(6.01\pm0.03)-\log(Ti,\mu g/g)}$ − 273 [44] of three spots showed the temperatures of about 890–920 ± 30 °C.

Zircon/Chondrites

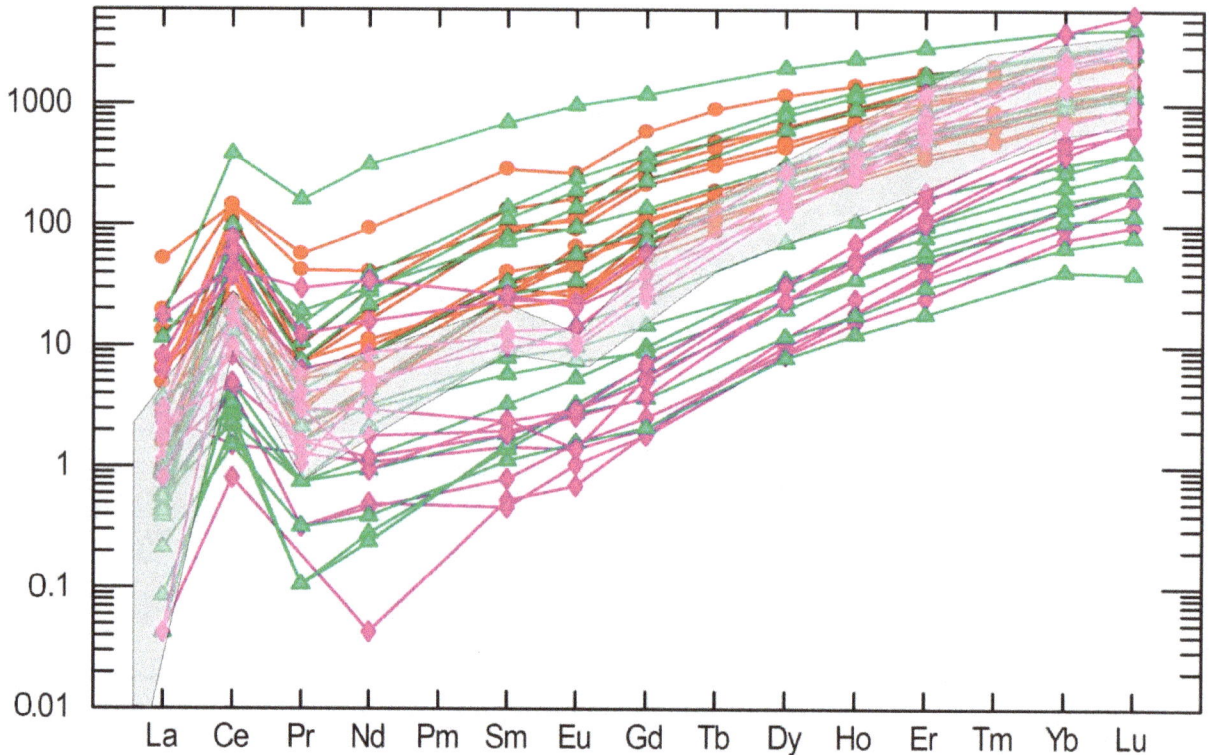

Figure 14. Chondrite-normalized concentration of REE in studied zircons from kyshtymites (samples 8-K—bold red circles), miascites (bold purple diamonds), and carbonatites (bold green triangles) of the Ilmenogorsky-Vishnevogorsky complex modified after (REE data on miascites and carbonatites are after [45]). Data on REE in chondrite are after [41]. Grey area is common REE distribution in magmatic zircons by Belousova et al. (2002) [43].

Twenty spots ablated on zircons during in situ LA-ICP-MS U-Pb geochronological measurements showed the stable ablation signal (cf. Table S8) indicating a homogeneous isotopic composition of the ablated volume. Eleven spots gave the Concordia age at 294 ± 6 Ma. Six spots showed the older Concordia age at 334 ± 10 Ma, which is common for the core areas of magmatic zircons (Figure S1). This older age was previously detected in zircons from carbonatites and nepheline syenites (miascites) of Ilmenogorsky-Vishnevogorsky complex [21–23,45] and is synchronous with the Hercynian metamorphism in this area [20,24] (Figure 15). Three spots were found to be in the Discordia, which is likely due to the later higher temperature event.

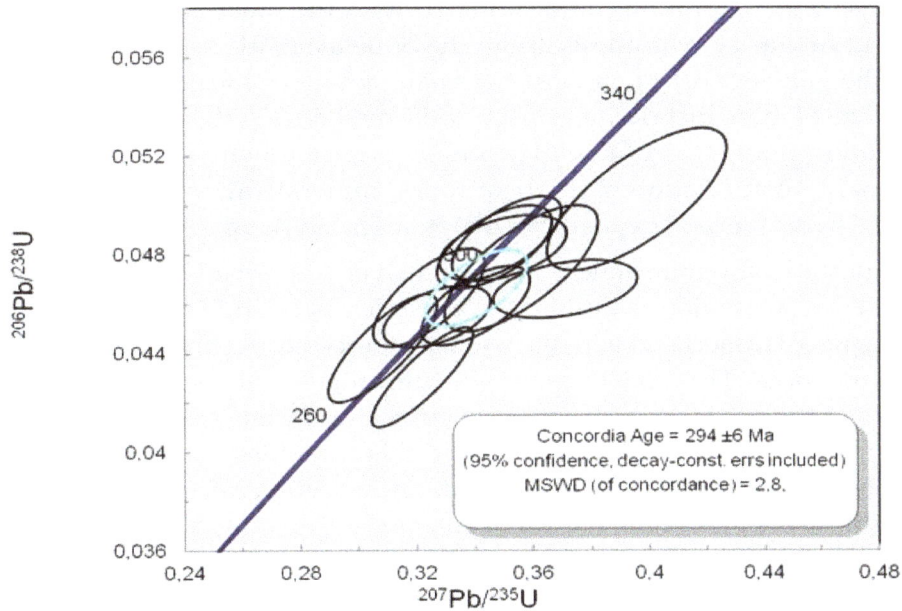

a.

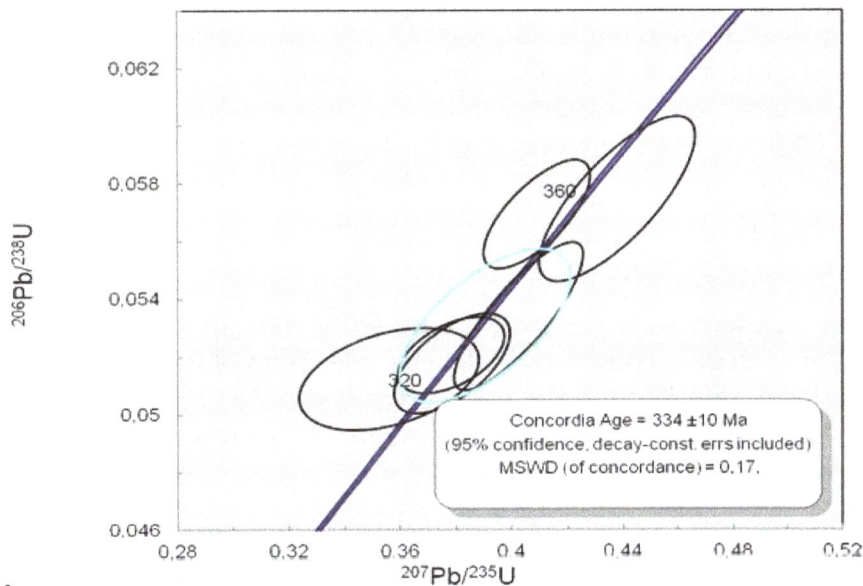

b.

Figure 15. Concordia diagram for zircons from kyshtymite sample 8-K: (**a**) Concordia ages at 294 ± 6 Ma; and (**b**) 334 ± 10 Ma.

5. Discussion

5.1. Genetic Models of Corundum Anorthosites-Kyshtymites

There are several hypotheses about the origin of anorthosite-kyshtymites in South Urals, however, most of them are controversial. For instance, Fersman [46] considered kyshtymites as a product of granitic pegmatite desilication due to interaction with meta-ultramafic host rocks. According to Fersman's theory, an excess of alumina in melt crystalized in the form of corundum, and the reactive rims with biotite (phlogopite), actinolite, chlorite, and talc formed at a contact zone.

Fersman's hypothesis was criticized by Lodochnikov [47], pointing out the similar reaction rims at a contact of meta-ultramafites to the older sedimentary or igneous rocks with the absence of kyshtymites. According to Lodochnikov's idea, corundum anorthosites are hydrothermal formations, while active mineral-forming gas–liquid solutions are associated with the meta-ultramafites themselves. Later, Korzhinsky [48] put forward a hypothesis about the bimetasomatic origin of anorthositic veins with corundum under the influence of granite post-magmatic solutions intruded later than meta-ultramafic rocks. However, none of these theories have been confirmed during the exploration of deposit as there was no clear indication of any hydrothermal processes occurred here (e.g., the presence of dispersion halos, hydrothermal transformation of the host rocks, mineralization that determines the primary mineral zonation of hydrothermal deposits were not observed there).

The problem of the kyshtymite genesis was extensively described by Kolesnik for the Borzovsky kyshtymite occurrence (an analogue of the 5th versta deposit) [16,17]. Kolesnik suggested that the formation of kyshtymites is associated with metasomatic processes occurring during intrusion of granite dykes into ultramafic rocks. The development of corundum anorthosite replacing aplite-like granite dyke with sections of the pegmatoid structure at contact with the granite gneisses was considered as a possible formation mechanism [14]. However, this hypothesis was not supported during the exploration of deposits, because the source of aluminum, as well as calcium, required for the formation of corundum anorthosite -kyshtymites, is still debatable.

Our previous studies have shown a possible genetic link between kyshtymites (corundum anorthosites), miascites (nepheline syenites), and carbonatites of the Ilmenogorsky-Vishnevogorsky complex [25]: similar REE patterns, i.e., enrichment of LREE compared to HREE (Figure 13); anomalies in U, Nb, P;, Sr, and Ti (Figure S3); moderate and highly fractionated distributions of REE $(La/Yb)_N$ = 4.20–48.12, and a small Eu maximum (Eu/Eu* = 1.02–1.32). Both miascites and kyshtymites are extremely enriched in Al_2O_3 (up to 42.94 wt. % in kyshtymites and up to 22.76 wt. % in miascites [25]). Moreover, these rocks have similar accessory minerals (i.e., Y-bearing phases, apatite group minerals, and monazite-(Ce) [49]). One elder zircon age of kyshtymites and those determined in zircons from carbonatites show similar Concordia age of ca. 334 Ma [50] (Figure S4).

5.2. Magmatic vs. Metamorphic Origin of Kyshtymites

Assuming that sapphires in kyshtymites represent in situ minerals of the primary rock, their geochemical signatures can be used to decode the possible origin of the sapphire-bearing anorthosites. Several discrimination diagrams from the literature presented in Figures 6–11 provide constraints on the metamorphic vs. magmatic genesis of sapphires. According to the diagrams in Figures 6–8, sapphires from kyshtymites demonstrate metamorphic imprint, whereas plots in Figures 9–11 show that they have magmatic and metamorphic signature. Compositional profiles across sapphires (Table 1 and Figure 4) indicate existence of the core and rim zones and point to at least two events of sapphire formation. However, trace element compositions of both the rim and the core are within the variations between compositions of different sapphire grains. Hence, observed zonation does not provide any additional key to unravel sapphire crystallization environment.

The ambiguity in the interpretation of sapphire origin is most probably caused by the criteria used in discriminating between magmatic and metamorphic trace element signatures proposed previously (as in Figures 6–11). Recent studies have discovered transitional groups of sapphires having trace element compositions which are located between the proposed end-members on the discrimination diagrams and which are difficult to classify [35,38,51]. For instance, recently published data on geochemistry of sapphires with obvious magmatic origin like those from Gortva syenite xenoliths within alkali basalts (Slovakia) [38], sapphires from lamphrophiric dyke in Yogo Gulch (USA) [39], sapphires from syenite pegmatites of Ilmen Mountains (South Urals of Russia) [2] indicate that the compositions of "syenitic/magmatic" sapphires lie within the nominally "metasomatic" field (Figure 9).

The magmatic origin of those rocks is also confirmed by the $\delta^{18}O$ data. The $\delta^{18}O$ value of the blue sapphire from xenolith of Gortva is 5.1 ± 0.1‰ [38] fits with the $\delta^{18}O$ range of sapphires associated with syenites/anorthoclasites [52]. The $\delta^{18}O$ values of sapphires from lamphrophiric dyke in Yogo Gulch (Montana, USA) showed 5.4–6.8‰ [39] overlapping the field defined for sapphires from lamprophyres (4.5–7.0‰) and sapphires from syenites (5.2–7.8‰) [52]. Sapphires from syenite pegmatite of the Ilmen Mountains showed $\delta^{18}O$ about 4.3‰, i.e., in range defined for magmatic rocks (lamprophyre, basalt, and syenite) [53]. Magmatic origin of sapphires from Yogo Gulch, Gortva, and Ilmen syenite pegmatites is also confirmed by presence of syngenetic inclusions of primary magmatic minerals (Table S10). In fact, the presence of mainly Ca-plagioclase solid inclusions within sapphires from Yogo and in the lamprophyre xenoliths, as in case of kyshtymites, as well as trachytic melt inclusions indicates that a slab-related troctolitic or anorthositic protolith may be involved in their formation [39]. Plagioclase, as in case of kyshtymites, and alkali feldspar, as in case of Ilmen syenite pegmatites, were found in mineral association, while zircon, spinel, monazite-(Ce), ilmenite, and Y-REE phase were identified as solid inclusions within sapphires from Gortva. Columbite-(Fe), zircon, minerals of alkali feldspar group, monazite-(Ce), sub-micron grains of uraninite, muscovite, diaspore, and ilmenite were identified as syngenetic solid inclusions within the blue sapphire from syenite pegmatites of Ilmenogorsky complex (mines 298 and 349) [2,35]. Thus, the "magmatic/syenitic" field in Figure 9 could be extended toward the low boundary of plot with Ilmen sapphires within syenite pegmatites.

On the Fe vs. Ga/Mg diagram and Fe–Mg*100–Ti*10 ternary plot (Figures 6 and 7), sapphires from kyshtymites plotted in the "metamorphic" field overlapping those from lamphrophiric dyke of Yogo Gulch and Gortva syenite xenoliths, the igneous nature of which was shown above. Besides they are also overlap Pailin (Cambodia) placer sapphire xenocrysts trapped by alkali basalts. The $\delta^{18}O$ of last is 7.1–7.8‰, which is also in range defined for magmatic sapphire in syenite [53]. Moreover, sapphires from Pailin deposit [54] contain inclusions of pyrochlore, columbite-(Fe), goethite, zircon, monazite-(Ce), and rutile [55,56] as in case of Ilmen syenite pegmatites [2,35], while plagioclase identified there as well is also the case of kyshtymites. Sapphires from kyshtymites and Ilmen syenite pegmatites overlap with those from alluvial placer deposit in Montana (USA). Besides, anatase, Ca-rich plagioclase (Raman spectra of them match with anorthite and bytownite) and alkali feldspar (Raman spectra of them match with orthoclase) as in the case of Ilmen syenite pegmatites, along with rutile, ilmenite, monazite-(Ce), apatite group minerals, etc., were identified as solid inclusions within Montana sapphires (see Table S10).

These observations indicate that the existing discrimination diagrams do not provide a clear answer on the sapphire origin. In this sense, the "metamorphic" signature of the sapphires from kyshtymites could be only apparent and incorrect. Thus, based on the available classification of trace element compositions in sapphires, it is currently impossible to unambiguously determine the origin of kyshtymites.

Another possible genetic tool is the geochemistry of zircons from kyshtymites. The zircons demonstrate a clear REE pattern typical for magmatic zircons and similar to that found in zircons from syenites and carbonaties which are obviously magmatic rocks [45] (Figure 14 and Figure S2). Furthermore, both syenites and carbonatites contain also other types of zircons, which are interpreted as metamorphic, with significantly different trace element patterns (Figures 14 and 15). Since kyshtymites are syngenetic to magmatic syenites and carbonaties (see Figure 13), their REE signature of zircons can imply a magmatic origin of anorthosites. Further analysis of trace element compositions of zircons also shows magmatic origin within the continental crust (Figures 16 and 17).

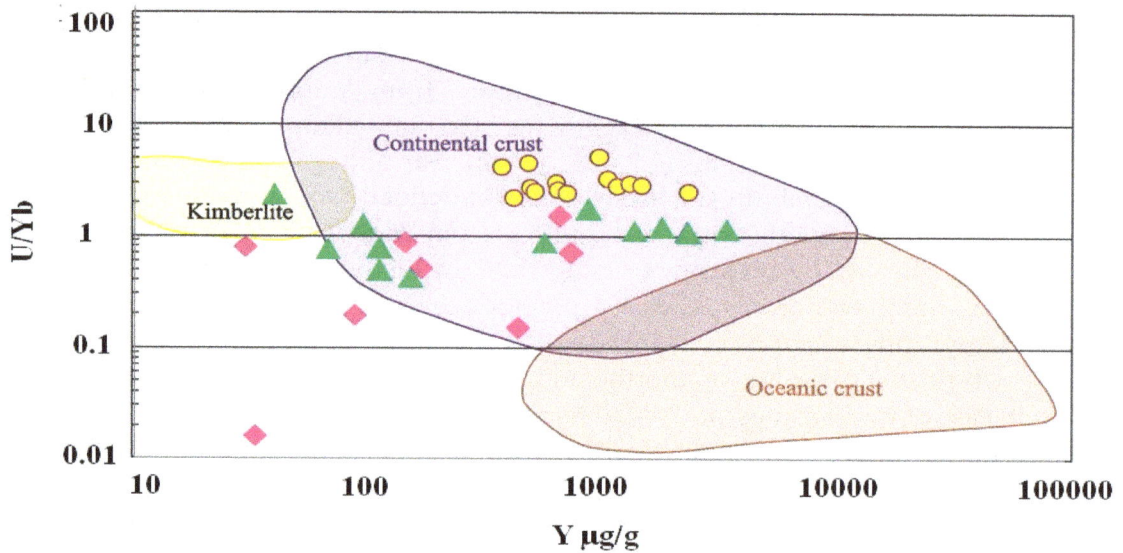

Figure 16. U/Nb vs. Y discriminant diagram of zircons from kyshtimtes (bold yellow circles), miascites (bold purple diamonds), and carbonatites (bold green triangles) of the Ilmenogorsky-Vishnevogorsky complex (the chemistry on miascites and carbonatites is modified after [45]). Fields for continental and oceanic crust, and kimberlite are from Grimes et al. (2007) [57].

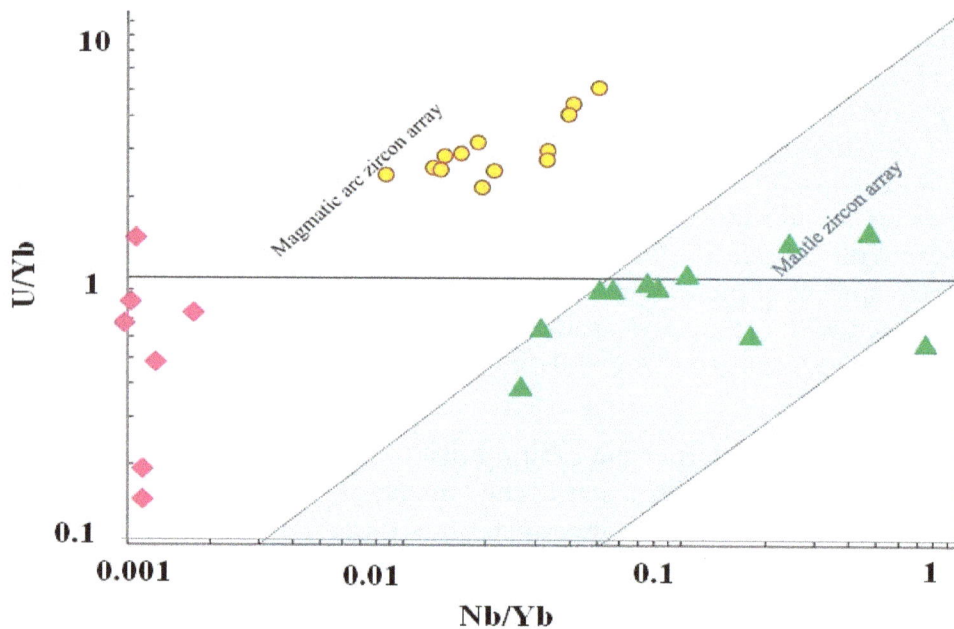

Figure 17. U/Nb vs. Nb/Yb discriminant diagram of zircons from kyshtimtes (bold yellow circles), miascites (bold purple diamonds), and carbonatites (bold green triangles) of the Ilmenogorsky-Vishnevogorsky complex (the chemistry on miascites and carbonatites is modified after [45]); fields for mantle and magmatic arc arrays are from Grimes et al. (2015) [58].

On the chondrite-normalized REE diagram for zircons (Figure 16), REE distributions show the main trend of enrichment in HREE compared to LREE with Ce pattern, which is similar to those observed in miascites and carbotanites of Ilmenogorsky-Vishnevogorsky complex. Positive patterns of Th, U, Hf, and Ta are also observed in chondrite-normalized REE diagrams (Figure S3).

Thus, formation of "magmatic" corundum-blue sapphire anorthosites could occur, likely during at least two stages. In the first stage, about 450–420 Ma [21–23,50], primary anorthosites formed as cumulates in the magmatic chamber, along with miascites and carbonatites of the Ilmenogorsky-Vishnevogorsky complex, with likely more Ca-rich plagioclase in mineral association. Formation of

Corundum Anorthosites-Kyshtymites from the South Urals, Russia: A Combined...

57

cumulates in magmatic chamber on the later stage in excess of H_2O was confirmed by experimental studies [59], whereas another study showed formation of cumulates at the top of magmatic chamber due to the lower density than the rest of magma [60]. Later, at 300–285 Ma, primary anorthositic cumulates probably could be re-melted during the collision process. Plagioclase recrystallized to more Ca-poor member, whereas the excess of aluminum crystallized in form of corundum, while sodium probably came from nepheline syenites (miascites).

However, even though zircons demonstrate clear evidence of magmatic origin, their signature may not represent the genesis of anothosites. Zircons are known as very stable and inert minerals and, hence, they could be inherited in anorthosites from another primary rock. Since both syngenetic syenites and carbonataites contain zircons, they can be a source of magmatic zircons in the system. On U/Nb vs. Y discriminant diagram (Figure 16), zircons from nepheline syenites (miascites) and carbonatites are plotted on the "continental crust" fields despite on their mantle origin [45] overlapping zircons from kyshtymites. On U/Nb vs. Nb/Yb discriminant diagram, zircons from miascites and kyshtymites are also plotted to "magmatic arc" field (Figure 17). In other words, the zircons in kyshtymites can be magmatic, but the host rock can be produced by a metamorphic process. Such a process must be a relatively low temperature process to preserve original zircons unaffected. The Ti-in-zircon thermomentry provides temperature estimates of about 890–920 °C, which is close to or below the closure temperature of the U-Pb geochronometer in zircons [61]. On the other hand, the U-Pb age determinations show only few discordant zircons (Figure 16) which could be affected by high-temperature events (either magmatic or metamorphic). Most zircons demonstrate concordant ages indicating that after their formation they were not exposed to high temperatures. Thus, compositions of zircons are also not an ultimate tool to reconstruct the origin of kyshtymites.

Typically, it is believed that anorthosites are formed by differentiation of mafic magmas when plagioclases become buoyant due to density contrast and accumulate at the roof of the magma chamber [58]. Such a process would require a relatively large volume of magma to produce plagioclase cumulates. The dimensions of kyshtymite anorthositic veins are quite small with the length from 15 to 70 m and thickness from 0.1 to 3 m, which are not realistic for anorthositic body of magmatic origin. Furthermore, formation of anorthosites by re-melting processes during orogenic events is also questionable as temperatures recorded by zircons are relatively low and obviously not enough to melt anorthite-rich substrate/rock.

The ages of 294 ± 6 Ma as determined by zircon geochronology correspond to the ages of Hercynian metamorphism during collision events in the area of Ilmenogorsky-Vishnevogorsky complex. Thus, formation of kyshtymites could also happen due to metamorphic processes. The kyshtymite vein shown in Figure 2 is located within the meta-ultramafic rocks separated by reaction rim consisting mostly of chrysotile-asbestos [62]. The ultra-mafic rocks originally having an orthopyroxenitic composition are strongly metamorphosed and enriched in silica (see Figure 12). This silica-enrichment process is likely linked to the Caledonian metamorphism [63] accompanied by the silicon metasomatism and partial removal of magnesium. Moreover, the excess of Si could also occur likely during the metamorphic re-working and desilication of carbonatites (see the discussion below). The importance of buffering capacity of ultramafic host rocks with respect to silica has been shown in ruby deposit from the Greenland [64]. The similar process observed on another corundum-blue sapphire-bearing deposit within meta-ultramafic host rocks (mine 418) in Ilmenogorsky complex [3].

However, the genesis of the contact zone is difficult to reconstruct since there is no clear evidence for a magmatic contact and there is no indication for any metasomatic zonation. Kyshtymites demonstrate Rb-Sr and Sm-Nd isotopic signatures ($^{87}Sr/^{86}Sr$ of 0.70637–0.706936 and εNd from −5.3 to −10.7) corresponding to the signature of the low-crustal material [25], whereas miascites and carbonatites show mantle-derived values [18]. The data lie on the miascites-carbonatites trend of the Ilmenogorsky-Vishnevogorsky complex and show increasing contribution from the crust (Figure S5). The age and isotopic data could be indicative of the late stage process of kyshtymite formation related to the Hercynian metamorphic event. The main question is the initial protolith to form anorthosites.

The geochemical features of syenites and carbonatites suggest that one of these rocks could serve as a substrate for metamorphic reactions to form anorthosites. Assuming that Al is mostly immobile major element and that carbonates can be decomposed by metamorphic fluids, it can be suggested that metamorphism of carbonatites can be responsible for the formation of Ca-Al-rich anorthosites. Carbonaties from Ilmenogorsky-Vishnevogorsky complex have in addition to CaO, about 5–30 wt. % of SiO_2 and 1–11 wt. % of Al_2O_3 on the CO_2-free basis [18]. If Ca and CO_2 are removed by the metamorphic fluids, the residue will be enriched in Al and Si. Assuming that host meta-ultramafic rocks can be effectively silicified and hence, consume Si, the concentration of Al in former carbonatite vein can reach very high values resulting in the formation of corundum and Ca-rich plagioclases. At the later stages with decreasing temperature, muscovite and clinozoisite, which particulary replace plagioclase, could form in kyshtymite veins as a lower-temperature phases. Clinochlore crystallized at the final stage. Its formation occurred due to replacement of muscovite and the introduction of Mg into the system from the meta-ultramaifc host rocks by metasomatic fluids. One evidence for such a process could be a UV-Vis-NIR spectra of sapphires in one kyshtymite sample which are similar to the spectra observed for metamorphic/metasomatic sapphires.

Such a scenario would explain the metamorphic signature of newly-formed sapphires and magmatic signature of zircons inherited from parental carbonatitic rock. On the other hand, such petrogenetic history of kyshtymites is not visible in the host rocks. It is expected that intensive metamorphism and decomposition of carbonates would result in the development of metamorphic/metasomatic zonation around kyshtymite veins and will cause significant volume changes, which is not observed.

Thus, the new geochemical data presented in this study still do not provide a final and unambiguous answer on the origin of kyshtymites and further investigations are required, e.g., oxygen isotopy of corundum is promising for clarifying the genesis of the kyshtymites.

6. Conclusions

Sapphires were found in situ within primary rock—anorthosites (kyshtymites)—in the 20th century. Until now, the genesis of kyshtymites remains controversial. They are located at a boundary of the Ilmenogorsky-Vishnevogorsky alkaline complex of Russia's South Urals to the north of blue sapphires in syenite pegmatites (mines 298 and 349) and of metasomatites within meta-ultramafic host rocks (mine 418).

Measured trace elements are in the range for metamorphic sapphires: 10,000 Ga/Al > 0.60–0.80, Ga/Mg is 0.11–0.80, Fe/Mg is 6.15–43.32, Cr/Ga is 0.01–0.32, and Fe/Ti is 1.08–22.95. Sapphires are plotted to the "metamorphic" field in Fe versus Ga/Mg and Fe–Mg*100–Ti*10 diagrams. Besides, they also overlap "metasomatic" and extended "magmatic/syenitic" fields in $FeO – Cr_2O_3 – MgO – V_2O_3$ versus $FeO + TiO_2 + Ga_2O_3$ discriminant diagrams.

Two possible magmatic and metamorphic-metasomatic scenarios were proposed for the formation of anorthosites-kyshtymites. The magmatic formation of corundum anorthosites should take place during at least two stages. At about 450–420 Ma, primary anorthosites formed as cumulates in the magmatic chamber, along with miascites and carbonatites of the Ilmenogorsky-Vishnevogorsky complex. Further at 300–285 Ma, primary anorthositic cumulates could be re-melted during the collision process. Plagioclase recrystallized to more Ca-poor member, whereas the excess of aluminum crystallized in form of corundum. However, small sizes of anorthositic veins and obtained sub-solidus temperatures of 890–920 ± 30 °C by Ti-in-zircon thermometry do not support the hypothesis of magmatic origin.

The second possible scenario is the formation of kyshtymites through the metamorphic-metasomatic re-working of initial carbonatites. Calcium and CO_2 could be removed by metamorphic fluids, the residue would be enriched in Al and Si. While meta-ultramafic host rocks consume Si, the concentration of Al in former carbonatite vein can reach very high values resulting in the formation of corundum and Ca-rich plagioclases.

Trace-element chemistry and solid inclusions identified in sapphires from kyshtymites, along with those of syenite pegmatites of Ilmenogorsky complex, provide possible genetic links with sapphires from Montana (USA), Gortva (Slovakia), and Pailin (Cambodia). However, more research is required to unravel the nature of anorthositic-syenitic protoliths that could be involved in their formation.

Supplementary Materials

Table S1. The samples and methods used in the research; Table S2. Minerals identified in association with the sapphire; Table S3. Representative compositions (wt. %) of major minerals found in association with sapphire; Table S4. Representative compositions (wt. %) of minor minerals found in association with sapphire; Table S5. Representative compositions (wt. %) of the zircon inclusions in sapphire; Table S6. Chemical composition of sapphires from Ilmenogorsky-Vishnevogorsky complex and sapphire occurrences with possible anorthositic–syenitic origin. Table S7. Representative analyses of corundum-blue sapphire anorthosites-kyshtymites, meta-ultramafic host rocks, and reaction rim; Table S8. LA-ICP-MS trace-elements measurements of zircons from sample 8-K; Table S9 In situ LA-ICP-MS geochronology of zircons from sample 8-K; Table S10. Frequently detected solid inclusions and common minerals found in association with sapphires from Ilmenogorsky-Vishnevogorsky complex and sapphire occurrences with possible anorthositic-syenitic origin. Figure S1. Zircon CL images and maps in average weighted atomic numbers with the spots positions for trace-elements (No. of spots inside of circles) and U-Pb geochronological measurements (U-Pb ages inside of circles). Black, Concordia kyshtymite age; red, 6 Concordia elder ages; purple, 3 ages in Discordia. Figure S2. Chondrite-normalized concentration of REE and trace elements in studied zircons from kyshtymites (**a**), miascites (**b**), and carbonatites (**c**), of the Ilmenogorsky-Vishnevogorsky complex modified after [45]. Data on REE and trace elements in chondrite are after [41]. Figure S3. Trace-elements and REE distribution normalized to primitive mantle (the data on REE and trace-elements in primitive mantle are from [41]) in reaction rim (1) between kyshymites and meta-ultramafic host rock, meta-ultramafic host rocks (2), kyshtymites (3), miascites (4), and carbonatites (5) of the Ilmenogorsky-Vishnevogorsky complex modified after [18] and Medvedeva E.V (unpublished data). Figure S4. Concordia diagram for zircons from kyshtymite sample 8-K and carbonatites of Vishnevogorsky complex (sample 354 by Nedosekova [50]). Figure S5. Diagram εSr (T) vs. εNd (T) for kyshtimte and miascite of the Ilmenogorsky-Vishnevogorsky complex modified after [18,25], the diagram shows mantle reservoirs DMM, HIMU, EM1, EM2, MORB and OBI [65].

Author Contributions: M.I.F. formulated the idea of a paper, performed the petrography of kyshtymte and host rocks, designed experiments on EDXRF and WDS EMPA, assembled most of tables and figures, and wrote the manuscript; E.S.S. formulated the idea of a paper, designed experiments on Raman spectroscopy, WDS EMPA, ICP-MS, LA-ICP-MS sapphire and zircon geochemistry, and LA-ICP-MS U-Pb zircon geochronology, performed LA-ICP-MS sapphire geochemistry and its following data-reduction process, interpreted the zircon dating data and constructed Concordia diagrams, assembled some tables and figures, wrote the part of manuscript, and edited the final text of paper; R.B. supervised the LA-ICP-MS zircon geochemical and U-Pb zircon geochronological measurements, formulated the part of possible scenario for kyshtymites origin, and edited the manuscript; S.K. helped with data interpretation and editing of the manuscript; M.A.R. collected samples for research and provided the geological setting data; N.N.K. performed WSD EMPA, and produced zircon CL and BSE images and maps of average weighted atomic numbers; A.G.N. provided absorption spectra of sapphires and their interpretation; J.B. performed LA-ICP-MS zircon geochemistry and U-Pb zircon geochronology, and their following data-reduction process; and W.H. supervised the LA-ICP-MS zircon geochemical and U-Pb zircon geochronological measurements.

Acknowledgments: The authors are grateful to Elena V. Medvedeva (Ilmen State Reserve, Russia) for the provided chemical analyses of miascites. The authors are grateful to the colleagues from GEOKHI RAS (T.G. Kuzmina, V.N. Ermolaeva, T.V. Romashova, V.A. Turkov, B.S. Semiannikov), Ya.V. Bychkova (Moscow State University, Russia), E.A. Minervina (IGEM RAS), K.Ponkratov (Renishaw Moscow), Delia Rösel (Technische Universität Bergakademie Freiberg, Germany), as well as Stephan Buhre and Tobias Häger (Johannes Gutenberg Universität Mainz, Germany) for their assistance in sample preparation and analytical studies of kyshmymites and miascites, as well as following data interpretation.

References

1. Giuliani, G.; Ohnenstetter, D.; Fallick, A.E.; Groat, L.; Fagan, A.G. The geology and genesis of gem corundum deposits. In *Geology of Gem Deposits*, 2nd ed.; Groat, L.A., Ed.; Mineralogical Association of Canada Short Course Series; Mineralogical Association of Canada: Tucson, AZ, USA, 2014; Volume 44, pp. 29–112.

2. Sorokina, E.S.; Karampelas, S.; Nishanbaev, T.P.; Nikandrov, S.N.; Semiannikov, B.S. Sapphire Megacrysts in Syenite Pegmatites from the Ilmen Mountains, South Urals, Russia: New Mineralogical Data. *Can. Mineral.* **2017**, *55*, 823–843. [CrossRef]

3.   Sorokina, E.S.; Rassomakhin, M.A.; Nikandrov, S.N.; Karampelas, S.; Kononkova, N.N.; Nikolaev, A.G.; Anosova, M.O.; Orlova, A.V.; Kostitsyn, Y.A.; Kotlyarov, V.A. Origin of blue sapphire in newly discovered spinel–chlorite–muscovite rocks within meta-ultramafites of Ilmen Mountains, South Urals of Russia: Evidence from mineralogy, geochemistry, Rb-Sr and Sm-Nd isotopic data. *Minerals* **2019**, *9*, 36. [CrossRef]

4.   Simonet, C.; Paquette, J.L.; Pin, C.; Lansner, B.; Fritsch, E. The Dusi (Garba Tula) sapphire deposit, Central Kenya–a unique Pan-African corundum-bearing monzonite. *J. Afr. Earth Sci.* **2004**, *38*, 401–410. [CrossRef]

5.   Monchoux, P.; Fontan, F.; De Parseval, P.; Martin, R.F.; Wang, R.C. Igneous albitite dikes in orogenic lherzolites, western Pyrenees, France: a possible source for corundum and alkali feldspar xenocrysts in basaltic terranes. I. Mineralogical associations. *Can. Mineral.* **2006**, *44*, 817–842. [CrossRef]

6.   Kan-Nyunt, H.P.; Karampelas, S.; Link, K.; Thu, K.; Kiefert, L.; Hardy, P. Blue sapphires from the Baw Mar Mine in Mogok. *Gems Gemol.* **2013**, *49*, 223–232. [CrossRef]

7.   Khoi, N.N.; Hauzenberger, C.A.; Sutthirat, C.; Tuan, D.A.; Häger, T.; Van Nam, N. Corundum with Spinel Corona from the Tan Huong–Truc Lau Area in Northern Vietnam. *Gems Gemol.* **2018**, *54*, 4. [CrossRef]

8.   Voudouris, P.; Mavrogonatos, C.; Graham, I.; Giuliani, G.; Melfos, V.; Karampelas, S.; Karantoni, V.; Wang, K.; Tarantola, A.; Zaw, K.; et al. Gem Corundum Deposits of Greece: Geology, Mineralogy and Genesis. *Minerals* **2018**, *9*, 49. [CrossRef]

9.   Keulen, N.; Kalvig, P. Fingerprinting of corundum (ruby) from Fiskenæsset, West Greenland. *Geol. Surv. Denmark Greenland* **2013**, *25*, 53–56.

10.   Karmakar, S.; Mukherjee, S.; Sanyal, S.; Sengupta, P. Origin of peraluminous minerals (corundum, spinel, and sapphirine) in a highly calcic anorthosite from the Sittampundi Layered Complex, Tamil Nadu, India. *Contrib. Mineral. Petrol.* **2017**, *172*, 67. [CrossRef]

11.   Gibson, G.M. Margarite in Kyanite- and Corundum-Bearing Anorthosite, Amphibolite, and Hornblendite from Central Fiordland, New Zealand. *Contrib. Miner. Petrol.* **1979**, *68*, 171–179. [CrossRef]

12.   Pratt, G.H. *Corundum and Its Occurrence and Distribution in the United States*; US Government Printing Office: Washington, DC, USA, 1906.

13.   McElhaney, M.S.; McSween, H.Y. Petrology of the Chunky Gal Mountain mafic-ultramafic complex, North Carolina. *GSA Bull.* **1983**, *94*, 855–874. [CrossRef]

14.   Koptev-Dvornikov, V.S.; Kuznetsov, E.A. *Borzovskoe Corundum Deposit: Petrological Study*; State Technical Publishing House: Moscow, Russia, 1931; p. 320. (In Russian)

15.   Claire, M.O. Corundum and emery on the Urals. *Uralskiy Technik.* **1918**, *7*, 1–17. (In Russian)

16.   Kolesnik, N.Y. *High-Temperature Metasomatism in Ultrabasic Massifs*; Science Publishing House: Novosibirsk, Russia, 1976; p. 240. (In Russian)

17.   Kolesnik, N.Y.; Korolyuk, V.N.; Lavrent'ev, Y.G. Spinels and ore minerals of the Borzovsk deposit of corundum plagioclasites. *Notes Russian Mineral. Soc.* **1974**, *103*, 373–378. (In Russian)

18.   Nedosekova, I.L.; Vladykin, N.V.; Pribakin, S.V.; Bayanova, T.B. The structure of the Ilmenogorsky-Vishnevogorsky Miaskite-Carbonatite Complex: Origin, ore-bearing. Sources of matter (Ural, Russia). *Geol. Ore Depos.* **2009**, *51*, 157–181. (In Russian) [CrossRef]

19.   Rusin, A.I.; Krasnobaev, A.A.; Valizer, P.M. Geology of the Ilmen Mountains: situation and problems. In *Geology and Mineralogy of the Ilmenogorsky Complex: Situation and Problems*; IGZ UB RAS: Miass, Russia, 2006; pp. 3–19. (In Russian)

20.   Krasnobaev, A.A.; Puzhakov, B.A.; Petrov, V.I.; Busharina, S.V. Zirconology of metamorphites of the Kyshtym-Arakulian strata of the Sysert-Ilmenogorsky complex. *Proc. Zavaritsky Inst. Geol. Geochem. (Trudy Instituta Geologii i Geokhimii im. Akademika A.N. Zavaritskogo)* **2009**, *156*, 264–268. (In Russian)

21.   Kramm, U.; Blaxland, A.B.; Kononova, V.A.; Grauert, B. Origin of the Ilmenogorsk-Vishnevogorsk nepheline syenites, Urals, USSR, and their time of emplacement during the history of the Ural fold belt: A Rb-Sr study. *J. Geol.* **1983**, *91*, 427–435. [CrossRef]

22.   Kramm, U.; Chernyshev, I.V.; Grauert, S. Zircon typology and U-Pb systematics: A case study of zircons from nefeline syenite of the Il'meny Mountains, Ural. *Petrology* **1993**, *1*, 474–485.

23.   Chernyshev, I.V.; Kononova, V.A.; Kramm, U. Isotope geochronology of alkaline rocks of the Urals in the light of zircon uranium-lead data. *Geochemistry* **1987**, *3*, 323–338.

24.   Ivanov, K.S.; Erokhin, Y.V. About the Age and nature of the metamorphic complexes of the Ilmenogorsk zone of the Urals. *Rep. Acad. Sci.* **2015**, *461*, 312–315.

25. Filina, M.I.; Sorokina, E.S.; Rassomakhin, M.A.; Kononkova, N.N.; Kostitsyn, Y.A.; Orlova, A.V. Genetic linkage of corundum plagioclazite-kyshtymite and miaskites of Ilmensky-Vishnevogorsky complex, Southern Urals, Russia: new data on Rb-Sr and Sm-Nd isotopic composition, geochemistry and mineralogy. *Geochem. Int.* **2019**. (accepted).

26. Arslanova, K.A.; Golubchina, M.N.; Iskanderova, A.D. *Geological Dictionary*; Nedra Publishing House: Moscow, Russia, 1978; Volume 2, p. 456. (In Russian)

27. Jochum, K.P.; Weis, U.; Stoll, B.; Kuzmin, D.; Yang, Q.; Raczek, I.; Jacob, D.E.; Stracke, A.; Birbaum, K.; Frick, D.A.; et al. Determination of reference values for NIST SRM 610-617 glasses following ISO Guidelines. *Geostand. Geoanal. Res.* **2011**, *35*, 397–429. [CrossRef]

28. Jochum, K.P.; Scholz, D.; Stoll, B.; Weis, U.; Wilson, S.A.; Yang, Q.; Schwalb, A.; Börner, N.; Jacob, D.E.; Andreae, M.O. Accurate trace element analysis of speleothems and biogenic calcium carbonates by LA-ICP-MS. *Chem. Geol.* **2012**, *318*, 31–44. [CrossRef]

29. Jackson, S.E.; Pearson, N.J.; Griffin, W.L.; Belousova, E.A. The application of laser ablation-inductively coupled plasma-mass spectrometry to in situ U–Pb zircon geochronology. *Chem. Geol.* **2004**, *211*, 47–69. [CrossRef]

30. Kooijman, E.; Berndt, J.; Mezger, K. U-Pb dating of zircon by laser ablation ICP-MS: Recent improvements and new insights. *Eur. J. Miner.* **2012**, *24*, 5–21. [CrossRef]

31. Wiedenbeck, M.; Alle, P.; Corfu, F.; Griffin, W.L.; Meier, M.; Oberli, F.; von Quadt, A.; Roddick, J.C.; Spiegel, W. Three natural zircon standards for U-Th-Pb, Lu-Hf, trace-element and REE analyses. *Geostand. Newsl.* **1995**, *19*, 1–23. [CrossRef]

32. Ludwig, K.R. *A User's Manual*; Barkeley Geochonology Center: Berkeley, CA, USA, 2009; p. 100.

33. Platonov, A.N.; Taran, M.N.; Balitsky, V.S. *The Nature of the Coloring of Gems*; Nedra Publishing House: Moscow, Russia, 1984; p. 196. (In Russian)

34. Sorokina, E.S.; Koivula, J.I.; Muyal, J.; Karampelas, S. Multiphase fluid inclusions in blue sapphires from the Ilmen Mountains, southern Urals. *Gems Gemol.* **2016**, *52*, 209–211.

35. Peucat, J.J.; Ruffault, P.; Fritch, E.; Bouhnik-Le Coz, M.; Simonet, C.; Lasnier, B. Ga/Mg ratio as a new geochemical tool to differentiate magmatic from metamorphic blue sapphires. *Lithos* **2007**, *98*, 261–274. [CrossRef]

36. Sutherland, F.L.; Zaw, K.; Meffre, S.; Giuliani, G.; Fallick, A.E.; Graham, I.T.; Webb, G.B. Gem-corundum megacrysts from east Australian basalt fields: trace elements, oxygen isotopes and origins. *Aust. J. Earth Sci.* **2009**, *56*, 1003–1022. [CrossRef]

37. Zwaan, J.C.; Buter, E.; Merty-Kraus, R.; Kane, R.E. The origin of Montana's alluvial sapphires. *Gems Gemol.* **2015**, *51*, 370–391.

38. Uher, P.; Giuliani, G.; Szaka, L.L.S.; Fallick, A.; Strunga, V.; Vaculovic, T.; Ozdin, D.; Greganova, M. Sapphires related to alkali basalts from the Cerov'a Highlands, Western Carpathians (southern Slovakia): Composition and origin. *Geologica Carpathica* **2012**, *63*, 71–82. [CrossRef]

39. Palke, A.C.; Wong, J.; Verdel, C.; Avila, J.N. A common origin for Thai/Cambodian rubies and blue and violet sapphires from Yogo Gulch, Montana, U.S.A.? *Am. Mineral.* **2018**, *103*, 469–479. [CrossRef]

40. Giuliani, G.; Caumon, G.; Rakotosamizanany, S.; Ohnenstetter, D.; Rakototondrazafy, M. Classification chimique des corindons par analyse factorielle discriminante: application a la typologie des gisements de rubis et saphirs. Chapter mineralogy, physical properties and geochemistry. *Revue Gemmol.* **2014**, *188*, 14–22.

41. Sun, S.S.; McDonough, W.F. Chemical and isotopic systematics of oceanic basalts: Implications for mantle composition and processes. *Geol. Soc. London Spec. Publ.* **1989**, *42*, 313–345. [CrossRef]

42. Fowler, A.; Prokoph, A.; Stenr, R.; Dupuis, C. Organization of oscillatory zoning in zircon: Analysis, scaling, geochemistry, and model of a zircon from Kipawa, Quebec, Canada. *Geochim. Cosmochim. Acta.* **2002**, *66*, 311–328. [CrossRef]

43. Belousova, E.A.; Griffin, W.L.; O'Reilly, S.Y.; Fisher, N.I. Igneous zircon: trace element compositon as an indicator of source rock type. *Contrib. Mineral. Petrol.* **2002**, *143*, 602–622. [CrossRef]

44. Watson, E.B.; Wark, D.A.; Thomas, J.B. Crystallization thermometers for zircon and rutile. *Contrib. Mineral. Petrol.* **2006**, *151*, 413–433. [CrossRef]

45. Nedosekova, I.L.; Belyatsky, B.V.; Belousova, E.A. Trace elements and Hf isotope composition as indicator of zircon genesis due to the evolution of alkaline-carbonatite magmatic system (Il'meny–Vishnevogorsky complex, Urals, Russia). *Geol. Geophys.* **2016**, *57*, 1135–1154. [CrossRef]

46. Fersman, L.E. *Pegmatites*; Publishing House of the Academy of Sciences of the USSR: Moscow, Russia, 1940; p. 712. (In Russian)

47. Lodochnikov, V.N. *Serpentines and Serpentinites Ilchirsk and Other Petrological Issues Associated with Them*; United Scientific and Technical Publishing: Moscow, Russia, 1936; p. 817. (In Russian)

48. Korzhinskiy, D.S. Essay on metasomatic processes. In *The Main Problems in the Theory of Magmatic Ore Deposits*; Publishing House of the Academy of Sciences of the USSR: Moscow, Russia, 1953; pp. 332–450. (In Russian)

49. Eskova, E.M.; Zhabin, A.G.; Mukhitdinov, G.N. *Mineralogy and Geochemistry of Rare Elements of the Vishnevogorsky Mountains*; Science Publishing House: Moscow, Russia, 1964; p. 318. (In Russian)

50. Nedosekova, I.L. U-Pb age and Lu-Hf isotopic systems of zircons Ilmenogorsky-Vishnevogorsky alkaline-carbonatitic complex, South Ural. *Lithosphere* **2014**, *5*, 19–31. (In Russian)

51. Sutherland, F.L.; Abduriym, A. Geographic typing of gem corundum: a test case from Australia. *J. Gemmol.* **2009**, *31*, 203–210. [CrossRef]

52. Giuliani, G.; Fallick, A.E.; Garnier, V.; France-Lanord, C.; Ohnenstetter, D.; Schwarz, D. Oxygen isotope composition as a tracer for the origins of rubies and sapphires. *Geology* **2005**, *33*, 249–252. [CrossRef]

53. Vysotsky, S.V.; Nechaev, V.P.; Kissin, A.Y.; Yakovlenko, V.V.; Velivetskaya, T.A.; Sutherland, F.L.; Agoshkov, A.I. Oxygen isotopic composition as an indicator of ruby and sapphire origin: A review of Russian occurrences. *Ore Geol. Rev.* **2015**, *68*, 164–170. [CrossRef]

54. Sutherland, F.L.; Giuliani, G.; Fallick, A.E.; Garland, M.; Webb, G. Sapphire-ruby characteristics, West Pailin, Cambodia: Clues to their origin based on trace element and O isotope analysis. *Aust. Gemmol.* **2008**, *23*, 329–368.

55. Sutherland, F.L.; Schwarz, D.; Jobbins, E.A.; Coenraads, R.R.; Webb, G. Distinctive gem corundum suites from discrete basalt fields: a comparative study of Barrington, Australia, and west Pailin, Cambodia, gemfields. *J. Gemmol.* **1998**, *26*, 65–85. [CrossRef]

56. Saeseaw, S.; Sangsawong, S.; Vertriest, W.; Atikarnsakul, U. *An In-Depth Study of Blue Sapphires from Pailin, Cambodia*; Gemological Institute of America report: Carlsbad, CA, USA, 2017; p. 45.

57. Grimes, C.B.; John, B.E.; Kelemen, P.B.; Mazdab, F.K.; Wooden, J.L.; Cheadle, M.J.; Hanghoj, K.; Schwartz, J.J. Trace element chemistry of zircons from oceanic crust: A method for distinguishing detrital zircon provenance. *Geology* **2007**, *35*, 643–646. [CrossRef]

58. Grimes, C.B.; Wooden, J.L.; Cheadle, M.J.; John, B.E. "Fingerprinting" tectono-magmatic provenance using trace elements in igneous zircon. *Contrib. Mineral. Petrol.* **2015**, *170*, 46. [CrossRef]

59. Botcharnikov, R.E.; Almeev, R.R.; Koepke, J.; Holtz, F. Phase relations and liquid lines of descent in hydrous ferrobasalt—Implications for the Skaergaard Intrusion and Columbia River flood basalts. *J. Petrol.* **2008**, *49*, 1687–1727. [CrossRef]

60. Arndt, N. The formation of massif anorthosite: Petrology in reverse. *Geosci. Front.* **2013**, *7*, 875–889. [CrossRef]

61. Leet, J.K.; Williams, I.S.; Ellis, D.J. Pb, U and Th diffusion in natural zircon. *Nature* **1997**, *390*, 159–162.

62. Biondi, J.C. Neoproterozoic Cana Brava chrysotile deposit (Goiás, Brazil): Geology and geochemistry of chrysotile vein formation. *Lithos* **2014**, *184*, 132–154. [CrossRef]

63. Varlakov, A.S.; Kuznetsov, G.P.; Korablev, G.G. *Hyperbasites of the Ilmenogorsky-Vishnevogorsky complex Complex (Southern Urals)*; Publishing House of the Institute of Mineralogy, Ural Branch RAS: Miass, Russia, 1998; p. 195. (In Russian)

64. Yakymchuk, C.; Kristoffer, S. Corundum formation by metasomatic reactions in Archean metapelite, SW Greenland: Exploration vectors for ruby deposits within high-grade greenstone belts. *Geosci. Front.* **2017**, *9*, 1–24. [CrossRef]

65. Hofmann, A.W. Mantle geochemistry: The message from oceanic volcanism. *Nature* **1997**, *385*, 219–229. [CrossRef]

4

Chemical Characteristics of Freshwater and Saltwater Natural and Cultured Pearls from Different Bivalves

Stefanos Karampelas *, Fatima Mohamed, Hasan Abdulla, Fatema Almahmood, Latifa Flamarzi, Supharart Sangsawong and Abeer Alalawi

Bahrain Institute for Pearls & Gemstones (DANAT), WTC East Tower, P.O. Box 17236 Manama, Bahrain;
Fatima.Mohamed@danat.bh (F.M.); Hasan.Abdulla@danat.bh (H.A.); Fatema.Almahmood@danat.bh (F.A.);
Latifa.Flamarzi@danat.bh (L.F.); Supharart.Sangsawong@danat.bh (S.S.); Abeer.Alalawi@danat.bh (A.A.)

* Correspondence: Stefanos.Karampelas@danat.bh

Abstract: The present study applied Laser Ablation-Inductively Coupled Plasma-Mass Spectrometry (LA-ICP-MS) on a large number of natural and cultured pearls from saltwater and freshwater environments, which revealed that freshwater (natural and cultured) pearls contain relatively higher quantities of manganese (Mn) and barium (Ba) and lower sodium (Na), magnesium (Mg) and strontium (Sr) than saltwater (natural and cultured) pearls. A few correlations between the host animal's species and chemical elements were found; some samples from *Pinctada maxima* (*P. maxima*) are the only studied saltwater samples with ^{55}Mn >20 ppmw, while some *P. radiata* are the only studied saltwater samples with ^{24}Mg <65 ppmw and some of the *P. imbricata* are the only studied saltwater samples with ^{137}Ba >4.5 ppmw. X-ray luminescence reactions of the studied samples has confirmed a correlation between its yellow-green intensity and manganese content in aragonite, where the higher Mn^{2+} content, the more intense the yellow-green luminescence becomes. Luminescence intensity in some cases is lower even if manganese increases, either because of pigments or because of manganese self-quenching. X-ray luminescence can be applied in most cases to separate saltwater from freshwater samples; only samples with low manganese content (^{55}Mn <50 ppmw) might be challenging to identify. One of the studied natural freshwater pearls contained vaterite sections which react by turning orange under X-ray due to a different coordination of Mn^{2+} in vaterite than that in aragonite.

Keywords: pearls; freshwater; saltwater; LA-ICP-MS; X-ray luminescence

1. Introduction

Pearls are probably the most appreciated organogenic gems and they are either natural or cultured. Natural pearls (NPs) are secreted accidentally, without human intervention within naturally formed sacs (cysts made of epithelium cells), by molluscs such as bivalves or gastropods and very rarely also by cephalopods. Cultured pearls (CPs) are formed by molluscs within a pearl sac (cyst) produced with human intervention; e.g., after transplantation of epithelial cells cut from the mantle—a.k.a. tissue—(with or without the implantation of a bead) by human. Pearls are also classified, following their external appearance, into nacreous and non-nacreous forms. Under an optical microscope, nacreous (natural and cultured) pearls show terrace like structures (sometimes looking like fingerprints) composed of aragonite and organic matter (mixture of beta-chitin and acidic glycoproteins) stacked in a "brick-wall" pattern (i.e., sheet nacre). All pearls without nacreous appearance are considered non-nacreous. The vast majority of natural and cultured pearls used in jewellery are found in bivalves and they have a nacreous surface which is entirely made out of aragonite [1].

Natural and cultured pearls may be separated following the growth environment of their host mollusc; the environment is either freshwater (i.e., living in rivers or lakes; FW) or saltwater (i.e., living in sea; SW). Usually, the implanted bead in cultured pearls (both SW and FW) is cut from a freshwater bivalve shell [2].

Freshwater cultured and natural pearls contain more manganese (Mn) than their saltwater counterparts. As a consequence, eye visible (nowadays captured with a digital camera in most cases) reactions under X-rays (a.k.a. X-ray luminescence) are different for samples found in different water environments. Freshwater (natural and cultured) pearls luminate a green-yellow colour and saltwater pearls (natural and cultured without bead) remain inert [3–5]. However, some cultured saltwater pearls with a bead (made out of freshwater shell) can also give a green–yellow form of luminescence under X-rays. This reaction is due to the manganese content of freshwater bead and the relatively thin nacre (i.e., thickness of nacre material covering the bead) [5]. Cathodoluminescence (CL) microscopy and spectroscopy were also used for pearl characterization, and it was suggested that Mn^{2+} differs from cultured freshwater to natural freshwater and to natural saltwater pearls [6].

Energy dispersive X–ray fluorescence (EDXRF), a non-destructive method commonly used on gems, is applied to separate freshwater from saltwater samples [7–9]. Saltwater samples contain more Sr and less Mn than freshwater samples. The SrO/MnO ratio was suggested to be used for freshwater and saltwater pearls separation, as it is >12 for saltwater pearls and <12 for freshwater pearls [8]. EDXRF is also used to detect treatments used on natural and cultured pearls to improve their colour—with inorganic substances such as silver, iodine, bromine etc. [10,11].

Few examples of Laser Ablation-Inductively Coupled Plasma-Mass Spectrometry (LA-ICP-MS) analysis on cultured and natural freshwater and saltwater pearls from different regions can be found in literature [12–16]. However, this method is widely applied on other biogenic calcium carbonates (of freshwater and saltwater) such as molluscs shells and corals, as they are considered to be valuable environmental monitors and archives of paleoclimates [17–22]. This is because of their immobility and their trace elements content, which could be linked with the water conditions they grew in, making them a useful indicator of climate pollution and ecosystem changes. Biological factors such as growth rate, age etc. can also influence biogenic carbonate chemistry, making the interpretation of chemical elements incorporation in calcium carbonate challenging [23].

For the present work, 1113 natural and cultured nacreous pearls were studied using X-ray luminescence and LA-ICP-MS. The studied samples were collected from various bivalves and different geographic areas. This study was carried out in order to better study natural and cultured pearls' chemical characteristics and look for potential differences linked to their origin (environment, host animal and/or geographic). This is the first study that combines these methods on such a large group of (natural and cultured) pearls.

2. Materials and Methods

All studied 1113 samples are listed in Table 1; 999 samples were natural saltwater pearls found in three different bivalves and areas. Two hundred forty-eight were natural saltwater pearls from *P. radiata*, fished in the early 1980s in the Arabian Gulf, around 15 km NNE off Bahrain ("main heirats") for a project conducted by the Bahrain Centre for Studies and Research, headed by Dr. Hashim Al-Sayed (ex-dean of College of Science, University of Bahrain), principally for environmental studies. The samples currently belong to the Bahrain National Museum. The pearls have various shapes, sized from 2.3 to 8.8 mm and weigh from 0.2 to 4.3 carats (1 carat = 0.2 grams). Their colours vary from light cream to yellow.

Eighty natural saltwater pearls were from *P. margaritifera*, fished early 1990s from the Red Sea, off Hurghada (Egypt) were also studied. All samples were reportedly found in the same animal. The samples have various shapes, are sized from 1.5 to 5.4 mm, weigh from 0.1 to 1.1 carats and their colours vary from white to light grey.

The rest of the studied natural saltwater pearls were six hundred seventy-one samples from Venezuela, part of a private collection. These samples were reportedly collected in the Pre to Early-Colombian Era from *P. imbricata* and kept in a jar for several years [9]. The pearls also have

Chemical Characteristics of Freshwater and Saltwater Natural and Cultured Pearls from Different Bivalves

various shapes, sized from 2.9 to 19.5 mm, weigh from 0.2 to 47.5 carats and their colours vary from white to cream and sometimes light grey and light yellow.

Table 1. List of studied samples.

Environment	Bivalve	Area	No. of Samples	
Saltwater	P. radiata *	Arabian Gulf (Heirats, Bahrain)	248 ****	
	P. margaritifera *	Red Sea (Hurghada, Egypt)	80 (****)	
	P. imbricata *	Venezuela	671 ****	
	P. maxima **/***	Indonesia	53 (****)	1058
	P. maxima ***	Burma	3	
	P. fucata ***	Vietnam (Halong Bay)	3	
Freshwater	Margaritifera margaritifera *	Scotland (Spey river)	12	
	Unionidae indet. *	North American rivers	26	55
	Hyriopsis sp. ***	Chinese rivers and lakes	17	

* Natural pearls; ** Cultured pearls without bead; *** Cultured pearls with bead; **** 1 spot was analysed using LA-ICP-MS, (****) one spot was analysed using LA-ICP-MS in most pearls.

Fifty-nine cultured saltwater pearls from two different animals and three different areas were also studied. These are of *P. maxima*; where 21 samples with bead and 32 samples without bead are from different farms off Indonesia and 3 samples with bead from a farm off Burma as well as 3 samples (with bead) of *P. fucata* cultivated off Vietnam (Halong Bay). The samples have various shapes, from 3.5 to 13.3 mm in size, weigh from 0.3 to 28.2 carats and their colours vary from light cream to light yellow. Thus, a total number of 1058 of saltwater pearls (natural and cultured) were studied for the present work.

Twelve natural freshwater pearls found in *Margaritifera margaritifera* in the Spey river (Scotland) during two different expeditions at the same season in the late 1980s were also studied. The samples have various shapes, are 4.6–9.1 mm in size, weigh from 0.4 to 3.8 carats and their colours vary from white to light grey to light yellow.

Twenty-six natural freshwater pearls reportedly found in USA freshwaters and from various animals belonging to Unionidae (Unionidae indet.) were studied. The samples have various shapes, sized from 2.8 to 8.9 mm, weigh from 0.2 to 4.9 carats and their colours vary from white to light yellow to light grey to light purple to brown.

Seventeen cultured freshwater pearls, without bead, were all reportedly cultivated in Chinese freshwaters into animals from *Hyriopsis* sp.; which today dominate the gem market. The samples have various shapes, sized from 4.4 to 10.6 mm, weigh from 0.4 to 8.3 carats and their colours vary from white to cream to purple and to grey. A total number of 55 freshwater pearls (natural and cultured) were studied for the present work.

X-ray luminescence was studied with a PXI GenX-100 (Pacific X-ray Imaging, San Diego, CA, USA) under 100 kV and 5 mA (500 W), the samples were placed around 20 cm from the X–ray tube. A Nikon D850 camera (Nikon, Tokyo, Japan) was used with an AF–S Micro-Nikkor 105 mm lens (Nikon, Tokyo, Japan), utilizing an exposure time of 8 seconds, F6.3 aperture and ISO Hi 0.7. Natural and cultured pearls may change their colour after exposure to X-ray irradiation, but no alteration of samples surface was noticed after the performed measurement. The samples used for more than 100 times as references for X-ray luminescence may turn darker in colour.

Laser Ablation-Induced Coupled Plasma-Mass Spectrometer (LA-ICP-MS) chemical analysis were performed using a iCAP Q (Thermo Fisher Scientific; Waltham, MA, USA) Induced Coupled Plasma-Mass Spectrometer (ICP-MS) coupled with a Q-switched Nd:YAG Laser Ablation (LA) device operating at a wavelength of 213 nm (Electro Scientific Industries/New Wave Research, Fremont,

CA, USA/San Diego, CA, USA). A laser spot of 40 μm in diameter was used, along with a fluence of around 5 J/cm^2 and a 10 Hz repetition rate. Laser warm up/background time was 20 s, its dwell time was 30 s, and its wash out time was 50 s. For the ICP-MS operations, the forward power was set at ~1550 W and the typical nebulizer gas (argon) flow was ~1.0 L/min and the carrier gas (helium) set at ~0.80 L/min. The criteria for the alignment and tuning sequence were to maximize Cobalt (Co), Lanthanum (La), Thorium (Th), and Uranium (U) counts and keep the ThO/Th ratio below 2%. A MACS-3 standard synthetic calcium carbonate (CaCO$_3$) pellet was used to minimize matrix effects [24]. The time-resolved signal was processed in Qtegra ISDS 2.10 software using calcium (^{43}Ca) as the internal standard applying 40.04 wt % theoretical value—calculated from pure aragonite (CaCO$_3$). Several isotopes were measured, but in this study only sodium (^{23}Na), strontium (^{88}Sr), barium (^{137}Ba) and lead (^{208}Pb) as well as manganese (^{55}Mn) and magnesium (^{24}Mg) are presented. These isotopes were selected as they present low to no matrix and gas blank related interferences (Mn and Mg are relatively high) [16,24]. 1022 samples were analysed with one spot only (all natural saltwater pearls from Bahrain and Venezuela, 59 out of 80 natural saltwater pearls from Egypt and 44 out of 53 cultured saltwater pearls from Indonesia) and in 91 samples three spots were analysed (Table 1). Chemical elements can be incorporated differently in various calcium carbonate polymorphs. All chemical analyses were acquired on spots made of aragonite (checked with Raman spectroscopy) with nacreous microstructure (checked with optical microscope). Limits of detection (LOD) and limits of quantification (LOQ) for each of the abovementioned elements are shown in Table 2. These limits differ from day to day (for every set of measurements) so they are presented as ranges from the lowest to the highest.

Table 2. Laser Ablation-Induced Coupled Plasma-Mass Spectrometer (LA-ICP-MS) detection limits and ranges in ppmw.

Limits	^{23}Na (ppmw)	^{24}Mg (ppmw)	^{55}Mn (ppmw)	^{88}Sr (ppmw)	^{137}Ba (ppmw)	^{208}Pb (ppmw)
LOD	1.62–70.84	0.02–0.57	0.09–0.65	0.01–0.05	0.01–0.39	0.01–0.02
LOQ	5.35–233.77	0.66–1.88	0.21–1.18	0.03-0.16	0.17–1.18	0.02–0.07

LOD: Limits of detection; LOQ: Limits of quantification.

Raman spectra were acquired in a Renishaw inVia spectrometer from 100 to 2000 cm^{-1}, coupled with an optical microscope, 514 nm excitation wavelength (diode-pumped solid-state laser), 1800 grooves/mm grating, notch filter, 40 microns slit, a spectral resolution of around 2 cm^{-1} and calibrated using a diamond at 1331.8 cm^{-1}. A laser power of 5 mW on the sample (to avoid any destruction of fragile organic matter) was used to acquire all Raman spectra, 50× long distance objective lens, an acquisition time was 30 seconds and 7 accumulations.

3. Results and Discussion

3.1. LA-ICP-MS Results

LA-ICP-MS data are presented in Table 3 where the results of saltwater and freshwater samples are listed together and in Table 4 the results by bivalve are listed. All acquired individual data-points are presented in Tables S1–S9. Measured freshwater samples contain higher ^{55}Mn than measured saltwater samples (see again Table 3). Manganese content of water is considered one of the important factors affecting molluscs and shells ^{55}Mn content. Rivers and lakes contain a higher content of ^{55}Mn than sea water [6,7,25].

Table 3. LA-ICP-MS analysis of saltwater and freshwater samples.

Samples	Element	Min–Max (ppmw)	Average (SD) (ppmw)	Median (ppmw)
Saltwater	^{23}Na	2130–7270	4711.27 (816.6)	4720
	^{24}Mg	29.5–950	241.29 (111.13)	240
	^{55}Mn	BQL–45.7	1.84 (5.01)	BQL
	^{88}Sr	518–1860	941.2 (197.31)	912
	^{137}Ba	BQL–11	1.18 (1.17)	0.84
	^{208}Pb	BQL–177	5.18 (14.88)	0.95
Freshwater	^{23}Na	1030–2450	1672.81 (303.89)	1680
	^{24}Mg	5.58–81.3	30.87 (16.45)	28.3
	^{55}Mn	17.4–1440	504.95 (386.34)	473
	^{88}Sr	70.6–2700	431.84 (389.41)	350
	^{137}Ba	13.2–249	72.76 (48.25)	61.6
	^{208}Pb	BQL–11.6	0.24 (1.24)	BQL

BQL: Below quantification limit; SD: Standard deviation.

Table 4. LA-ICP-MS analysis of the studied samples by mollusc.

Samples	Element	Min–Max (ppmw)	Average (SD) (ppmw)	Median (ppmw)
P. radiata (Heirats, Bahrain) Natural	^{23}Na	3340–7270	5249.52 (800.85)	5250
	^{24}Mg	29.5–477	147.45 (88.18)	131
	^{55}Mn	BQL–7.62	0.79 (1.38)	BQL
	^{88}Sr	518–1650	943.9 (219.42)	902
	^{137}Ba	BQL–4.32	0.78 (0.6)	0.58
	^{208}Pb	BQL–3.91	0.17 (0.4)	0.08
P. margaritifera (Red Sea, Egypt) Natural	^{23}Na	2630–7240	4442.05 (1204.07)	4595
	^{24}Mg	67.6–675	362.72 (109.6)	365
	^{55}Mn	BQL–5.5	2.24 (1)	2.31
	^{88}Sr	663–1560	1052.57 (180.98)	1030
	^{137}Ba	BQL–2.02	0.26 (0.26)	0.28
	^{208}Pb	BQL–1.74	0.3 (0.36)	0.17
P. imbricata (Venezuela) Natural	^{23}Na	3200–6240	4621.64 (564.29)	4600
	^{24}Mg	108–531	264.34 (72.01)	258
	^{55}Mn	BQL–10.3	0.33 (0.95)	BQL
	^{88}Sr	572–1620	898.07 (164.56)	880
	^{137}Ba	BQL–11	1.56 (1.32)	1.18
	^{208}Pb	0.35–177	8.58 (18.68)	2.42
P. maxima (Indonesia) Cultured	^{23}Na	2150–5930	4430.28 (860.33)	4460
	^{24}Mg	71.1–279	125.76 (55.38)	110
	^{55}Mn	2.6–37.4	15.37 (8.65)	13.5
	^{88}Sr	791–1540	1061.54 (151.92)	1050
	^{137}Ba	0.09–1.85	0.53 (0.38)	0.38
	^{208}Pb	BQL–1.27	0.17 (0.22)	0.11
P. maxima (Burma) Cultured	^{23}Na	2130–5000	3317.78 (1170.44)	2790
	^{24}Mg	107–172	139.67 (23.58)	139
	^{55}Mn	7.04–45.7	25.82 (14.42)	30.90
	^{88}Sr	840–1700	1279.78 (364.37)	1380
	^{137}Ba	0.86–2.39	1.58 (0.47)	1.55
	^{208}Pb	0.2–1.16	0.43 (0.32)	0.3
P. fucata (Halong Bay, Vietnam) Cultured	^{23}Na	2370–4950	3821.11 (1092.03)	4150
	^{24}Mg	237–950	476.11 (298.01)	306
	^{55}Mn	1.29–15.9	7.2 (6.31)	4.38
	^{88}Sr	888–1860	1284.22 (403.18)	1020
	^{137}Ba	0.36–1.47	0.88 (0.41)	0.78
	^{208}Pb	0.35–1.08	0.7 (0.24)	0.70
Margaritifera margaritifera (Spey river, Scotland) Natural	^{23}Na	1390–2220	1733.61 (202.26)	1725
	^{24}Mg	11.1–60.6	31.37 (11.79)	29.15
	^{55}Mn	58–896	355.25 (234.92)	343.5
	^{88}Sr	175–1030	559.94 (244.16)	599
	^{137}Ba	15.30–233	85.96 (56.2)	90.45
	^{208}Pb	BQL–0.39	0.08 (0.12)	BQL
Unionidae indet. (North American rivers and lakes) Natural	^{23}Na	1030–2300	1519.48 (BQL)	1510
	^{24}Mg	9.21–81.3	34.06 (16.96)	33.45
	^{55}Mn	17.40–1020	382.07 (298.5)	427.5
	^{88}Sr	70.6–2700	389.12 (527.14)	260.5
	^{137}Ba	13.2–249	77.04 (48.49)	65.9
	^{208}Pb	BQL–11.6	0.44 (1.79)	BQL
Hyriopsis sp. (Chinese rivers and lakes) Cultured	^{23}Na	1260–2450	1880.39 (BQL)	1870
	^{24}Mg	5.58–76.9	25.65 (17.51)	19.20
	^{55}Mn	45–1440	798.54 (431.41)	898
	^{88}Sr	241–690	406.75 (111.39)	378
	^{137}Ba	13.6–169	56.88 (37.33)	52
	^{208}Pb	BQL–0.35	0.04 (0.09)	BQL

BQL: Below quantification limit; SD: Standard deviation.

Most (40 out of 55) of the studied freshwater samples present ^{55}Mn >200 ppmw. In some samples ^{55}Mn is higher than 1000 ppmw and up to 1440 ppmw. Large amount of Mn^{2+} incorporation into biogenic aragonite has been attributed to local crystallographic alterations [26]. On the other hand, some freshwater samples (15 out of 55) present relatively low ^{55}Mn concentrations (<150 ppmw). It has been suggested that ^{55}Mn found in nacreous aragonitic freshwater mollusc shells is linked with the availability of Mn^{2+} in the sediment-water interface [27]. Manganese variations of shells were also linked to local geology, anthropogenic factors, animal's age and growth rates as well as phytoplankton blooms, water temperature and pH as well as others [28,29]. However, measured natural pearls from *Margaritifera margaritifera* bivalve fished off the same season (two consecutive years) at the same location are presenting great variability of ^{55}Mn (58–896 ppmw; Table 4).

All studied saltwater cultured pearls, as well as around half of the natural saltwater pearls present detectable ^{55}Mn using LA-ICP-MS reaching up to 45.7 ppmw (see again Tables 3 and 4). Noteworthy, 1/3 of the cultured saltwater samples present ^{55}Mn >15 ppmw and all studied natural saltwater pearls contained ^{55}Mn <15 ppmw. Half of the studied natural saltwater pearls did not present any detectable amounts of ^{55}Mn. Moreover, three natural freshwater pearls from US presented amounts of ^{55}Mn similar to those presented to some cultured saltwater pearls from *P. maxima* bivalve from Indonesia and Burma (Table 4).

^{88}Sr in studied saltwater samples is relatively higher than in freshwater samples (Table 3). Most studied freshwater samples present concentrations from 70.6 ppmw to 800 ppmw, with only four natural freshwater pearls presenting ^{88}Sr >800 ppmw. There are two natural pearls from *Margaritifera margaritifera* presenting ^{88}Sr values of 874 and 1150 ppmw and two natural samples from Unionidae indet. with about 1487 ppmw and 2663 ppmw. The other 24 studied natural samples from Unionidae indet. presented ^{88}Sr <470 ppmw. It was suggested that the variation of ^{88}Sr in aragonitic (nacreous and non-nacreous) bivalves is influenced by the animal's growth rates (which is linked with water temperature as well as other parameters) and physiological processes; ^{88}Sr variations are not under environmental control [30–32]. Strontium content measured in nacreous aragonitic freshwater bivalve inner shells has been recently suggested as a good proxy of water's ^{88}Sr [27].

Figure 1 is a binary plot of manganese oxide (MnO) and strontium oxide (SrO). This plot is similar to those previously presented using EDXRF chemical analysis [7–9], taking into account that the limit of detection for MnO is higher for EDXRF than for LA-ICP-MS. It was formerly published that the SrO/MnO ratio, measured with EDXRF, is separating freshwater (here presented as filled squares in the figure) from saltwater pearls, as it is >12 for saltwater pearls and <12 for freshwater pearls [8]. Most of saltwater samples can be separated using this ratio as they present higher strontium and lower manganese with a SrO/MnO ratio of >100. However, for the freshwater samples, most present lower strontium and higher manganese samples with a ratio of <1. Chemical analysis of some freshwater samples might lead to wrong conclusions if the previously published plots and ratio are used [8,9], these are the freshwater (natural and cultured) pearls with ^{55}Mn <150 ppmw (i.e., MnO <ca. 194 ppmw) or ^{88}Sr >950ppmw (i.e., SrO >ca. 1124 ppmw). There are two natural freshwater pearls presenting SrO/MnO ratio >12. The SrO/MnO ratio of the studied samples is still not overlapping; all the studied saltwater samples present SrO/MnO ratio >17 and freshwater samples <16.

Interestingly, all studied freshwater samples presented higher ^{137}Ba concentration than the saltwater samples; no overlapping values between the studied freshwater and saltwater samples were observed (Table 3). Barium is present in various concentrations in freshwater and saltwater [33] and in some studies barium uptake of a nacreous and non-nacreous shell was principally related to water's barium content [34–36]. A strong inverse correlation was also found between salinity and barium in an aragonitic non-nacreous shell [30]. Other factors are also playing a role as the content is highly variable from a freshwater sample to another; even for those collected at the same region. For instance, measured ^{137}Ba of natural pearls from *Margaritifera margaritifera* bivalves collected from the Spey river

(Scotland) varied from 15.3 to 233 ppmw (Table 4). Food supply and shell growth have also been suggested as influencing factors [37,38] along with another yet undetermined environmental factor [35]. Animal age might also play a role. The vast majority of saltwater samples present a detectable [137]Ba, while only 30 samples out of 1058 sample present undetectable samples. Also, only 120 samples out of the studied 1058 samples present [137]Ba above 2.5 ppmw. All these samples are natural pearls; 116 sample found in *P. imbricata*. Few of these are samples with [137]Ba >4.5 ppmw.

Figure 2 is the binary plot of [55]Mn and [137]Ba of the studied samples. All freshwater samples present higher [55]Mn and [137]Ba (shown as coloured filled squares in the plot) than the studied saltwater samples (shown as coloured empty circles in the figure). The population fields of these two group of samples are well apart, and the plot can be used to efficiently separate saltwater and freshwater (natural and cultured) pearls. In Figure 3, freshwater and saltwater samples are well separated. Moreover, freshwater samples present a somehow linear trend for [137]Ba vs. [88]Sr plots. This is not observed for saltwater samples and it further supports previous studies mentioning that [88]Sr incorporation differs in freshwater and saltwater bivalves [27]. This should be further studied on samples from known molluscs and areas like the studied natural pearls founded in *Margaritifera margaritifera* bivalve from the river Spey (Scotland).

[23]Na is the most abundant element present in the studied samples in both saltwater and freshwater (Table 3) with higher concentrations for the first than for the latter. Sodium concentration of biogenic carbonate is linked with water salinity and pH [25,36]. All studied samples present diverse concentrations, not linked to specific bivalves or regions. Even the natural saltwater pearls found into the same *P. margaritifera* bivalve present a great variability of [23]Na from 2630 to 7240 ppmw (Table 4) and the natural freshwater pearls from Spey river with [23]Na from 1030 to 2450 ppmw. This may indicate that other factors can also influence cultured and natural pearls' sodium concentration, such as the age of the pearl sac's epithelial cells where it was found into. In the [23]Na and [137]Ba diagram presented in Figure 4, saltwater samples are more highly separated than the freshwater samples (filled squares).

Measured [24]Mg content was also lower on the studied freshwater samples than on the saltwater (Table 3). However, some saltwater samples present relatively low manganese content, similar to the content of freshwater samples. The lowest content on saltwater samples was measured for *P. radiata*, where 45 samples out of 284 sample has [24]Mg <65 ppmw (Table 3). Samples found in the same animal (see again Table 4 for *P. margaritifera*) have a great variation of [24]Mg (from 67.6 to 675 ppmw). It has been suggested that Mg content is not (or is little) related to water conditions (e.g., temperature) as previously believed, but instead is related to physiological processes [30,34,39].

[208]Pb content was relatively low on all the studied samples, sometimes including BQL. All studied freshwater samples present below 0.4 ppmw of [208]Pb. Solely two freshwater samples, both natural from Unionidae indet., presented concentrations of >0.4 pppmw, one presented around 1.5 ppmw and the other around 9 ppmw. Most of the studied saltwater samples presented below 1.5 ppmw [208]Pb. Only five samples from *P. radiata* containing above 1.5 ppmw [208]Pb and up to 3.91 ppmw (Table 4). However, around 2/3 of the studied *P. imbricata* samples contained above 1.5 ppmw [208]Pb, with around 1/3 above 3.91 ppmw and up to 177 ppmw (see also Figure 5). It has been suggested that [208]Pb measured in aragonite bivalve shells can be linked with anthropogenic lead pollution [40]. Anthropogenic small-scale lead pollution might be the reason of [208]Pb >1 ppmw concentration in eight natural freshwater samples from *P. radiata*. Relatively high concentrations of [208]Pb observed in *P. imbricata* natural pearls are probably superficial due to samples storage, as some natural pearls fished near Central America around 16th century were stored in lead boxes (e.g., natural pearls of Santa Margarita shipwreck treasures were found [41]). However, a possible link with polluted water caused by human and lead content cannot be excluded.

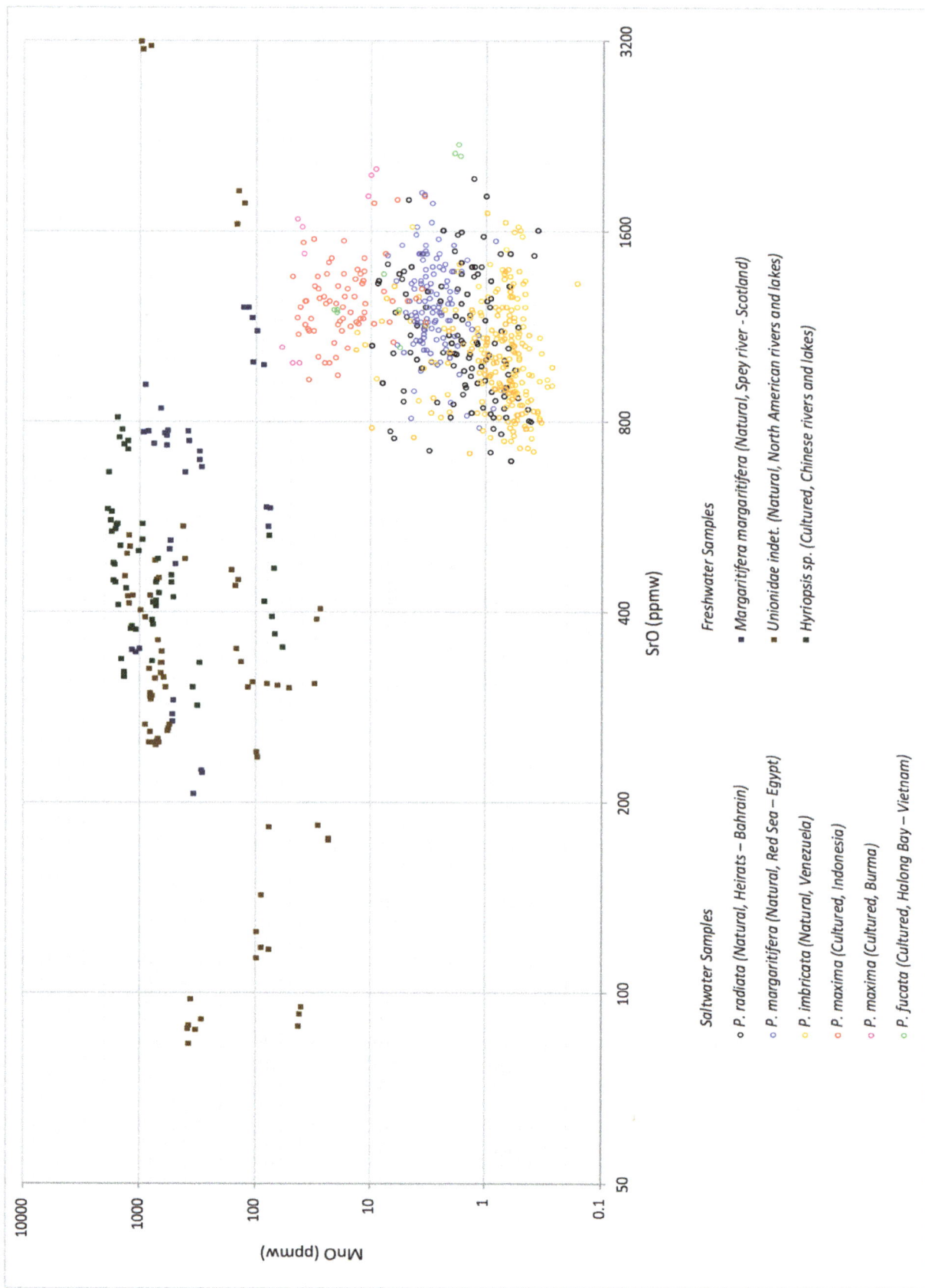

Figure 1. Binary plot of SrO vs. MnO.

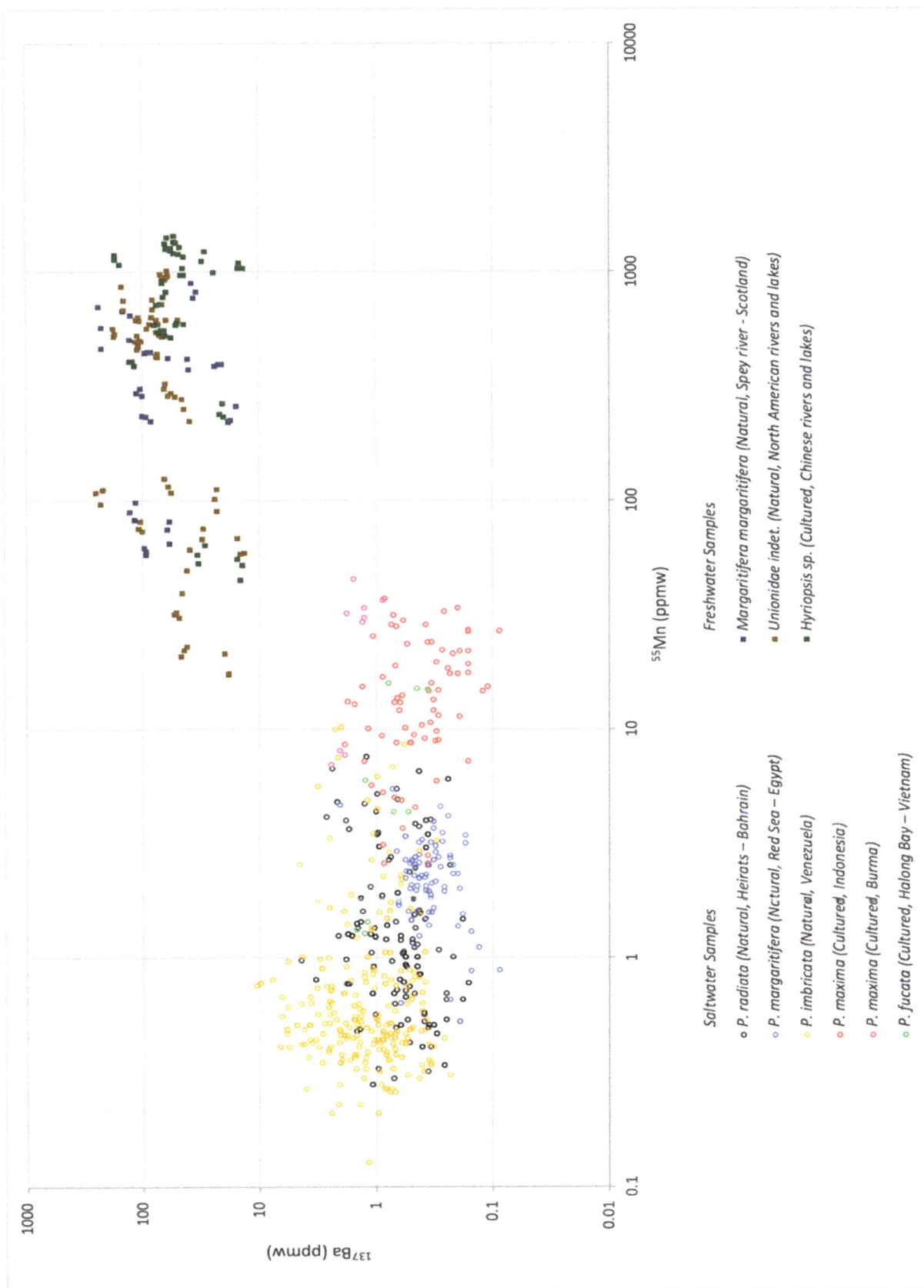

Figure 2. Binary plot of ^{55}Mn vs. ^{137}Ba of the studied samples.

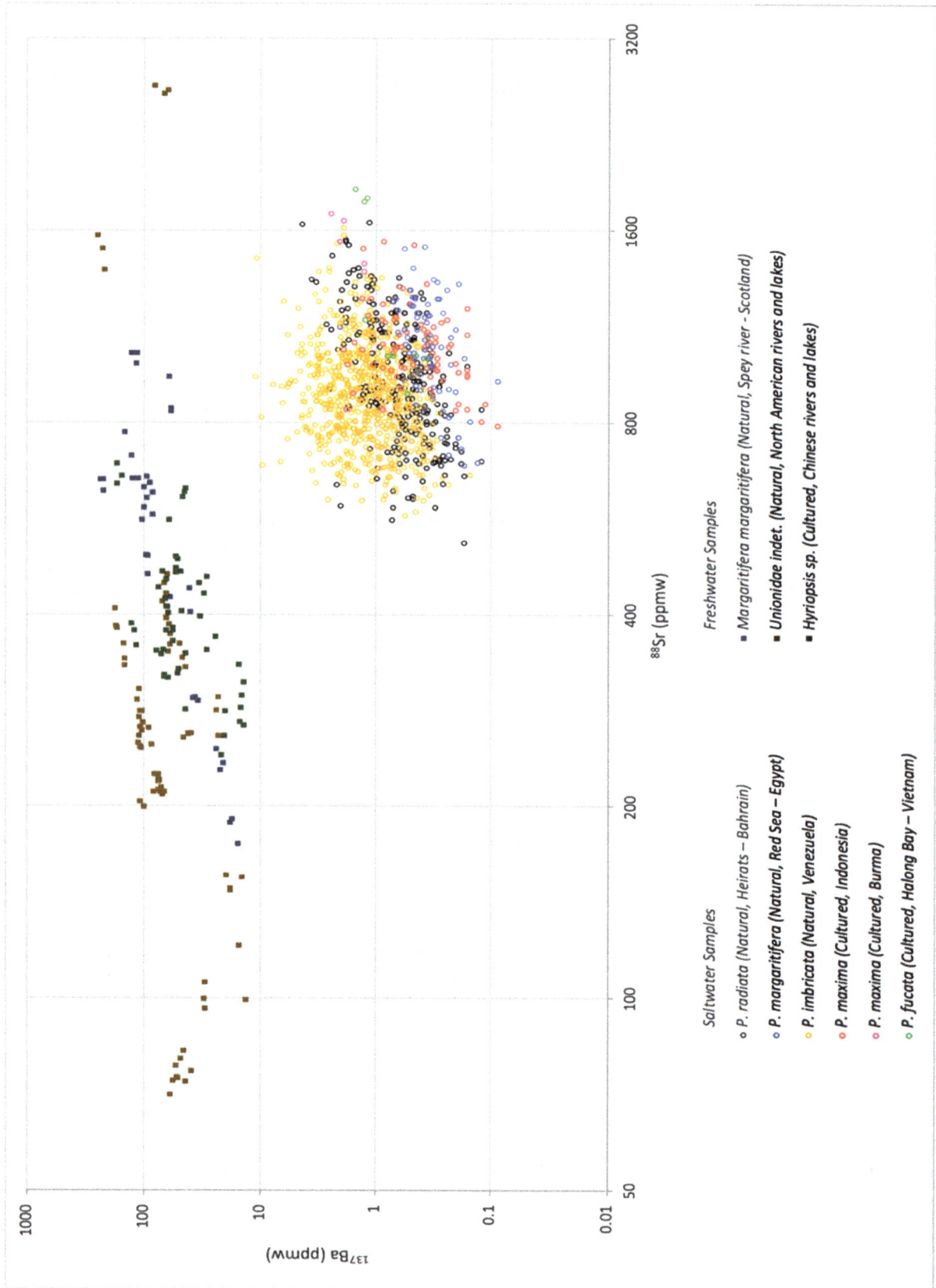

Figure 3. Binary plot of ^{88}Sr vs. ^{137}Ba of the studied samples.

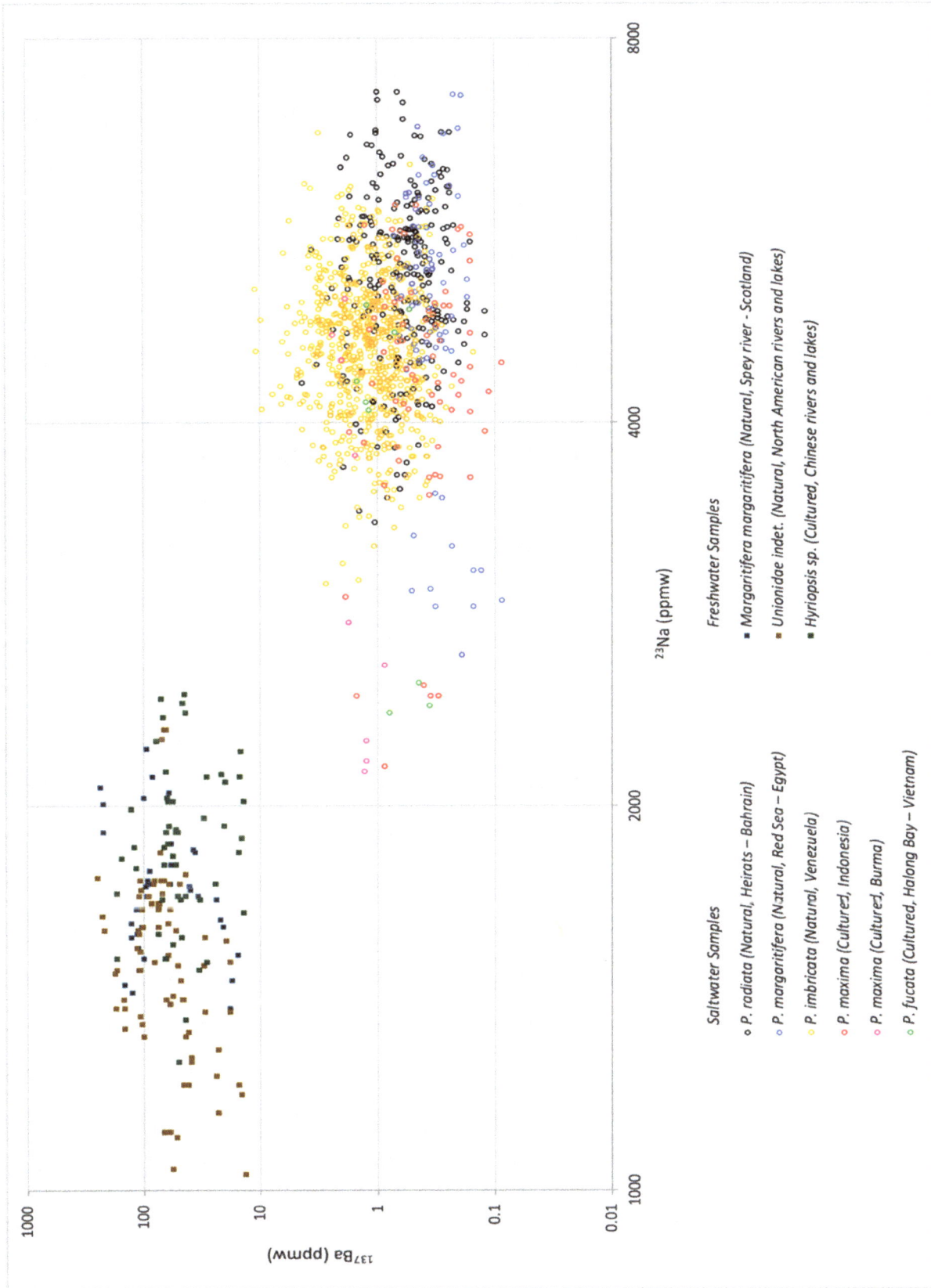

Figure 4. Binary plot of ^{137}Ba vs. ^{23}Na of the studied samples.

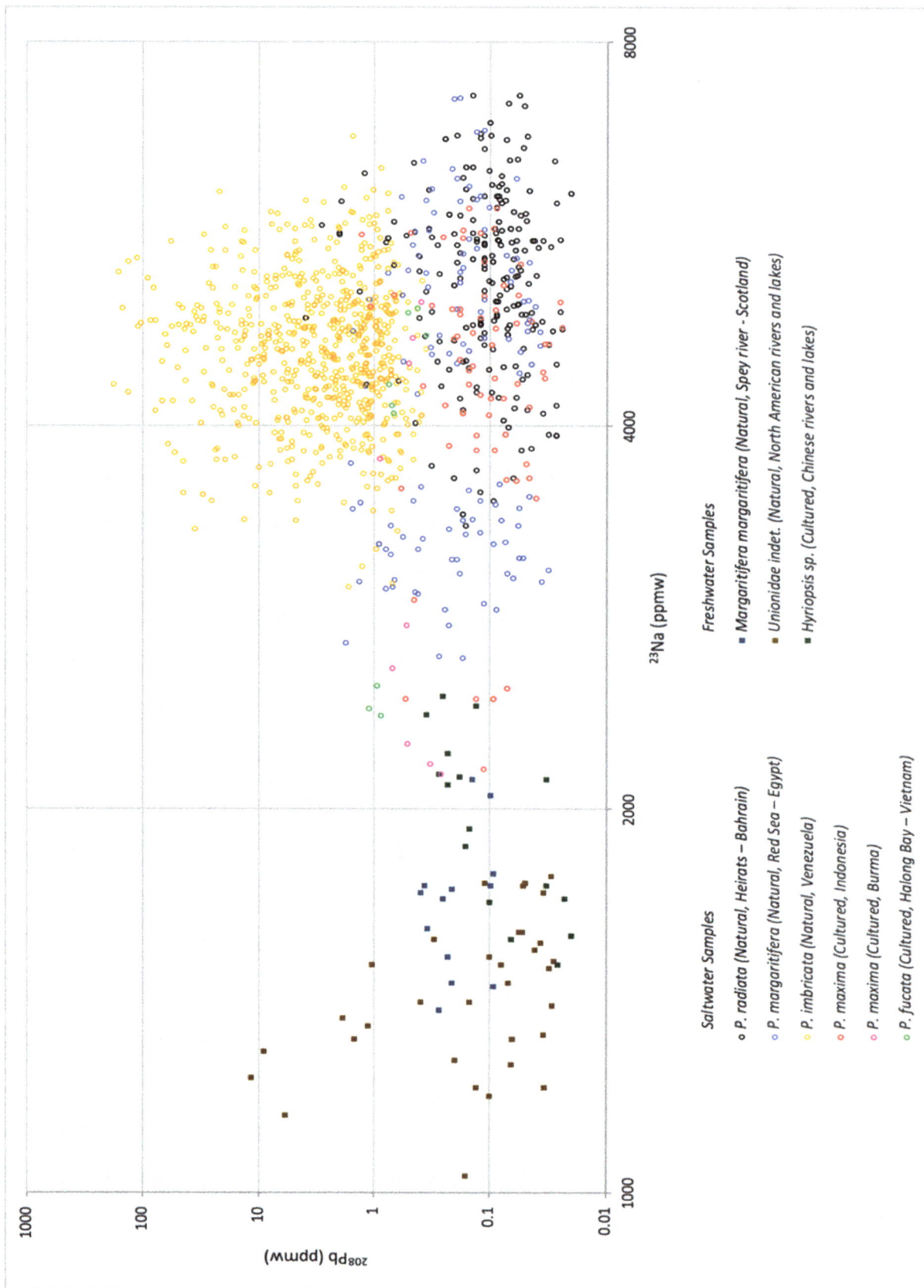

Figure 5. Binary plot of ^{23}Na vs. ^{208}Pb of the studied samples.

3.2. X-ray Luminescence and Raman Spectroscopy

Under X-rays, most natural and cultured (without bead) saltwater pearls remain inert and most natural and cultured freshwater pearls presented yellow-green luminescence of medium to high intensity (Figures 6–9). This luminescence is attributed to 9-fold coordination of oxygen around Mn^{2+} in aragonite [42]. Luminescence intensity is correlated with an Mn^{2+} content, as the higher Mn^{2+} content, the more intense the yellow-green luminescence appears to be [42,43] (Figure 6). Meanwhile, it has been noticed that a higher Mn^{2+} content could cause lower luminescence (see again Figure 6). Similar to what was previously observed with CL spectroscopy (and X-ray induced luminescence) of biogenic aragonites, this is probably due to self-quenching of manganese. The presence of Fe^{2+} (as well as Cu^{2+}, Co^{2+} and Ni^{2+}) could quench Mn^{2+} luminescence as well [42,43].

Some coloured natural and cultured pearls present photoluminescence phenomena [44]. None of the studied samples presented any reaction under X-rays related to pigments. It was previously mentioned that natural (and artificial) pigments of freshwater samples suppress luminescence under X-rays [5]. The studied freshwater samples confirm that coloured samples with higher ^{55}Mn (measured with LA-ICP-MS) present lower luminescence than less coloured samples with lower manganese (measured with LA-ICP-MS); see Figure 7.

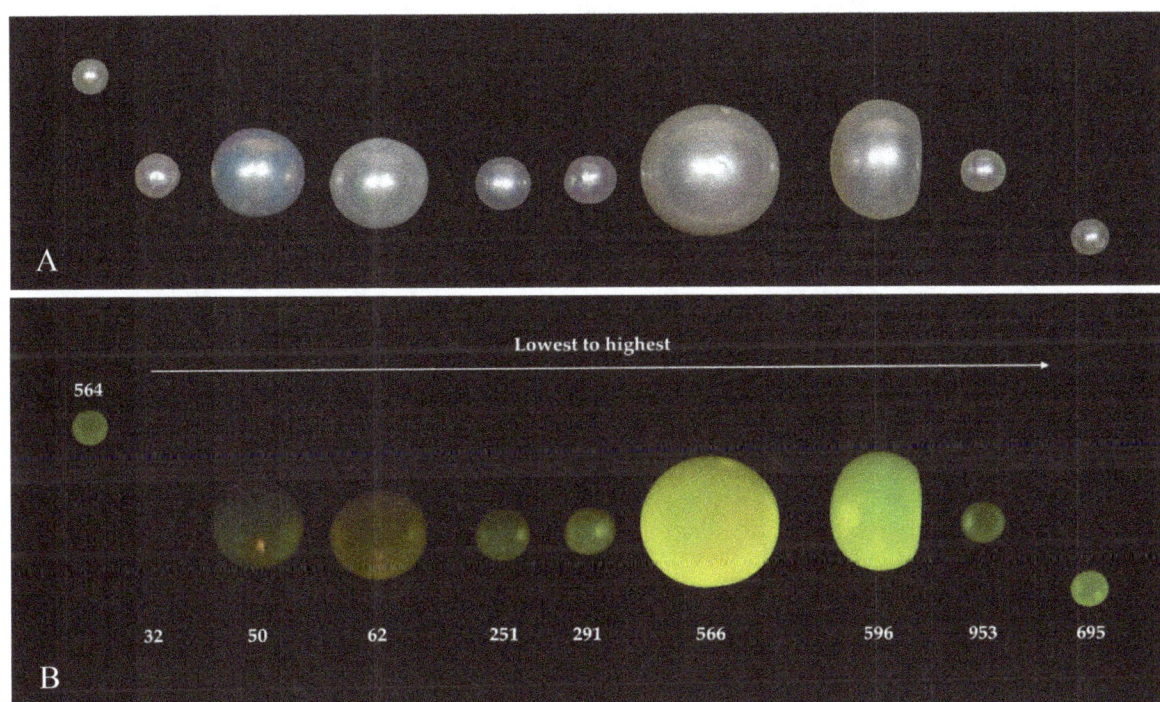

Figure 6. Freshwater samples under daylight (**A**) and under X-rays (**B**). The two freshwater samples upper left (diameter: 2.85 mm) and bottom right (diameter: 2.95 mm) used as reference. The concentrations of ^{55}Mn in ppmw are mentioned below for the studied samples and above for the reference samples. For the samples 3 analyses with LA-ICP-MS were acquired, the average values are presented (rounded to the nearest one).

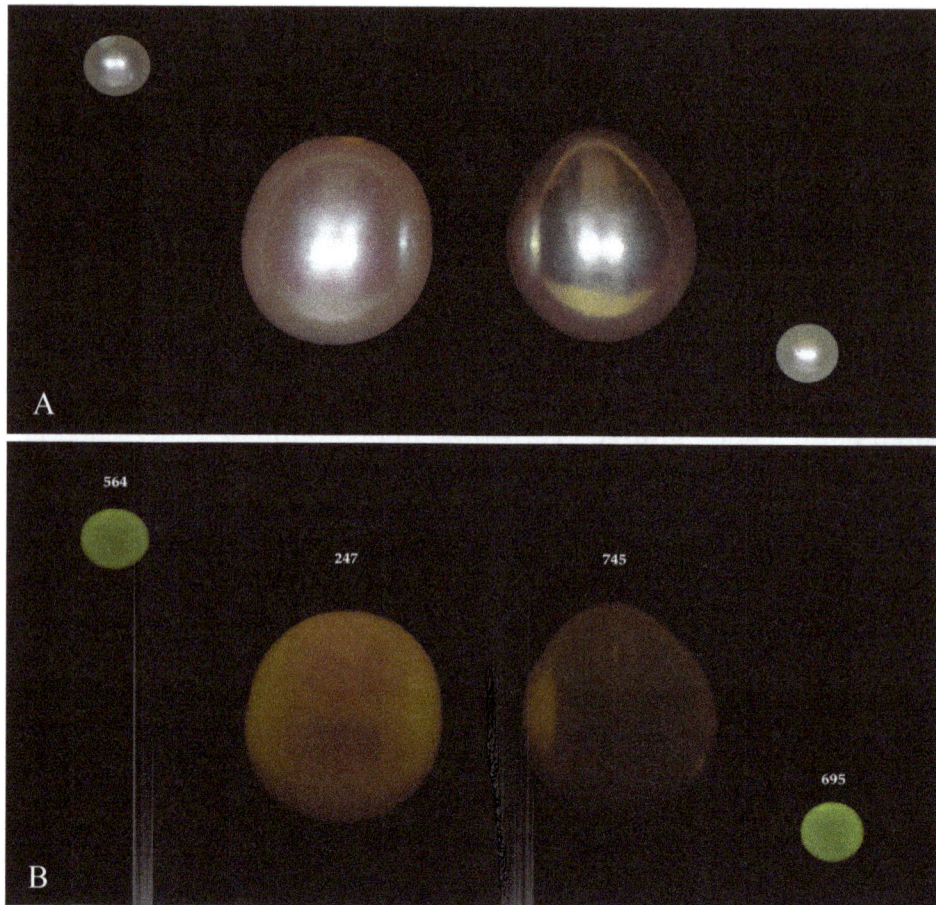

Figure 7. Freshwater samples under daylight (**A**) and under X-rays (**B**). Two natural-coloured freshwater samples (left sample of pink colour and right samples of grey colour) with the amount ^{55}Mn in ppmw. The two freshwater samples upper left (diameter: 2.85 mm) and bottom right (diameter: 2.95 mm) used as reference. For the samples 3 analyses with LA-ICP-MS were acquired, the average values are presented (rounded to the nearest one).

Some saltwater cultured pearls, with freshwater bead, also present intense yellow-green luminescence (Figure 8). The amount of ^{55}Mn measured on these samples is from 8 to 38 ppmw and it is not directly linked to the luminescence. The reaction under X-rays is due to the content of Mn^{2+} of the bead of these saltwater cultured samples and the relatively thin nacre thickness (i.e., thickness of the layer of saltwater nacre deposited around the freshwater bead) of the studied samples [5]. In Figure 8, the samples with the highest fluorescence has the thinnest nacre thickness (0.65 mm; measured using X-ray microradiography) and the other two samples have similar relatively thick nacre thickness (1.8 mm).

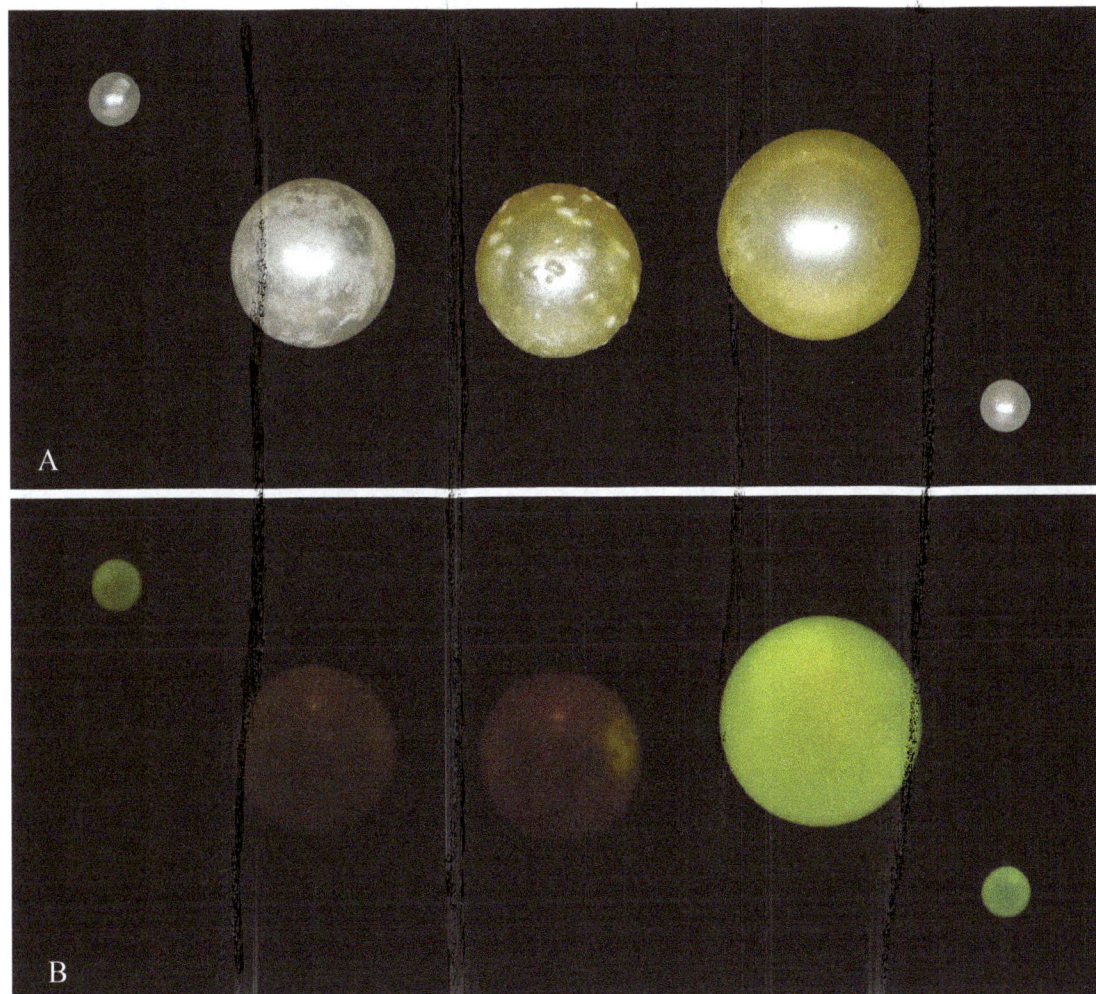

Figure 8. Saltwater cultured pearls (with freshwater bead) under daylight (**A**) and under X-rays (**B**). The first and second pearls from the left has a similar nacre thickness that equals 1.8 mm, while the third one has a nacre thickness of 0.65 mm (measured using X-ray microradiography). The two freshwater samples upper left (diameter: 2.85 mm) and bottom right (diameter: 2.95 mm) used as reference.

As mentioned above, some of the studied freshwater samples present ^{55}Mn <150 ppmw. This samples present luminescence of low to medium intensity under X-rays (see examples in Figures 6 and 9). On the other hand, some of the studied saltwater samples present detectable manganese measured with LA-ICP-MS. On our samples, X-ray luminescence appeared with low intensity at the studied samples only when ^{55}Mn was higher than 10 ppmw. However, this will be different, if different parameters (e.g., less X-ray power) are used to acquire luminescence images. All the studied freshwater samples presented luminescence as well as some of the studied saltwater samples of low intensity. The latter were sometimes difficult to separate them from the freshwater samples containing low manganese (Figure 9). The amount of manganese in freshwater samples is low and below the detection limit of some EDXRF instruments and might lead to wrong conclusions. This samples might be identified by EDXRF if barium is taken into consideration or with LA-ICP-MS using the plots presented from Figures 2–4.

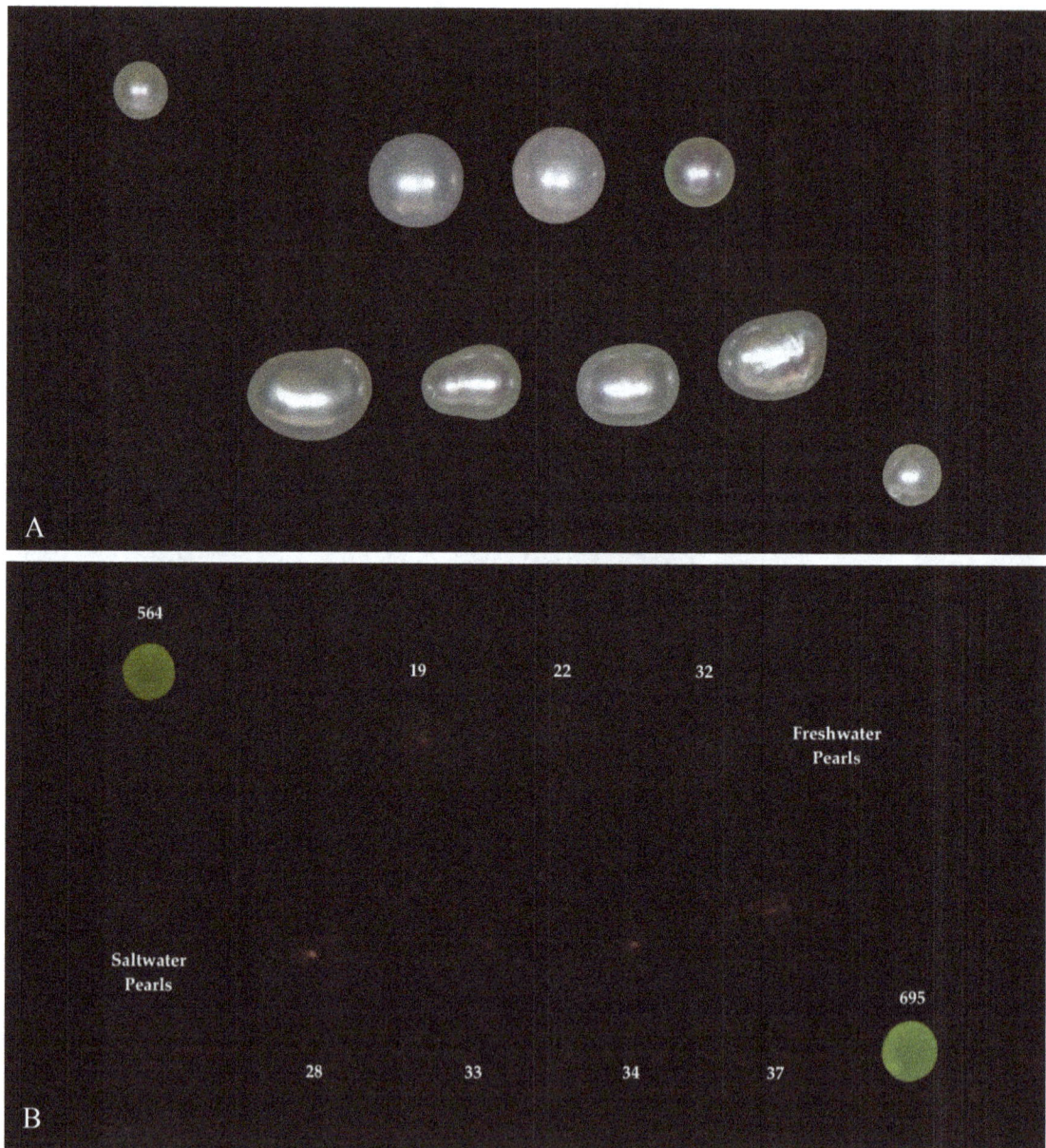

Figure 9. Three freshwater (upper row) and four saltwater (bottom row) samples under daylight (**A**) and under X-rays (**B**). The two freshwater samples upper left (diameter: 2.85 mm) and bottom right (diameter: 2.95 mm) used as reference. All samples present similar luminescence intensity and separation of freshwater samples from saltwater is challenging. The concentrations of ^{55}Mn in ppmw are mentioned. For the samples 3 analyses with LA-ICP-MS that were acquired, the average values are presented (rounded to the nearest one).

One of the studied natural freshwater pearls presented a section with intense orange luminescence under X-rays; and phosphorescence which lasted for about 3 seconds (Figure 10). Orange luminescence activated by CL was previously observed on some cultured freshwater pearls, without bead, from China and this was attributed to calcite emitting Mn^{2+} [6]. Raman spectra acquired on the section where orange luminescence presents (and which, macroscopically and under optical microscope, appears with different lustre than the rest of the sample; refer to Figure 10A) have shown Raman bands at around 1091, 1081 and 1075 cm^{-1} due to ν_1 symmetric stretching of carbonate ions in vaterite (Figure 11) and not at 1086 cm^{-1} due to ν_1 symmetric stretching of carbonate ions in aragonite as on the section where medium yellow-green luminescence was observed [45]. CL and X-ray induced luminescence spectra of

Mn^{2+} in calcite and vaterite are very close, which suggest a 6-fold coordination of oxygen around both calcite and vaterite [42]. Sections of vaterite were already previously identified in freshwater cultured pearls without bead from Japan and China [45–47]; but not in natural freshwater pearls.

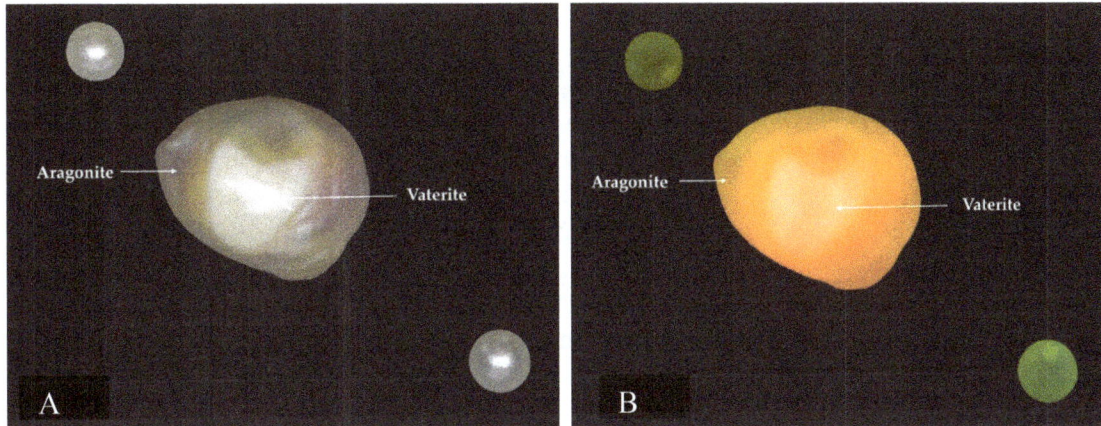

Figure 10. Natural freshwater pearl under daylight (**A**) and under X-rays (**B**); orange coloured part is made of vaterite (verified by Raman; see Figure 11). The two freshwater samples upper left (diameter: 2.85 mm) and bottom right (diameter: 2.95 mm) were used as references.

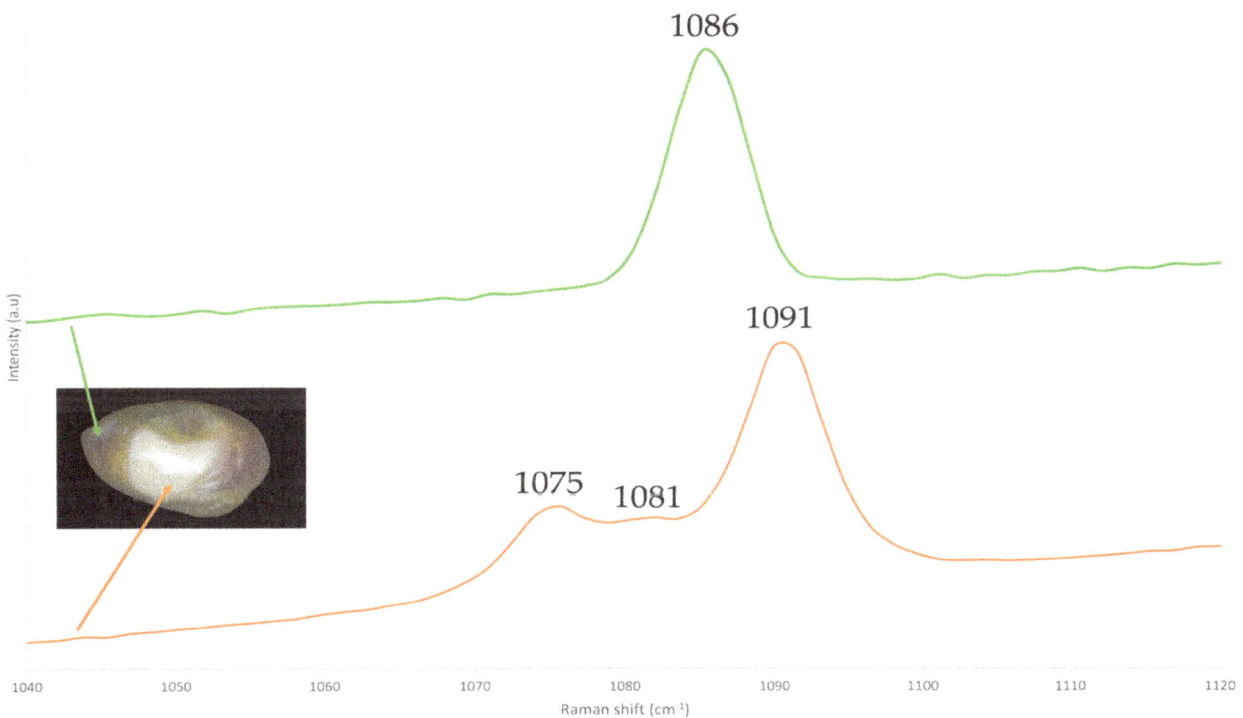

Figure 11. Raman spectra on the white part with low lustre (bottom spectrum) and on the cream coloured part with medium lustre (top spectrum) showing respectively bands due to vaterite and calcite. Note that the spectra are stacked and shifted for clarity.

4. Conclusions

LA-ICP-MS measurements on freshwater samples reveal higher ^{55}Mn and ^{137}Ba and lower ^{23}Na, ^{24}Mg and ^{88}Sr concentrations than the studied saltwater samples. Only ^{137}Ba concentrations of the studied samples do not present any overlap; the other concentrations slightly overlap. Plots combining ^{137}Ba with ^{55}Mn, ^{23}Na and ^{88}Sr clearly separate freshwater samples from saltwater samples. Very little correlation between the studied chemical elements of natural and cultured pearls and host animal's species was found. Saltwater samples from *P. maxima* are the only studied saltwater samples which contain ^{55}Mn >20 ppmw. On the other hand, only some saltwater samples from *P. radiata* could present ^{24}Mg <65 ppmw and saltwater samples from *P. imbricata* could present ^{137}Ba >2.5 ppmw. It seems that all the chemical elements studied here are linked to a combination of environmental and physiological factors.

Most studied freshwater samples can be separated from the saltwater samples as X-ray luminescence intensity is correlated with Mn^{2+} content in aragonite. Some of the studied freshwater samples present similar manganese content with some saltwater samples as well as similar low intensity luminescence under X-rays. The growth environment of these samples could only inadequately be identified by LA-ICP-MS or a well calibrated EDXRF with a good limit of detection for manganese and barium.

Studies on a bigger number of samples from known animals and various regions (with known geographic coordinates) that take into account the local environmental factors, along with statistical analysis (e.g., principal components), should be performed in order to confirm the abovementioned differences and better understand the link between the chemical elements and the natural and cultured pearls. Studying natural pearls is not an easy task because they are very rarely found and in parallel rarely used for scientific purposes. Additional measurements of isotopes might shed more light on a possible link between environmental factors and natural (or cultured) pearls. In order to inspect the possible role of organic matter, TGA analysis and chemical analysis would also be useful. It is also important to further study the possible role of elements such us Fe^{2+}, Cu^{2+}, Co^{2+} and Ni^{2+} which could quench Mn^{2+} luminescence and their effect on the separation of saltwater from freshwater (natural and cultured) pearls. Orange coloured X-ray luminescence of freshwater samples should also be further investigated and its link with vaterite should be confirmed with spectroscopic means.

Supplementary Materials
Table S1: LA-ICP-MS analysis in ppmw of the studied natural pearls samples from P. radiata (Heirats, Bahrain), Table S2: LA-ICP-MS analysis in ppmw of the studied natural pearls samples from *P. margaritifera* (Red Sea, Egypt), Table S3: LA-ICP-MS analysis in ppmw of the studied natural pearls samples from *P. imbricata* (Venezuela), Table S4: LA-ICP-MS analysis in ppmw of the studied cultured pearls samples from *P. maxima* (Indonesia), Table S5: LA-ICP-MS analysis in ppmw of the studied cultured pearls samples from *P. maxima* (Burma), Table S6: LA-ICP-MS analysis in ppmw of the studied cultured pearls samples from *P. fucata* (Halong Bay, Vietnam), Table S7: LA-ICP-MS analysis in ppmw of the studied natural pearls samples from *Margaritifera margaritifera* (Spey river, Scotland), Table S8: LA-ICP-MS analysis in ppmw of the studied natural pearls samples from Unionidae indet. (North American rivers and lakes), Table S9: LA-ICP-MS analysis in ppmw of the studied cultured pearls samples from *Hyriopsis* sp. (Chinese rivers and lakes).

Author Contributions: S.K. formulated the paper designed the experiments, participated to the data interpretation and wrote the manuscript. F.M. prepared the experiments for X-ray luminescence analysis, did part of LA-ICP-MS analysis, data reduction, edited the manuscript and draw its figures and tables. H.A. prepared all the images of the manuscript. F.A. and L.F. performed most of LA-ICP-MS analysis on the samples and S.S. has designed, supervised and evaluated LA-ICP-MS analysis. A.A. selected the samples, designed the experiments and edited the manuscript.

Acknowledgments: The authors wish to thank the National Museum of Bahrain which loaned the natural saltwater pearls from *P. radiata* fished off Bahrain, Peter Balogh for the natural saltwater pearls from *Pinctada imbricata* fished off Venezuela and Peter Piva for the natural saltwater pearls from *Pinctada margaritifera* fished off Egypt. The rest of the samples are from DANAT reference collection.

References

1.   Gauthier, J.P.; Karampelas, S. Pearls and corals: "Trendy biomineralizations". *Elements* **2009**, *5*, 179–180. [CrossRef]

2.   Cartier, L.; Krzemnicki, M.S. New developments in cultured pearls production: Use of organic and baroque shell nuclei. *Aust. Gemmol.* **2013**, *25*, 6–13.

3. Schiffmann, C.A. Pearl identification: Some laboratory experiments. *J. Gemmol.* **1971**, *12*, 284–296. [CrossRef]

4. Lorenz, I.; Schmetzer, K. Possibilities and limitations in radiographic determination of pearls. *J. Gemmol.* **1986**, *20*, 114–123. [CrossRef]

5. Hänni, H.A.; Kiefert, L.; Giese, P. X-ray luminescence, a valuable test in pearl identification. *J. Gemmol.* **2005**, *29*, 325–329. [CrossRef]

6. Habermann, D.; Banerjee, A.; Meijer, J.; Stephan, A. Investigation of manganese in salt- and freshwater pearls. *Nucl. Instrum. Methods* **2001**, *181*, 739–743. [CrossRef]

7. Gütmannsbauer, W.; Hänni, H.A. Structural and chemical investigations on shell and pearls of nacre forming salt- and freshwater bivalve mollusks. *J. Gemmol.* **1994**, *24*, 241–252. [CrossRef]

8. Karampelas, S.; Kiefert, L. Gemstones and minerals. In *Analytical Archaeometry: Selected Topics*, 1st ed.; Edwards, H.G.M., Vandenabeele, P., Eds.; Royal Society of Chemistry: London, UK, 2012; pp. 291–317. ISBN 978-1-84973-162-1.

9. Zhou, C.; Hodgins, G.; Lange, T.; Saruwatari, K.; Sturman, N.; Kiefert, L.; Schollenbruch, K. Saltwater pearls from the pre- to early Columbian era: A gemological and radiocarbon dating study. *Gems Gemol.* **2017**, *53*, 286–295. [CrossRef]

10. Goebel, M.; Dirlam, D.M. Polynesian black pearls. *Gems Gemol.* **1989**, *25*, 130–148. [CrossRef]

11. Elen, S. Spectral reflectance and fluorescence characteristics of natural-color and heat-treated "Golden" south sea cultured pearls. *Gems Gemol.* **2001**, *37*, 114–123. [CrossRef]

12. Jacob, D.E.; Wehrmeister, U.; Hager, T.; Hofmeister, W. Identifying Japanese freshwater cultured pearls from Lake Kasumigaura. *J. Gemmol.* **2006**, *22*, 539–541.

13. Scarratt, K.; Bracher, P.; Bracher, M.; Atawi, A.; Safar, A.; Saeseaw, S.; Homkrajae, A.; Sturman, N. Natural pearls from Australian *Pinctada maxima*. *Gems Gemol.* **2012**, *48*, 236–261. [CrossRef]

14. Hänni, H.A.; Cartier, L.E. Tracing cultured pearls from farm to consumer: A review of potential methods and solutions. *J. Gemmol.* **2013**, *33*, 185–191. [CrossRef]

15. Sturman, N.; Homkrajae, A.; Manustrong, A.; Somsaard, N. Observations on pearls reportedly from the Pinnidae family (pen pearls). *Gems Gemol.* **2014**, *50*, 202–215. [CrossRef]

16. Homkrajae, A.; Sun, Z.; Blodgett, T.; Zhou, C. Provenance discrimination of freshwater pearls by laser ablation–inductively coupled plasma–mass spectrometry (LA-ICP-MS) and linear discriminant analysis (LDA). *Gems Gemol.* **2019**, *55*, 47–60.

17. Perkins, W.T.; Fuge, R.; Pearce, N.J.G. Quantitative analysis of trace elements in carbonates using laser ablation inductively coupled plasma mass spectrometry. *J. Anal. Atom. Spectr.* **1991**, *6*, 445–449. [CrossRef]

18. Sinclair, D.J.; Kinsley, L.P.J.; McCulloch, M.T. High resolution analysis of trace elements in corals by laser-ablation ICP-MS. *Geochim. Cosmochim. Acta* **1998**, *62*, 1889–1901. [CrossRef]

19. Gillikin, D.P. Geochemistry of Marine Bivalve Shells: The Potential for Paleoenvironmental Reconstruction. Ph.D. Thesis, Vrije University, Brussel, Belgium, 2005; 265p.

20. Jacob, D.E.; Soldati, A.L.; Wirth, R.; Huth, J.; Wehrmeister, U.; Hofmeister, W. Nanostructure, composition and mechanisms of bivalve shell growth. *Geochim. Cosmochim. Acta* **2009**, *72*, 5401–5415. [CrossRef]

21. Mertz-Kraus, R.; Brachert, T.C.; Jochum, K.P.; Reuter, M.; Stoll, B. LA-ICP-MS analyses on coral growth increments reveal heavy winter rain in the eastern Mediterranean at 9 Ma. *Palaeogeogr. Palaeoclimatol. Palaeoecol.* **2009**, *273*, 25–40. [CrossRef]

22. Phung, A.T.; Baeyens, W.; Leermakers, M.; Goderis, S.; Vanhaecke, F.; Gao, Y. Reproducibility of laser ablation–inductively coupled plasma–mass spectrometry (LA–ICP–MS) measurements in mussel shells and comparison with micro-drill sampling and solution ICP–MS. *Talanta* **2013**, *396*, 42–50. [CrossRef]

23. Strasser, C.A.; Mullineaux, L.S.; Walther, B.D. Growth rate and age effects on Mya Arenaria shell chemistry: Implications for biogeochemical studies. *J. Exp. Mar. Biol. Ecol.* **2008**, *355*, 153–163. [CrossRef]

24. Jochum, K.P.; Scholz, D.; Stoll, B.; Weis, U.; Wilson, S.A.; Yang, Q.; Schwalb, A.; Borner, N.; Jacob, D.E.; Andreae, M.O. Accurate trace element analysis of speleothems and biogenic calcium carbonates by LA-ICP-MS. *Chem. Geol.* **2012**, *318–319*, 31–44. [CrossRef]

25. Gordon, C.M.; Carr, R.A.; Larson, R.E. The influence of environmental factors on the sodium and manganese content of barnacle shells. *Limnol. Oceanogr.* **1970**, *15*, 461–466. [CrossRef]

26. Soldati, A.L.; Jacob, D.E.; Glatzel, P.; Swarbrick, J.C.; Geck, J. Element substitution by living organisms: The case of manganese in mollusc shell aragonite. *Sci. Rep.* **2016**, *6*, 22514. [CrossRef] [PubMed]

27. Geeza, T.J.; Gillikin, D.P.; Goodwin, D.H.; Evans, S.D.; Watters, T.; Warner, N.R. Controls on magnesium, manganese, strontium, and barium concentrations recorded in freshwater mussel shells from Ohio. *Chem. Geol.* **2019**. [CrossRef]

28. Carroll, M.; Romanek, C.S. Shell layer variation in trace element concentration for the freshwater bivalve *Elliptio complanta*. *Geo-Mar. Lett.* **2008**, *28*, 369–381. [CrossRef]

29. Kelemen, Z.; Gillikin, D.P.; Bouillon, S. Relationship between river water chemistry and shell chemistry of two tropical African freshwater bivalve species. *Chem. Geol.* **2019**. [CrossRef]

30. Poulain, C.; Gillikin, D.P.; Thebault, J.; Munaron, J.M.; Bohn, M.; Robert, R.; Paulet, Y.M.; Lorrain, A. An evaluation of Mg/Ca, Sr/Ca, and Ba/Ca ratios as environmental proxies bivalve shells. *Chem. Geol.* **2015**, *396*, 42–50. [CrossRef]

31. Wanamaker, D.; Gillikin, D.P. Strontium, magnesium and barium incorporation in aragonitic shells of junevile *Arctica islandica*: Insights from temperature controlled experiments. *Chem. Geol.* **2019**. [CrossRef]

32. Stecher, H.A.; Krantz, D.E.; Lord, C.J.; Luther, G.W.; Bock, K.W. Profiles of strontium and barium in *Mercenaria mercenaria* and *Spisula solidisima* shells. *Geochim. Cosmochim. Acta* **1996**, *60*, 3445–3456. [CrossRef]

33. Schroeder, H.A.; Tipton, I.H.; Nason, A.P. Trace metals in man: Strontium and barium. *J. Chronic Dis.* **1972**, *25*, 491–517. [CrossRef]

34. Gillikin, D.P.; Dehairs, F.; Lorrain, A.; Steenmans, D.; Baeyens, W.; Andre, L. Barium uptake into the shells of the common mussel (*Mytilus edulis*) and the potentials for estuarine paleo-chemistry reconstruction. *Geochim. Cosmochim. Acta* **2006**, *70*, 395–407. [CrossRef]

35. Gillikin, D.P.; Lorrain, A.; Paulet, Y.M. Synchronous barium peaks in high-resolution profiles of calcite and aragonite marine bivalve shells. *Geo-Mar. Lett.* **2008**, *28*, 351–358. [CrossRef]

36. Zhao, L.; Schone, B.R.; Mertz-Kraus, R.; Yang, F. Controls on strontium and barium incorporation into freshwater bivalve shells (*Corbicula fluminea*). *Palaeogeogr. Palaeoclimatol. Palaeoecol.* **2017**, *465*, 386–394. [CrossRef]

37. Zhao, L.; Schone, B.R.; Mertz-Kraus, R. Sodium provides unique insights into transgenerational effects of ocean acidification on bivalve shell formation. *Sci. Total Environ.* **2017**, *577*, 360–366. [CrossRef] [PubMed]

38. Herath, D.; Jacob, D.E.; Jones, H.; Fallon, S.J. Potential of shells of three species of Eastern Australian freshwater mussels (Bivalvia: Hyriidae) as environmental proxy archives. *Mar. Freshw. Res.* **2018**, *70*, 255–269. [CrossRef]

39. Foster, L.C.; Finch, A.A.; Allison, N.; Andersson, C.; Clarke, L.J. Mg in aragonitic bivalve shells: Seasonal variations and mode of incorporation in *Arctica islandica*. *Chem. Geol.* **2008**, *254*, 113–119. [CrossRef]

40. Gillikin, D.P.; Dehairs, F.; Baeyens, W.; Navez, J.; Lorrain, A.; André, L. Inter- and intra-annual variations of Pb/Ca ratios in clam shells (*Mercenaria mercenaria*): A record of anthropogenic lead pollution. *Mar. Pollut. Bull.* **2005**, *50*, 1530–1540. [CrossRef]

41. Collins, D. Thousands of Pearls Found in Shipwreck. Available online: https://www.cbsnews.com/news/thousands-of-pearls-found-in-shipwreck/ (accessed on 20 March 2019).

42. Sommer, S.E. Cathodoluminescence of carbonates 1. Characterization of cathodoluminescence from carbonate solid solutions. *Chem. Geol.* **1972**, *9*, 257–273. [CrossRef]

43. Götte, T.; Richter, D.K. Quantitative aspects of Mn activated cathodoluminescence of natural and synthetic aragonite. *Sedimentology* **2009**, *56*, 483–492. [CrossRef]

44. Hainschwang, T.; Karampelas, S.; Fritsch, E.; Notari, F. Luminescence spectroscopy and microscopy applied to study gem materials: A case study of C centre containing diamonds. *Mineral. Petrol.* **2013**, *107*, 393–413. [CrossRef]

45. Wehrmeister, U.; Jacob, D.E.; Soldati, A.L.; Häger, T.; Hofmeister, W. Vaterite in freshwater cultured pearls from China and Japan. *J. Gemmol.* **2007**, *31*, 269–276. [CrossRef]

46. Qiao, L.; Feng, Q.L.; Li, Z. Special vaterite in freshwater lackluster pearls. *Cryst. Growth Des.* **2006**, *7*, 275–279. [CrossRef]

47. Ma, H.; Su, A.; Zhang, B.; Li, R.K.; Zhou, L.; Wang, B. Vaterite or aragonite observed in the prismatic layer of freshwater-cultured pearls from South China. *Prog. Nat. Sci.* **2009**, *19*, 817–820. [CrossRef]

5

A Review of the Classification of Opal with Reference to Recent New Localities

Neville J. Curtis [1,2], **Jason R. Gascooke** [2], **Martin R. Johnston** [2] and **Allan Pring** [1,2,*]

[1]   South Australian Museum, North Terrace, Adelaide, SA 5000, Australia; neville.curtis@flinders.edu.au

[2]   College of Science and Engineering, Flinders University, Sturt Rd, Bedford Park, SA 5042, Australia; jason.gascooke@flinders.edu.au (J.R.G.); martin.johnston@flinders.edu.au (M.R.J.)

*   Correspondence: Allan.Pring@flinders.edu.au

Abstract: Our examination of over 230 worldwide opal samples shows that X-ray diffraction (XRD) remains the best primary method for delineation and classification of opal-A, opal-CT and opal-C, though we found that mid-range infra-red spectroscopy provides an acceptable alternative. Raman, infra-red and nuclear magnetic resonance spectroscopy may also provide additional information to assist in classification and provenance. The corpus of results indicated that the opal-CT group covers a range of structural states and will benefit from further multi-technique analysis. At the one end are the opal-CTs that provide a simple XRD pattern ("simple" opal-CT) that includes Ethiopian play-of-colour samples, which are not opal-A. At the other end of the range are those opal-CTs that give a complex XRD pattern ("complex" opal-CT). The majority of opal-CT samples fall at this end of the range, though some show play-of-colour. Raman spectra provide some correlation. Specimens from new opal finds were examined. Those from Ethiopia, Kazakhstan, Madagascar, Peru, Tanzania and Turkey all proved to be opal-CT. Of the three specimens examined from Indonesian localities, one proved to be opal-A, while a second sample and the play-of-colour opal from West Java was a "simple" Opal-CT. Evidence for two transitional types having characteristics of opal-A and opal-CT, and "simple" opal-CT and opal-C are presented.

Keywords: opal; hyalite; silica; X-ray diffraction; Raman; Infrared; ^{29}Si nuclear magnetic resonance; SEM; provenance

1. Introduction

Opal is a generic term for a group of amorphous and paracrystalline silica species, containing up to 20% "water" as molecular H_2O or silanol (R_3SiOH) or both [1,2]. Opals, both common opal and precious opal, have been the subject of considerable study over the last five decades or so [3]. Australia has long been the dominant sources of precious opal, exhibiting play-of-colour (POC) [4], but recently, opals from new fields, such as Ethiopia [5], Madagascar [6], Indonesia [7], Tanzania [8] and Turkey [9], have appeared on the market. This influx of new material on the market makes a re-examination of the classification opal timely.

Conventionally, opals are classified into three types [10]: opal-A (further divided into opal-AG (opal) and opal-AN (hyalite)), opal-CT and opal-C [11,12]. For nearly 50 years, the classification of opals proposed by Jones and Segnit [1], based on X-ray powder diffraction (XRD), has been widely adopted, e.g., [13–15]. The pertinent features of the XRD classification are:

- Opal-A (both AG and AN): broad absorption only, centred on 4.0 Å.
- Opal-CT: two prominent peaks at ~4.1 Å and 2.5 Å with a further peak showing variable degrees of separation at ~4.27 Å.
- Opal-C: prominent peaks at 4.04 Å and 2.5 Å.

The notation refers to "amorphous" (A) or is based on the similarity of the XRD reflection positions for α-cristobalite (C) and α-tridymite (T) [16,17]. The exact nature of atomic structures of opal-CT and opal-C remain largely unresolved. Opal-CT appears to be not just a fine-grained intergrowth of layers of cristobalite and tridymite, but a paracrystalline form of silica that has some structural characteristics of these minerals. Separate from the structural features revealed by XRD, some opals exhibit a play-of-colour. This is a textural feature related to the regular packing of silica spheres which are of a size to diffract visible light [4,10,12,15,18]. Both opal-A and opal-CT can show play-of-colour, but the atomic structure of the silica in the spheres is clearly different.

Other techniques such as Raman spectroscopy [13,19–21], ^{29}Si nuclear magnetic resonance (NMR) [22–27], near infra-red spectroscopy [28–30] and neutron scattering [31] have also been used to try to unravel the complex structural relationships between the types of opal. Trace chemical analysis both of opal and associated minerals [7,15,18,32–36] may also provide evidence of provenance [37].

In this paper, we explore whether the XRD classification system is still valid for the newer-sourced material, particularly the precious or play-of-colour opals and that the terms opal-AG, opal-AN, opal-CT and opal-C represent homogenous structural groups. Transitions from opal-A to opal-CT to opals-C to quartz have been reported [38–43] and we take the opportunity to try to identify specimens showing evidence of intermediate forms. In addition to XRD, we will present results of techniques that focus on Si–O bonding to see if they provide evidence for homogeneity of the opal types. Techniques comprise Raman spectroscopy, far and medium infra-red (IR) spectroscopy, and single-pulse ^{29}Si NMR. The key to this approach is the large suite of samples with all opals measured under similar conditions so that trends or differences readily come to light. We hope that this suite of opal samples will also be used by other researchers in their investigations.

The form of this paper is to focus on the results of the characterization of some 48 samples from new or unusual localities out of a total sample suite of some 230 samples that we have examined. A primary classification of all samples into groups according to the XRD methodology of Jones and Segnit [1] (opal-A, opal-CT and opal-C) was undertaken. Then selected samples were subjected to further study using Raman spectroscopy, infra-red spectroscopy and ^{29}Si nuclear magnetic resonance spectroscopy.

2. Materials and Methods

Over 230 opal samples (opal-A, 67 samples; opal-CT, 161 samples; opal-C, 4 samples and 4 samples which appear to be intermediates between forms) were sourced from the South Australian Museum (G prefix), Flinders University (E), the Tate Collection (T) of the University of Adelaide, Museum Victoria (M), the Smithsonian National Museum of Natural History (NMNH) and through recent acquisitions (G NEW, OOC and SO). Where ambiguity exists, such as multiple samples in a single catalogued specimen lot, obvious differences between subsamples are indicated. About 250 individual specimens were analysed. We used well-documented specimens where possible and assumed that locality details were correct. A full list of the specimens examined is given in Supplementary Materials.

X-ray powder diffraction patterns (Bruker D8 Advance machine, Co source $K_\alpha = 1.78897$ Å) were recorded with a scan speed of 0.0195° per second over the 2θ range 10 to 65°. Samples were ground under acetone before use. At least half of the samples were free from any obvious impurity. Literature [25,44,45] d-spacings of the major lines in the XRD patterns of the crystalline silica polymorphs were as follows: quartz: 4.25 Å and 3.33 Å, moganite: 4.43 Å, 4.38 Å, 3.39 Å (obscuring 3.33 Å), 3.10 Å and 2.86 Å, α-cristobalite: 4.04 Å, 3.12 Å, 2.83 Å and 2.48 Å and α-tridymite: 4.38 Å, 4.14 Å, 3.75 Å, 2.98 Å and 2.51 Å.

The major "impurity" was quartz, and this varied from minute traces to overwhelming amounts. A range of clays and related layer silicates was also noted. Only specimens with no or only trace amounts of impurities in the XRD pattern were included in the study.

Curve fitting for XRD data (d-spacing) gave peak positions, full-width half-maximum (FWHM) and relative proportions. A baseline spline was calculated using data read from the relatively flat portions of the pattern and the Microsoft Excel Solver software was used to calculate the minimised least squares fit (about 1000 data points up to 45° 2θ). Since the major peaks spanned a wide range of 2θ, baselines showed variability (particularly for opal-A and some of the opal-CT samples) and only a generalised curve fitting regime was followed. The literature suggests a mix of Lorentzian and Gaussian types (pseudo-Voigt formalization) [46]. Fits could be obtained using either form or a mixture of forms, though this introduced extra variables but showed no obvious gain, and we adopted a pure Lorentzian peak shape for this initial study. Patterns were not corrected for $K_{\alpha 2}$. Consistent results from the large number of samples analysed for opal-A and opal-CT gave confidence in this approach.

Raman spectra were collected in the 100 to 1500 wavenumber (cm^{-1}) region, using a XplorRA Horiba Scientific Confocal Raman microscope. Spectra were acquired using a 50× objective (numerical aperture 0.6) at an excitation wavelength of 786 nm (27 mW measured at the sample) and spectrometer resolution of 4.5 cm^{-1} FWHM. Typical integrations times for the spectra were 30 s and averaged from 6 (for chip samples) to 60 (for powdered samples). The instrument was calibrated using the 520.7 cm^{-1} line of silicon and spectra were corrected to account for absorptions by the edge filter used to suppress the Rayleigh scattering peak. The spectrum of each sample was confirmed at several locations on the sample. Additional runs were made with laser wavelengths of 640 nm (3.8 mW) and 532 nm (7.3 mW) to confirm the Raman spectra recorded at 786 nm. Raman spectra of reference compounds are found in References [21,47–51]. There is, however, inconsistency in reported values for tridymite [52] with several types of Raman spectra despite having similar XRD patterns. Silanole [53] may also present a complicating Raman band at around 500 cm^{-1}. Several samples showed overwhelming fluorescence or gave a weak and largely featureless spectrum indicating that Raman is not as universal as XRD for characterisation. Because of the relatively low Raman scattering cross-section of opal, small amounts of "impurities" could produce misleadingly large peaks. All opals showed a significant baseline component that was most intense below 500 cm^{-1}. Baseline correction was not undertaken since the observed peaks were broad and the exact form of the baseline is unknown.

Attenuated total reflectance (ATR) infra-red (IR) spectra using a diamond ATR crystal were collected in two wavenumber ranges (100–600 cm^{-1} and 400–4000 cm^{-1}). Spectra in the low wavenumber region were recorded at the Australian Synchrotron's THz-Far Infrared beamline using a Bruker IFS 125/HR spectrometer at a spectral resolution of 4 cm^{-1}. Four-hundred to five-hundred scans of the powdered samples were averaged to generate the final spectrum. For mid-IR, ATR spectra were collected using a Perkin–Elmer Frontier spectrometer and recorded at a 2 cm^{-1} resolution. Overlapping wavenumber regions were examined and found to be comparable. The extended range provides full cover of the Si–O bonding derived peaks. Peaks consistent with water stretching (3000–3500 cm^{-1}) and bending (1600–1650 cm^{-1}) were noted but not further investigated for the current study. A simple intensity correction was applied by assuming the ATR penetration depth was directly proportional to the wavelength. We note that the spectra will give slightly different peak positions and band shapes to those obtained via transmission IR spectra or ATR spectra recorded with different crystals due to the anomalous dispersion (change of refractive index) across an absorption band.

The ^{29}Si solid-state Magic Angle Spinning (MAS) NMR spectra were obtained using a Bruker Avance III 400 MHz spectrometer equipped with either a Bruker 4- or 7-mm probe with rotors spinning at 5 or 2.5 kHz, respectively. All spectra were collected at ambient temperature. Single-pulse (SP) experiments were typically carried out using a 90° pulse, high-power decoupling during acquisition (TPPM or SPINAL-64), followed by a recycle delay of 60 s. Spectra were referenced to 4,4-dimethyl-4-silapentane-1-sulfonic acid (DSS) at 0 ppm. It was noted that the opal-A samples required less sample or time to achieve good ^{29}Si NMR spectra compared to other opal forms. The combination of the insensitivity of ^{29}Si and the requirement for relatively large amounts of powdered sample (100 mg) limited our ability to measure numerous samples, thus only selected samples were used. The Q_4 (RO$_4$Si), Q_3 (RO$_3$SiOH) and Q_2 (RO$_2$Si(OH)$_2$) peaks were centred at around −112, −102 and −93 ppm [22] and overlapped owing to the broadness of the resonances (up to 10 ppm). The presence of a comparatively large Q_3 peak will affect the Q_4 peak position and, for this reason, the spectra were deconvoluted. We have also measured ^{1}H and ^{29}Si [22] cross-polarisation

(CP) MAS NMR spectra at various contact times allowing analysis of CP dynamics, and we will report these in a further paper (Curtis, Gascooke, Johnston and Pring, to be published). Reported chemical shifts (Q_4) [27,28] were: quartz −107.2 ppm to −107.1 ppm, cristobalite: −108.1 ppm and tridymite: −111.4 ppm and −109.3/−110.7/−114.0 ppm.

3. Results

The results are presented in two parts. The first is a summary of the results in terms of the overall classification of the opal family from the various techniques. The second presents the results for each opal group, focusing on the variation within the opal-CT group.

3.1. Overview

Figure 1 shows an overall appreciation of the XRD patterns, Raman, mid-IR spectra and NMR of the four groups of opal-AG, opal-AN (hyalite), opal-CT and opal-C using a set of exemplar specimens. As can be seen there are clear differences between opal-A, opal-CT and opal-C, though as will be discussed below the situation is more complex than this figure might suggest. Table 1 presents specimen data for the samples we have designated as exemplars (or typical examples) with a fuller list in the Supplementary Materials.

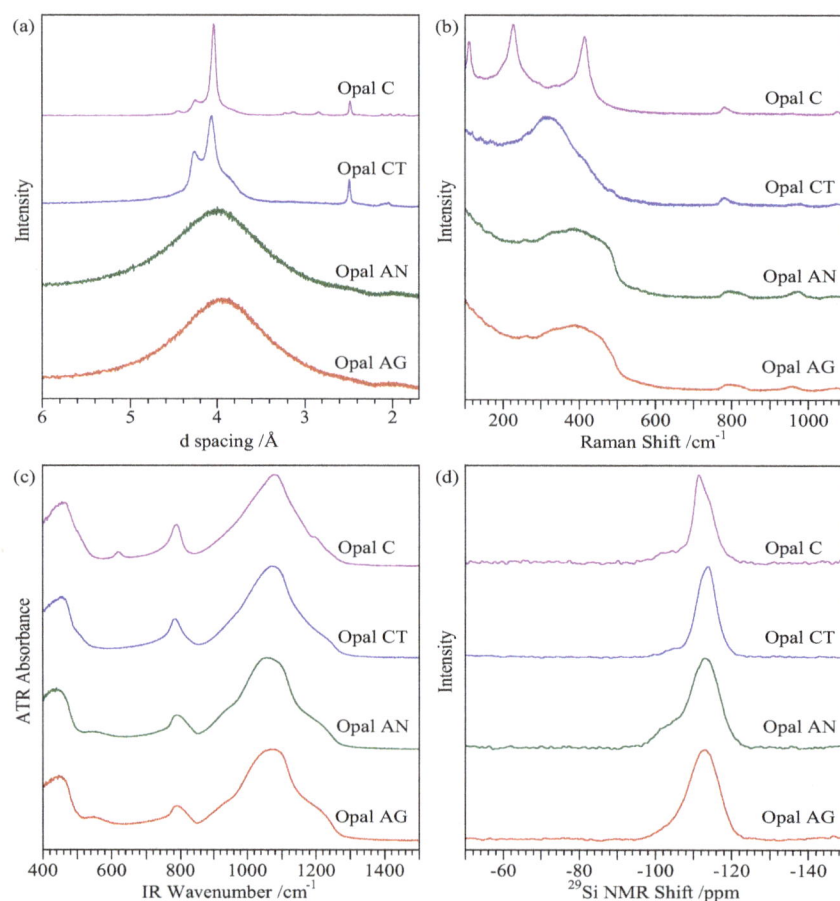

Figure 1. Patterns and spectra of typical samples showing (**a**) XRD patterns, (**b**) Raman spectra, (**c**) attenuated total reflectance mid-IR spectra and (**d**) single-pulse ^{29}Si MAS NMR spectra. In ascending order, the samples are: opal-AG (White Cliffs, Australia G13771) (red); opal-AN/hyalite (Valec, Czech Republic G32740) (green); opal-CT (Angaston, Australia, G9942) (blue) and opal-C (Iceland M5081) (purple). Spectra were scaled and offset for comparison.

Table 1. List of exemplar opal samples examined in detail in the study [a].

Country	Location	Sample ID	Appearance	Type
Australia	White Cliffs, NSW	G1401	Translucent no POC	A
Australia	Eurolowie, NSW	G1425	Translucent pale brown glossy	CT
Australia	Iron Monarch, SA	G9620	White glossy opaque	CT
Australia	White Cliffs, NSW	G8608	White opaque POC	A
Australia	Unknown	G9260	White to grey opaque	A
Australia	Two mile Coober Pedy, SA	G9594	Translucent milky glossy minor POC	A
Australia	Four miles S of Angaston, SA	G9942	Translucent white glossy	CT
Australia	Near Murwillumbah, NSW	G9964	Slightly cloudy, clear POC	CT
Australia	Lightning Ridge, NSW	G13769	Black, glassy band in matrix	A
Australia	White Cliffs, NSW	G13771	Bag of samples	A
Australia	Angaston, SA	G24346	Brown opaque	CT
Australia	Springsure, Qld	M8736	Glassy (hyalite like)	A
Australia	Yinnar, Vic	T19006	Glassy grey-brown	CT
Czech Rep	Valec, Bohemia	OOC11	Glassy clear	A
Czech Rep	Valec, Bohemia	G32740	Hyalite outgrowth, colourless	A
Ethiopia	Mezezo	G25374	Deep-brown translucent	CT
Ethiopia	Afar	G32752	Brown glass, some with POC	CT
Ethiopia	Mezezo	NMNH Eth1	Pinkish POC on white	CT
Ethiopia	Yita Ridge, Menz-Gishe	G31892	Nodules with clear orange centres	CT
Ethiopia	Mezezo	NMNH Eth 2	Transparent brown POC	CT
Ethiopia	Wello	NMNH Eth 3	Milky transparent POC	CT
Honduras	Unknown	G1441	Milky transparent some POC	CT
Iceland	Unknown	M5081	Opaque white	C
Indonesia	Cilayang Village, West Java	G34240	Colourless with POC	CT
Indonesia	Mangarrai Prov, Flores	OOC6	Translucent white	CT
Indonesia	Mamuju, West Sulawesi	OOC13	Blue-green matrix of "grape agate"	A
Kazakhstan	Voznesenovka, Martuk	M53407	Orange glass	CT
Kazakhstan	Zelinograd	G32925	Translucent vermilion glassy	CT
Madagascar	Bemi, Befotaka District	G NEW05	Clear yellow	CT
Madagascar	Bemi, Befotaka District	G NEW07	Translucent pale brown	CT
Mexico	La Trinidad Queretaro	G31851	Single piece with opal inclusions	CT
Namibia	Khorixas district	G NEW29	Blue to white opaque	CT
Peru	Acari	G33912	Massive blue	CT
Spain	Mazarron, Murcia	OOC4	Composite with green zones	CT
Tanzania	Kigoma, Region	G NEW19	Pale orange shades glassy	CT
Tanzania	Haneti	G NEW03	Opaque green	CT
Tanzania	Haneti	G NEW04	Opaque green, some glassy zones	CT
Tanzania	Arusha	G34238	Transparent green layer	CT
Turkey	Kutahya	G NEW24	Translucent green and brown	CT
Turkey	Eskisehir	G NEW25	Opaque white with indigo speckles	CT
Turkey	Anatolia	G NEW26	Opaque white transparent green inside	CT
Turkey	Yozgat, Anatolia	G NEW27	Blue-green transparent glass	CT
Turkey	Yozgat, Anatolia	G NEW28	Olive-green transparent glass	CT
USA	Opal Butte Mine, Oregon	G NEW18	Glassy white	CT-C
USA	Manzano Mtns. New Mexico	G NEW30	White opaque mass	CT
USA	Virgin Valley, Nevada	G31852	Milky and translucent zones	CT
USA	Virgin Valley, Nevada	G32263	Translucent brown	CT
USA	Virgin Valley, Nevada	M19717	Opaque glassy POC	CT
USA	Virgin Valley, Nevada	OOC5	White and POC zones	CT

[a] Some samples yielded more than one experimental sample.

The form of the XRD patterns shown in Figure 1a matches those given in the original opal classification by Jones and Segnit [1] published in 1971. We found that for the most part samples taken from the same specimen, though differing in colour or texture, showed similar, but not necessarily identical, XRD patterns. We note that the difference between opal-CT and opal-C may be subtle when faced with a single sample, as assignment to one group or the other can be difficult without detailed comparison.

The major absorption for the Raman spectra for exemplar specimens (Figure 1b) was in the 200–500 cm^{-1} region with isolated smaller peaks up to 1100 cm^{-1}, and the spectra for opal-A, opal-CT and opal-C were distinct. Spectra were consistent between the unground and the powdered samples (which was also used in XRD). Generally, samples were mostly homogenous but, in some cases, extra peaks were observed in the spectra. These were probably non-opal impurities such as inclusions of silicate minerals. The most likely "impurity", quartz, has a characteristic spectrum of a relatively strong, isolated, sharp peak at 459 cm^{-1} and lesser ones at 108 cm^{-1} (sharp) and 227 cm^{-1} (broad).

All ATR-IR spectra (Figure 1c) for opals show a common set of peaks at around 470, 790 and 1080 cm^{-1}. Below 400 cm^{-1}, peaks were either non-existent or very weak, and thus measurement of spectra in the 400-1600 cm^{-1} range is needed to provide unambiguous differentiation of opal-A, opal-CT and opal-C. The IR spectra of reference compounds are found in References [48–50,54]. Quartz may be identified by peaks at 697 and 780 cm^{-1} which are at lower energy to the common peak at 795 cm^{-1}, seen for both quartz and all the opal samples. Both opal-A and opal-C showed distinct peaks that were absent for opal-CT (see later). A complication occurred with the far-IR as a range of spectra were seen (Figure 2a). The peak at around 470 cm^{-1} showed a progressive trend of broadening, shifting to lower wavenumber and the appearance of a second peak at around 440 cm^{-1}. This was found to be common for all types of opal. There was no obvious correlation with the other spectral methods or with the behaviour in the mid-IR. Figure 2b also shows variation in the mid-IR range (in this case for opal-A for which it is most pronounced).

The ^{29}Si MAS NMR spectra are potentially discriminating for the opal types, although the average Q_4 peak positions were close for all the forms. Some difference was seen for FWHM and the positions of the Q_3 peaks. Peaks may be readily curve-fitted to differentiate the Q_4 and Q_3 peaks (Figure 3).

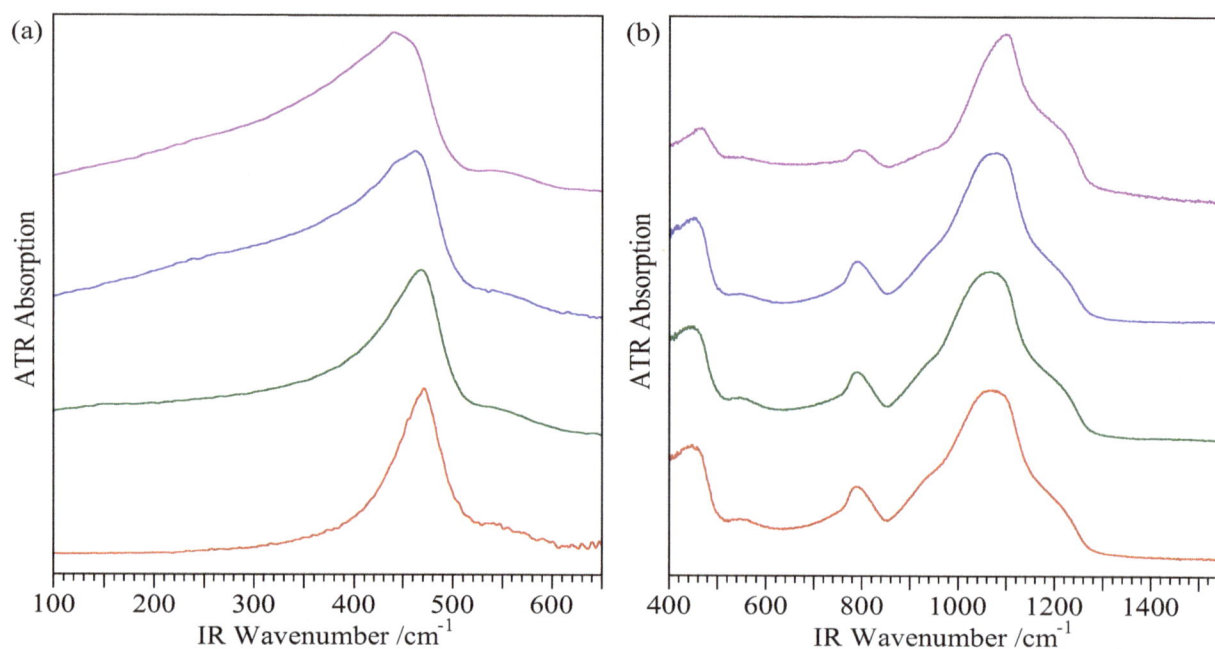

Figure 2. Far IR spectra of opal-AG samples showing the range of types. In ascending order: (**a**) White Cliffs, Australia (G8608), Lightening Ridge, Australia (G13769), Coober Pedy Australia (G9594) and Iron Monarch, Australia (G9260). (**b**) Mid-IR spectra of opal-A samples showing the range of types. In ascending order: Valec, Czech Republic (OOC11) (opal-AN), Coober Pedy Australia (G9594) (opal-AG), Springsure, Australia (M8736) (opal-AN) and White Cliffs, Australia (G1401) (opal-AG). Spectra were scaled and offset (y-axis) for comparison.

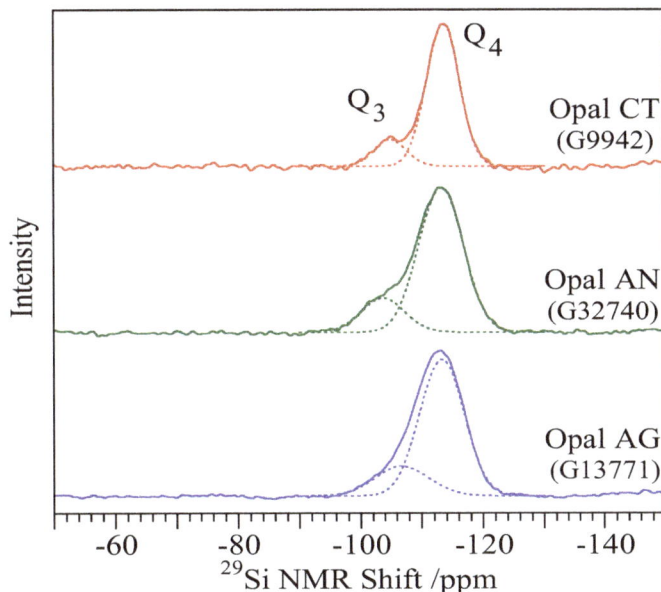

Figure 3. Experimental and fitted curves for ^{29}Si MAS NMR spectra. The exemplar opal-AG, opal-AN and opal-CT samples: opal-AG (White Cliffs, Australia G13771) (red), opal-AN/hyalite (Valec, Czech Republic G32740) (green) and opal-CT (Angaston, Australia, G9942) (blue).

3.2. Opal-A

All samples showed broad and weak XRD patterns with absorption centred on ~4.0 Å (Figure 1a) and had a poorly defined baseline compared to the other opal types. There was also a weak broad band at around 2.0 Å. Consistency was demonstrated through curve fitting with peak maxima of 3.98 ± 0.04 Å and FWHM 0.53 ± 0.02 Å. There was little, if any, difference between the opal-AG and opal-AN samples.

Opal-A Raman spectra were distinct from the spectra for opal-C and opal-CT in the 100–600 cm^{-1} range, so this provided a secondary delineation of the opal type. There were no obvious differences between Raman spectra for opal-AG and opal-AN in the 100–600 cm^{-1} range. Raman spectra between 700 and 1200 cm^{-1} (Figure 4) showed several weak peaks and a broad peak (probably two peaks) in the range 760–860 cm^{-1} that was characteristic of both opal-AG and opal-AN and which also separated these from most examples of opal-CT and opal-C. In general, the middle peak in Figure 4 was at 960–965 cm^{-1} for opal-AG and 970–975 cm^{-1} for opal-AN, though the peaks were weak. The IR spectra (Figure 1c) for opal-A were characteristic and distinct to those for opal-CT and opal-C with the peak at around 550 cm^{-1} being specific for opal-A, but there were no obvious differences between opal-AG and opal-AN.

The ^{29}Si MAS NMR spectra (Figure 1d) were potentially discriminating for opal-A, as the FWHMs tended to be larger for it than for other types. The Q$_4$ peak positions were −113.3 ± 0.2 ppm with FWHM of 8.5 ± 0.2 ppm for opal-AG and −113.4 ± 0.3 ppm and 8.7 ± 0.6 ppm for opal-AN. These FWHMs were higher than the 6.5 ± 0.6 ppm seen for opal-CT. All samples showed significant amounts of Q$_3$ peaks within the range 10–40%. The visual difference of opal-AG and opal-AN in Figure 1d can be traced to the placement of the Q$_3$ peaks with the former at −106.5 ± 0.9 ppm and the latter at −103.8 ± 0.8 ppm (see Figure 3).

Of the more recent finds, the massive blue-green opal that occurred as the base of some of the purple "grape agate" from near Mamuju, West Sulawesi, Indonesia was an opal-AG. A previous report [37] suggested that this material might have been a clay.

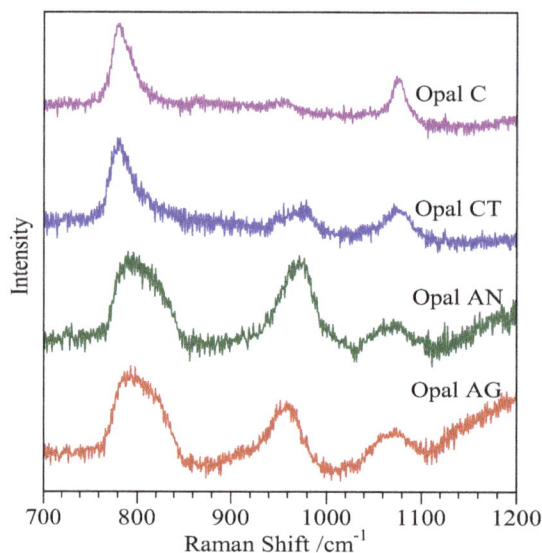

Figure 4. Raman spectra (700 to 1200 cm^{-1}) of opal samples. In ascending order: White Cliffs, Australia (G13771) (opal-AG), Valec, Czech Republic (G32740) (opal-AN), Angaston, Australia (G9942) (opal-CT) and Iceland (M5081) (opal-C). Spectra were scaled and offset for comparison.

3.3. Opal-CT

All samples showing substantial XRD peaks at ~4.1 Å and 2.5 Å may be considered as opal-CT according to the Jones and Segnit [1] classification. However, we believe that this is an oversimplification. The maximum was distinct from that for the opal-C grouping (~4.1 Å to 4.04 Å, respectively), though this was only apparent when two samples were directly compared. The increasing complexity of the main peak ~4.1 Å in the XRD pattern (Figure 5), presented in terms of increasing sharpness of the peak at 2.5 Å) suggests that the term opal-CT does not represent a homogenous group. We noted a trend whereby a peak at 4.28 Å was sometimes absent, sometimes present as a shoulder and sometimes as a separate peak. This was noted in a number of previous studies [55–57]. Complexity of the peak at ~4.1 Å appears associated with the peak width at 2.5Å. It is also apparent from the large number of samples measured here (161 specimens) that the patterns varied according to the relative peak heights and widths in the composite at around ~4.1 Å (see later). Sample G32752 (Afar, Ethiopia) (Figure 5) showed a simple, near symmetric XRD peak at 4.1 Å, as well a broad peak at 2.5 Å. At the other extreme, sample G NEW19 (Kigoma, Tanzania) showed two sharp peaks and a prominent shoulder in the composite at around ~4.1 Å and a sharp peak at 2.5 Å.

We note that the more "simple" opal-CT specimens include POC specimens from Ethiopia (G25374, G31892, G32752 (three distinct samples from the same specimen), NMNH Ethiopia samples 1, 2 and 3), Honduras (G1441), the USA (M19717, OOC5), Mexico (NMNH 117414) and Australia (G9964). The group also includes mostly transparent samples from worldwide localities such as Turkey (G NEW24, G NEW26, G NEW27, G NEW28), Madagascar (G NEW05), Mexico (G34738, NMNH115816, NMNHR1694), Australia (T19006, T23363, M12495, M20970), Peru (G33912), Brazil (T1152), Indonesia (OOC6), Spain (G NEW12) and the USA (G9116, G31852, G34243). The converse is not, however, true, with for instance G NEW07 from the same site in Madagascar as G NEW05 showing a more complex pattern. Some samples showing POC also have more complex XRD patterns.

We also noted a distinct change in the shape of the background of the XRD pattern. In the simpler opal-CT patterns, the background was distinctly lower on the low-angle side and higher on the high-angle side of the peak at 4.1 Å. In contrast, the background around this peak became more uniform as the peak at 4.1 Å became more complex.

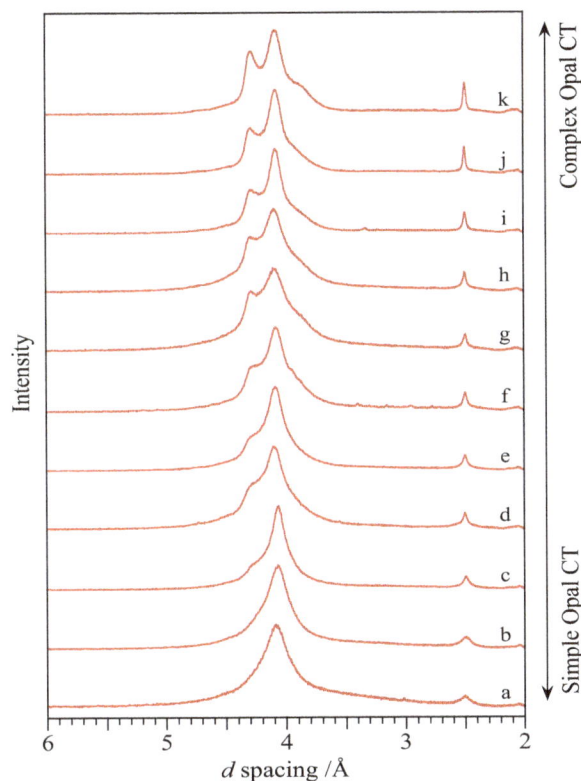

Figure 5. A series of XRD patterns illustrating the structural changes across the opal-CT group. The patterns were arranged in order of increasing complexity of the main peak at ~4.1 Å. Any subdivision between "simple" opal-CT and "complex" opal-CT is arbitrary, but patterns *a* and *b* are clearly distinct from patterns *i*, *j* and *k*. Note the progressive change in the sharpness and shape of the reflections at 4.1 and 2.5 Å. The specimens are (in ascending order from simplest to most complex): (**a**) Afar, Ethiopia (G32752), (**b**) Mezezo, Ethiopia (NMNH Eth 1), (**c**) Murwillumbah, Australia (G9964), (**d**) Acari, Peru (G33912), (**e**) Honduras (G1441), (**f**) Indonesia (OOC6), (**g**) Kazakhstan (M53407), (**h**) Nevada, USA (G32263), (**i**) Tanzania (G34238), (**j**) Nevada, USA (G31851) and (**k**) Tanzania (G NEW19).

We explored peak fitting to investigate the heterogeneity seen in opal-CT to see if quantifiable support may be gained for the continuum of "simple" and "complex" opal-CT. As Figure 6 shows, a reasonable proposition is that there were three peaks present (P1 to P3) in the 4.1 Å complex with a further one at 2.5 Å (P4). Well-fitting deconvolutions were obtained for all relatively pure (i.e., no visible or only a very minor amounts of quartz or other impurities) samples and confirms three peaks with positions at 4.27 ± 0.01 Å (P1), 4.08 ± 0.01 Å (P2) and 3.89 ± 0.03 Å (P3). The peak at 2.50 ± 0.01 Å (P4) was also constant.

The peak areas were normalized to P2 and the FWHM examined for each peak used to explore trends. A strong correlation in FWHM was for P1 and P4 (Figure 7a) where the linear trend between "simple" and "complex" extremes is manifest. The "simple" opal-CTs plot in the top right-hand corner are Ethiopian opals with POC. Other POC opals are distributed though the plot suggesting that this textural effect is independent of crystal–chemical features. It is also worth noting that other textural features such as the alignment and size of spheres are also implicated in POC.

There is also a linear correlation for the relative areas of P1 and P3 peaks as shown in Figure 7b, implying that, in simplistic terms, that this could be considered to represent the increase in the "tridymite" component comparted to the "cristobalite" (P2 coinciding with the overlap of the C and T contribution). The presence of trace amounts of quartz does not seem to affect this trend, but the problem associated with fitting caused by the asymmetric baseline may affect the ratio and over-represent P1 or P3.

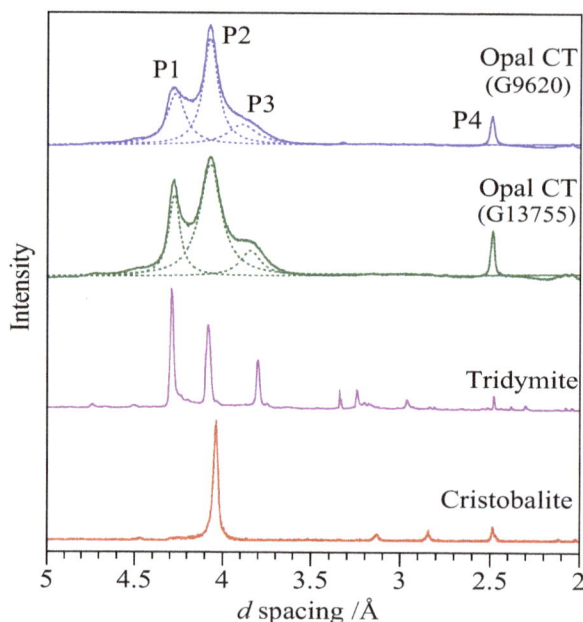

Figure 6. Comparison of mixed XRD patterns for mineral samples of tridymite (G1395) and cristobalite (RRUFF Database ID R060648) with Ross, Tasmania, Australia (G13755) and Iron Monarch, Australia (G9620) showing curve-fitting elements and actual pattern. Spectra were scaled and offset for comparison.

The implication of the separate relationships shown in Figure 7 is that the XRD pattern for any particular opal-CT will be hard to predict. Peaks may be large or small and sharp or narrow. We think it unlikely that a single parameter may be derived to quantify an opal-CT. Principal component analysis may provide a solution, but this would need to have a chemical or structural basis rather than mathematical manipulation. We noted that the two points at the top right of Figure 7b (G9960 from Iron Monarch, Australia and G NEW17 from Turkey) lie in the middle of the plot in Figure 7a. Similarly, G13755 (Ross, Tasmania, Australia, bottom left of Figure 7a) which shows very prominent and sharp XRD peaks (see Figure 6) lies in the centre of Figure 7b. It is also unlikely that the XRD pattern will allow prediction for the potential of POC from that locality.

The Raman spectra Figure 8 also demonstrated that opal-CT is not a homogenous group as shown by the spectra in Figure 5. An arbitrary classification may be made according to the rationales: (i) featureless spectrum with a maximum above 300 cm^{-1}, (ii) signs of structure with maximum at or below 300 cm^{-1} and (iii) more developed spectrum with partially delineated peaks. It is not clear if the trend was due to the change in structure with different spectra or merely a sharpening of existing peaks. We noted that those samples with few features in the Raman were mostly coincident with the "simple" opal-CT points at the upper right of Figure 7b (see Figure 8b). The most Raman-structured samples tended to lie at the lower left of Figure 8b and thus correlated with the more complex XRD patterns.

The 700 cm^{-1} to 1000 cm^{-1} Raman region showed potential differences between the "simple" and "complex" opal-CT examples in the 850–1000 cm^{-1} region. Unfortunately, the spectra were generally weak and were indicative rather than definitive. This is further complicated as not all samples were amenable to Raman due to the fluorescence, and this limits the applicability of the technique.

The ^{29}Si MAS NMR data showed that SP opal-CT Q$_4$ peak positions were marginally downfield of those for opal-A at −113.7 ± 0.3 ppm with FWHM 6.5 ± 0.6 ppm. In general, the Q$_3$ peaks were visible and were at −105.3 ± 0.8 ppm. The FWHMs of 6.5 ± 0.6 ppm were smaller than for opal-A at 8.4 ± 0.4 ppm. We also found that some of the "simple" opal-CT types gave visually different spectra that could not be deconvoluted satisfactorily.

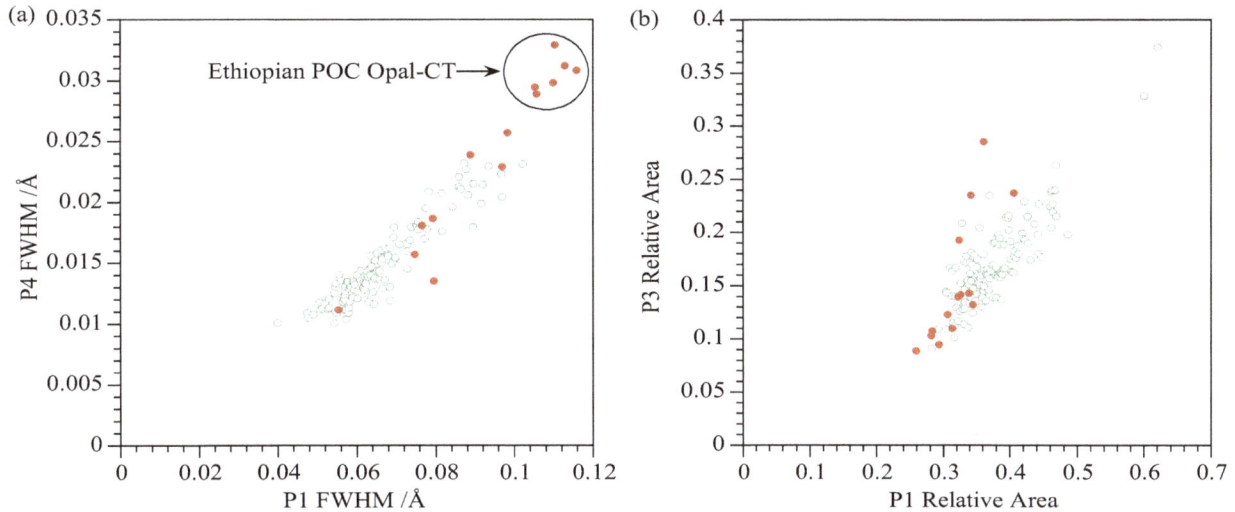

Figure 7. Correlations of XRD curve fitting data: (**a**) FWHM for P1 (*x*-axis) and P4 (*y*-axis), and (**b**) relative amounts of the P1 (*x*-axis) and P3 (*y*-axis) peaks (P2 is set at unity). Samples showing play-of-colour (POC) are shown as red-filled circles, whereas non-POC samples are represented by green-open circles. The subset of samples from Ethiopia displaying POC are circled in panel (**a**).

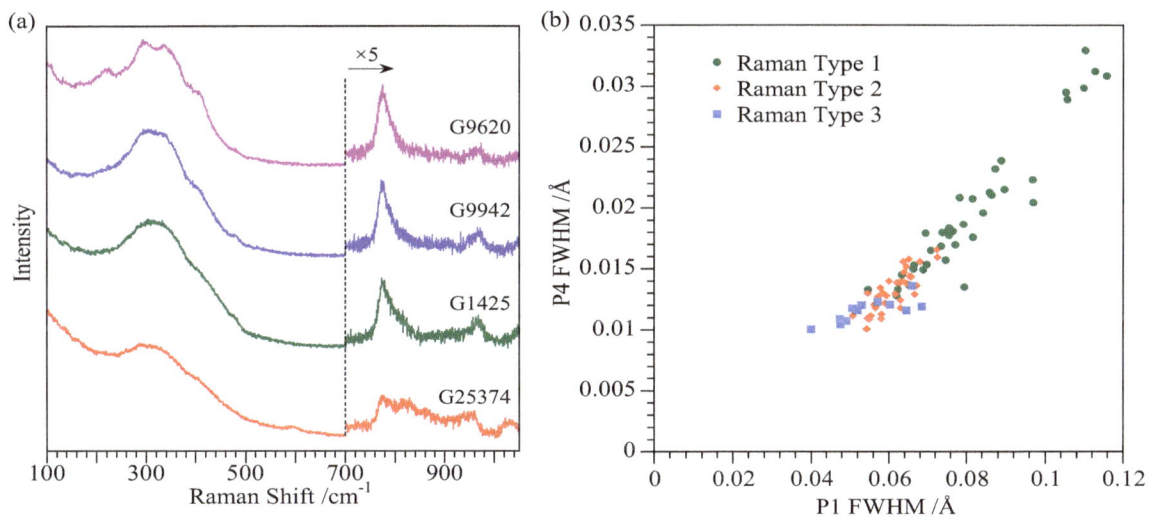

Figure 8. Raman spectra of opal-CT showing progressive structure. In ascending order: (**a**) "simple" opal-CT from Mezezo Ethiopia (G25374), and increasingly complex forms from Eurolowie, Australia (G1425), Angaston, Australia (G9942) and Iron Monarch Australia (G9620). (**b**) Plot of the XRD pattern FWHM of P4 versus P1 separated into different Raman types. See text for details regarding the definitions of the Raman types observed in this study. Not all samples yielded a Raman spectrum due to the problems with fluorescence.

3.4. Opal-C

Only four discrete samples were identified as opal-C in this study. Two of these had zones of both transparency and opaqueness though having identical XRD patterns. All showed very small peaks for quartz. The common feature was a large peak at 4.04 Å with a smaller one at 4.28 Å. The small pair of equally sized peaks at 3.11 Å and 2.84 Å was diagnostic of cristobalite [44].

Raman spectra (Figure 1b) showed characteristic spectra similar to that of cristobalite [51] at low wavenumber (peaks at 107, 222 and 409 cm^{-1}), while the medium wavenumber spectrum shown in Figure 4 is possibly diagnostic but again weak. The combined far- and mid-IR for the opal-C samples showed these features: 300 cm^{-1} (sh), 385 cm^{-1} (sh), 480 cm^{-1} (m), 625 cm^{-1} (w), 795 cm^{-1}

(m), 1090 cm^{-1} (s) and a shoulder at 1200 cm^{-1}. The most delineating feature was the band at around 625 cm^{-1} which was visible both via far- and mid-IR. We did not, however, find any evidence of an IR band at 145 cm^{-1} as was reported in a previous study [50].

The ^{29}Si MAS NMR spectra of the opal-C samples showed variability and the Iceland sample (M5081) shown in Figure 1d should be treated as an example rather than a typical spectrum. This had characteristics of a peak position of -114.4 ppm and an FWHM of 5.7 ppm. Spectra suggest more complexity than those seen for opal-A and opal-CT with the likelihood of additional peaks.

3.5. Transitional Samples

3.5.1. Samples Showing Opal-A and Opal-CT Characteristics

Three samples had XRD patterns which had characteristics of both opal-A and the simpler form of opal-CT, with a very broad peak at 4.1 Å and a relatively small and broad peak at 2.5 Å. Sample OOC4 from Mazarron, Spain had a major peak FWHM estimate of about 6.3°, while that for T22824, from Megyasro, Hungary, was 4.3°. Neither gave a satisfactory XRD curve fitting. Not shown is E1950 (Canungra Mts. Australia) which had an FWHM of around 3° and which could be XRD curve fitted (and plottted as a "simple" opal-CT). This contrasts with a value of about 8° for a typical opal-A and less than 2° for the POC opal-CT from Mezezo, Ethiopia (G25374) and the remainder of the simple opal-CT samples. Only T22824 yielded a Raman spectrum and this was consistent at both low and medium wavenumbers with opal-A. The orange opal-CT from Voznesenovka, Kazakhstan (T22824) and OOC4 from Mazarron, Spain showed a peak at 550 cm^{-1} in the IR though this was not readily apparent in E1950 (Figure 9).

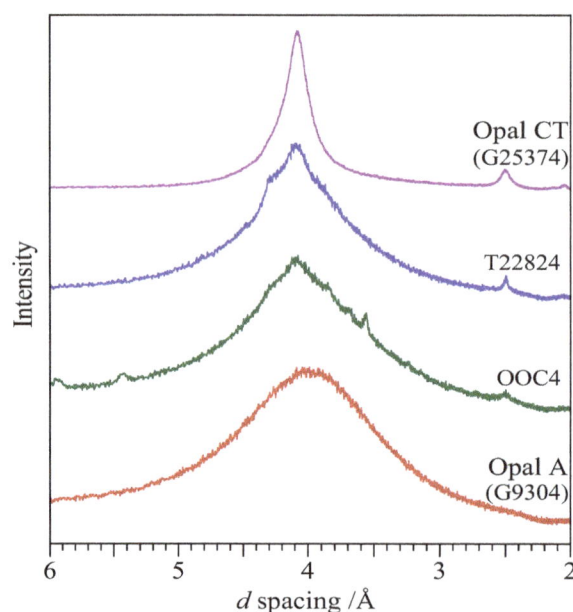

Figure 9. Transitional opal XRD pattern (lower-middle OOC4 from Mazarron, Murcia, Spain, upper-middle T22824 from Megyasro, Hungary). Shown with G9304 (opal-A, lower) and G25374 (simple opal-CT, upper). Scaled and offset (y-axis) for comparison.

Raman spectra showed broad peaks at 760–860 cm^{-1} (as in opal-A) and 975 cm^{-1} (as in hyalite). While there were coincidences in position, it is difficult to assert that the XRD peak at 2.5 Å represents a narrowing of the broad and weak second peak in opal-A. Critically, the XRD patterns and Raman spectra showed no evidence of cristobalite, so a reasonable assumption was that these represent transitional opal-A/opal-CT. The SEM images (Figure 10) were not consistent with opal-AG.

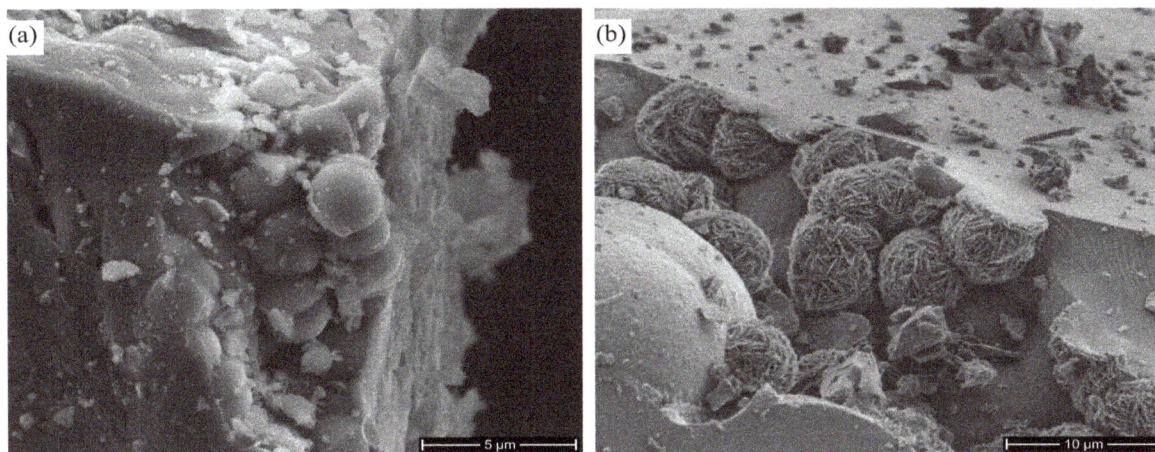

Figure 10. SEM of: (**a**) OOC4 AND (**b**) T22824 (RHS) showing large spheres and bundles of plates.

3.5.2. Samples Showing Opal-CT and Opal-C Characteristics

The sample from the Opal Butte Mine, Oregon USA (G NEW18) appears to be a transitional form of opal-CT to opal-C as the characteristic cristobalite peaks are present in both the XRD pattern and Raman spectrum (Figure 11). The XRD maximum is at 4.02 Å which is consistent with the cristobalite-like patterns of opal-C as are the two small peaks between 3.14 Å and 2.85 Å. The form and position of the Raman absorption at low wavenumbers are not characteristic of opal-A and are more like that seen for the "simple" opal-CT samples with an overlay of cristobalite peaks. At medium wavenumbers only one peak is seen at around 800 cm^{-1} while the peak at just below 1000 cm^{-1} is perhaps more significant than is seen for the opal-C samples. A weak response at 625 cm^{-1} in the IR is also consistent with opal-C as are the patterns at lower wavenumber. Overall this suggests a transitional "simple" opal-CT to opal-C species.

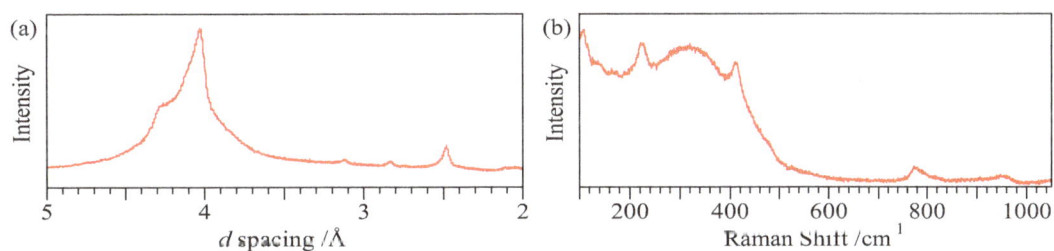

Figure 11. Transitional opal CT to C from Opal Butte Mine, Oregon USA (G NEW18). (**a**) XRD pattern and (**b**) Raman spectrum.

4. Discussion

4.1. Applicability of XRD for Primary Classification

We found that all opal samples, including those from "newer" sources, including Ethiopia, Indonesia, Kazakhstan, Madagascar, Peru and Tanzania could be readily classified using XRD into one of the Jones and Segnit [1] groups. All of the opals for newer localities were opal-CT, except for two of the Indonesia samples which were Opal-A (Table 1). XRD provides a ready and informative delineation of opal-A, opal-CT and opal-C but the identification of transitional forms requires additional data from spectroscopic techniques. Differentiation of opal-AG and opal-AN is, however, not possible from XRD, but can usually be readily seen visually. The two types of precious opal (opal-A and "simple" opal-CT) can be readily separated by XRD.

Further detail may be gleaned, particularly of opal-CT, through semi-quantitative curve fitting.

The work described here provides an alternative analysis to previous studies [14,17,55,58] where the maximum and width of the composite peak at 4.1 Å were interpreted in terms of contributions from cristobalite-like and tridymite-like components.

4.2. Homogeneity and Characterisation of Opal Groups

4.2.1. Opal-A

This group was clearly identified by the distinctive broad XRD pattern, the featureless but broad Raman pattern centred at 370 cm^{-1} (uncorrected) and a specific IR peak at around 530 cm^{-1}. The ^{29}Si NMR will show a relatively broad (8.5 ppm) and symmetric Q_4 peak at around -113.3 ppm hiding the Q_3 component. Hyalites (opal-AN) were visually different from the opal-AG group being botryoidal and gel-like though with similar XRD patterns and Raman and IR. The major ^{29}Si NMR spectra will show resolved peaks for Q_4 and Q_3 unlike for opal-AG.

4.2.2. Opal-CT

The XRD patterns for this group are characterized by peaks at 4.1 Å and 2.5 Å. With the 4.1 Å reflection range from a "simple" asymmetric peak accompanied by broad weak reflections at 2.5 Å, to a "complex" broad group of three peaks centred on 4.1 Å and a sharp reflection at 2.5 Å. The opal-CT that gave the simple X-ray patterns were more or less transparent, though they may have been coloured and showed POC to some extent, while those with complex patterns tended to be opaque and included material that was generally considered common opal, though there were examples with POC. The Raman spectra for those that gave "simple" XRD patterns were generally weak and featureless showing only a peak around 300 cm^{-1}, while those for the opal-CTs with more complex XRD patterns (if obtainable) showed broad absorption in the 200 to 500 cm^{-1} range and varying degrees of structure, possibly with discrete peaks at 220, 295, 340 and 410 cm^{-1}. The IR spectra and ^{29}Si MAS NMR across the range of opal-CTs did not provide differentiation.

These "simple" opal-CTs included the Ethiopian play-of-colour opals as well as some Australian specimens, but were not from the current active opal fields. For instance, G9964 is labelled as "jelly opal" from Murwillumbah in New South Wales and has a cloudy but transparent appearance. It has been proposed that Australian POC opal is sedimentary if opal-A and volcanic if opal-CT [59]. Other sources for "simple" opal-CT without play-of-colour include: Brazil, France, Honduras, Indonesia, Madagascar, Mexico, Peru, USA (Nevada). Published spectra show further examples of these simple XRD patterns, e.g., [7,13,19,60].

Recent measurements suggest that POC in opal-CT is also caused by the presence of diffracting patterns of spheres [5,61] similar in concept to those for opal-A [4]. Our work on the chemical characteristics of opal-CT, however, can provide no light for prediction of POC, though we note that an Ethiopian opal with a wide peak at 2.5Å is likely to show the effect. Other POC examples do not show any correlation with XRD patterns or spectroscopic measurements.

4.2.3. Opal-C

There is no visible difference between these and opal-CT samples. Some show the feature of transparent specimens merged with white translucent areas [49] which have the same XRD patterns, Raman and IR spectra. The distinct features are XRD pattern peaks at 4.04 Å and 2.5 Å coupled with a pair of small peaks between the major ones. Raman shows a characteristic pattern with peaks 107, 222 and 409 cm^{-1}. The IR spectra were distinct with a small but clearly defined peak at 625 cm^{-1}. The ^{29}Si MAS NMR showed 6 ppm FWHM peak at -113 ppm.

4.3. Spectroscopic Characterisation Techniques

Raman spectroscopy [6,7,13,20,32] can be used in non-destructive mode with data gained in the far- and mid-range providing information related to Si–O bonding, although direct assignment of bands to the opal structure is difficult. The spectra for opal-A, opal-CT and opal-C were distinct and provided ready differentiation. With some care, hyalite may also be identified if it is not already apparent from its distinctive appearance. Raman spectroscopy is not as viable a technique as XRD for two major reasons [62]. Firstly, opal has a relatively low cross-section for Raman scattering resulting in weak signals, thus requiring long collection times. Second, many samples exhibited a large degree of fluorescence which swamps the signal. The "complex" opal-CT samples were more likely to be fluorescent than the "simple" types, indicating possibly higher content of metal impurities in "complex" opal-CT. Opal-C samples, and some proposed transitional forms, had similar peaks to that of α-cristobalite and this is probably the preferred means for identification. The current work is consistent with previous studies.

ATR IR has been used previously to examine silica species [1,54] but has been little exploited recently for opal. It does, however, represent an alternative to XRD as opal-A, opal-CT and opal-C can be readily discriminated, though without the additional information relating to "simple" or "complex" forms. While we used ground samples, it could be a non-destructive method in ATR mode. Although not performed here, we suggest that reflectance IR spectroscopy may also be a valuable tool for non-destructive examination of opal samples. The ^{29}Si NMR is of more interest for investigation of the chemical structure of opal rather than as a differentiation technique.

4.4. Comments on Nature of Opal-CT

This has been the subject of controversy for some years. The term opal-CT can be interpreted as zones of cristobalite and tridymite or as an intimate intergrowth, whether regular or disordered. In our opinion, we feel that the term "opal-CT" is a misnomer. The notation was based on a similarity of XRD peak positions of α-cristobalite and α-tridymite with those in opal samples [1,25,44]. This has been complemented by modelling studies of XRD patterns [63] and Raman spectra [52,64]. The XRD patterns have been analysed in terms of α-cristobalite to α-tridymite ratio [14]. The TEM images have been interpreted in terms of the intergrowth of domains of cristobalite and tridymite and as tridymite-like stacking faults in cristobalite [55,56]. As Figure 6 implies, a reasonable proposal based on peak positions of reference compounds is that the opal-CT peaks derive from cristobalite and tridymite, with P1 due to tridymite, P2 due to cristobalite and tridymite and P3 due to tridymite (and possibly cristobalite). While the XRD curve fitting suggests complexity in the structure, it is not clear how many discrete species may be involved, as correlation was noted between the intensities of the P1 and P3 peaks in one instance and the sharpness of the P1 and P4 peaks in another. The trending evidence (Figure 7) did not imply this. If tridymite was present, then we might expect the P1, P2 and P3 peaks to be linked in some way. P3 did not correlate in sharpness with P1 and P4 and was shifted compared to α-tridymite. We also believe that the Raman evidence was equivocal, as while we find (baseline uncorrected) peaks at 220, 295, 340 and 410 cm^{-1} in the most structured opal-CT samples, we do not believe that the spectra were of sufficient quality to add anything significant to the issue.

The notion that opal-CT is a disordered, intimate mix of cristobalite and tridymite has also been questioned on the basis on XRD and Raman data [52,57,64,65]. The lack of evidence for the presence of cristobalite led to the proposal for "opal-T" based on interpretation of the Raman data [52,57,65]. Whether this is a "not cristobalite" rather than a "positive tridymite" assignment is a moot point. Supporting evidence for tridymite, however, presents a major problem with the potential multitude of stacking variations of this structure [44]. We found no evidence for the triplet reported for tridymite in ^{29}Si studies [24,27]. The topology of the silica structural frameworks in opal-CT remain a matter for debate, and the changes we observed between "simple" and "complex" opal-CTs may represent different structural states rather than different structural intergrowths.

4.5. Comments on Opal Formation and Transitions Between Opal-A, Opal-CT, Opal-C and Quartz

A recent paper [39] has proposed that temperature of formation (≤45° for opal-AG and >160° for opal-CT) is the prime determinant of opal type rather than type of deposit, i.e., volcanic versus sedimentary [40,49] sources for opal-CT and opal-A, respectively. Our results are consistent with this proposition, our simple opal-CT showing a play-of-colour all come from deposits associated with volcanism.

It is possible that the transition between the different forms of opal derives from a similar process as the initial formation: for example, opal-A dissolution (partially hydrated silica to silicic acid) followed by deposition of opal-CT [41]. Analysis of dated sinter samples from hot springs sites [42] show the presence of opals with XRD patterns consistent with opal-A, a transitional form between opal-A and opal-CT, opal-CT, opal-C and quartz. The SEM images also show a change in form. While the opal-CT sample was probably consistent with "simple" opal-CT, verification is difficult owing to a significant amount of quartz. The SEM evidence for transition is also noted for geysers [38]. Changes in XRD pattern and near-IR have also been noted in accelerated aging at 300 °C of deep sea deposits [43].

We see two types of transitional forms that could be interpreted as opal-A to opal-CT and "simple" opal-CT to opal-C. However, these transitional samples are very uncommon and possibly far less than would be expected if this was a routine occurrence, but this could depend on the kinetics of the process and the geological age of the samples. The transformation of one form of opal to another is most probably a dissolution/reprecipitation reaction, given that opal is associated with flow of aqueous crustal fluids and these processes are known to be relatively rapid in terms of geological time [66,67].

Table 2 gives a brief summary of the defining characteristics found for the different opal types described in this work. The presence of "impurities" may cause misidentification for single samples.

Table 2. Summary of differentiating opal properties (this work).

Opal Type	XRD	Raman [a]	IR [b] (ATR)	^{29}Si NMR [b] (Single Pulse)
Opal-AG	Very broad peak between ~2.2 Å and ~6.5 Å with maximum at 3.9–4.0 Å	Broad peak between ~230 and ~530 cm^{-1} with maximum at ~370 cm^{-1}; 760–860 cm^{-1} 970–975 cm^{-1}	Peak at 530 cm^{-1}	Q_4 FWHM 8.5 ppm Q_3 peak(s) not prominent
Opal-AN (hyalite)	Very broad peak between ~2.2 Å and ~6.5 Å with maximum at 3.9–4.0 Å	Broad peak between ~230 and ~530 cm^{-1} with maximum at ~370 cm^{-1} 760–860 cm^{-1} 960–965 cm^{-1}	Peak at 530 cm^{-1}	Q_4 FWHM 8.3 ppm Q_3 peak visible as a shoulder
Opal-CT [c]	All have peak at 2.50 Å. Simpler types have a single peak at 4.08 Å. More complex types also show a peak or shoulder at 4.28 Å and a shoulder at 3.89 Å	Broad peak between ~180 and ~500 cm^{-1} with maximum at ~300 cm^{-1} to more defined maxima at 220, 295, 340 and 410 cm^{-1}	*Absence of peaks at 530 and 625 cm^{-1}*	Q_4 FWHM 6.5 ppm Q_3 peak visible
Opal-C	4.04 Å and 2.50 Å	Sharp peaks at 107, 222 and 409 cm^{-1}	Peaks at 300, 385, 470 and 625 cm^{-1}	*No common feature*

[a] Without baseline correction. [b] Only unique features are noted. [c] Trend discussed in text.

4.6. Summary

This study provides a classification of examples from many sites, both gem quality and other samples, and incorporates a number of techniques. The large number of spectra and XRD patterns presented here illustrate the range of opals that may be found under each heading.

XRD remains the primary analytic method of choice as all samples, including those from newer sources, can be readily classified as opal-A, opal-CT, opal-C or a reasonable case may be made for a transitional form. We note that mid-IR ATR spectroscopy also fulfils this role. The XRD patterns for Opal-CT exhibited a range of forms from quite simple patterns to more complex ones, but this range was a continuum. At the "simple" end, the Ethiopian POC occurred while the "complex" forms included the common opals. Play-of-colour opals may belong to either the opal-A or opal-CT groups. Thus, the terms "precious opal", "play-of-colour", "potch", "common opal" and "fire opal" are best treated with caution, possibly only to be used within the trade, rather than in scientific studies.

The large body of samples examined in this work has allowed us to identify exemplars or typical specimens for the various opal groups. These provide authentic references that could be used in provenance authentication, in comparing purity and for providing characterised samples for further studies, such as chemical or thermal modification. New samples can be compared against these established and well-characterised examples. They may also be used for other studies such as geological (e.g., volcanic versus sedimentary) settings or trace element content.

Author Contributions: Conceptualization, A.P. and N.J.C.; Data curation, N.J.C.; Formal analysis: N.J.C., J.R.G. and M.R.J.; Investigation, N.J.C., J.R.G. and M.R.J.; Methodology N.J.C., J.R.G. and M.R.J.; Validation, N.J.C., J.R.G., M.R.J. and A.P.; Visualization, N.J.C. and A.P.; Writing—original draft preparation, N.J.C.; Writing—review and preparation, N.J.C., J.R.G. and A.P.

Acknowledgments: The authors thank Ben McHenry of the South Australian Museum and Tony Milne of the Tate collection at the University of Adelaide for their unstinting assistance in locating samples for this study. The collection mangers of the Flinders University of South Australia, Museum Victoria and the Smithsonian National Museum of Natural History are thanked for the provision of samples and useful conversations. The authors acknowledge the expertise, equipment and support provided by Microscopy Australia and the Australian National Fabrication Facility (ANFF) at the South Australian nodes under the National Collaborative Research Infrastructure Strategy. We acknowledge access to the facilities at the Australian Synchrotron and the technical support provided by Dominique Appadoo for the Far-IR data collection. The constructive comments of the two anonymous referees and the associate editor are gratefully acknowledged.

References

1. Jones, J.B.; Segnit, E.R. The Nature of Opal I. Nomenclature and Constituent Phases. *J. Geol. Soc. Aust.* **1971**, *18*, 57–68. [CrossRef]

2. Murray, M.J.; Sanders, J.V. Close packed structures of spheres of two different sizes II. The packing densities of likely arrangements. *Philos. Mag. A* **1980**, *42*, 721–740. [CrossRef]

3. Caucia, F.; Ghisoli, C.; Marinoni, L.; Bordoni, V. Opal, a beautiful gem between myth and reality. *Neues Jahrbuch für Mineralogie Abhandlungen J. Miner. Geochem.* **2013**, *190*, 1–9. [CrossRef]

4. Sanders, J.V. Colour of precious opal. *Nature* **1964**, *204*, 1151–1153. [CrossRef]

5. Chauvire, B.; Rondeau, B.; Mazzero, F.; Ayalew, D. The precious opal deposit at Wegel Tena, Ethiopia: Formation via successive pedogenesis events. *Can. Miner.* **2017**, *55*, 701–723. [CrossRef]

6. Simoni, M.; Caucia, F.; Adamo, I.; Galinetto, P. New occurence of fire opal from Bemia, Madagascar. *Gems Gemol.* **2010**, *46*, 114–121. [CrossRef]

7. Ansori, C. Model mineralisasi pembentukan opal banten. *Jurnal Geol. Indones.* **2010**, *5*, 151–170. [CrossRef]

8. Shigley, J.E.; Laurs, B.M.; Renfro, N.D. Chrysoprase and prase opal from Haneti, Central Tanzania. *Gems Gemol.* **2009**, *45*, 271–279. [CrossRef]

9. Hatipoglu, M.; Kibici, Y.; Yanik, G.; Ozkul, C.; Demirbilek, M.; Yardmici, Y. Nano-structure of the Cristobalite and Tridymite Staking Sequences in the Common Purple Opal from the Gevrekseydi Deposit, Seyitomer-Kutahka, Turkey. *Orient. J. Chem.* **2015**, *31*, 35–49. [CrossRef]

10. Pewkliang, B.; Pring, A.; Brugger, J. The formation of precious opal: Clues from the opalization of bone. *Can. Miner.* **2008**, *46*, 139–149. [CrossRef]

11. Gauthier, J.-P.; Fritsch, E.; Aguilar-Reyes, B.; Barreau, A.; Lasnier, B. Phase de Laves dans la première opale CR disperse. *Mineralogie* **2004**, *336*, 187–196.

12. Sanders, J.V. Close-packed structures of spheres of two different sizes I. Observations of natural opal. *Philos. Mag. A* **1980**, *42*, 705–720. [CrossRef]

13. Sodo, A.; Municchia, A.C.; Barucca, S.; Bellatreccia, F.; Ventura, G.D.; Butini, F.; Ricci, M.A. Raman, FT-IR and XRD investigation of natural opals. *J. Raman Spectrosc.* **2016**, *47*, 1444–1451. [CrossRef]

14. Ghisoli, C.; Caucia, F.; Marinoni, L. XRPD patterns of opals: A brief review and new results from recent studies. *Powder Diffr.* **2010**, *25*, 274–282. [CrossRef]

15. Liesegang, M.; Milke, R. Australian sedimentary opal-A and its associated minerals: Implications for natural silica sphere formation. *Am. Miner.* **2014**, *99*, 1488–1499. [CrossRef]

16. Jones, J.B.; Segnit, E.R. Genesis of Cristobalite and Tridymite at Low Temperature. *J. Geol. Soc. Aust.* **1972**, *18*, 419–422. [CrossRef]

17. Elzea, E.L.; Odom, I.E.; Miles, W.J. Distinguishing well ordered opal-CT and opal-C from high temperature cristoblite by x-ray diffraction. *Anal. Chim. Acta* **1994**, *286*, 106–116. [CrossRef]

18. Gaillou, E.; Delaunay, A.; Rondeau, B.; Bouhnik-le-Coz, M.; Fritsch, E.; Corren, G.; Monnier, C. The geochemistry of gem opals as evidence of their origin. *Ore Geol. Rev.* **2008**, *34*, 113–126. [CrossRef]

19. Ostrooumov, N.; Fritsch, E.; Lasnier, B.; Lefrant, S. Spectres Raman des opales: Aspect diagnostique et aide a la classification. *Eur. J. Miner.* **1999**, *11*, 899–908. [CrossRef]

20. Smallwood, A.G.; Thomas, P.S.; Ray, A.S. Characterisation of sedimentary opals by Fourier transfrom Raman spectroscopy. *Spectrochim. Acta Part A* **1997**, *53*, 2341–2345. [CrossRef]

21. Kingma, K.J.; Hemley, R.J. Raman spectroscopic study of microcrystalline silica. *Am. Miner.* **1994**, *79*, 269–273.

22. Paris, M.; Fritsch, E.; Aguilar-Reyes, B. ^{1}H, ^{29}Si and ^{27}Al NMR study of the destabilization process of a paracrystalline opal from Mexico. *J. Non-Cryst. Solids* **2007**, *353*, 1650–1656. [CrossRef]

23. Brown, L.D.; Ray, A.S.; Thomas, P.S. ^{29}Si and ^{27}Al NMR study of amorphous and paracrystalline opals from Australia. *J. Non-Cryst. Solids* **2003**, *332*, 242–248. [CrossRef]

24. de Jong, B.W.H.S.; van Hoek, J.; Veeeman, W.S.; Manson, D.V. X-ray diffraction and ^{29}Si magic-angle-spinning NMR of opals: Incoherent long- and short-range order in opal-CT. *Am. Miner.* **1987**, *72*, 1195–1203.

25. Graetsch, H.; Gies, H.; Topalovic, I. NMR, XRD and IR study on microcrstalline opal. *Phys. Chem. Miner.* **1994**, *21*, 166–175. [CrossRef]

26. Graetsch, H.; Mosset, A.; Gies, H. XRD and ^{29}Si MAS-NMR study of some non-crystalline silica minerals. *J. Non-Cryst. Solids* **1990**, *119*, 173–190. [CrossRef]

27. Smith, J.V.; Blackwell, C.S. Nuclear magnetic resonance of silica polymorphs. *Nature* **1983**, *303*, 223–225. [CrossRef]

28. Day, R.; Jones, B. Variations in water content in opal-A and opal-CT from geyser discharge aprons. *J. Sediment. Res.* **2008**, *78*, 301–315. [CrossRef]

29. Bobon, M.; Christy, A.A.; Kluvanec, D.; Illasova, L. State of water molecules and silanol groups in opal minerals: A near infrared spectrscopic study of opals from Slovakia. *Phys. Chem. Miner.* **2011**, *38*, 809–818. [CrossRef]

30. Chauviré, B.; Rondeau, B.; Mangold, N. Near infrared signature of opal and chalcedony as a proxy for their structure and formation conditions. *Eur. J. Miner.* **2017**, *29*, 409–421. [CrossRef]

31. Eckert, J.; Gourdon, O.; Jacob, D.E.; Meral, C.; Monteiro, P.J.M.; Vogel, S.C.; Wirth, R.; Wenk, H.-R. Ordering of water in opals with different microstructures. *Eur. J. Miner.* **2015**, *27*, 203–213. [CrossRef]

32. Rondeau, B.; Fritz, E.; Mazzero, F.; Gauthier, J.-P.; Cencki-Tok, B.; Bekele, E.; Gaillou, E. Play-of-color opal from Wegel Tena, Wollo Province, Ethiopia. *Gems Gemol.* **2010**, *46*, 90–105. [CrossRef]

33. McOrist, G.D.; Smallwood, A. Trace elements in precious and common opals using neutron activation analysis. *J. Radioanal. Nucl. Chem.* **1997**, *223*, 9–15. [CrossRef]

34. Brown, L.D.; Ray, A.S.; Thomas, P.S. Elemental Analysis of Australian amorphous banded opals by laser-ablation ICP-MS. *Neues Jahrbuch für Minerologie Monatshefte* **2004**, *2004*, 411–424. [CrossRef]

35. Dutkiewicz, A.; Landgrebe, T.C.W.; Rey, P.F. Origin of silica and fingerprinting of Australian sedimentary opals. *Gondwana Res.* **2015**, *27*, 786–795. [CrossRef]

36. Rondeau, B.; Cenki-Tok, B.; Fritsch, E.; Mazzero, F.; Gauthier, J.-P.; Bodeur, Y.; Bekele, E.; Gaillou, E.; Ayalew, D. Geochemical and petrological characterizarion of gem opals from Wegel Tena, Wolo, Ethiopia: Opal formation in an Oligocene soil. *Geochem. Explan. Environ. Anal.* **2012**, *12*, 93–104. [CrossRef]

37. Ivey, J. Grape agate from West Sulawesi, Indonesia. *Miner. Rec.* **2018**, *49*, 827–836.

38. Jones, B.; Renaut, R.W. Microstructural changes accompanying the opal-A to opal-CT transition: New evidence from the siliceous sinters of Geysir, Haukadalur, Iceland. *Sedimentology* **2007**, *54*, 921–948. [CrossRef]

39. Martin, E.; Gaillou, E. Insight on gem opal formation in volcanic ash deposits from a supereruption: A case study through oxygen and hydrogen isotopic composition of opals from Lake Tecopa, California, U.S.A. *Am. Miner.* **2018**, *103*, 803–811. [CrossRef]

40. Fritsch, E.; Gaillou, E.; Rondeau, B.; Barreau, A.; Albertini, D.; Ostroumov, M. The nanostructure of fire opal. *J. Non-Cryst. Solids* **2006**, *352*, 3957–3960. [CrossRef]

41. Williams, L.A.; Crerar, D.A. Silica diagenesis, II. General mechanisms. *J. Sediment. Petrol.* **1985**, *55*, 312–321.

42. Lynne, B.Y.; Campbell, K.A.; James, B.; Browne, P.R.L.; Moore, J. Siliceous sinter diagenesis: Order among the randomness. In Proceedings of the 28th NZ Geothermal Workshop, Auckland, New Zealand, 15–17 November 2006.

43. Rice, S.B.; Freund, H.; Huang, W.-L.; Clouse, J.A.; Isaacs, C.M. Application of Fourier transform infrared spectrscopy to silica diagenis: The opal-A to opal-CT transformation. *J. Sediment. Res.* **1995**, *A65*, 639–647.

44. Smith, D.K. Opal, cristobalite and tridymite: Noncrystallinity versus crystallinity, nomenclature of the silica minerals and bibliography. *Powder Diffr.* **1998**, *13*, 2–19. [CrossRef]

45. Anthony, J.W.; Bideaux, R.A.; Bladh, K.W.; Nichols, M.C. *Handbook of Mineralogy, vol 2 Silica, Silicates*; Mineral Data Publishing: Tucson, AZ, USA, 1995.

46. Pecharsky, V.L.; Zavalij, P.Y. *Fundamentals of Powder Diffraction and Structural Characterization of Materials*; Springer Verlag: Secaucus, NJ, USA, 2005.

47. Kihara, K.; Hirose, T.; Shinoda, K. Raman spectra, normal modes and disorder in monoclinic tridymite and its higher temperature orthorhombic modification. *J. Miner. Petrol. Sci.* **2005**, *100*, 91–103. [CrossRef]

48. Schmidt, P.; Bellot-Gurlet, L.; Sciau, P. Moganite detection in silica rocks using Raman and inrared spectroscopy. *Eur. J. Miner.* **2013**, *25*, 797–805. [CrossRef]

49. Ostrooumov, M. A Raman, infrared and XRD analysis of the instability in volcanic opals from Mexico. *Spectrochim. Acta Part A* **2007**, *68*, 1070–1076. [CrossRef]

50. Etchepare, J.; Merian, M.; Kaplan, P. Vibrational normal modes of SiO_2, II Cristobalite and tridymite. *J. Chem. Phys.* **1978**, *68*, 1531–1537. [CrossRef]

51. Bates, J.B. Raman Spectra of alpha and beta cristobalite. *J. Chem. Phys.* **1972**, *57*, 4042–4047. [CrossRef]

52. Ilieva, A.; Mihailova, B.; Tsintov, Z.; Petrov, O. Structural state of microcrystalline opals: A Raman spectroscopic study. *Am. Miner.* **2007**, *92*, 1325–1333. [CrossRef]

53. Schmidt, P.; Bellot-Gurlet, L.; Slodczyk, A.; Froehlich, F. A hitherto unrecognised band ub the Raman spectra of silica rocks: Influence of hydroxylated Si-O bonds (silanole) on the Raman moganite band in chalcedony and flint (SiO_2). *Phys. Chem. Miner.* **2012**, *39*, 455–464. [CrossRef]

54. Lippincott, E.R.; Valkenbur, A.v.; Weir, C.E.; Bunting, E.N. Infrared studies of polymorphs of silicon dioxide and germaniun dioxide. *J. Res. Natl. Bur. Stand.* **1958**, *61*, 61–70. [CrossRef]

55. Elzea, J.M.; Rice, S.B. TEM and X-Ray diffraction evidence for cristobalite and tridymite stacking sequences in opal. *Clays Clay Miner.* **1996**, *44*, 492–500. [CrossRef]

56. Nagase, T.; Akizura, M. Texture and structure of opal-CT and opal-C in volcanic rocks. *Can. J. Miner.* **2007**, *35*, 947–958.

57. Wilson, M.J. The structure of opal-CT revisited. *J. Non-Cryst. Solids* **2014**, *405*, 68–75. [CrossRef]

58. Esenli, F.; Sans, B.E. XRD studies of opals (4A peak) in bentonites from Turkey: Implications for the origin of bentonites. *Neues Jahrbuch für Minerologie Abhandlungen* **2013**, *191*, 45–63. [CrossRef]

59. Smallwood, A.G.; Thomas, P.S.; Ray, A.S. Comparative Analysis of Sedimentary and Volcanic Opals from Australia. *J. Aust. Ceram. Soc.* **2008**, *44*, 17–22.

60. Thomas, P.S.; Ray, A. The thermophysical properties of australian opal. In Proceedings of the 9th International Congress for Appplied Mineralogy, Brisbane, Australia, 8–10 September 2008; pp. 557–565.

61. Gaillou, E. An overview of gem opals: From the geology to color and microstructure. In Proceedings of the Thirteenth Annual Sinkankas Symposium—Opal, Carlsbad, CA, USA, 18 April 2015.

62. Kiefert, L.; Karampelas, S. The use of the Raman spectrometer in gemmological laboratories: Review. *Spectrochim. Acta Part A* **2011**, *80*, 119–124. [CrossRef]

63. Guthrie, G.D.; Bish, D.L.; Reynolds, R.C. Modeling the X-ray diffraction pattern of opal-CT. *Am. Miner.* **1995**, *80*, 869–872. [CrossRef]

64. Ivanov, V.G.; Reyes, B.A.; Fritsch, E.; Faulques, E. Vibrational States in Opal Revisited. *J. Phys. Chem. C* **2011**, *115*, 11968–11975. [CrossRef]

65. Eversull, L.G.; Ferrell, R.E. Disordered silica with tridymite-like structure in the Twiggs clay. *Am. Miner.* **2008**, *93*, 565–572. [CrossRef]

66. Altree-Williams, A.; Pring, A.; Ngothai, Y.; Brugger, J. Textural and compositional complexities resulting from coupled dissolution–reprecipitation reactions in geomaterials. *Earth-Sci. Rev.* **2015**, *150*, 628–651. [CrossRef]

67. Xia, F.; Brugger, J.; Ngothai, Y.; O'Neill, B.; Chen, G.; Pring, A. Three-dimensional ordered arrays of zeolite nanocrystals with uniform size and orientation by a pseudomorphic coupled dissolution–reprecipitation replacement route. *Cryst. Growth Des.* **2009**, *9*, 4902–4906. [CrossRef]

6

Femtosecond Laser Ablation-ICP-Mass Spectrometry and CHNS Elemental Analyzer Reveal Trace Element Characteristics of Danburite from Mexico, Tanzania and Vietnam

Le Thi-Thu Huong [1,*], Laura M. Otter [2,3,*], Michael W. Förster [3], Christoph A. Hauzenberger [1], Kurt Krenn [1], Olivier Alard [3,4], Dorothea S. Macholdt [2], Ulrike Weis [2], Brigitte Stoll [2] and Klaus Peter Jochum [2]

[1] NAWI Graz Geocentre, University of Graz, 8010 Graz, Austria; christoph.hauzenberger@uni-graz.at (C.A.H.); kurt.krenn@uni-graz.at (K.K.)

[2] Climate Geochemistry Department, Max Planck Institute for Chemistry, 55128 Mainz, Germany; d.macholdt@mpic.de (D.S.M.); ulrike.weis@mpic.de (U.W.); Brigitte.stoll@mpic.de (B.S.); k.jochum@mpic.de (K.P.J.)

[3] Department of Earth and Planetary Sciences, Macquarie University, Sydney NSW 2109, Australia; michael.forster@hdr.mq.edu.au (M.W.F.); olivier.alard@mq.edu.au (O.A.)

[4] Géosciences Montpellier, UMR 5243, CNRS & Université Montpellier, 34095 Montpellier, France

[*] Correspondence: thi.le@uni-graz.at (L.T.-T.H.); laura.otter@hdr.mq.edu.au (L.M.O.)

Abstract: Danburite is a calcium borosilicate that forms within the transition zones of metacarbonates and pegmatites as a late magmatic accessory mineral. We present here trace element contents obtained by femtosecond laser ablation-inductively coupled plasma (ICP)-mass spectrometry for danburite from Mexico, Tanzania, and Vietnam. The Tanzanian and Vietnamese samples show high concentrations of rare earth elements ($\sum$REEs 1900 $\mu g \cdot g^{-1}$ and 1100 $\mu g \cdot g^{-1}$, respectively), whereas Mexican samples are depleted in REEs (<1.1 $\mu g \cdot g^{-1}$). Other traces include Al, Sr, and Be, with Al and Sr dominating in Mexican samples (325 and 1611 $\mu g \cdot g^{-1}$, respectively). Volatile elements, analyzed using a CHNS elemental analyzer, reach <3000 $\mu g \cdot g^{-1}$. Sr and Al are incorporated following $Ca^{2+} = Sr^{2+}$ and $2 B^{3+} + 3 O^{2-} = Al^{3+} + 3 OH^- + \square$ (vacancy). REEs replace Ca^{2+} with a coupled substitution of B^{3+} by Be^{2+}. Cerium is assumed to be present as Ce^{4+} in Tanzanian samples based on the observed Be/REE molar ratio of 1.5:1 following $2 Ca^{2+} + 3 B^{3+} = Ce^{4+} + REE^{3+} + 3 Be^{2+}$. In Vietnamese samples, Ce is present as Ce^{3+} seen in a Be/REE molar ratio of 1:1, indicating a substitution of $Ca^{2+} + B^{3+} = REE^{3+} + Be^{2+}$. Our results imply that the trace elements of danburite reflect different involvement of metacarbonates and pegmatites among the different locations.

Keywords: danburite; trace elements; REE; femtosecond LA-ICP-MS; CHNS elemental analyzer; pegmatites; skarn

1. Introduction

Danburite crystallizes in the orthorhombic system and has the formula $CaB_2Si_2O_8$. Its structure consists of a tetrahedral framework with boron and silicon orderly distributed in different tetrahedral sites. The framework of corner-sharing Si_2O_7 and B_2O_7 groups are interconnected by Ca atoms [1,2]. According to previous studies [3,4], the structural unit of danburite contains two tetrahedrally coordinated cations (T1: B and T2: Si), one calcium, and five oxygen atoms, among which O1, O2, and O3 are bonded to both B and Si, while O4 and O5 are bridging oxygens of the Si_2O_7 and B_2O_7 groups, respectively.

Danburite is one of the few boron minerals that are valued as gemstones. After its discovery in Danbury, Connecticut, USA, colorless gem-quality danburite has been subsequently found in Japan, Mexico, Russia, Sri Lanka, and Switzerland [5]. Exceptionally rare is yellow danburite, which so far has been reportedly found only in Madagascar, Tanzania, Myanmar, and Vietnam [6,7]. The important geological environments that are known to have produced gem-quality danburite specimens include pegmatites and metacarbonates associated with hydrothermal activity [8–10].

Previous studies presenting danburite compositions were generally limited to its major element geochemistry (e.g., [11,12]) due to the lack of microanalytical reference materials for boron minerals. While Huong et al. [7] overcame this issue by applying femtosecond laser ablation-inductively coupled plasma-mass spectrometry (LA-ICP-MS), which allows virtually matrix-independent calibration, their study was confined to a regional scale. Here, we present state-of-the-art femtosecond LA-ICP-MS determination of major and trace element concentrations in danburite from three distinct worldwide distributed occurrences (Tanzania, Vietnam, and Mexico). The aims of this study are (1) to investigate the geochemical differences of danburites from different locations (Tanzania, Vietnam, and Mexico) and rock types (pegmatites and skarn), (2) to elucidate their potential for further provenance discrimination, and (3) to understand the incorporation of trace elements into the danburite structure.

2. Materials and Methods

2.1. Sample Material

For this study, we selected 6 danburite samples from 3 deposits in Mexico, Tanzania, and Vietnam (2 samples from each deposit), representing different geological environments. The Mexican danburites were collected from the polymetallic skarn deposit (sulfides of Ag, Pb, Cu, and Zn) in the Charcas mining district, San Luis Potosi. The area is characterized by marine siliciclastic and volcaniclastic rocks, with 2 domains (east and west) separated by a regional fault [9]. In the region of San Luis Potosí, numerous volcanic systems and igneous rocks are associated with different mineral deposits. The Charcas deposit has a large Ca–B metasomatic envelope composed of early datolite and later danburite. Other minerals associated with danburite include calcite, apophyllite, stilbite, chalcopyrite, sphalerite, and citrine. The samples appear as colorless, transparent, prismatic euhedral crystals and are up to 6 cm in length.

The Tanzanian danburite originates from the central zone of a pegmatite mostly as yellowish, fine-grained, massive, opaque aggregates, but occasionally also as larger single crystals with color and transparency. The mine is referred to by the locals as "Munaraima" and is situated in Eastern Tanzania, at the edge of the Uluguru Mountains [10] near the village of Kivuma. The region around Kivuma is dominated by a metasedimentary sequence including metapelites, gneisses, and spinel and ruby-bearing marbles, which underwent granulite facies metamorphism during the East African Orogen at ~640 Ma [13,14]. Tonalitic dikes and pegmatites, commonly found in this area, intruded the basement rocks during slow cooling of the whole area. The contact zone of marble and pegmatite is dominated by a mineral assemblage consisting of microcline (variety amazonite), blue quartz, kyanite, and dravite, while the core complex is mainly composed of massive quartz and schörl [10]. For this study, small, anhedral, transparent yellow danburite crystals ranging from 0.5 to 1 cm in size were selected.

The Vietnamese danburite samples (1–1.5 cm) appear as yellow, transparent, broken, and slightly rounded crystals and have been found in a placer deposit (Bai Cat) in the Luc Yen mining area, Yen Bai province, Northern Vietnam [7,15]. The geology of Luc Yen is dominated by metamorphic rocks, mainly granulitic gneisses, mica schists, and marbles, which are associated with the large-scale Ailao Shan–Red River shear zone. Locally, aplitic and pegmatitic dykes occur [16]. Danburite crystals are associated with ruby, sapphire, spinel, topaz, and tourmaline in the Bai Cat placer deposit, which is surrounded by marble units. While the primary formations of ruby, sapphire, and spinel in Luc Yen are associated with metamorphosed limestones, those of tourmaline and topaz originate from pegmatite bodies. Besides tourmaline and topaz, these pegmatites contain orthoclase, smoky quartz, lepidolite, and beryl. Danburite crystals have not yet been discovered in situ, hence their genetic relationship with the Luc Yen pegmatites is not verified. However, fluid inclusion studies of Luc Yen danburites indicate a pegmatitic origin [15].

2.2. Analytical Methods

Chemical data for major elements were obtained by electron microprobe at the Institute of Geosciences, Johannes Gutenberg University Mainz, by laser ablation-inductively coupled plasma-mass spectrometry (LA-ICP-MS) at the Max Planck Institute for Chemistry, Mainz, and by CHNS Elemental Analyzer at Macquarie University.

Electron probe micro-analysis (EPMA) was performed at the University of Mainz with a JEOL JXA 8200 Superprobe instrument equipped with 5 wavelength-dispersive spectrometers, using 15 kV acceleration voltage and 12 nA filament current. Calcium and silicon were analyzed with wollastonite as a standard material.

LA-ICP-MS data for a total of 55 elements were obtained using an NWRFemto femtosecond laser operating at a wavelength of 200 nm in combination with a ThermoFisher Element2 single-collector sector-field ICP mass spectrometer (see Table 1). Pre-ablation cleaning was performed using a spot size of 65 μm, 80 μm/s scan speed, and 50 Hz pulse repetition rate at 100% energy output to remove any superficial surface residue. Thereafter, samples were ablated using line scans of 300 μm length at a spot size of 55 μm and a scan speed of 5 μm/s. These parameters resulted in an energy density of ca. 0.51 J/cm^2 at the sample surface, and the pulse repetition rate was set to 50 Hz. Since there is no matrix-matched calibration material for Ca–B silicates available, we applied a laser device that produces pulses at 150 fs, enabling virtually matrix-independent calibration [17]. The glass microanalytical reference material NIST SRM 610 was used as calibration material in the evaluation process, where ^{43}Ca was used as internal standard. Reduction of data and elimination of obvious outliers were performed following a programmed routine in Microsoft Excel described in Jochum et al. [18].

All samples were additionally analyzed for their H, C, N, and S contents in a vario EL cube elemental analyzer (Elementar, Langenselbold, Germany). For analysis, 50 to 100 mg samples were packed in Sn-foils (no flux added) and were ignited in an oxygen–He gas atmosphere furnace at around 1150 °C. The produced gases were then trapped and released in a set of chromatographic columns for the sequential analysis of N (no trapping), then C, H, and S. Each sample was measured for 9 min, and released gases were sequentially analyzed with a thermal conductivity detector. Sample measurements were repeated 3 times for each sampling location, and all values were calibrated against the reference materials BAM-U110, JP-1, and CRPG BE-N (Table 2). Analytical uncertainties were evaluated from reference material values, which were found to lie within 16% and 25% for C and H of the data tabulated in the GeoReM database [19].

Table 1. Operating conditions of the femtosecond laser ablation-inductively coupled plasma-mass spectrometry (fs-LA-ICP-MS) system.

Operating Conditions of NWRFemto200 Laser System	
Wavelength λ (nm)	200
Fluence (J·cm^{-2})	0.51
Pulse length (fs)	150
Pulse repetition rate (Hz)	50
Laser energy output (%)	100
Spot size (μm)	55
Line length (μm)	300
Scan speed (μm·s^{-1})	5
Warm-up time (s)	28
Dwell time (s)	60
Washout time (s)	30

Table 1. *Cont.*

Operating Conditions of the Element2 Mass Spectrometer	
RF power (W)	1055
Cooling gas (Ar) flow rate (L·min^{-1})	16
Auxiliary gas (Ar) flow rate (L·min^{-1})	1.19
Additional gas (He) flow rate (L·min^{-1})	0.7
Sample gas (Ar) flow rate (L·min^{-1})	0.7
Sample time (s)	0.002
Samples per peak	100
Mass window (%)	10
Time per pass (s)	2
Scan mode (Escan/Bscan)	both
Mass resolution	300

Table 2. Reference materials for CHNS analyzer.

BE-N Altered Basalts (SARM)	H TCD	C TCD	N TCD	S TCD	S IR
n	14	20	17	21	8
Average (µg·g^{-1})	2771 ± 534	2301 ± 147	197 ± 42	301 ± 37	298 ± 23
RSD %	19	6	21	12	8
BAM-U110					
n	13	18	18	17	–
Average (µg·g^{-1})	12,258 ± 1758	72,340 ± 2640	4237 ± 165	9114 ± 1082	–
RSD %	14	4	4	12	–
JP-1 Peridotite massif (JGS)					
n	4	12	14	14	14
Average (µg·g^{-1})	3195 ± 170	763 ± 82	91 ± 23	27 ± 14	26 ± 7
RSD %	5	11	26	51	27

n denotes the number of measurement performed; average refers to arithmetic means of the *n* values measured; RSD % is relative standard deviation expressed in %. "TCD" refers to the thermal conductivity detector and "IR" to the infrared detector devices.

3. Results

The chemical composition of the danburite samples from Mexico, Tanzania, and Vietnam are presented in Table 3. The major element mass fractions of B, Ca, and Si are close to the stoichiometric composition, i.e., 28.32 wt % B_2O_3 (calculated from 87,890 ppm B obtained by ICP-MS), 22.81 wt % CaO, and 48.88 wt % SiO_2, respectively.

The trace elements Li, Sc, Ga, Se, Rb, Zr, Nb, Ag, Cd, Sn, Cs, Hf, Ta, W, Ir, Pt, Au, Tl, Bi, and U have concentrations below the detection limit in all samples (see detection limits in footnote of Table 3). A C1-chondrite–normalized plot of rare earth element (REE) mass fractions (normalizing data from [20]) displays a strong enrichment of light rare earth elements (LREEs: La, Ce, Pr, Nd, Sm, Eu) compared to heavy rare earth elements (HREEs: Gd, Tb, Dy, Ho, Er, Tm, Yb, Lu) in danburite from Tanzania and Vietnam, while the samples from Mexico are mostly below detectability (Figure 1). Europium shows a negative anomaly of the same order for samples from Tanzania and Vietnam, while no other strong anomaly is observed (e.g., Ce). Total lanthanide content of Tanzanian samples is up to 1900 µg·g^{-1}, hence the mass fractions of LREE exceed those of HREE by a 500-fold enrichment. The Vietnamese samples are different, with a total REE content of around 1000 µg·g^{-1}, hence the mass fractions of LREE exceed those of HREE by a 200-fold enrichment. The Mexican danburites appear to be REE-poor, with total REE contents below 1 µg·g^{-1} for La and Ce, while the remaining REEs from Nd to Lu are below the detection limits. Possible quadrivalent trace elements such as Ti, Hf, and Zr were also below the detection limits. Thorium and Pb show low mass fractions of 0.1 and 11 µg·g^{-1} in the Mexican danburites, and 0.6 and 8 µg·g^{-1} in the Vietnamese and Tanzanian danburites, respectively.

The trivalent element Al shows highly varying concentrations among the different deposits and is anti-correlated with $\sum$REE, thus also with Be (see Figure 2A,B), while no correlation was observed between Al and Sr (Figure 2C). The highest mass fractions of Al are found in the Mexican samples ($325 \ \mu g \cdot g^{-1}$) and the lowest in the Tanzanian samples ($84 \ \mu g \cdot g^{-1}$) (Table 3). Strontium is highest ($1611 \ \mu g \cdot g^{-1}$) in the Mexican samples and lowest ($66 \ \mu g \cdot g^{-1}$) in the Vietnamese samples (Figure 2D). Manganese yields up to $18 \ \mu g \cdot g^{-1}$ in the Vietnamese danburites, while the Mexican and Tanzanian samples have Mn contents below the detection limits. In general, transition metals (e.g., Fe and Ni) have extremely low concentrations in all danburite samples. The three danburite origins can also be separated from each other using the mass fractions of Y (Figure 2E). Beryllium is found to be highest in the Tanzanian danburite (up to $178 \ \mu g \cdot g^{-1}$) and lowest in the Mexican samples ($3 \ \mu g \cdot g^{-1}$). In addition, low concentrations of less than 1, 9, and $2 \ \mu g \cdot g^{-1}$ of the elements Ba, Mg, and Cu, respectively, are identified in all samples regardless of their origin. The positive correlation between Be and $\sum$REE is exceptionally strong for all three deposits (Figure 3A), and due to varying mass fractions of all shown elements, this plot enables excellent discrimination among the three deposits. Univalent elements such as Li, Na, and, K fall below the detection limits.

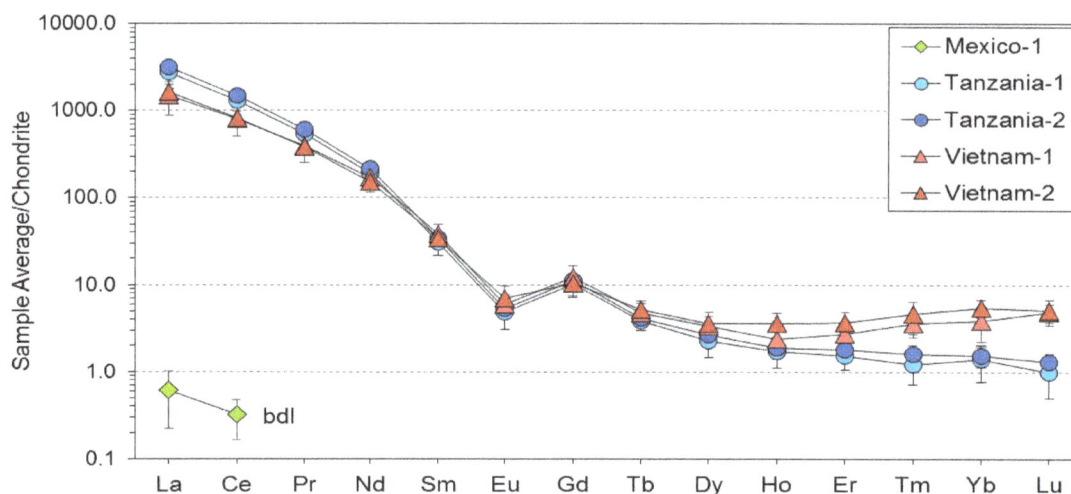

Figure 1. Rare earth element (REE) mass fractions in danburite from Mexico, Tanzania, and Vietnam. All values are presented as averages and normalized to C1-chondrite (data from [20]). Samples from Tanzania and Vietnam have a high abundance of REEs, especially LREEs, while samples from Mexico are largely devoid of these elements (bdl, below detection limit, from Pr to Lu).

The contents of the light volatile elements H, C, N, and S are generally low for both unpowdered and powdered samples (see Table 4). Samples from both Tanzania and Mexico exhibit mass fractions of H, C, N, and S of <10, <200, <100, and $<15 \ \mu g \cdot g^{-1}$, respectively. The samples from Vietnam show significantly higher mass fractions for H, C, N, and S in unpowdered specimens, reaching values of up to 300, 5600, 1000, and $15 \ \mu g \cdot g^{-1}$, respectively. However, most elements are present in lower concentrations in the powdered sample set (H, C, and N of 40, 3000, and $520 \ \mu g \cdot g^{-1}$, respectively). Higher values for powdered samples from Mexico and Tanzania are likely attributed to the significantly increased surface-to-volume ratio facilitating higher adhesion of atmospheric gases. Significantly higher values for H and C in all Vietnamese samples agree well with a previous study by Huong et al. [15], who characterized a high abundance of primary CO_2-bearing fluid inclusions in these samples, which likely accounts for the elevated concentrations of both elements, while fluid inclusion is not present in the Mexican or Tanzanian specimens. High N mass fractions are in the expected range of metamorphosed sediments, which are involved in danburite formation and are known to contain ~200–3000 $\mu g \cdot g^{-1}$ N for a typical metamorphic gradient of 500–700 °C (e.g., [21]). Overall lower values of H, C, and N in the powdered Vietnamese sample set support the observation that these elements are derived from fluid inclusions and were lost during crushing in the agate mortar. Nevertheless, powdered samples were still found to contain up to 40 $\mu g \cdot g^{-1}$ structurally bound H (equivalent to 0.036 wt % H_2O+), which is close to the value of 0.04 wt % H_2O+ determined by IR spectroscopy as published in [4].

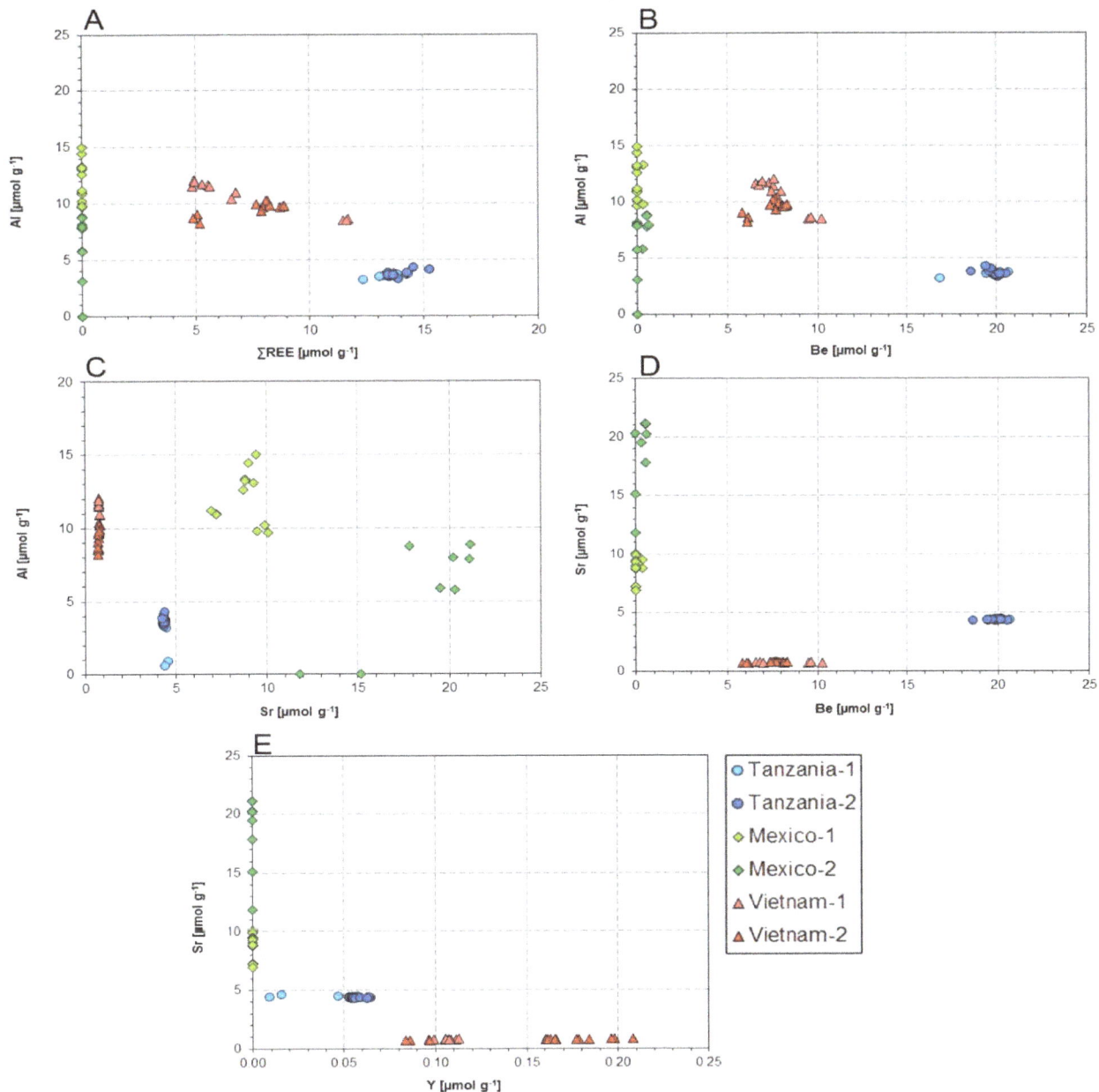

Figure 2. Molar abundance of Al versus (**A**) $\sum$REE, (**B**) Be, and (**C**) Sr, and Sr versus (**D**) Be and (**E**) Y. Aluminum is negatively correlated with (**A**) REEs, (**B**) Be, and (**D**) partly Sr, while Sr correlates negatively with (**D**) Be and (**E**) Y.

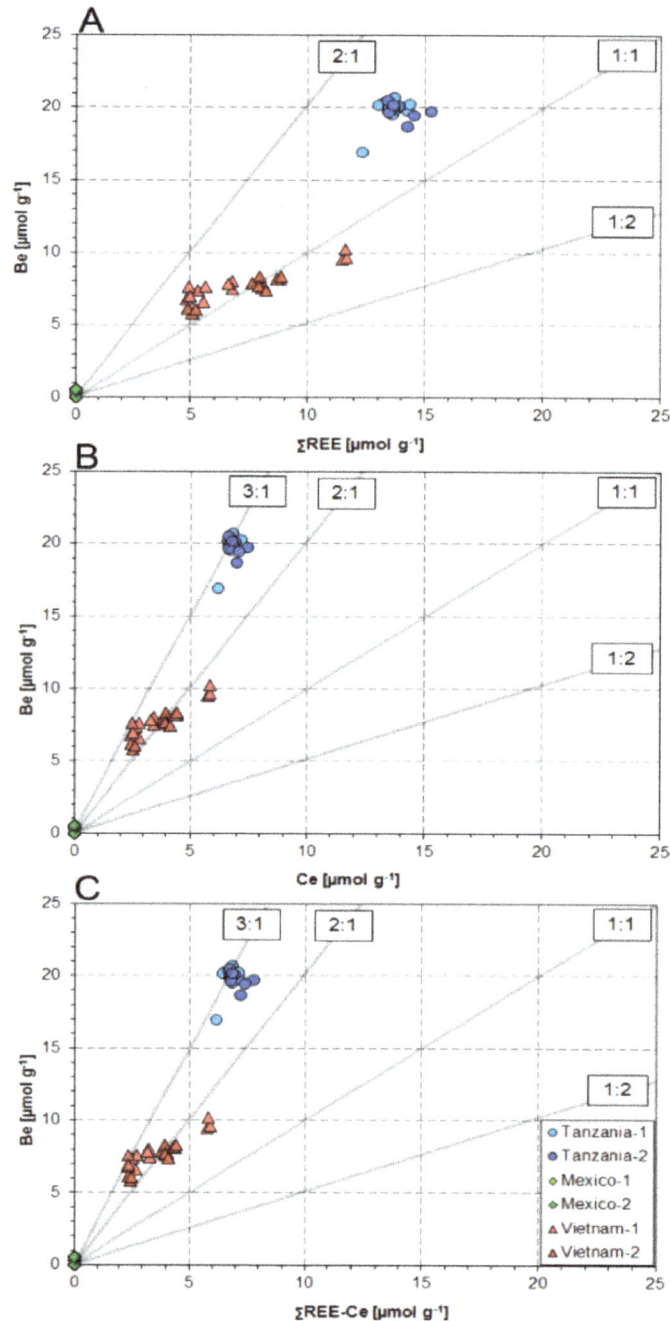

Figure 3. (A) Molar abundance of Be shows a linear correlation with $\sum$REE and illustrates that incorporation of REEs in the danburite structure is accompanied by Be. The Be/$\sum$REE ratio in Tanzanian samples is approximately 1.5:1, while it varies in Vietnamese samples from ca. 1:1 to ca. 1.5:1. Correlations between Be and $\sum$REE are ideally suited to distinguish danburite sampling locations. However, the simple substitution equation $Ca^{2+} + B^{3+} = REE^{3+} + Be^{2+}$, where the Be/$\sum$REE ratio is 1:1, is not sufficient to explain the varying ratios of Be and $\sum$REE. (B) Molar abundance of Be and Ce shows a linear correlation, implying that incorporation of Ce into the danburite structure is accompanied by Be. The Be/Ce ratio in Tanzanian samples is approx. 3:1, while in Vietnamese samples it varies from ca. 2:1 to ca. 3:1. The equation $2\,Ca^{2+} + 3\,B^{3+} = Ce^{4+} + REE^{3+} + 3\,Be^{2+}$, where the Be/Ce ratio is 3:1, explains the Tanzanian and most of the Vietnamese cases very well. Therefore, we argue here that Ce occurs not only as Ce^{3+}, but also as Ce^{4+} in Tanzanian and Vietnamese danburite and the substitution mechanism $2\,Ca^{2+} + 3\,B^{3+} = Ce^{4+} + REE^{3+} + 3\,Be^{2+}$ takes places in these samples. (C) Be/($\sum$REE $-$ Ce) ratios show similar behavior to Be/Ce ratios. As the Be/($\sum$REE $-$ Ce) ratios vary from 2:1 to 3:1, both substitution mechanisms should take place: $Ca^{2+} + B^{3+} = REE^{3+} + Be^{2+}$, $2\,Ca^{2+} + 3\,B^{3+} = Ce^{4+} + REE^{3+} + 3\,Be^{2+}$.

Table 3. Average chemical composition (in μg·g⁻¹ and μmol·g⁻¹) and relative standard deviation (RSD, %) obtained from danburite of three different occurrences. All concentrations were obtained by fs LA-ICP-MS (n = 12 line scans), except CaO and SiO₂, which were evaluated with Electron Probe Microanalyzer (EPMA) (averaged from n = 20 spot analyses per sample).

Element	Isotope Used	L.O.D.	Charcas, San Luis Potosi, Mexico						Morogoro, Tanzania						Luc Yen, Vietnam					
			Mex-1			Mex-2			Tanz-1			Tanz-2			Viet-1			Viet-2		
			ø (μg·g⁻¹)	ø (μmol·g⁻¹)	RSD (%)	ø (μg·g⁻¹)	ø (μmol·g⁻¹)	RSD (%)	ø (μg·g⁻¹)	ø (μmol·g⁻¹)	RSD (%)	ø (μg·g⁻¹)	ø (μmol·g⁻¹)	RSD (%)	ø (μg·g⁻¹)	ø (μmol·g⁻¹)	RSD (%)	ø (μg·g⁻¹)	ø (μmol·g⁻¹)	RSD (%)
CaO*	–	–	22.47	0.04	0.49	22.43	–	0.71	22.63	0.04	0.60	22.48	0.04	0.57	22.32	0.03	0.51	22.40	0.03	0.37
B	11	10	86218	7975	6.67	90471	8368	4.05	86287	7981	1.56	87270	8072	2.69	85698	7927	3.13	89138	8245	3.19
SiO₂*	–	–	48.31	0.8	0.40	48.81	0.8	0.33	48.80	0.8	0.36	48.88	0.8	0.38	48.81	0.8	0.43	48.43	0.8	0.37
La	139	0.01	0.15	0.0011	63.5	0.01	0.0001	124.0	675	4.9	35.3	807	5.8	4.7	368	2.7	42.3	400	2.9	22.0
Ce	140	0.01	0.20	0.0014	48.1	0.05	0.0004	10.00	827	5.9	34.7	964	6.9	3.4	508	3.6	36.9	525	3.8	19.6
Pr	141	0.03	<0.03	–	–	<0.03	–	–	51.4	0.37	35.1	59.5	0.42	4.0	37.4	0.27	35.3	36.5	0.26	17.5
Nd	143	0.08	<0.08	–	–	<0.08	–	–	91.1	0.63	34.5	106	0.74	4.9	80.9	0.56	32.2	72.8	0.51	16.2
Sm	147	0.05	<0.05	–	–	<0.05	–	–	4.65	0.03	28.7	5.68	0.04	9.6	5.94	0.04	27.1	5.23	0.03	18.7
Eu	151	0.05	<0.05	–	–	<0.05	–	–	0.28	0.002	37.7	0.31	0.002	14.1	0.34	0.002	28.5	0.41	0.003	38.2
Gd	157	0.04	<0.04	–	–	<0.04	–	–	2.05	0.013	28.1	2.38	0.015	10.9	2.46	0.016	37.1	2.13	0.014	28.3
Tb	159	0.01	<0.01	–	–	<0.01	–	–	0.15	0.001	22.4	0.16	0.001	15.5	0.18	0.001	29.5	0.20	0.001	26.1
Dy	163	0.05	<0.05	–	–	<0.05	–	–	0.58	0.004	35.7	0.68	0.004	7.80	0.85	0.005	30.5	0.92	0.006	33.7
Ho	165	0.01	<0.01	–	–	<0.01	–	–	0.10	0.001	35.9	0.11	0.001	11.3	0.13	0.001	40.3	0.20	0.001	31.1
Er	167	0.04	<0.04	–	–	<0.04	–	–	0.25	0.001	30.0	0.28	0.002	38.6	0.46	0.003	41.0	0.61	0.004	33.3
Tm	169	0.02	<0.02	–	–	<0.02	–	–	0.03	0.0002	40.5	0.04	0.0002	49.3	0.09	0.0005	31.4	0.12	0.0007	40.1
Yb	173	0.04	<0.04	–	–	<0.04	–	–	0.23	0.0013	44.4	0.26	0.0015	32.5	0.64	0.0037	42.1	0.90	0.0052	22.4
Lu	175	0.02	<0.02	–	–	<0.02	–	–	0.020	0.0001	50.3	0.03	0.0002	43.5	0.12	0.0007	23.7	0.13	0.0007	32.2
Al	27	10	325	12.0	14.9	171	6.4	40.5	84.3	3.1	36.2	101	3.7	7.58	288	10.7	12.8	257	9.5	6.41
As	69	1	46.6	0.6	8.74	7.05	0.1	19.0	<1	0.01	56.5	<1	0.01	67.8	1.13	0.02	103	1.10	0.01	46.0
Ba	135, 137	0.1	0.50	0.004	54.1	0.19	0.001	36.60	0.29	0.002	25.06	0.22	0.002	14.77	0.48	0.003	89.66	0.63	0.005	27.03
Be	9	3	3.06	0.3	12.5	4.94	0.5	26.0	156	17.4	32.8	178	19.8	2.42	71.7	8.0	14.09	67.0	7.4	12.2
Cr	53	5	<5	–	–	<5	–	–	<5	–	–	<5	–	–	6.44	0.12	43.8	<5	–	–
Cu	65	1	1.56	0.02	65.3	<1	–	–	<1	–	–	<1	–	–	1.10	0.02	78.2	2.00	0.03	56.4
Fe	57	20	<20	–	–	<20	–	–	<20	–	–	<20	–	–	<20	–	–	<20	–	–
K	39	7	<7	–	–	<7	–	–	<7	–	–	<7	–	–	<7	–	–	<7	–	–
Mg	25	9	<9	–	–	<9	–	–	<9	–	–	<9	–	–	<9	–	–	<9	–	–
Mn	55	1	<1	–	–	<1	–	–	<1	–	–	<1	–	–	18.4	0.34	8.93	14.57	0.27	14.3
Na	23	50	<50	–	–	<50	–	–	<50	–	–	<50	–	–	<50	–	–	<50	–	–
Ni	62	18	<18	–	–	<18	–	–	<18	–	–	<18	–	–	<18	–	–	<18	–	–
Pb	207, 208, 209	0.1	0.27	0.0013	49.2	0.23	0.0011	40.0	7.46	0.0360	33.8	8.05	0.0389	8.94	11.1	0.0534	5.05	10.6	0.0514	5.60

Table 3. *Cont.*

Element	Isotope Used	L.O.D.	Charcas, San Luis Potosi, Mexico						Morogoro, Tanzania						Luc Yen, Vietnam					
			Mex-1			Mex-2			Tanz-1			Tanz-2			Viet-1			Viet-2		
			Ø (µg·g⁻¹)	Ø (µmol·g⁻¹)	RSD (%)	Ø (µg·g⁻¹)	Ø (µmol·g⁻¹)	RSD (%)	Ø (µg·g⁻¹)	Ø (µmol·g⁻¹)	RSD (%)	Ø (µg·g⁻¹)	Ø (µmol·g⁻¹)	RSD (%)	Ø (µg·g⁻¹)	Ø (µmol·g⁻¹)	RSD (%)	Ø (µg·g⁻¹)	Ø (µmol·g⁻¹)	RSD (%)
Sb	121, 123	1	13.0	0.1	81.9	35.8	0.3	75.6	<1	-	-	<1	-	-	<1	-	-	<1	-	-
Sr	88	0.1	767	8.8	12.09	1611	14.8	40.6	387	4.4	1.44	381	4.4	1.02	66.5	0.8	2.81	66.1	0.8	5.08
Th	232	0.01	<0.01	-	-	<0.01	-	-	0.47	0.0020	41.3	0.66	0.0028	8.45	0.11	0.0005	51.7	0.11	0.0005	44.6
Ti	49	3	<3	-	-	<3	-	-	<3	-	-	<3	-	-	<3	-	-	<3	-	-
V	51	0.5	0.71	0.01	36.54	0.98	0.02	19.1	0.84	0.02	10.8	0.83	0.02	11.3	0.73	0.01	33.8	0.95	0.02	22.5
Y	89	0.1	<0.1	-	-	<0.1	-	-	4.21	0.05	35.1	5.23	0.1	6.19	10.6	0.1	21.2	14.1	0.2	27.9
Zn	67	10	59.6	0.9	34.6	27.5	0.4	53.1	15.8	0.2	32.5	15.5	0.2	41.7	63.2	1.0	39.3	89.9	1.4	49.0

* (wt %), excluded due to concentrations below limits of detection (L.O.D.) in µg·g⁻¹ in all samples: Li (<5), Sc (<1), Ga (<0.5), Se (<10), Rb (<0.5), Zr (<0.1), Nb (<0.01), Ag (<0.1), Cd (<1), Sn (<1), Cs (<0.1), Hf (<0.01), Ta (<0.01), W (<0.01), Ir (<0.01), Pt (<0.01), Au (<0.01), Tl (<0.1), Bi (<0.01), and U (<0.01).

Table 4. Light volatile elements in unpowdered and powdered danburite samples provided as $\mu g \cdot g^{-1}$; calculated H$_2$O+ values are given in wt %.

Sample Location	Unpowdered Samples					Powdered Samples				
	H	eq. H$_2$O+ (wt %)	C	N	S	H	eq. H$_2$O+ (wt %)	C	N	S
Mexico	<10	<0.01	200 ± 5	50 ± 10	13 ± 1	31 ± 10	0.028	130 ± 50	230 ± 130	25 ± 7
Tanzania	<10	<0.01	120 ± 40	100 ± 50	8 ± 1	14 ± 1	0.012	70 ± 40	100 ± 50	22 ± 3
Vietnam	300 ± 50	0.244	5600 ± 20	1000 ± 10	15 ± 3	40 ± 10	0.036	3000 ± 40	520 ± 20	19 ± 11

4. Discussion

4.1. Substitution Mechanisms of REEs, Be, and Sr in Danburite Structure

Referring to the similarity in ionic size and charge, eightfold-coordinated Ca^{2+} (1.12 Å) can be replaced to a certain extent by Sr^{2+} (1.26 Å). This substitution commonly takes place in danburite from all deposits and is mostly observed in the Mexican samples, where the concentration of Sr reaches 1611 $\mu g \cdot g^{-1}$, followed by the Tanzanian and Vietnamese samples, with Sr concentrations up to 387 $\mu g \cdot g^{-1}$ and 66 $\mu g \cdot g^{-1}$, respectively (Table 3).

$$Ca^{2+} = Sr^{2+} \tag{1}$$

Another substitution in danburite is the replacement of Ca^{2+} by REE^{3+} (here we presume that all REEs are trivalent, with the exception of Ce, which can be quadrivalent under strongly oxidizing conditions). It is obvious that danburite from all deposits prefer to incorporate LREE over HREE by a 200- to 500-fold enrichment. The REE^{3+} have decreasing radii with respect to increasing atomic number, i.e., from La (1.16 Å) to Ce (1.15 Å) to Lu (0.98 Å). Moreover, LREE radii are more compatible with the eightfold-coordinated Ca^{2+} lattice site. This explains why LREEs, especially La and Ce, are preferentially incorporated in the danburite lattice. The negative Eu anomaly observed in the Vietnamese and Tanzanian danburite is in accordance with a general depletion of Eu in highly oxidized magma, such as granites and pegmatites [22].

The substitution of Ca^{2+} by a REE^{3+} requires charge compensation and is therefore coupled with the substitution of B^{3+} (0.11 Å) by Be^{2+} (0.27 Å) and/or Si^{4+} (0.26 Å) by Al^{3+} (0.39 Å). These coupled substitutions theoretically allow all sites to be filled and charges to be balanced accordingly:

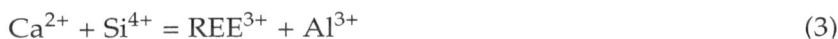

$$Ca^{2+} + B^{3+} = REE^{3+} + Be^{2+} \tag{2}$$

$$Ca^{2+} + Si^{4+} = REE^{3+} + Al^{3+} \tag{3}$$

An omission-style substitution of Ca^{2+} by a trivalent REE^{3+} is also suitable to gain charge balance:

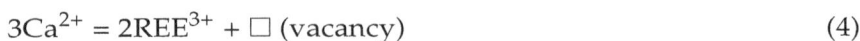

$$3Ca^{2+} = 2REE^{3+} + \square \text{ (vacancy)} \tag{4}$$

However, the positive correlation of molar abundance of REE and Be suggests that Equation (2) is the dominating process of REE incorporation into the danburite lattice (Figure 3A). The remaining substitution Equations (3) and (4) are theoretically possible, but are not supported by the datasets, which show, e.g., a negatively correlated relationship of Al with $\sum$REE (Figure 2A). In the Vietnamese samples, the Be/REE ratio is at an approximate 1:1 trend. In the Tanzanian samples, all the values are approximately equal to 1.5:1 (Figure 3A). A Be/REE ratio that is equal to or higher than 1:1 indicates that REEs are fully coupled with Be; subsequently, the two other forms of substitution, $Ca^{2+} + Si^{4+} = REE^{3+} + Al^{3+}$ (3) and $3Ca^{2+} = 2REE^{3+} + \square$ (4), are subordinate mechanisms. However, Equation (2) implies a Be/REE ratio equal to 1:1, rather than the 1.5:1 ratio measured in the Tanzanian samples. Hence, the excessive molar abundance of Be over REE in the Tanzanian samples needs to be explained by another substitution process with different ratios for Be and REEs or by an REE-independent substitution mechanism. This first hypothesis leads to the suggestion that

Ce occurs not only as Ce^{3+}, but also Ce^{4+} in the samples (with ionic sizes of 1.15 Å and 0.97 Å, respectively). The existence of Ce^{4+} in geological materials has been observed in various studies [23–26]. Hence, we extend substitution mechanism (2) to account for the probable presence of quadrivalent Ce:

$$2Ca^{2+} + 3B^{3+} = Ce^{4+} + REE^{3+} + 3Be^{2+} \tag{5}$$

Equation (5) is an example of a substitution mechanism where the ratio of $\sum REEs$ (all REE^{3+} and Ce^{4+}) to Be is equal to 3:2 (or 1.5:1). According to Equation (5), the ratios Be/Ce and Be/($\sum REEs - Ce$) are both 3:1. Our chemical data (Figure 3B,C) show that the Be/Ce and Be/($\sum REEs - Ce$) ratios in the Tanzanian samples are approximately 3:1 and 1.5:1. Therefore, we assume that mechanisms (2) and (5) take place predominantly in the Vietnamese and Tanzanian samples, respectively, with Ce likely present as Ce^{3+} and Ce^{4+}, respectively. This might suggest that REE uptake into Tanzanian danburite occurs at elevated oxygen fugacity compared to Mexican and Vietnamese danburite.

4.2. Substitution Mechanisms Involving OH and Al in the Danburite Lattice

The presence of Be may also be the result of a REE-independent substitution of B^{3+} by Be^{2+} coupled with the substitution of O^{2-} by OH^-:

$$B^{3+} + O^{2-} = Be^{2+} + OH^- \tag{6}$$

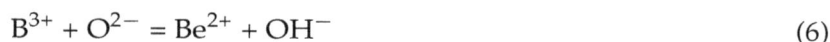

The presence of OH^- species in the danburite lattice was indicated in [4,7] by means of FTIR spectroscopy. The bridging oxygen O5 in the B_2O_7 group is an ideal candidate for partial OH^- replacement, which allows the presence of low amounts of OH^- in danburite. However, a coupled incorporation of Be^{2+} and OH^- was not observed in our study (Figure 4A). Beran [4] proposed a coupled 1:1 substitution of Si^{4+} and O^{2-} by Al^{3+} and OH^- to charge balance OH incorporation. However, a direct substitution (1:1) was not confirmed by either dataset in the present study. Instead, we observed a positive correlation of Al^{3+} with OH^- (Figure 4B) in a 1:3 ratio, which suggests a coupled incorporation of Al^{3+} and OH^-, substituting for B^{3+} and O^{2-}, respectively:

$$2B^{3+} + 3O^{2-} = Al^{3+} + 3OH^- + \square \text{ (vacancy)} \tag{7}$$

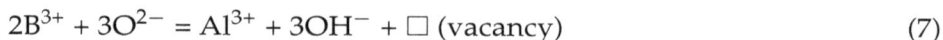

Regarding the possibility of Al incorporation into danburite, it should be noted that the geochemical behavior of B^{3+} and Al^{3+} is very similar; however, they differ in radius size, with 0.11 Å for B^{3+} and 0.39 Å for Al^{3+} when in a tetrahedral coordination environment. Hence, a simple substitution mechanism such as $B^{3+} = Al^{3+}$ would not be possible. Although a substitution of Al^{3+} with B^{3+} has been observed in the system albite $NaAlSi_3O_8$ - $NaBSi_3O_8$ reedmergnerite [27], as well as in a synthetic phlogopite $KMg_3(BSi_3)O_{10}(OH)_2$ [28], this process seems not to be valid for the danburite datasets (Figure 4A,B). Substitution mechanism (7) explains the Mexican samples as well, where the Al concentration is high and both REE and Be concentrations are low.

In general, the four main substitution mechanisms discussed above take place with different priority in the three studied locations. The substitutions of Ca^{2+} by Sr^{2+} and $2B^{3+}$ by Al^{3+} are more common in the Mexican samples, while the substitutions of Ca^{2+} by REEs in the forms $Ca^{2+} + B^{3+} = REE^{3+} + Be^{2+}$ and $2Ca^{2+} + 3B^{3+} = Ce^{4+} + REE^{3+} + 3Be^{2+}$ are more common in the Vietnamese and Tanzanian samples, respectively.

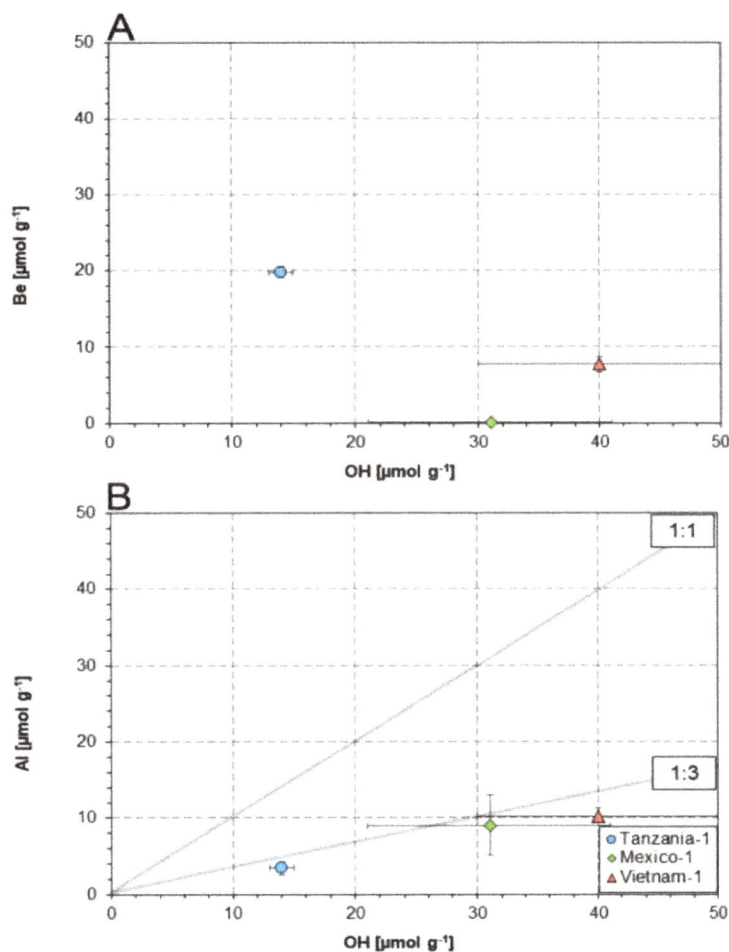

Figure 4. (**A**) Beryllium and OH correlate negatively, which suggests that incorporation into the danburite structure is not coupled. (**B**) However, Al shows a positive correlation with OH at a 1:3 ratio for all sampling locations, making a coupled incorporation of Al with OH likely.

4.3. Constraints on the Geochemical Formation Environment of Danburite

Since danburite samples from Tanzania and Vietnam show a similar strong enrichment in LREE, the reported low values in the Mexican samples must therefore mean either a deficit of REE in the source material or REEs were already sequestered in datolite, which co-occurs with Mexican danburite. However, the depletion of REEs in the Mexican samples is accompanied by exceedingly high amounts of Sr (ca. 1000 $\mu g \cdot g^{-1}$ on average), which is an independent indicator of a different source composition with a higher component of biogenic calcareous sediments in the source material of this location. Biogenic limestone is known to contain high Sr values by being virtually free of REE [29,30]. In comparison, samples from Vietnam and Tanzania exhibit high REE, Y, and Be coupled with low Sr, thereby representing a composition that results from the involvement of highly differentiated late-stage silicic magmas (i.e., pegmatites). This is in agreement with the observed negative Eu anomalies in these samples, which are characteristic for late magmas, where the depletion in Eu is driven by fractional crystallization of plagioclase, which is commonly found to incorporate high amounts of Eu^{2+} [22]. This is in agreement with the nature of the outcrop and mineral assemblage in which the danburite was found. The high compatibility of LREE in the danburite lattice must therefore only be limited by the availability from the source material (i.e., highest in Tanzanian and lowest in Mexican samples). High contents of REE are in accordance with the significantly low amounts of N in the Tanzanian samples, due to the incompatibility of N in highly fractionated magmatic rocks, indicating a strong pegmatite component in the Tanzanian samples. Charge compensation and a

Be/$\sum$REEs ratio of 1.5:1 indicate the presence of Ce^{4+} (Equation (5)), additionally supporting a highly oxidized pegmatitic source for the Tanzanian samples. Nitrogen mass fractions generally increase with decreasing temperature and increasing involvement of metamorphic rocks (e.g., [21]), which suggests that the Vietnamese and Mexican samples were either formed at greater distance from the pegmatite or sourced from a higher proportion of recycled metamorphic rocks. Hence, we conclude that even though danburite always forms in a transition zone of metacarbonates and pegmatites, there are significant geochemical differences, i.e., $\sum$REE, Be, Sr, Al, and OH in danburite, that directly reflect the different proportions and compositions of these source materials. Thus, our results suggest that trace element concentrations are suitable for determining the origins and locations of danburite crystals (i.e., for gem-testing laboratories).

5. Conclusions

In this study, we investigated trace element variations of danburite from three different locations in Mexico, Tanzania, and Vietnam. The most important trace elements in danburite that reflect their provenance include REEs, Sr, Al, Be, and, to a lesser extent, Mn, Zn, and Y. Mexican samples are fairly devoid of REEs, while Tanzanian samples contain up to 1900 $\mu g \cdot g^{-1}$ and Vietnamese samples have intermediate total values of around 1100 $\mu g \cdot g^{-1}$. LREEs are more abundant than HREEs in all danburite samples, showing a 200- to 500-fold relative enrichment. Strontium and Al are more enriched in Mexican danburite than in Tanzanian and Vietnamese danburite, with mass fractions up to 1611 and 325 $\mu g \cdot g^{-1}$, respectively.

Based on fs-LA-ICP-MS and CHNS analysis, we identified four mechanisms of trace element substitution in danburite: Two replacements of Ca^{2+} by Sr^{2+} ($Ca^{2+} = Sr^{2+}$) and of B^{3+} by Al^{3+}, which are coupled with an incorporation of OH^- for O^{2-}: $2B^{3+} + 3O^{2-} = Al^{3+} + 3\,OH^- + \square$, are dominant in the Mexican samples. The incorporation of REE^{3+} for Ca^{2+} coupled with a simultaneous replacement of B^{3+} by Be^{2+} is present in the Vietnamese and Tanzanian samples. Different valance states of Ce are present in the Vietnamese and Tanzanian samples, leading to two different substitutions: $Ca^{2+} + B^{3+} = REE^{3+} + Be^{2+}$ and $2Ca^{2+} + 3B^{3+} = Ce^{4+} + REE^{3+} + 3Be^{2+}$, respectively.

The observed significant differences in trace element abundance not only suggest a high potential for provenance discrimination, but also provide information on the contrasting source compositions of the three deposits. The formation of danburite generally involves both metacarbonates and pegmatites as source materials. Different proportions of these two source components were involved in the formation of danburite at the three locations, and likely explain the observed trace element variations. Low REE and Be coupled with high Sr, Al, N, and OH in the Mexican samples indicate a dominant biogenic metacarbonate component, while the Vietnamese and Tanzanian samples show high REE and Be coupled with low Sr, Al, N, and OH, characteristic of a predominantly pegmatitic source. The 200- to 500-fold enrichment of LREE over HREE in the Tanzanian and Vietnamese samples results from the preferential replacement of Ca ions by similarly sized LREE ions. The negative Eu anomaly, which is characteristic of highly fractionated igneous rocks, is characteristic of Vietnamese and Tanzanian danburite and supports the predominance of the pegmatitic source at these locations.

Author Contributions: L.T.-T.H., L.M.O., K.P.J., C.A.H., and K.K. designed and coordinated the study. K.P.J. and C.A.H. supervised the project. L.T.-T.H., L.M.O., and K.K. collected and prepared the samples. M.W.F. prepared epoxy mounts and performed electron probe microanalyses. L.M.O., D.S.M., B.S., and U.W. collected and evaluated trace element concentrations. M.W.F. and O.A. collected and evaluated light volatile element concentrations and calibrated the CHNS analyzer. L.T.-T.H., L.M.O., and M.W.F. wrote the first drafts of the manuscript. K.P.J., C.A.H., K.K., and O.A. carefully edited the final version. All authors contributed to the final version and gave their approval for submission.

Acknowledgments: Lauren Gorojovsky is acknowledged for preparation and analyses of reference materials for the CHNS elemental analyzer, and we kindly acknowledge the insightful comments by two anonymous reviewers.

References

1. Lindbloom, J.T.; Gibbs, G.V.; Ribbe, P.H. Crystal Structure of Hurlbutite—Comparison with Danburite and Anorthite. *Am. Mineral.* **1974**, *59*, 1267–1271.
2. Best, S.P.; Clark, R.J.H.; Hayward, C.L.; Withnall, R. Polarized single-crystal Raman spectroscopy of danburite, $CaB_2Si_2O_8$. *J. Raman Spectrosc.* **1994**, *25*, 557–563. [CrossRef]
3. Sugiyama, K.; Takéuchi, Y. Unusual thermal expansion of a B–O bond in the structure of danburite $CaB_2Si_2O_8$. *Z. Kristallogr. -Cryst. Mater.* **1985**, *173*, 293–304. [CrossRef]
4. Beran, A. OH groups in nominally anhydrous framework structures: An infrared spectroscopic investigation of danburite and labradorite. *Phys. Chem. Miner.* **1987**, *14*, 441–445. [CrossRef]
5. Hurwit, K.N. Gem Trade Lab Notes: Golden yellow danburite from Sri Lanka. *Gems Gemol.* **1986**, *22*, 47.
6. Chadwick, K.M.; Laurs, B.M. Gem News International: Yellow danburite from 346 Tanzania. *Gems Gemol.* **2008**, *44*, 169–171.
7. Huong, L.T.-T.; Otter, L.M.; Häger, T.; Ullmann, T.; Hofmeister, W.; Weis, U.; Jochum, K.P. A New Find of Danburite in the Luc Yen Mining Area, Vietnam. *Gems Gemol.* **2016**, *52*. [CrossRef]
8. De Vito, C.; Pezzotta, F.; Ferrini, V.; Aurisicchio, C. Nb–Ti–Ta oxides in the gem-mineralized and "hybrid" Anjanabonoina granitic pegmatite, central Madagascar: A record of magmatic and postmagmatic events. *Can. Mineral.* **2006**, *44*, 87–103. [CrossRef]
9. Cook, R.B. Connoisseur's: Danburite, Charcas, San Luis Potosí, Mexico. *Rocks Miner.* **2003**, *78*, 400–403. [CrossRef]
10. Hintze, J. Safari njema—AFRIKANISCHES TAGEBUCH (I): Reise zu den gelben Danburiten von Morogoro, Tansania. *Lapis Die Aktuelle Monatsschrift Fuer Liebhaber Und Sammler Von Mineralien Und* **2010**, *35*, 25.
11. Dyar, M.D.; Wiedenbeck, M.; Robertson, D.; Cross, L.R.; Delaney, J.S.; Ferguson, K.; Francis, C.A.; Grew, E.S.; Guidotti, C.V.; Hervig, R.L. Reference minerals for the microanalysis of light elements. *Geostand. Geoanal. Res.* **2001**, *25*, 441–463. [CrossRef]
12. Ottolini, L.; Cámara, F.; Hawthorne, F.C.; Stirling, J. SIMS matrix effects in the analysis of light elements in silicate minerals: Comparison with SREF and EMPA data. *Am. Mineral.* **2002**, *87*, 1477–1485. [CrossRef]
13. Balmer, W.A.; Hauzenberger, C.A.; Fritz, H.; Sutthirat, C. Marble-hosted ruby deposits of the Morogoro Region, Tanzania. *J. Afr. Earth Sci.* **2017**, *134*, 626–643. [CrossRef]
14. Möller, A.; Mezger, K.; Schenk, V. U–Pb dating of metamorphic minerals: Pan-African metamorphism and prolonged slow cooling of high pressure granulites in Tanzania, East Africa. *Precambrian Res.* **2000**, *104*, 123–146. [CrossRef]
15. Huong, L.T.-T.; Krenn, K.; Hauzenberger, C. Sassolite- and CO_2-H_2O-bearing Fluid Inclusions in Yellow Danburite from Luc Yen, Vietnam. *J. Gemmol.* **2017**, *35*, 544–550. [CrossRef]
16. Garnier, V.; Ohnenstetter, D.; Giuliani, G.; Maluski, H.; Deloule, E.; Trong, T.P.; Van, L.P.; Quang, V.H. Age and significance of ruby-bearing marble from the Red River Shear Zone, northern Vietnam. *Can. Mineral.* **2005**, *43*, 1315–1329. [CrossRef]
17. Jochum, K.P.; Stoll, B.; Weis, U.; Jacob, D.E.; Mertz-Kraus, R.; Andreae, M.O. Non-Matrix-Matched Calibration for the Multi-Element Analysis of Geological and Environmental Samples Using 200 nm Femtosecond LA-ICP-MS: A Comparison with Nanosecond Lasers. *Geostand. Geoanal. Res.* **2014**, *38*, 265–292. [CrossRef]
18. Jochum, K.P.; Stoll, B.; Herwig, K.; Willbold, M. Validation of LA-ICP-MS trace element analysis of geological glasses using a new solid-state 193 nm Nd: YAG laser and matrix-matched calibration. *J. Anal. At. Spectrome.* **2007**, *22*, 112–121. [CrossRef]
19. Jochum, K.P.; Nohl, U.; Herwig, K.; Lammel, E.; Stoll, B.; Hofmann, A.W. GeoReM: A new geochemical database for reference materials and isotopic standards. *Geostand. Geoanal. Res.* **2005**, *29*, 333–338. [CrossRef]
20. Palme, H.; Jones, A. Solar system abundances of the elements. *Treat. Geochem.* **2003**, *1*, 711.
21. Plessen, B.; Harlov, D.E.; Henry, D.; Guidotti, C.V. Ammonium loss and nitrogen isotopic fractionation in biotite as a function of metamorphic grade in metapelites from western Maine, USA. *Geochim. Cosmochim. Acta* **2010**, *74*, 4759–4771. [CrossRef]
22. Fowler, A.D.; Doig, R. The significance of europium anomalies in the REE spectra of granites and pegmatites, Mont Laurier, Quebec. *Geochim. Cosmochim. Acta* **1983**, *47*, 1131–1137. [CrossRef]
23. Braun, J.-J.; Pagel, M.; Muller, J.-P.; Bilong, P.; Michard, A.; Guillet, B. Cerium anomalies in lateritic profiles. *Geochim. Cosmochim. Acta* **1990**, *54*, 781–795. [CrossRef]

24. Takahashi, Y.; Shimizu, H.; Kagi, H.; Yoshida, H.; Usui, A.; Nomura, M. A new method for the determination of CeIII/CeIV ratios in geological materials; application for weathering, sedimentary and diagenetic processes. *Earth Planet. Sci. Lett.* **2000**, *182*, 201–207. [CrossRef]

25. Taunton, A.E.; Welch, S.A.; Banfield, J.F. Microbial controls on phosphate and lanthanide distributions during granite weathering and soil formation. *Chem. Geol.* **2000**, *169*, 371–382. [CrossRef]

26. Bao, Z.; Zhao, Z. Geochemistry of mineralization with exchangeable REY in the weathering crusts of granitic rocks in South China. *Ore Geol. Rev.* **2008**, *33*, 519–535. [CrossRef]

27. Wunder, B.; Stefanski, J.; Wirth, R.; Gottschalk, M. Al-B substitution in the system albite ($NaAlSi_3O_8$)-reedmergnerite ($NaBSi_3O_8$). *Eur. J. Mineral.* **2013**, *25*, 499–508. [CrossRef]

28. Stubican, V.; Roy, R. Boron substitution in synthetic micas and clays. *Am. Mineral.* **1962**, *47*, 1166.

29. Chen, C.; Liu, Y.; Foley, S.F.; Ducea, M.N.; He, D.; Hu, Z.; Chen, W.; Zong, K. Paleo-Asian oceanic slab under the North China craton revealed by carbonatites derived from subducted limestones. *Geology* **2016**, *44*, 1039–1042. [CrossRef]

30. Gozzi, F.; Gaeta, M.; Freda, C.; Mollo, S.; Di Rocco, T.; Marra, F.; Dallai, L.; Pack, A. Primary magmatic calcite reveals origin from crustal carbonate. *Lithos* **2014**, *190*, 191–203. [CrossRef]

7

Fingerprinting Paranesti Rubies through Oxygen Isotopes

Kandy K. Wang [1,*], **Ian T. Graham** [1], **Laure Martin** [2], **Panagiotis Voudouris** [3], **Gaston Giuliani** [4], **Angela Lay** [1], **Stephen J. Harris** [1] and **Anthony Fallick** [5]

[1]   PANGEA Research Centre, School of Biological, Earth and Environmental Sciences, University of NSW, 2052 Sydney, Australia; i.graham@unsw.edu.au (I.T.G.); angela.lay@unsw.edu.au (A.L.); s.j.harris@student.unsw.edu.au (S.J.H.)

[2]   Centre for Microscopy Characterisation and Analysis, The University of Western Australia, 6009 Perth, Australia; laure.martin@uwa.edu.au

[3]   Faculty of Geology and Geoenvironment, National and Kapodistrian University of Athens, 157 84 Athens, Greece; voudouris@geol.uoa.gr

[4]   Université de Lorraine, IRD and CRPG UMR 7358 CNRS-UL, BP 20, 15 rue Notre-Dame-des-Pauvres, 54501 Vandœuvre-lès-Nancy, France; giuliani@crpg.cnrs-nancy.fr

[5]   Isotope Geosciences Unit, S.U.E.R.C., Rankine Avenue, East Kilbride, Glasgow G75 0QF, UK; anthony.fallick@glasgow.ac.uk

*   Correspondence: kandy.wang@student.unsw.edu.au

Abstract: In this study, the oxygen isotope ($\delta^{18}O$) composition of pink to red gem-quality rubies from Paranesti, Greece was investigated using in-situ secondary ionization mass spectrometry (SIMS) and laser-fluorination techniques. Paranesti rubies have a narrow range of $\delta^{18}O$ values between ~0 and +1‰ and represent one of only a few cases worldwide where $\delta^{18}O$ signatures can be used to distinguish them from other localities. SIMS analyses from this study and previous work by the authors suggests that the rubies formed under metamorphic/metasomatic conditions involving deeply penetrating meteoric waters along major crustal structures associated with the Nestos Shear Zone. SIMS analyses also revealed slight variations in $\delta^{18}O$ composition for two outcrops located just ~500 m apart: PAR-1 with a mean value of 1.0‰ $\pm$ 0.42‰ and PAR-5 with a mean value of 0.14‰ $\pm$ 0.24‰. This work adds to the growing use of in-situ methods to determine the origin of gem-quality corundum and re-confirms its usefulness in geographic "fingerprinting".

Keywords: rubies; corundum; in-situ oxygen isotopes; Paranesti Greece; Nestos Shear Zone; Secondary ion mass spectrometry (SIMS)

1. Introduction

1.1. Oxygen Isotopic Studies in Corundums

Oxygen is an abundant element in the Earth's crust, mantle and fluids. Oxygen consists of three naturally-occurring stable isotopes: ^{16}O (99.76%), ^{17}O (0.04%) and ^{18}O (0.2%). $\delta^{18}O$ expressed as Vienna standard mean ocean water (VSMOW) in per mil is the standard for the oxygen isotopic composition which is a measure of the ratio of the stable isotopes oxygen-18 (^{18}O) and oxygen-16 (^{16}O). There are numerous applications of oxygen isotope geochemistry including paleoclimatology, urban forensics, geological genesis and many more [1–3]. Oxygen isotope fractionation is a function of the initial Rayleigh evaporation-precipitation cycle, temperature of the system and degree of water-rock interaction and therefore great care must be taken when interpreting oxygen isotope values [4–7].

Although worldwide corundum oxygen isotope values have been found in a wide range from -27‰ (Khitostrov, Russia) to +23‰ (Mong Hsu, Myanmar), most are in the range of +3‰ to +21‰ [8–10].

This criterion has often been used to determine the geological origin of coloured corundum and especially the gem corundums, rubies and sapphires. $\delta^{18}O$ has been particularly useful in determining the likely primary geological origin of placer corundums where the primary origin is uncertain [11]. As isotopic fractionation is a function of both temperature and geological processes, oxygen isotope data need to be treated with some degree of caution Thus, there are very few examples where oxygen isotopes have been used to "fingerprint" the geographic location [12].

1.2. Geological Setting and Sample Background

The Paranesti rubies are found within the Nestos Shear Zone (NSZ) of the Rhodope Mountain Complex (RMC) in north-eastern Greece (Figure 1). The tectonic and polymetamorphic record of this Northern Aegean region (including the RMC) reflects the Middle Jurassic to Neogene northeast dipping subduction and convergence of the African-Eurasian plates which resulted in the closure of the Tethys Ocean [13,14]. The NSZ is thought to be a one of the syn-metamorphic thrusts in the RMC that are responsible for regional metamorphic inversion, placing higher amphibolite-facies intermediate terranes onto upper-greenschist to lower amphibolite-facies rocks of the lower terrane [15,16].

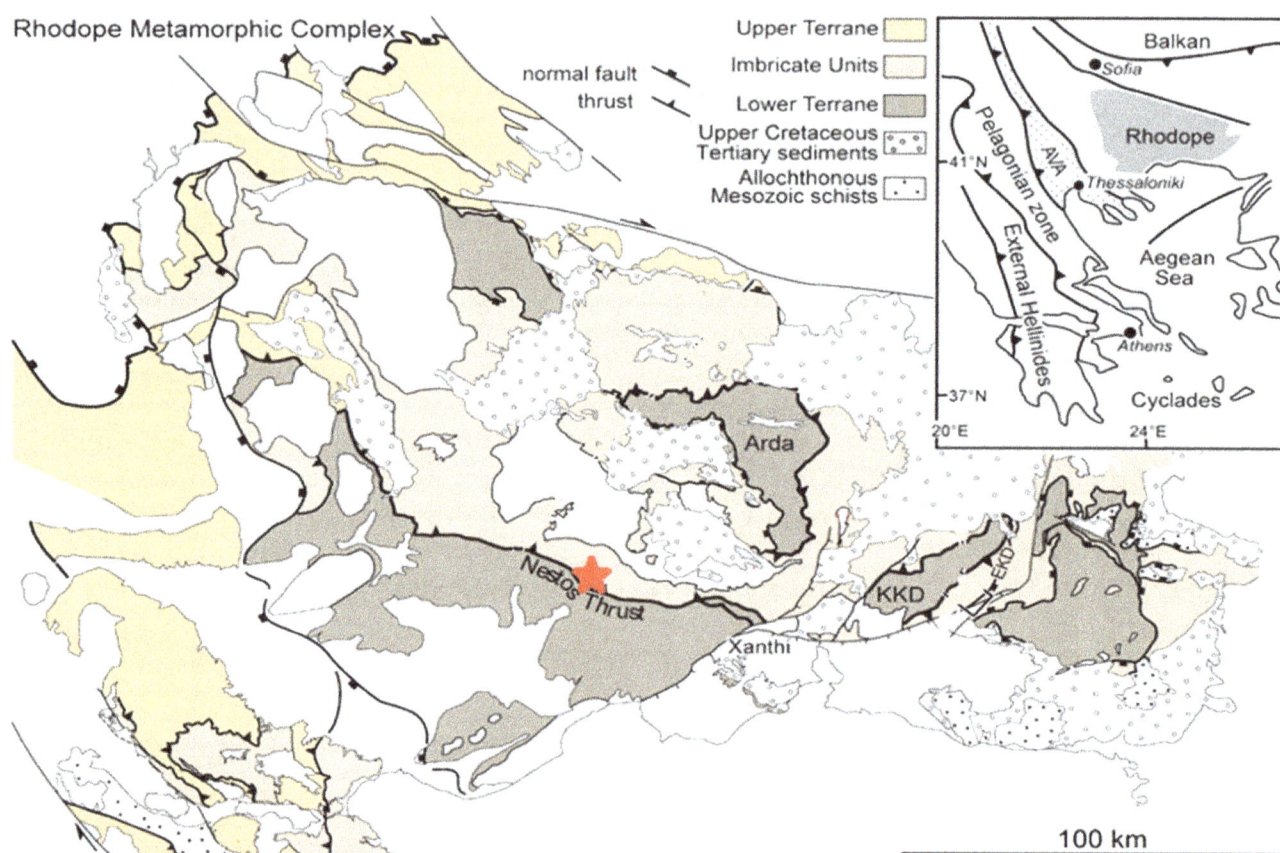

Figure 1. Geological map of the Rhodope Mountain Complex, with Paranesti located within the Nestos Shear Zone (red star) (Adapted from Moulas et al, 2017 [17]).

Based on an earlier systematic study on Paranesti [18], the ruby-bearing occurrences were found to be hosted in pargasite schist with a mafic/ultramafic protolith. The surrounding non-corundum-bearing chlorite schist was found to mainly be comprised of clinochlore. The ruby-bearing occurrence found on the hillside is referred to as PAR-1 (Figure 2a) and the road-side occurrence is termed PAR-5 (Figure 2b). Not all of the pargasite boudins nor the pegmatite intrusion found within the vicinity of the two sites contained corundum (Figure 2c,d).

(a)

(b)

(c)

(d)

Figure 2. Locality diagram of the ruby occurrences. (**a**) PAR-1 location on top of the hill. (**b**) PAR-5 location on the roadside. (**c**) Pargasite schist boudin found approx. 500 m north of PAR-5 without any corundum. (**d**) Pegmatite on top of the ruby-bearing pargasite schist at PAR-1.

A summary of the main findings from this previous study is listed in Table 1. Detailed LA-ICP-MS trace element analyses showed that the rubies are of metamorphic origin (Figure 3a) with minor partial metasomatic influences (Figure 3b). The high R^2 value based on the Fe/Mg vs Ga/Mg elemental discrimination diagram shows both PAR-1 and PAR-5 rubies to contain highly consistent trace element compositions (Figure 3a).

120

120

Handbook of Gemmology

Table 1. Summary of prior Paranesti ruby results (Wang et al. 2017 [18]).

Attributes	PAR-1	PAR-5
Physical Characteristics		
Site	Hillside surface outcrop	Roadside surface outcrop—500 m east of PAR1
Grain-size	10 mm–20 mm	5 mm–10 mm
Colour	Deeper red than PAR-5 (generally)	Medium red
Inclusions	Spinels	None
Microscope view	More fractured, finer-grained	-
Host rock	Pargasite schist	Pargasite schist
EMPA Analyses (wt. %)		
Cr_2O_3	0.11–1.68	0.13–0.29
FeO	0.19–0.73	0.18–0.36
TiO_2	0–0.01	0–0.06
Ga_2O_3	0–0.04	0–0.04
LA-ICP-MS: Trace Element Analysis (ppm)		
Cr	360–2856	4–8627
Fe	1572–2664	1833–3822
V	1–3	2–5
Mg	7–42	8–376
Ti	6–184	10–190
Ga	14-23	13–29
Si	781–2456	837–2123
Ca	769–2119	653–1903

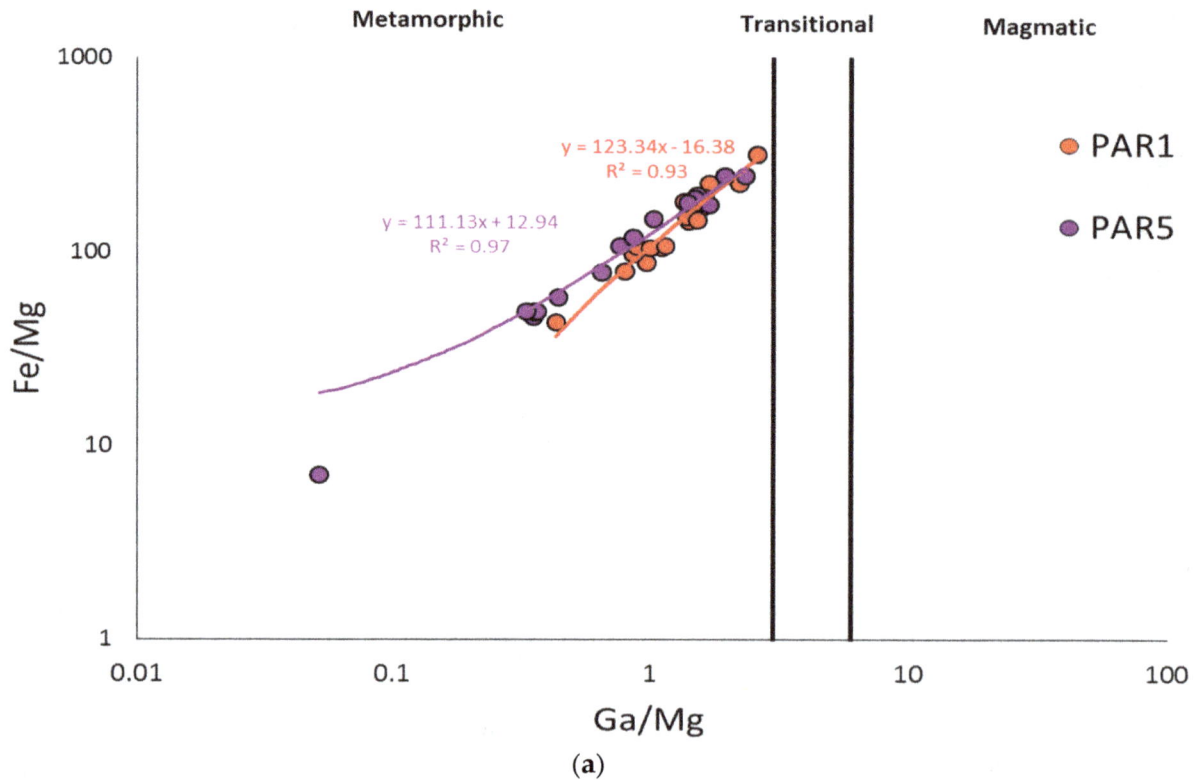

(a)

Figure 3. *Cont.*

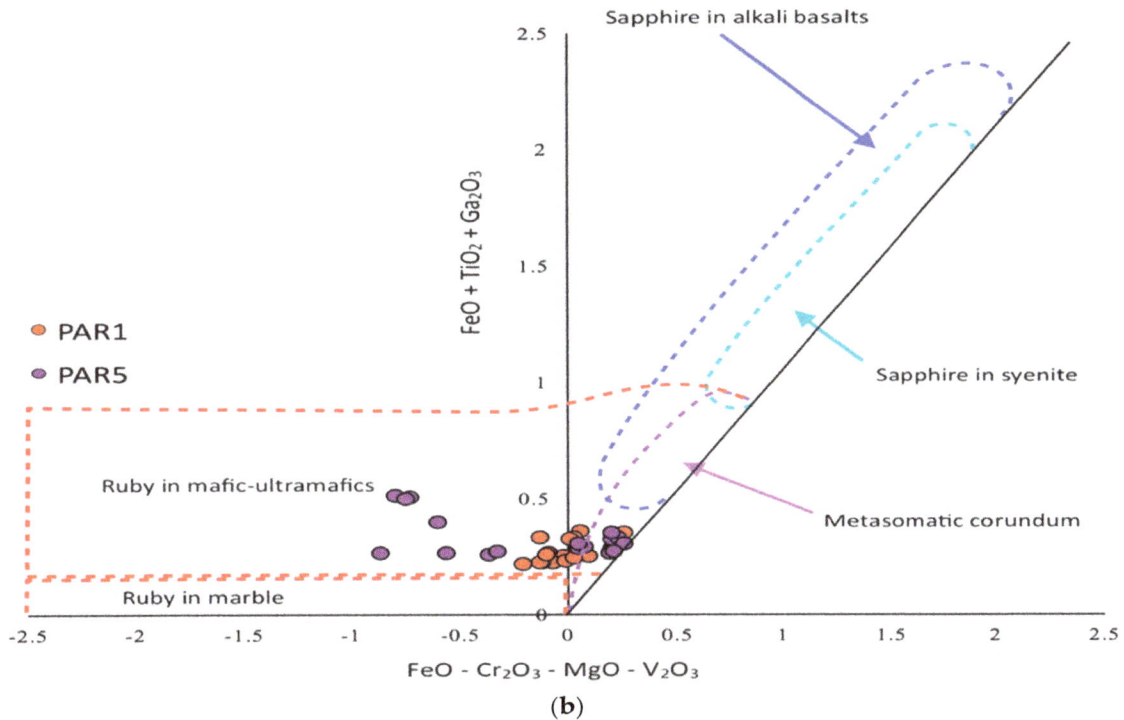

(b)

Figure 3. Trace element discrimination diagrams showing the fields for magmatic, metamorphic and metasomatic corundums, along with the plots for the Paranesti rubies. (a) FeMg vs. GaMg elemental diagram showing the metamorphic vs magmatic fields and SD lines with Paranesti ruby plots. Adapted with permission from Sutherland et al. 2014 [19]. (b) FeO + TiO_2 + Ga_2O_3 vs. FeO-Cr_2O_3-MgO-V_2O_3 elemental diagram showing a metasomatic origin as well as a mafic-ultramafic influence on the Paranesti rubies. Adapted with permission from Giuliani et al. 2014 [20].

2. Materials and Methods

Two different oxygen isotopic analytical methods have been used in this study in order to verify the oxygen isotope values of the Paranesti rubies. The rubies were mechanically extracted from the pargasite host matrix and carefully cleaned prior to being sent for analysis. In many samples, the ruby crystals occur in clusters of platy crystals where the grain sizes generally range between 0.5–1.5 cm (Figure 4a–c). Importantly, in the previous study [18] the ruby grains were generally found to be free of inclusions and thus amenable to in-situ analysis.

(a)   (b)

Figure 4. *Cont.*

Figure 4. Images of ruby samples from Paranesti. (**a**) Dark red ruby samples from PAR-1 in pargasite schist host rock 0.5–1.0 cm; (**b**) Cluster of pale ruby samples from PAR-5 in pargasite schist host rock 0.5–1.5 cm; (**c**) Clean PAR-1 ruby sample free from inclusions used for the 2009 Oxygen Isotope analysis.

2.1. Laser-Fluorination Method (2009)

In 2009, a reconnaissance study was conducted, whereby five individual grains, one each from different corundum localities/geological environments in Greece, were studied for their $\delta^{18}O$ composition. These included two colourless to blue sapphires in desilicified pegmatite from Naxos, one pink marble-hosted ruby from Kimi and one purple marble-hosted ruby from Xanthi. One medium red intensity ruby in pargasite schist from Paranesti (PAR-1) was also included. Oxygen isotope analyses were performed using a modification of the laser-fluorination technique described by Sharp [21] that was similar to that applied by Giuliani et al. in 2005 [11].

The method involves the complete reaction of ~1 mg of ground corundum. This powder is then heated by a CO_2 laser, with ClF_3 as the fluorine reagent. The released oxygen is passed through an in-line Hg-diffusion pump before conversion to CO_2 on platinized graphite. The yield is then measured by a capacitance manometer. The gas-handling vacuum line is connected to the inlet system of a dedicated VG PRISM 3 dual inlet isotope-ratio mass spectrometer. All oxygen isotope ratios are reported in $\delta^{18}O$ (‰) relative to Vienna standard mean ocean water (VSMOW). The secondary standard used for the laser-fluorination method was an internal quartz standard, NBS28 quartz that gave an average $\delta^{18}O$ value of 9.6‰. Oxygen yields differing significantly from the theoretical value of 14.07 μmol. per mg were taken as likely evidence of analytical artefact. Precision and accuracy on the internal quartz standard are ± 0.1‰ (1σ). Duplicate and triplicate analyses of sapphire and ruby suggested that this is appropriate for such materials.

2.2. Secondary Ion Mass Spectrometry (SIMS) Method (2017)

The 2017 analyses were performed exclusively on a range of coloured ruby grains from the two distinct Paranesti locations described in the previous study [13]. Unlike the 2009 analyses, these analyses were made using secondary ionisation mass spectrometry (SIMS) to analyse different areas of selected ruby grains in-situ to measure oxygen isotope ratios with less than one per mil (‰) level precision. Oxygen isotope ratios ($^{18}O/^{16}O$) in ruby were determined using a Cameca IMS 1280 multi-collector ion microprobe within the Centre for Microscopy, Characterisation and Analysis (CMCA), University of Western Australia (UWA). The materials examined in 2017 included 3 ruby grains from 3 different samples (57 analyses) from PAR-1 and 5 ruby grains from 5 different samples (44 analyses) from PAR-5. Each analysis point is shown in Figure 5a,b.

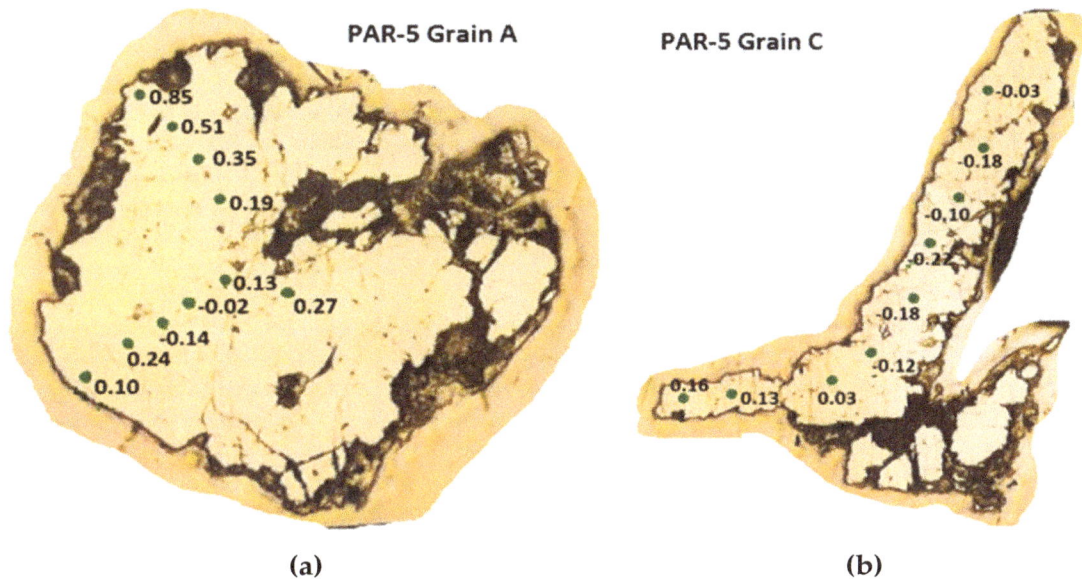

Figure 5. (a) SIMS in-situ analysis spot location individually marked on the ruby grain–PAR-5 Grain A; (b) SIMS in-situ analysis spot location individually marked on the ruby grain–PAR-5 Grain C.

The sample mounts were carefully cleaned with detergent, distilled water and ethanol in an ultrasonic bath and then coated with gold (30 nm in thickness) prior to SIMS O isotope analyses. For oxygen isotopic analyses, secondary ions were sputtered from the sample by bombarding its surface with a Gaussian Cs^+ beam and a total impact energy of 20 keV. The surface of the sample was rastered with a 2.5 nA primary beam over a 15×15 µm area. An electron gun was used to ensure charge compensation during the analyses. Secondary ions were admitted in the double focusing mass spectrometer within a 100 µm entrance slit and focused in the centre of a 4000 µm field aperture ($\times 100$ magnification). They were energy filtered using a 30 eV band pass with a 5 eV gap toward the high-energy side. ^{16}O and ^{18}O were collected simultaneously in multicollection mode in Faraday Cup detectors fitted with 10^{10} Ω and 10^{11} Ω, respectively. Each analysis includes a pre-sputtering over a 20×20 µm area during 30 s and the automatic centring of the secondary ions in the field aperture, contrast aperture and entrance slit and consisted of 20 four-second cycles which give an average internal precision of ~0.16‰ (2 SE).

External reproducibility during the analytical sessions was evaluated by repeating analyses in one single fragment of PAR-1. External reproducibility in this fragment was 0.3 and 0.4 per mil (2SD) during the two analytical sessions. In total, three large fragments of PAR-1 were analysed for their oxygen isotope composition, altogether yielding an average value of 0.9 ± 0.6 per mil (2SD, n = 57, Table 2). Raw oxygen isotope ratios were corrected for instrumental mass fractionation using the $\delta^{18}O$ composition of PAR-1, which oxygen isotope composition was obtained by laser fluorination method (from the 2009 study). Uncertainty on each $\delta^{18}O$ spot has been calculated by propagating the errors on instrumental mass fractionation determination, which include the standard deviation of the mean oxygen isotope ratio measured on the primary standard during the session and internal error on each sample data point. Corrected $\delta^{18}O$ (quoted with respect to Vienna Standard Mean Ocean Water or VSMOW) are presented in Supplementary Material Table S1.

Table 2. Oxygen isotope results from 2017 using the SIMS method.

Grain	$\delta^{18}O$ Min	$\delta^{18}O$ Max	$\delta^{18}O$ Mean	Number of Analyses
PAR-1a	0.64	1.62	1.00 ± 0.42	31
PAR-1b	0.44	1.17	0.67 ± 0.37	13
PAR-1c	0.77	1.68	1.27 ± 0.47	13
PAR-1 Total	0.44	1.68	1.00 ± 0.42	57
PAR-5central	-0.04	0.51	0.27	12
PAR-5a	-0.14	0.85	0.25	10
PAR-5b	-0.31	0.42	0.03	8
PAR-5c	-0.22	0.16	-0.06	9
PAR-5d	0.08	0.27	0.17	5
PAR-5 Total	-0.31	0.85	0.14 ± 0.24	44
Combined PAR-1 and PAR-5	-0.31	1.31	0.60	93

3. Results

3.1. Laser-Fluorination Results

Using the laser-fluorination method, the oxygen isotope ratio for the pargasite schist hosted PAR-1 ruby was found to be $\delta^{18}O$ 1.0‰. This analytical run also included a number of rubies and sapphires from different geological environments. Sapphires from desilicified pegmatites were found to range from 4.8‰ to 5.0‰ and rubies from marble-hosted deposits were found to range from 20‰ to 22‰ (Table 3).

Table 3. Oxygen isotope results from the 2009 reconnaissance using the laser-fluorination method, n = 1.

Sample	Location	Sample Type	Deposit Type	$\delta^{18}O$
NAX2	Naxos, Greece	Colourless sapphire	Desilicified pegmatite	4.80
NAX3	Naxos, Greece	Colourless to blue sapphire	Desilicified pegmatite	5.05
PAR-1	Paranesti, Greece	Red ruby	Pargasite schist	1.00
KIM2	Kimi, Greece	Pink ruby	Marble-hosted	20.50
Xanthi	Xanthi, Greece	Purple-pink ruby	Marble-hosted	22.09

3.2. Secondary Ionisation Mass Spectrometry (SIMS) Results

The oxygen isotope ratios $\delta^{18}O$ (VSMOW) are presented in Table 2. PAR-5 results show values of -0.31‰ to 0.85‰ (0.14 ± 0.24), on average slightly lower compared to PAR-1 results 0.44‰ to 1.68‰ (1.00 ± 0.42) even though the two occurrences are only 500 m apart.

4. Discussion

4.1. Corundum Oxygen Isotopes as An Identifier for Geological Origin

A framework on the interpretation of the geological origin of gem corundums using the $\delta^{18}O$ ratio proposed by Giuliani et al is now widely adopted [11]. Based on this framework, rubies and pink sapphires can be classified into 5 categories based on their $\delta^{18}O$ value range.

1. Mafic gneiss hosted from 2.9‰ to 3.8‰;
2. Mafic-ultramafic rocks (amphibolite, serpentinite) from 3.2‰ to 6.8‰;
3. Desilicated pegmatites from 4.2‰ to 7.5‰;
4. Shear zones cross-cutting ultramafic lenses and pegmatites within sillimanite gneisses 11.9‰–13.1‰;
5. Marble-hosted rubies 16.3‰–23‰. This framework has been further validated by numerous subsequent corundum oxygen isotope studies [22–25].

The reconnaissance 2009 laser fluorination results on the sapphires and rubies from different geological environments very closely fits the oxygen isotope value ranges from the model of Giuliani et al. where over 200 corundum samples were analysed under the same method [1]. That is the marble-hosted value from 20‰ to 22‰ is within the range of 16.3‰ to 23‰ and the sapphires from the desilicified pegmatites with a value from 4.8‰ to 5‰ fits within the framework range from 4.2‰ to 7.5‰. Therefore, the $\delta^{18}O$ results obtained using the laser-fluorination method in 2009 are further validated as accurate measurements.

4.2. PAR-1 vs. PAR-5 Variations

The oxygen isotope values obtained using SIMS indicates that the Paranesti rubies have a narrow defined band of oxygen isotope signatures with a mean on +1‰ (ranging from −0.31‰ to 1.31‰). This is lower than any ratios based on the existing framework for rubies. There are further distinctive constrained values between PAR-1 (+0.65‰ to 1.31‰) and PAR-5 (−0.31‰ to 0.85‰). There is a slight overlap of the individual highest value in PAR-5 to the lowest value in PAR-1. The average for PAR-1 is +1‰ whilst the average for PAR-5 is +0.14‰.

There may be some differences between core-rim oxygen isotope values observed in the PAR-5 SIMS results where the core average (−0.02) is lower than the rim average (0.29). However, this is within the range of uncertainty when the errors are taken into account. This narrow range within individual localities and between the two localities that are 500 m apart is in stark contrast to the findings of Bindeman et al (2010) [7] who found variation within single 10 cm samples of up to 3‰ and variation within single ruby grains of up to 1.5‰. It also does not rule-out variances due to partitioning in individual crystals during growth. Therefore, a detailed cathodoluminescence analysis to determine the homogeneity or heterogeneity of the sample grains is suggested for future studies. As the traditional laser fluorination method consumes the entire grain, such subtle zoning would not be seen using this technique. Thus, the greater spatial resolution of the SIMS technique enables us to analyse discrete isotopic domains (i.e., rims, cores, sectors) within single corundum crystals.

4.3. Global Low to Ultra-Low Oxygen Isotope Corundum Comparison

$\delta^{18}O$ (SMOW) values for gem corundums below 1‰ are very rare and not shown on the original systematic framework by Giuliani et al 2005 [11]. Other than the Paranesti rubies shown above, the only negative value for corundums are from Karelia in north-western Russia and sapphire from a secondary deposit in Madagascar. Table 4 lists the global low to ultra-low oxygen isotope analyses for corundums.

The Madagascar sapphire deposit of Ilakaka with $\delta^{18}O$ of −0.3‰ to 16.5‰ is a consolidated placer formed in a sandstone environment. The geological origin of the different ranges of isotopic values found for the sapphires corresponds to at least five different geological environments [26]. The low $\delta^{18}O$ delta values for some sapphires correspond up to an unknown geological sapphire type.

The PAR-1 result of +1.0‰ is lower than most known primary corundum oxygen isotope values other than the unique ultra-low values of corundums from Karelia [28,30] and one instance of ruby from the Soamiakatra area of Madagascar [26]. The Karelia corundum formed under unique circumstances (see discussion below) and can be easily distinguished from the Paranesti rubies. The Madagascar rubies show much higher average $\delta^{18}O$ values and the minimum value obtained corresponds to the maximum value from Paranesti. Therefore, oxygen isotope analysis is a valuable tool that can be used to fingerprint the Paranesti rubies Figure 6 from other worldwide occurrences.

Table 4. Global comparison of corundums with low oxygen isotope values.

Country	District	$\delta^{18}O‰$ (Min)	$\delta^{18}O‰$ (Max)	Host Rock	Primary vs. Secondary	Corundum Type
Greece [1]	Paranesti-1 *	0.65	1.31	Pargasite schist	Primary	Ruby
Greece [1]	Paranesti-5 *	−0.31	0.85	Pargasite schist	Primary	Ruby
Madagascar [2]	Soamiakatra *	1.25	4.70	Pyroxenitic enclaves in basalt	Primary	Ruby
Madagascar [2]	Ilakaka *	−0.30	16.5	Placer in sandstone	Secondary	Sapphire
Madagascar [2]	Andilamena *	0.50	3.9	Placer in basalt	Secondary	Ruby
Russia [3,4,7]	Khitostrov ^	−26.3	−17.7	plagiogneiss	Primary	Corundum
Russia [4,7]	Khitostrov *	−26	-	Crn-St-Gt-Bi-Prg-Pl rocks with coarse grained Crn	Primary	Corundum
Russia [7]	Khitostrov *	−18.6	-	Ky-Crn-Pl, leucocratic	Primary	Corundum
Russia [3]	Varastskoye ^	−19.2	−11.3	plagiogneiss	Primary	Corundum
Russia [4]	Varastskoye #	−17.3	-	Crn-Cam rock, coarse grained	Primary	Corundum
	Varastskoye #	−19.2	-	Crn and Crn-St-Pl substituting Ky	Primary	Corundum
Russia [4]	Dyadina #	0.49	-	Inclusion of Cam-Crn in giant Gt	Inclusion	Corundum
Russia [4]	Dyadina #	0.10	-	Crn-Cam rock, coarse grained	Primary	Corundum
Russia [5]	Dyadina *	0.4	0.8	Corundum amphibolite	Primary	Corundum
Russia [4]	Kulezhma #	0.31	-	Cam-Crn rock	Primary	Corundum
Russia [4]	Pulonga #	0.67	-	Crn-Gt-Ged rock	Primary	Corundum
Russia [4]	Perusel'ka *	0.26	3.45	Crn-Cam rock, coarse grained	Primary	Corundum
Russia [5]	Perusel'ka #	0.6	-	Corundum-kyanite amphibolite	Primary	Corundum
Russia [5]	Perusel'ka #	1.5	-	Corundum amphibolite	Primary	Corundum
Russia [5]	Notozero *	−1.7	−1.5	Ged-Gt rocks with Crn and St	Primary	Corundum
Russia [4]	Mironova Guba ^	(2.34)	-	Cam-Crn rock	Primary	Corundum
Thailand [6]	Bo Rai *	1.30	4.20	Placer in basalt	Secondary	Ruby

* Individual grain analysis ^ Whole-rock analysis # Only one analysis result, no range.   Bi—biotite, Cam—Ca-amphibole, Crn—corundum, Ged—gedrite amphibole, Gt—garnet, Ky—kyanite, Pl—plagioclase, Prg—pargasitic amphibole, St—staurolite. [1] Wang et al. (2017) [18]; [2] Giuliani et al. (2007) [26]; [3] Vysotskiy et al. (2015) [27]; [4] Bindeman and Serebryakov (2011) [28]; [5] Vysotskiy et al. (2014) [12]; [6] Yui et al. (2006) [29]; [7] Bindeman et al. (2010) [7].

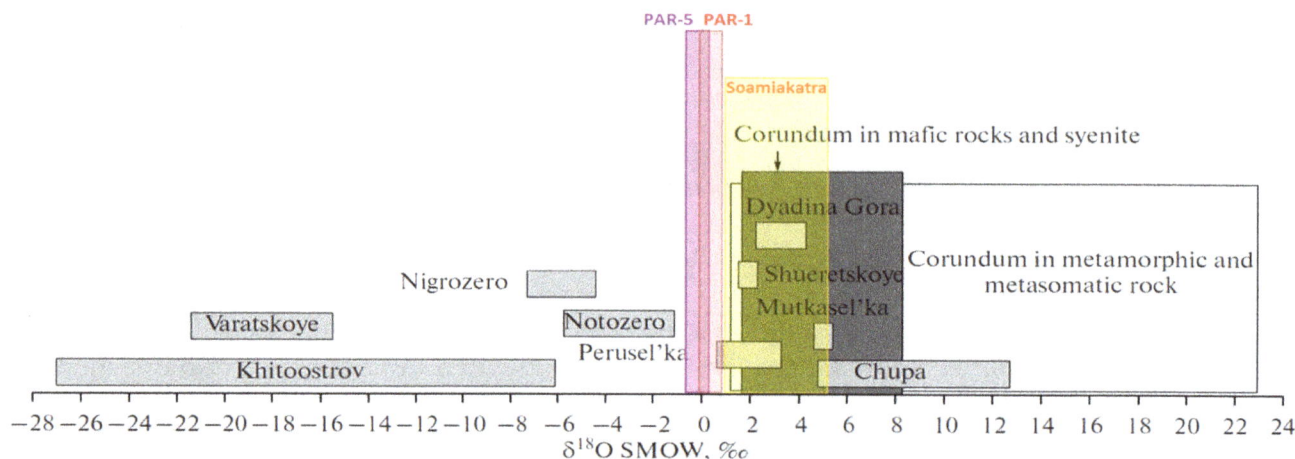

Figure 6. Oxygen isotopic comparison of rubies from Paranesti occurrences compared with low $\delta^{18}O$ corundums from Karelia in northwestern Russia and Soamiakatra in Madagascar.

4.4. Possible Causes for Low Oxygen Isotope Corundum Formation

There are several current hypotheses on how corundums can form with low $\delta^{18}O$ isotope ratios. These range from hydrothermal alteration of deeply penetrating surface meteoric waters to isotope separation by thermal diffusion during endogenous fluid flow [31,32].

4.4.1. Kinetic Isotope Fractionation

Kinetic isotope fractionation occurs when rapid thermal decomposition of hydrous phases results in isotope disproportionation into a high-$\delta^{18}O$ residue and a low-$\delta^{18}O$ fluid [33,34]. However, the high-δ^{18}O residue material should also be found within proximity of the studied samples for this hypothesis to apply. The water-rock interaction is kinetically restricted in supracrustal rocks and isotope fractionation factors are large at low temperatures, favouring higher-δ^{18}O solids [35]. In contrast, isotopic exchange is more rapid within a hydrothermal system. As the Paranesti rubies formed under amphibolite-facies conditions above 600 °C [18], significant kinetic isotope fractionation is highly unlikely and therefore rules out this hypothesis.

4.4.2. Thermal Diffusion

For thermal diffusion, the oxygen in a temperature gradient is redistributed with low δ^{18}O at the hotter end and high δ^{18}O at the colder end of melt or hydrous solution [31]. Akimova (2015) [32] has proposed a model of cascading thermo-diffusion within shear zones to explain the Karelian ultra-low δ^{18}O corundums. This scenario would require several individual thermal cells to align in the correct position simultaneously with a similar convection rate and timing. Given that only two locations have shown ruby-bearing pargasite schist with other very similar pargasite boudins nearby being ruby absent, this model appears to be less likely than the hydrothermal scenario.

4.4.3. Other Ultra-Low δ^{18}O Protoliths

Ultra-low δ^{18}O protoliths could potentially provide the low δ^{18}O during corundum syn-metamorphic formation [35]. However, a source for the ultra-low δ^{18}O protolith would be needed under this scenario such as a low δ^{18}O mantle reservoir or previously surface-exposed and then rapidly buried metamorphic rocks. Neither were observed at Paranesti. As oxygen isotope analyses were not performed on the whole-rock and associated mineral phases for the Paranesti occurrences, this hypothesis cannot be ruled out.

4.4.4. Hydrothermal Alteration Model

This hypothesis involves the conservation of the initial isotopic ratios of the protolith in the corundum-bearing rocks and then isotopic exchange between these rocks and meteoric waters before metamorphism [10]. Wang et al. [18] demonstrated that the rubies from Paranesti were syn-metamorphic and were largely free of inclusions. However, as shown by Bindeman et al (2010) [7], this does not preclude preservation of the initial ratios within the protolith for the Paranesti occurrence. Therefore, it is unlikely that the low isotopic ratios observed within the Paranesti rubies were due to preservation of the initial ratios within the protolith.

For granulite facies metamorphism, Wilson and Banksi (1983) [36] proposed three processes that could produce a low oxygen isotope value. These are (1) pre-granulite reaction between heated seawater and hot basic intrusives or an initial protolith such as a palaeosol for the sapphirine–spinel–(cordierite) assemblages; (2) syn-granulite depletion in ^{18}O related to dehydration during granulite metamorphism and removal of the resultant products of partial melting with a depletion in ^{18}O by up to 2‰ or 3‰ for the restite; and (3) post-granulite facies metamorphism with recrystallization under the effect of biotite and/or amphibole-metasomatism with depletion in δ^{18}O up to 4‰. Based on the previous study [18], the Paranesti rubies were found to have formed under amphibolite facies conditions and there is no evidence that they ever reached granulite facies within the specified zone. However, there are other locations within the Rhodope Mountain Complex (RMC) where regional metamorphism reached granulite facies conditions, though these are some distance away from Paranesti and no rubies are known from these locations.

The glacial meltwater influence during formation of corundums was proposed to explain the ultra-low δ^{18}O isotopic ratios observed for corundums from Karelia in north-western Russia [10,12,27]. However, there is no evidence suggesting the existence of glaciers in the Mediterranean region based on the reconstruction of the tectonic evolution of the East Mediterranean region since the late Cretaceous [37]. Could the RMC be a higher mountain with glaciers that have melted during ruby genesis? There is no evidence in the literature to suggest such and this would only be a remote possibility. Therefore, it is unlikely that a glacial meltwater source played a role during ruby formation at Paranesti. However, it is likely that meteoric water (but not glacial melt) interaction caused by downward flow of surface waters along deep crustal fractures/structures during the formation of the corundum would contribute in producing low $\delta^{18}O$ values for the Paranesti rubies.

5. Conclusions

The in-situ SIMS oxygen isotope analyses on the Paranesti rubies is the first time that a primary (and exclusively) ruby occurrence was found to have ~+1‰ for its $\delta^{18}O$-isotope composition. Based on the low $\delta^{18}O$ value and the local geology, it is most likely that the Paranesti rubies formed under metamorphic/metasomatic conditions involving deeply penetrating meteoric waters along major crustal structures related to the Nestos Shear Zone. PAR-5 is potentially closer to the source of the hydrothermal influence during ruby formation compared to PAR-1 and thus has a lower $\delta^{18}O$. Importantly, this study shows that in-situ gem corundum oxygen isotope analysis using the SIMS method may be used to determine the likely geographic origin for corundums lacking any provenance details. Importantly, with the SIMS method being only minimally destructive, with the analysis spot (15×15 μm) amenable to repolishing, a wider adoption of this technique has important applications/implications for the international gem and jewellery industry. A future area of research would be to apply this methodology for more gem mineral varieties other than corundum and emeralds. The aim of such future work would be to determine if in-situ SIMS oxygen isotope analysis can be used to both better understand gem formation and to see if it can be used to clearly separate the same gem mineral from different geographic locations.

Author Contributions: K.K.W. wrote the manuscript and interpreted the results of the analyses. I.T.G. collected the samples, provided technical input, funding and supervised the project. L.M. provided technical input and ran the SIMS analyses. P.V. collected the samples and provided geological expertise on the Paranesti region. G.G. and A.F. coordinated the laser fluorination analyses and provided technical input. A.L. and S.J.H. provided technical input.

Acknowledgments: The authors would like to thank Joanne Wilde, formerly of the School of Biological, Earth and Environmental Sciences, UNSW Sydney, for making the polished mounts required for SIMS analysis. The authors acknowledge the facilities and the scientific and technical assistance of the Australian Microscopy & Microanalysis Research Facility at the Centre for Microscopy, Characterisation & Analysis, The University of Western Australia, a facility funded by the University, State and Commonwealth Governments.

References

1. Hoefs, J. *Stable Isotope Geochemistry*; Springer: Berlin/Heidelberg, Germany, 1997; ISBN 9783662033791.

2. Bowen, G.J.; Wassenaar, L.I.; Hobson, K.A. Global application of stable hydrogen and oxygen isotopes to wildlife forensics. *Oecologia* **2005**, *143*, 337–348. [CrossRef] [PubMed]

3. Miller, K.G.; Fairbanks, R.G.; Mountain, G.S. Tertiary oxygen isotope synthesis, sea level history and continental margin erosion. *Paleoceanography* **1987**, *2*, 1–19. [CrossRef]

4. Criss, R.E.; Taylor, H.P., Jr. Meteoric-hydrothermal systems. *Rev. Mineral.* **1986**, *16*, 373–424.

5. Valley, J.W. Stable Isotope Thermometry at High Temperatures. *Rev. Mineral. Geochem.* **2001**, *43*, 365–413. [CrossRef]

6. Bindeman, I. Oxygen Isotopes in Mantle and Crustal Magmas as Revealed by Single Crystal Analysis. *Rev. Mineral. Geochem.* **2008**, *69*, 445–478. [CrossRef]

7. Bindeman, I.N.; Schmitt, A.K.; Evans, D.A.D. Limits of hydrosphere-lithosphere interaction: Origin of the lowest-known $\delta^{18}O$ silicate rock on Earth in the Paleoproterozoic Karelian rift. *Geology* **2010**, *38*, 631–634. [CrossRef]

8. Giuliani, G.; Ohnenstetter, D.; Garnier, V.; Fallick, A.E.; Rakotondrazafy, M.; Schwarz, D. The geology and

genesis of gem corundum deposits. In *Geology of Gem Deposits*; Groat, L.A., Ed.; Mineralogical Association of Canada: Québec, QC, Canada, 2007; Volume 37, pp. 23–78. ISBN 9780921294375.

9.   Yui, T.-F.; Khin, Z.; Limtrakun, P. Oxygen isotope composition of the Denchai sapphire, Thailand: A clue to its enigmatic origin. *Lithos* **2003**, *67*, 153–161. [CrossRef]

10. Krylov, D.P.; Glebovitsky, V.A. Oxygen isotopic composition and nature of fluid during the formation of high-Al corundum-bearing rocks of Mt. Dyadina, northern Karelia. *Doklady Earth Sci.* **2007**, *413*, 210–212. [CrossRef]

11. Giuliani, G.; Fallick, A.E.; Garnier, V.; France-Lanord, C.; Ohnenstetter, D.; Schwarz, D. Oxygen isotope composition as a tracer for the origins of rubies and sapphires. *Geology* **2005**, *33*, 249. [CrossRef]

12. Vysotskiy, S.V.; Ignat'ev, A.V.; Levitskii, V.I.; Nechaev, V.P.; Velivetskaya, T.A.; Yakovenko, V.V. Geochemistry of stable oxygen and hydrogen isotopes in minerals and corundum-bearing rocks in northern Karelia as an indicator of their unusual genesis. *Geochem. Int.* **2014**, *52*, 773–782. [CrossRef]

13. Bonev, N.; Burg, J.-P.; Ivanov, Z. Mesozoic–Tertiary structural evolution of an extensional gneiss dome—The Kesebir–Kardamos dome, eastern Rhodope (Bulgaria–Greece). *Int. J. Earth Sci.* **2006**, *95*, 318–340. [CrossRef]

14. Krenn, K.; Bauer, C.; Proyer, A.; Mposkos, E.; Hoinkes, G. Fluid entrapment and reequilibration during subduction and exhumation: A case study from the high-grade Nestos shear zone, Central Rhodope, Greece. *Lithos* **2008**, *104*, 33–53. [CrossRef]

15. Barr, S.R.; Temperley, S.; Tarney, J. Lateral growth of the continental crust through deep level subduction–accretion: A re-evaluation of central Greek Rhodope. *Lithos* **1999**, *46*, 69–94. [CrossRef]

16. Nagel, T.J.; Schmidt, S.; Janák, M.; Froitzheim, N.; Jahn-Awe, S.; Georgiev, N. The exposed base of a collapsing wedge: The Nestos Shear Zone (Rhodope Metamorphic Province, Greece): The Nestos Shear Zone. *Tectonics* **2011**, *30*. [CrossRef]

17. Moulas, E.; Schenker, F.L.; Burg, J.-P.; Kostopoulos, D. Metamorphic conditions and structural evolution of the Kesebir-Kardamos dome: Rhodope metamorphic complex (Greece-Bulgaria). *Int. J. Earth Sci.* **2017**, *106*, 2667–2685. [CrossRef]

18. Wang, K.K.; Graham, I.T.; Lay, A.; Harris, S.J.; Cohen, D.R.; Voudouris, P.; Belousova, E.; Giuliani, G.; Fallick, A.E.; Greig, A. The Origin of a New Pargasite-Schist Hosted Ruby Deposit from Paranesti, Northern Greece. *Can. Mineral.* **2017**, *55*, 535–560. [CrossRef]

19. Sutherland, F.; Zaw, K.; Meffre, S.; Yui, T.-F.; Thu, K. Advances in Trace Element "Fingerprinting" of Gem Corundum, Ruby and Sapphire, Mogok Area, Myanmar. *Minerals* **2014**, *5*, 61–79. [CrossRef]

20. Giuliani, G.; Ohnenstetter, D.; Fallick, A.E. The geology and genesis of gem corundum deposits. In *Geology of Gem Deposits. Series: Mineralogical Association of Canada Short Course Series, 44*; Groat, L.A., Ed.; Mineralogical Association of Canada: Québec, QC, Canada, 2014; pp. 35–112. ISBN 9780921294542.

21. Sharp, Z.D. In situ laser microprobe techniques for stable isotope analysis. *Chem. Geol.* **1992**, *101*, 3–19. [CrossRef]

22. Garnier, V.; Giuliani, G.; Ohnenstetter, D.; Fallick, A.E.; Dubessy, J.; Banks, D.; Vinh, H.Q.; Lhomme, T.; Maluski, H.; Pêcher, A.; et al. Marble-hosted ruby deposits from Central and Southeast Asia: Towards a new genetic model. *Ore Geol. Rev.* **2008**, *34*, 169–191. [CrossRef]

23. Zaw, K.; Sutherland, L.; Yui, T.-F.; Meffre, S.; Thu, K. Vanadium-rich ruby and sapphire within Mogok Gemfield, Myanmar: Implications for gem color and genesis. *Miner. Deposita* **2015**, *50*, 25–39. [CrossRef]

24. Graham, I.; Sutherland, L.; Zaw, K.; Nechaev, V.; Khanchuk, A. Advances in our understanding of the gem corundum deposits of the West Pacific continental margins intraplate basaltic fields. *Ore Geol. Rev.* **2008**, *34*, 200–215. [CrossRef]

25. Sutherland, F.L.; Duroc-Danner, J.M.; Meffre, S. Age and origin of gem corundum and zircon megacrysts from the Mercaderes–Rio Mayo area, South-west Colombia, South America. *Ore Geol. Rev.* **2008**, *34*, 155–168. [CrossRef]

26. Giuliani, G.; Fallick, A.; Rakotondrazafy, M.; Ohnenstetter, D.; Andriamamonjy, A.; Ralantoarison, T.; Rakotosamizanany, S.; Razanatseheno, M.; Offant, Y.; Garnier, V.; et al. Oxygen isotope systematics of gem corundum deposits in Madagascar: Relevance for their geological origin. *Miner. Deposita* **2007**, *42*, 251–270. [CrossRef]

27. Vysotskiy, S.V.; Nechaev, V.P.; Kissin, A.Y.; Yakovenko, V.V.; Ignat'ev, A.V.; Velivetskaya, T.A.; Sutherland, F.L.; Agoshkov, A.I. Oxygen isotopic composition as an indicator of ruby and sapphire origin: A review of Russian occurrences. *Ore Geol. Rev.* **2015** *68*, 164–170. [CrossRef]

28. Bindeman, I.N.; Serebryakov, N.S. Geology, Petrology and O and H isotope geochemistry of remarkably ^{18}O depleted Paleoproterozoic rocks of the Belomorian Belt, Karelia, Russia, attributed to global glaciation 2.4 Ga. *Earth Planet. Sci. Lett.* **2011**, *306*, 163–174. [CrossRef]

29. Yui, T.-F.; Wu, C.-M.; Limtrakun, P.; Sricharn, W.; Boonsoong, A. Oxygen isotope studies on placer sapphire and ruby in the Chanthaburi-Trat alkali basaltic gemfield, Thailand. *Lithos* **2006**, *86*, 197–211. [CrossRef]

30. Krylov, D.P. Anomalous ^{18}O/^{16}O ratios in the corundum-bearing rocks of Khitostrov, northern Karelia. *Doklady Earth Sci.* **2008**, *419*, 453–456. [CrossRef]

31. Bindeman, I.N.; Lundstrom, C.C.; Bopp, C.; Huang, F. Stable isotope fractionation by thermal diffusion through partially molten wet and dry silicate rocks. *Earth Planet. Sci. Lett.* **2013**, *365*, 51–62. [CrossRef]

32. Akimova, E.Y.; Lokhov, K.I. Ultralight Oxygen in Corundum-Bearing Rocks of North Karelia, Russia, as a Result of Isotope Separation by Thermal Diffusion (Soret Effect) in Endogenous Fluid Flow. *J. Mater. Sci. Chem. Eng.* **2015**, *3*, 42–47. [CrossRef]

33. Clayton, R.N.; Mayeda, T.K. Kinetic Isotope Effects in Oxygen in the Laboratory Dehydration of Magnesian Minerals. *J. Phys. Chem.* **2009**, *113*, 2212–2217. [CrossRef]

34. Mendybayev, R.A.; Richter, F.M.; Spicuzza, M.J.; Davis, A.M. Oxygen isotope fractionation during evaporation of Mg- and Si- rich CMAS-liquid in vacuum. *Lunar Planet. Inst. Sci. Conf. Abstr.* **2010**, *41*, 2725.

35. Bindeman, I.N.; Serebryakov, N.S.; Schmitt, A.K.; Vazquez, J.A.; Guan, Y.; Azimov, P.Y.; Astafiev, B.Y.; Palandri, J.; Dobrzhinetskaya, L. Field and microanalytical isotopic investigation of ultradepleted in ^{18}O Paleoproterozoic "Slushball Earth" rocks from Karelia, Russia. *Geosphere* **2014**, *10*, 308–339. [CrossRef]

36. Wilson, A.F.; Baksi, A.K. Widespread ^{18}O depletion in some precambrian granulites of Australia. *Precambrian Res.* **1983**, *23*, 33–56. [CrossRef]

37. Menant, A.; Jolivet, L.; Vrielynck, B. Kinematic reconstructions and magmatic evolution illuminating crustal and mantle dynamics of the eastern Mediterranean region since the late Cretaceous. *Tectonophysics* **2016**, *675*, 103–140. [CrossRef]

8

Metamorphic and Metasomatic Kyanite-Bearing Mineral Assemblages of Thassos Island (Rhodope, Greece)

Alexandre Tarantola [1,*], Panagiotis Voudouris [2], Aurélien Eglinger [1], Christophe Scheffer [1,3], Kimberly Trebus [1], Marie Bitte [1], Benjamin Rondeau [4], Constantinos Mavrogonatos [2], Ian Graham [5], Marius Etienne [1] and Chantal Peiffert [1]

[1] GeoRessources, Faculté des Sciences et Technologies, Université de Lorraine, CNRS, F-54506 Vandœuvre-lès-Nancy, France; aurelien.eglinger@univ-lorraine.fr (A.E.); christophe.scheffer.1@ulaval.ca (C.S.); kimberly.trebus5@etu.univ-lorraine.fr (K.T.); marie.bitte8@etu.univ-lorraine.fr (M.B.); marius.etienne5@gmail.com (M.E.); chantal.peiffert@univ-lorraine.fr (C.P.)

[2] Department of Geology & Geoenvironment, National and Kapodistrian University of Athens, 15784 Athens, Greece; voudouris@geol.uoa.gr (P.V.); kmavrogon@geol.uoa.gr (C.M.)

[3] Département de Géologie et de Génie Géologique, Université Laval, Québec, QC G1V 0A6, Canada

[4] Laboratoire de Planétologie et Géodynamique, Université de Nantes, CNRS UMR 6112, 44322 Nantes, France; benjamin.rondeau@univ-nantes.fr

[5] PANGEA Research Centre, School of Biological, Earth and Environmental Sciences, University of New South Wales, Sydney, NSW 2052 Australia; i.graham@unsw.edu.au

* Correspondence: alexandre.tarantola@univ-lorraine.fr

Abstract: The Trikorfo area (Thassos Island, Rhodope massif, Northern Greece) represents a unique mineralogical locality with Mn-rich minerals including kyanite, andalusite, garnet and epidote. Their vivid colors and large crystal size make them good indicators of gem-quality materials, although crystals found up to now are too fractured to be considered as marketable gems. The dominant lithology is represented by a garnet–kyanite–biotite–hematite–plagioclase ± staurolite ± sillimanite paragneiss. Thermodynamic Perple_X modeling indicates conditions of ca. 630–710 °C and 7.8–10.4 kbars. Post-metamorphic metasomatic silicate and calc-silicate (Mn-rich)-minerals are found within (i) green-red horizons with a mineralogical zonation from diopside, hornblende, epidote and grossular, (ii) mica schists containing spessartine, kyanite, andalusite and piemontite, and (iii) weakly deformed quartz-feldspar coarse-grained veins with kyanite at the interface with the metamorphic gneiss. The transition towards brittle conditions is shown by Alpine-type tension gashes, including spessartine–epidote–clinochlore–hornblende-quartz veins, cross-cutting the metamorphic foliation. Kyanite is of particular interest because it is present in the metamorphic paragenesis and locally in metasomatic assemblages with a large variety of colors (zoned blue to green/yellow-transparent and orange). Element analyses and UV-near infrared spectroscopy analyses indicate that the variation in color is due to a combination of Ti^{4+}–Fe^{2+}, Fe^{3+} and Mn^{3+} substitutions with Al^{3+}. Structural and mineralogical observations point to a two-stage evolution of the Trikorfo area, where post-metamorphic hydrothermal fluid circulation lead locally to metasomatic reactions from ductile to brittle conditions during Miocene exhumation of the high-grade host-rocks. The large variety of mineral compositions and assemblages points to a local control of the mineralogy and fO_2 conditions during metasomatic reactions and interactions between hydrothermal active fluids and surrounding rocks.

Keywords: kyanite; Mn-rich silicates; Rhodope; Thassos; amphibolite facies; metasomatism

1. Introduction

Dispersed metal ions in substitution for Al and Si may lead to a large variety of colors in calc- and alumino-silicate minerals, e.g., [1,2]. For instance, the most common natural color of kyanite is blue but it can also be green, yellow, orange, white, black, grey or colorless as a function of the nature of elemental substitutions [3,4]. Kyanite is composed of usually >98 wt. % Al_2O_3 and SiO_2. The ~2 wt. % left is generally dominated by Fe, Ti, Mg, Mn, Cr and V. The variation in blue colors is attributed to Ti and/or Fe substitution and intervalence charge transfers within the crystal lattice [5], while the orange color of kyanite is generally attributed to the presence of Mn^{3+} [1,6–9]. The incorporation of trace elements is interpreted as the result of (i) oxygen fugacity during crystal growth (Fe), (ii) metamorphic grade, temperature of formation (Ti), or the nature of the protolith (Cr and V) [4]. In the same way, Mn^{3+} may substitute for Al^{3+} in andalusite [10,11] resulting in the dark green Mn-rich variety of andalusite, formerly known as viridine [2,12]. This may also be the case for epidotes e.g., [13], garnets e.g., [14] and other metamorphic and metasomatic Ca-Al silicates. Among the interesting trace elements to produce gem quality calc- and alumino-silicate minerals, Mn seems to play an important role. High-PT manganian silicate and calc-silicate metamorphic assemblages are very rare on Earth and require very specific conditions such as the preservation of high fO_2 during the entire metamorphic cycle, e.g., [15,16]. The presence in the protolith of Mn-oxides capable of buffering fO_2 to high levels during metamorphism as well as low fluid/rock ratio during the whole geodynamic evolution are necessary [17,18]. The best way to produce gem-quality Mn-rich silicates seems thus hydrothermal fluid circulation through an already Mn-rich protolith.

The geology of Greece is marked by a recent alpine belt, which results in outstanding localities for minerals and gemstones owing to high PT metamorphic conditions and subsequent magmatism and hydrothermalism, e.g., [19–23]. Among these localities, the island of Thassos, in the Rhodope Belt, represents a unique mineralogical locality for kyanite and uncommon varieties of manganiferous/manganian and magnesian silicate and calc-silicate minerals, including garnet, andalusite and epidote among others [19,24–31]. These crystals are often large (several centimeters long) and vividly colored, translucent but commonly fractured, which makes them usually inappropriate for faceting gemstones, but more proper for cabochon-shaped material. However, considering the extent of the geological units, their occurrence is promising for gemstones exploration in the area. In this article, we use the term "near gem-quality" as a reference to their aesthetic aspect (large size and vivid color) rather than to their potential as facetable material.

Thassos Island is part of the Southern Rhodope Metamorphic Core Complex (SRCC), with PT conditions recorded by the garnet–kyanite–biotite–hematite–plagioclase ± staurolite ± sillimanite assemblage of the paragneisses of the Trikorfo area (intermediate unit) in the range 600–650 °C, 4–7 kbars [24]. At the Trikorfo area, four main distinct lithologies, often bearing Ca-, Mg- and Mn-rich silicates, are distinguished with (i) the dominant metamorphic lithology composed of metasedimentary rocks, mostly mica schists and paragneisses, with intercalation of carbonate layers especially towards the base and top of the formation [24], (ii) green to red horizons parallel to the regional foliation enriched in silicate and calc-silicate minerals always showing a typical mineralogical zonation from the contact with the paragneisses/mica schists to the center of the layers, (iii) kyanite-andalusite bearing mica schists, and (iv) kyanite–quartz ± feldspar ± andalusite weakly deformed coarse-grained veins. The transition to brittle deformation conditions is shown by quartz–clinochlore–ilmenite–adularia–albite alpine-type tension gashes cross-cutting the metamorphic foliation [25]. Locally, when cross-cutting Mn-rich layers, these brittle veins can host spessartine and Mn-bearing clinochlore, epidote and hornblende.

On the basis of the description made by Voudouris et al. [20], the aim of the present study is (i) to obtain a complete inventory of the metamorphic and metasomatic minerals and their assemblages found at the Trikorfo site, (ii) to evaluate the cation substitution responsible for the color variations, (iii) to reevaluate the PT conditions attained by the gneiss host-rock on the basis of a comprehensive petrological phase equilibria using the Perple_X program, and (iv) to discuss the conditions of formation, metamorphic versus metasomatic, of these exceptional parageneses and assemblages of

(Mn-rich)-silicate and calc-silicate minerals. Kyanite, as a ubiquitous silicate found in both metamorphic and localized metasomatic assemblages, is the referent mineral described in the manuscript.

2. Geological Context

2.1. The Southern Rhodope Core Complex and Thassos Island

The Aegean domain, which extends from the Rhodope Massif in the north to the island of Crete in the south of the Hellenides, is a broad metamorphic domain whose formation started with the subduction of the Adriatic microplate (the northern part of the Africa Plate) below the Eurasia Plate, in late Cretaceous to Eocene times, e.g., [32–35]. The progressive southward retreat of the downgoing slab since the Oligocene is responsible for the exhumation of metamorphic core complexes (MCC) along low-angle detachment faults [36]. The Southern Rhodope Core Complex (SRCC) [37–41] and the Attic-Cycladic Metamorphic Complex (ACMC) [42–44] are the two main metamorphic domains resulting from Aegean extension during the Oligocene to Miocene. The Aegean domain is now in a back-arc position and the subduction front is currently located to the south of the island of Crete [35]. The SRCC, located in the northern part of Greece, displays an almost triangular shape and extends to south to southern Bulgaria [40]. It is mainly constituted by the Vertiskos Gneiss Complex from the Serbo-Macedonian Massif and to the west by the Chalkidiki Peninsula dominated by gneisses from the Sideroneron Massif (Figure 1). The area is marked by Jurassic and Cretaceous age ultra-high-pressure metamorphic events, including kyanite eclogites, followed by Barrovian type metamorphism, e.g., [37–41].

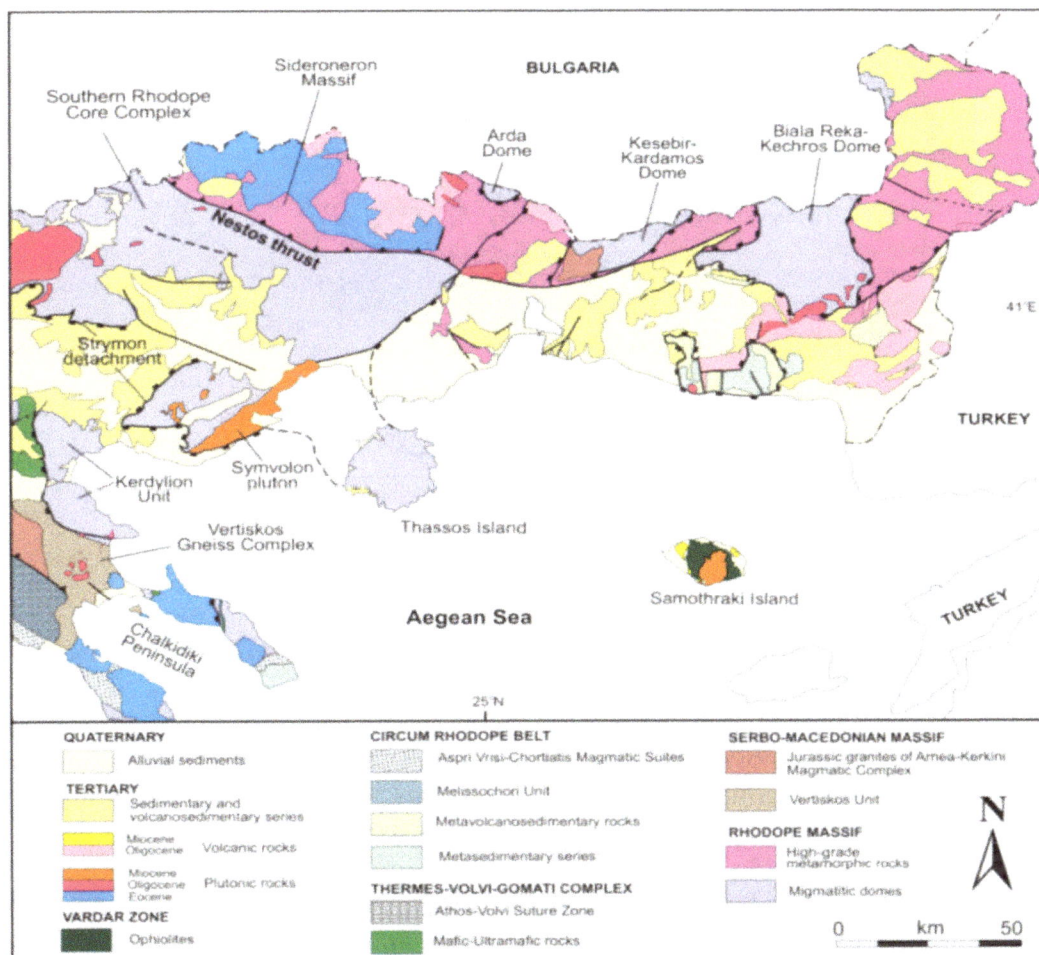

Figure 1. Geological map and regional tectonic framework of northern Greece, modified after Melfos and Voudouris [45].

Thassos Island, with an area of 400 km^2, exposes the south-east end of the SRCC. The geology of the island is dominated by the juxtaposition of shallow-dipping units mainly made of marbles, orthogneisses and paragneisses/mica schists (Figure 2). It is assumed that orthogneisses are the result of metamorphism of Hercynian plutonic units of the Greek Rhodope Massif [36,46] and that the protoliths of marbles and mica schists correspond to the Mesozoic sedimentary cover overlying the Hercynian plutonic basement.

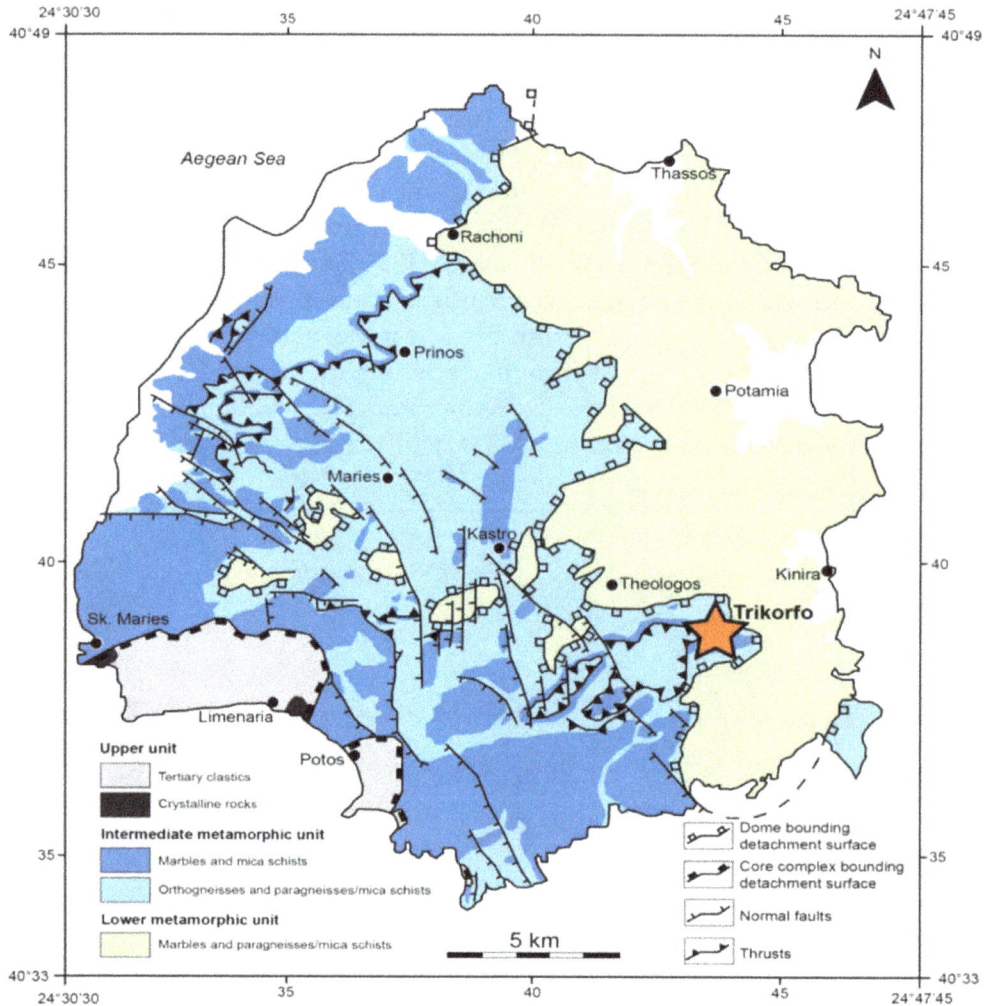

Figure 2. Geological map of Thassos Island with main units, modified from Wawrzenitz and Krohe [36] and Brun and Sokoutis [40]. The location of the Trikorfo area is indicated with an orange star.

Thassos Island is represented by three main metamorphic units with, from base to top, (i) the lower metamorphic unit dominated by mylonitic marbles with intercalations of mica schists and paragneisses forming metamorphic domes to the east of the island, (ii) an intermediate metamorphic unit composed at the base of gneisses overlapped by marbles with paragneiss, mica schist and amphibolite intercalations, (iii) the upper unit is considered as the hanging wall, consisting of non-mylonitic metamorphic rocks dominated by gneisses, migmatites, pegmatites and locally ultramafic rocks and marbles [36,40]. Metamorphic rocks from the lower and intermediate units reached amphibolite facies conditions [24,40,47,48], achieved during crustal thickening [49] at 21–18 Ma and 26–23 Ma (Rb–Sr mica ages from the main metamorphic fabric) for the lower and intermediate units, respectively [36].

Retrograde metamorphism is marked by the development of mylonite marbles, common boudinage along a shallow-dipping foliation with NE–SW stretching lineation with a top to SW sense of shear, and widespread brittle tension gashes [36,40,48,50]. Rb–Sr (on white mica and biotite) and U–Pb (on monazite and xenotime) dating by the above authors revealed a continuous cooling history under ductile to brittle conditions from 700 °C to 300 °C from 26 Ma to 12 Ma. The exhumation *PT* path is interpreted as the result of the progressive uplift of the metamorphic rocks during the formation of the metamorphic core complex [36,51]. The emplacement of the low-angle detachment fault in the Oligocene-Miocene is also associated with the intrusion of syntectonic plutons at the SRCC scale, where the Symvolon pluton crops out 30 km to the NW of Thassos Island [52–54] (Figure 1).

2.2. The Trikorfo Area

The study area at Trikorfo is located along the major SW-dipping low-angle normal detachment separating the lower from the intermediate unit of Thassos Island. Mica schists and paragneisses predominate and tectonically overlay the marbles of Theologos (i.e. lower unit). Muscovite, phlogopite, or two mica quartzo-feldspathic schists (mica schists or paragneisses) with plagioclase (oligoclase) porphyroblasts are the most common lithologies, locally interbedded with carbonate/dolomitic rocks towards the base and top of the formation [24]. In many occurrences, these rocks are rich in minerals from the epidote and garnet groups together with common tourmaline (schorl mainly) and opaque minerals, mostly hematite and ilmenite. Metamorphic Al-silicate minerals (kyanite, fibrolitic sillimanite, andalusite, staurolite) are also documented. These mica schists are thus likely derived from calcareous pelitic sediments [24]. Interestingly, this unit also locally presents uncommon Mn-rich mineral assemblages attributed to post-metamorphic hydrothermal recrystallization [20,24,26]. The garnet–kyanite–biotite–hematite–plagioclase ± staurolite ± sillimanite paragenesis of the mica schists, contemporaneous with the ductile fabric related to extensional deformation, has been specifically interpreted as having attained *PT* conditions of 5.5 ± 1.5 kbars and 600 ± 50 °C, without any evidence of partial melting [24]. Voudouris et al. [20] documented at the top of the formation the presence of an orthogneiss at the contact with the paragneisses/mica schists. Although not exposed, a granitoid is considered to be genetically related to the widespread Miocene gold mineralization in the area [27].

3. Materials and Methods

Representative samples of host-rocks, mineral assemblages and veins were selected for bulk rock analyses and conventional 30 μm polished sections for petrographic observations and electron probe microanalyses (EPMA). Individual kyanite crystals were prepared for bulk rock analyses (powder) and for EPMA and Laser Ablation-Inductively Coupled Plasma Mass Spectroscopy (LA-ICPMS) on 200–300 μm thick sections, and UV-visible-near infrared absorption spectroscopy on 1 mm thick sections. The metamorphic paragenesis of the dominant gneiss of the area was modeled by Perple_X for *PT* conditions estimation. This work represents a compilation of samples collected by different authors during successive field campaigns, resulting in a large variety of sample labelling. All samples come from the Trikorfo area (intermediate unit of Thassos Island) (Figure 2).

3.1. Bulk-Rock Analyses

Host-rock samples CS16_292b and CS16_297 and individual kyanite crystals were powderized so as to obtain a grain-size of about 80 μm. Kyanite crystals were cut as to not contain any apparent alteration minerals, mainly micas, and cleaned with deionized water. Bulk compositions (oxides and trace elements) were obtained by ICP-Optical Emission Spectrometry and ICP-Mass Spectrometry (LiBO$_2$ fusion) at the CRPG-CNRS laboratory (Nancy, France). Sample preparation, analytical conditions and limits of detection are given in Carignan et al. [55].

3.2. Electron Probe MicroAnalyses (EPMA)

The chemical composition of minerals (samples CS16_292b, CS297, KT01-03, KT07 and MB13) was determined using a Cameca SX100 electron microprobe analyzer (EPMA) equipped only with wavelength dispersive spectrometers at the GeoRessources laboratory (Nancy, France) operating with an emission current of 20 nA, an acceleration voltage of 15 kV, and a beam diameter of 1 μm.

The following elements, monochromators, standards, and limits of detection were used: Na (TAP, albite, 515 ppm), Si (TAP, albite, 330 ppm), Mg (TAP, olivine, 265 ppm), Al (TAP, Al_2O_3, 300 ppm), K (LPET, orthoclase, 200 ppm), Ca (PET, andradite, 370 ppm), Ti (LPET, $MnTiO_3$, 250 ppm), Mn (LIF, $MnTiO_3$, 635 ppm), Fe (LIF, Fe_2O_3, 1003 ppm). Three representative samples of blue-green-colorless, yellow and orange kyanite crystals from the same locality as the ones analyzed by bulk geochemistry were selected for EPMA (profiles) analyses and X-ray maps for the elements Al, Si, Fe, Ti and Mn. All analyses were performed with FeO and MnO and converted to Fe_2O_3 and Mn_2O_3 for epidote and andalusite group minerals and Ti-hematite.

Mineral analyses (samples Th01 to Th06) carried-out at Hamburg used a Cameca-SX 100 WDS, accelerating voltage of 20 kV, a beam current of 20 nA and counting time of 20 s. Standards used were andradite (Si, Fe and Ca), Al_2O_3 (Al), MgO (Mg), albite (Na), orthoclase (K), and $MnTiO_3$ (Ti and Mn). Corrections were made using the PAP online program [56].

3.3. Laser Ablation-Inductively Coupled Plasma Mass Spectroscopy (LA-ICPMS)

Kyanite in-situ quantitative LA-ICPMS elemental analyses were performed at GeoRessources laboratory (Nancy, France) using an Agilent 7500c quadrupole ICPMS coupled with a 193 nm GeoLas ArF Excimer laser (MicroLas, Göttingen, Germany). Laser ablation was performed continuously with a speed of 2 μm/s with a constant 5 Hz pulse frequency and a constant fluence of 4.20 J/cm^2 by focussing the beam at the sample surface, from edge to edge of the crystals, along the same profiles analyzed by EPMA. Diameter of the ablation spot was 32 μm. Helium was used as carrier gas to transport the laser-generated particles from the ablation cell to the ICPMS and argon was added as an auxiliary gas via a flow adapter before the ICP torch. Typical flow rates of 0.5 L/min for He and 1 L/min for Ar were used. EPMA analyses of Si and Al served as internal standards. The certified reference materials NIST SRM 610 and 612 (concentrations from Jochum et al. [57]) were used as external standards for calibration of all analyses; they were analyzed twice at the beginning and at the end for each kyanite sample, following a bracketing standardization procedure. ^{27}Al and ^{29}Si were measured with a dwell-time of 10 ms; the following 17 isotopes were measured with a dwell-time of 20 ms for each: ^7Li (limit of detection of 20 ppm), ^{23}Na (70 ppm), ^{24}Mg (1 ppm), ^{39}K (50 ppm), ^{43}Ca (6000 ppm), ^{47}Ti (20 ppm), ^{51}V (1 ppm), ^{53}Cr (15 ppm), ^{55}Mn (15 ppm), ^{57}Fe (150 ppm), ^{63}Cu (10 ppm), ^{66}Zn (10 ppm), ^{88}Sr (0.5 ppm), ^{107}Ag (2 ppm), ^{137}Ba (3 ppm), ^{197}Au (1 ppm), and ^{208}Pb (0.5 ppm). All data were reduced off-line with the limits of detection calculated using the commercial version of Iolite (Version 3.71, https://iolite-software.com) data reduction software [58] running with Igor Pro. The uncertainty is calculated for each element for each analysis as a function of the uncertainty of the reference standards, the absolute element content and the time of integration [58]. The resulting value of uncertainty is then varying for a single element between each analysis and is in the order of 10% for high concentration to 80% when the element concentration is measured close to the limit of detection.

3.4. UV-Visible-Near Infrared Absorption Spectra

UV-visible-near infrared absorption spectra were measured at room temperature using a Perkin-Elmer 1050 dual beam spectrophotometer at the Institut des Matériaux Jean Rouxel (University of Nantes, France) on blue-green (sample THA16), yellow (THA08), and orange (THA13) kyanite crystals. Spectra were acquired at a resolution of 1 nm at a rate of 120 nm/min. As our samples were significantly fractured and hence translucent but not transparent, we prepared them as 1 mm thick slices. At this thickness, they became sufficiently transparent so that the light beam could go through, and they also remained colored enough to characterize spectral features related to color.

3.5. Phase Diagram Calculation

PT conditions of the main paragneiss host-rock (sample CS16_292b) were modeled on the basis of the paragenesis described in Section 4.2. The phase diagram was calculated using Perple_X Version 6.8.5 (http://www.perplex.ethz.ch) [59] and the internally consistent end-member data set of Holland

and Powell [60]. Calculations were undertaken in a set of chemical systems ranging from NCKFMASH to MnNCKFMASHT(O) (MnO–Na$_2$O–CaO–K$_2$O–FeO–MgO–Al$_2$O$_3$–SiO$_2$–H$_2$O–TiO$_2$–O$_2$) in order to constrain the effect of TiO$_2$, Fe$_2$O$_3$ and MnO-bearing minerals.

Phases involved in modeling are biotite (Bt), chlorite (Chl), cordierite (Crd), epidote (Ep), garnet (Gt), ilmenite (Ilm), kyanite (Ky), melt (L), plagioclase (Pl), quartz (Qtz), rutile (Rt), sillimanite (Sil), staurolite (St) and white mica (Ms). The solution models utilized for metamorphic minerals and the bulk rock composition are presented in the Table 1 and Table S1, respectively. We used the Ilm(WPH) solution model which enables the calculation of hematite, geikielite and ilmenite end-member abundance [61]. Water was assumed to be saturated.

Table 1. Solution models used for the pseudo-sections (See Figure 15a). See Perple_X documentation (http://www.perplex.ethz.ch) for detailed information.

Phase	Solution Model Label in Perple_X	References
Biotite	Bi(W)	[62]
Chlorite	Chl(W)	[62]
Chloritoid	Ctd(W)	[62]
Cordierite	Crd(W)	[62]
Garnet	Gt(W)	[62]
Ilmenite	Ilm(WPH)	[62]
Melt	melt(W)	[62]
Orthopyroxene	Opx(W)	[62]
Plagioclase	Pl(h)	[63]
Staurolite	St(W)	[62]
White mica	Mica(W)	[62]

4. Results

4.1. Tectonic and Structural Setting of Al-Silicates

The main lithology of the Trikorfo area is dominated by garnet–kyanite–biotite–hematite–plagioclase ± staurolite ± sillimanite bearing paragneisses/mica schists. At the scale of Thassos Island, the orientation of the main foliation varies widely due to post-foliation tectonic events, including large-scale folding. In the Trikorfo area, rare relics of an early foliation S_{n-1} are observed. The paragneisses/mica schists mainly show a sub-horizontal (azimuth N140-300, dip up to 20° W) S_n foliation, bearing kyanite mineral lineation, oriented N255 ± 15°, and gently dipping to the WSW. Late brittle deformation is evidenced by tension gashes that cross-cut S_n. These veins are oriented N330 ± 15° and are generally close to vertical. The variation in the orientation of the tension gashes is the same as the regional lineation underlined by synmetamorphic kyanite crystals within 30° that may reflect their continuous formation during counter-clockwise rotation of Thassos Island [40].

4.2. Lithological Units and Rock Sampling

Samples CS16_292b (Figure 3a) and CS16_292a represent the main paragneiss/schist host-rock, apparently unaltered at the outcrop scale. An increasing number of green layers parallel to the gneiss foliation are observed towards the top of the formation (samples CS16_297 (Figure 3b) and Th06). Although not analyzed in our samples, the presence of braunite in the mica schists of the Trikorfo area was shown by previous studies [26,28]. Green to red calc-silicate rocks (up to 80 cm wide) are observed, at the top of the formation, interbedded or at the contacts between quartzo-feldspathic mica schists and metacarbonate rocks (Figure 3c). The rocks typically show a mineral zonation with tremolitic hornblende, diopside, quartz, anorthite/bytownite, epidote, grossular, titanite and Mn-bearing clinozoisite from an outer band adjacent to host mica schists towards the center of the layer (samples Th01, Th02, Th03, KT01 and KT02 (Figure 3c–e)). Sample Th04 is a spessartine–epidote–braunite–hornblende mica schist.

Figure 3. (**a**) Main garnet–kyanite–biotite–hematite–plagioclase ± staurolite ± sillimanite bearing gneiss of the Trikorfo area (sample CS16_292b, N290/65N). (**b**) Green epidote-rich horizon parallel to the main gneiss foliation (sample CS16_297, N270/11N). (**c**) Lense slightly oblique to the subhorizontal gneiss foliation showing a zonation of green epidote and pink clinozoisite together with grossular in a geode (sample KT02). (**d**) Mineralogical zonation with diopside, hornblende, epidote (green), grossular and clinozoisite (pink-red), sample KT02 equivalent to Th04. (**e**) Close up to the core of the geode to near gem-quality pink clinozoisite (clinothulite) and grossular association within locality sample KT02. (**f**) Deep green Mn-rich andalusite (viridine) and kyanite (deep blue and orange) association in sample KT07.

Kyanite is found as a metamorphic mineral, underlining the mineral lineation of the main paragneisses/mica schists rocks as in sample CS16_292b. Locally at the top of the formation, kyanite–piemontite ± andalusite ± muscovite ± spessartine ± braunite mica schists (where kyanite is distributed without any specific orientation) are intercalated within the spessartine–kyanite-bearing mica schists (samples Th05, KT03 (Figure 4a), KT05, and KT07 (Figure 3f)).

Figure 4. Different kyanite color types encountered in the Trikorfo area (Thassos Island). (**a**) Unoriented dark blue kyanite in a mica schist collected towards the top of the area. (**b**) Zoned blue (core) to green-transparent (rim) kyanite in a deformed and folded quartz vein. The pluricentimetric kyanite crystals are found at the interface between the gneiss and the vein. The crystals are weakly deformed and generally oriented parallel to the rock lineation. (**c**) Close-up on zoned-blue kyanite mostly aligned parallel to the rock foliation. (**d**) Field photograph of zoned dark blue-yellow isotropically distributed kyanite crystals in a quartz vein. (**e**) Orange kyanite crystals, generally oriented parallel to gneiss lineation, associated with green Mn-rich andalusite (viridine) and quartz. (**f**) Aggregate of centimetric green Mn-bearing andalusite crystals.

Towards the top of the formation, pluricentimetric kyanite crystals are observed at the contact between metamorphic quartzo-feldspathic kyanite-bearing rocks and coarse-grained quartz-feldspar veins, which are transposed within or secant to the main foliation. In this case, kyanite is generally weakly oriented parallel to the regional lineation. In general the crystals are zoned with dark blue cores and green-transparent (Figure 4b,c) to yellow rims (Figure 4d). Less common are orange kyanite-bearing samples showing association with green andalusite, spessartine and piemontite (Figure 4e,f).

An orthogneiss exposed in the area is composed of orthoclase, oligoclase/andesine along with phlogopite, quartz, hematite, epidote and allanite and is surrounded by a transitional zone with the same mineralogy as the above mentioned calc-silicate layers (i.e. amphibole, plagioclase, and epidote mineral groups from granitoid outwards) [20]. The presence of pluricentimetric crystals of tourmaline (schorl mainly) was also shown in quartz veins and late brittle fissures [20].

Late brittle veins cross-cutting the metamorphic foliation are dominated by euhedral quartz, clinochlore, ilmenite, albite and adularia [20]. Rare veins are enriched in Mn as evidenced by centimetric euhedral crystals of spessartine (Sample MB13; Figure 5).

Figure 5. Late brittle fracture sample MB13. (**a**) Hand specimen showing mainly spessartine (orange), Mn-bearing clinochlore and hornblende (green) and quartz. (**b**) Plane polarized light (PPL) photomicrograph with euhedral spessartine (Grt), Mn-rich epidote (Ep) and Mn-bearing clinochlore (Clc) assemblage. (**c**) Crossed-polarized light (XPL) photomicrograph with secondary spessartine (Grt) and hornblende (Hbl) overprinting the metamorphic foliation underlined by Mn-rich epidote (Ep). (**d**) XPL photomicrograph showing Mn-rich epidote (Ep), Mn-bearing clinochlore (Clc), Ti-hematite (Ti-Hem) and quartz (Qtz) association.

4.3. Mineralogy and Mineral Chemistry of Paragneisses/Mica Schists of the Metamorphic Unit at Trikorfo

The composition of two host-rock samples (CS16_292a and _292b) is shown in Table S1a. The structure of the rock is granolepidoblastic with quartz-plagioclase-dominated layers and a schistosity underlined by biotite locally retrogressed to chlorite. C'/S structures indicate top N228 sense of shear. The mineralogy is made of biotite, garnet, quartz, kyanite, staurolite, plagioclase, apatite and Ti-hematite (Figure 6a). Rare late muscovite and epidote are observed without any preferential orientation (Figure 6a).

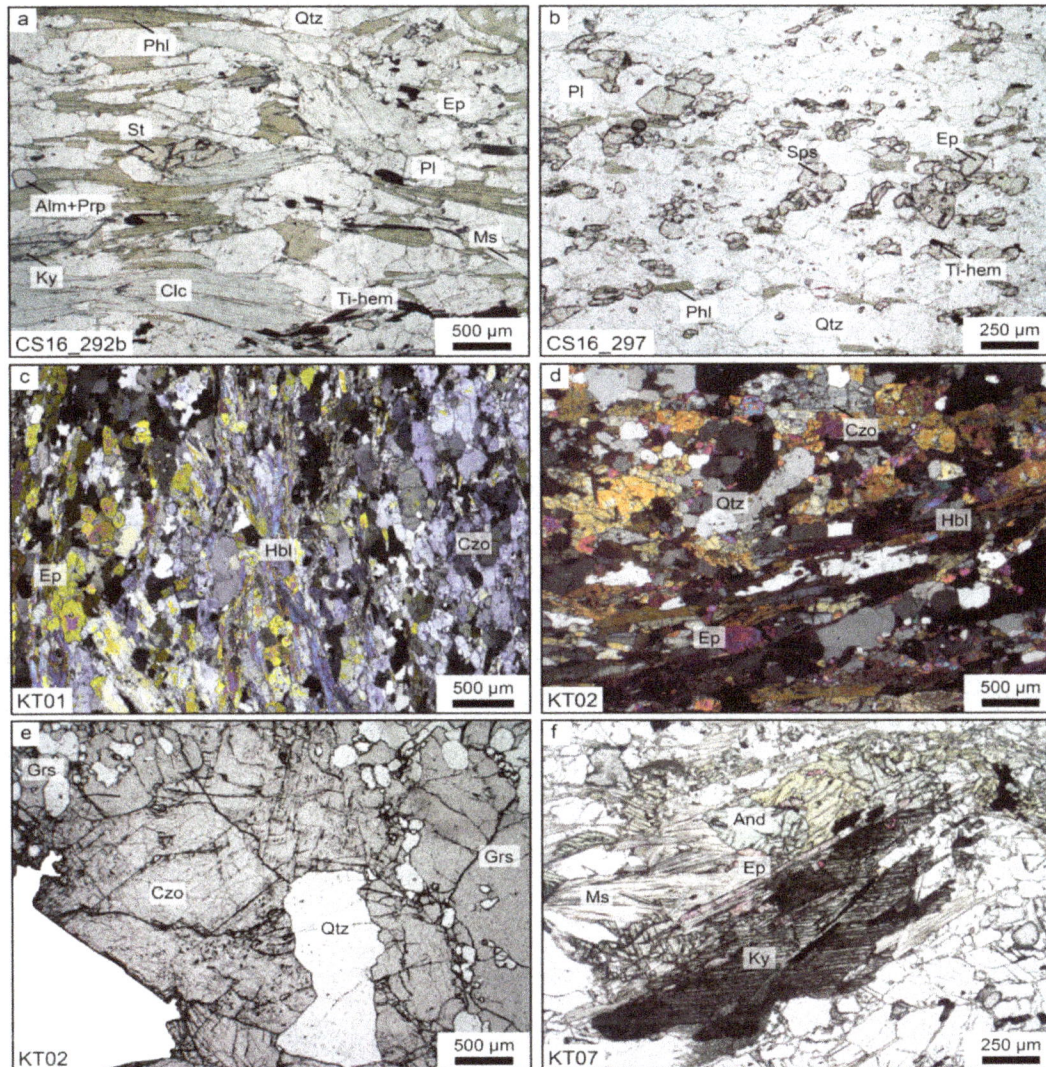

Figure 6. Main mineralogical assemblages from the Trikorfo area. (**a**) Biotite–garnet–kyanite–staurolite–plagioclase-quartz–Ti–hematite paragenesis of the main paragneiss unit modeled with Perple_X. Chlorite (clinochlore/amesite) is a retrogressed phase of phlogopite. Scarce late muscovite and epidote are observed. (**b**) Phlogopite–spessartine–epidote association of the quartz-dominated green horizon parallel to the gneiss foliation. (**c**) Clinozoisite and epidote layers and secondary amphibole in sample KT01 under XPL. (**d**) Clinozoisite–epidote–hornblende in sample KT02 under XPL. (**e**) Grossular-quartz–clinozoisite (clinothulite) microscopic view (PPL) of sample KT02. (**f**) Secondary Mn-rich epidote, orange kyanite and green andalusite assemblage of sample KT07 under PPL. Mineral abbreviation: Almandine + pyrope (Alm + Prp), Andalusite (And), Clinochlore/amesite (Clc), Clinozoisite (Czo), Epidote (Ep), Grossular (Grs), Hornblende (Hbl), Kyanite (Ky), Muscovite (Ms), Phlogopite (Phl), Plagioclase (Pl), Quartz (Qtz), Staurolite (St) and Ti-hematite (Ti-Hem).

Biotite has the composition of phlogopite with X_{Mg} in the range 0.73–0.75 (Table S2a). Plagioclase is oligoclase/andesine with a composition between An_{23} and An_{34} (Table S3a). Garnet and staurolite crystals are isolated in the mica-rich layers. Garnet grains are always corroded remnants of preexisting porphyroblasts (Figure 6a). Garnet chemistry (Table S4a) is dominated by X_{Fe} (up to 39.25) and X_{Mn} (up to 44.06); X_{Mg} is significant, up to 18.82, and X_{Ca} is never higher than 7.72. Garnet composition thus lies close to the border between almandine-(pyrope) and spessartine fields (Figure 7a). No significant and systematic chemical variation was noticed from border to border among all analyzed crystals (Table S4a). Staurolite, also corroded, contains significant amounts of MnO, up to 1.47 wt. % (Table S5). Although Dimitriadis [24] indicated that kyanite seems to postdate garnet growth (no kyanite inclusions in garnet and garnet early porphyroblast remnants might be enveloped by kyanite crystals), the relations are not so clear in sample CS16_292b because most of the crystals are isolated from each other. Our observations did not provide convincing evidence that kyanite postdates garnet growth and both minerals are thus considered as cogenetic. Fe-rich (up to 1.23 wt. % FeO) syn-metamorphic kyanite crystals are oriented parallel to the main foliation S_n. Chlorite analyses revealed a composition between clinochlore and amesite (Figure A1; Table S11). All opaque minerals analyzed by EPMA were Ti-rich hematite (Table S6a). Rare ilmenite was identified by optical microscopy.

4.4. Mineralogy and Mineral Chemistry of Green-Colored Horizons

Sample CS16_297 is representative of the many green-colored horizons parallel to the foliation of the main schist/paragneiss unit of the Trikorfo area (Figure 3b). Element analyses are reported in Table S1a. The structure is mainly granoblastic with the Sn foliation locally underlined by phlogopite alignment. Identified minerals include biotite, quartz, plagioclase, garnet, epidote, Ti-hematite, titanite, rutile and apatite (Figure 6b). Magnetite with hematite exsolution was noticed. The mineralogy is dominated by quartz with layers rich in secondary spessartine and/or epidote/clinozoisite. Biotite has the composition of phlogopite with X_{Mg} in the range 0.72–0.74 (Table S2b). Plagioclase is oligoclase with a composition of An_{15} to An_{21}, slightly depleted in CaO compared to CS16_292b (Table S3b). The content of TiO_2 in Ti-hematite does not exceed 8.15 wt. % (Table S6b). The composition of titanite and rutile is given in Table S9a. Chlorite, staurolite, kyanite and sillimanite are not found in this rock.

The shape of garnet crystals is usually irregular. X_{Fe} and X_{Mg} are much lower (up to 21.69 and 3.24, respectively), and X_{Mn} and X_{Ca} (up to 58.00 and 37.29, respectively) significantly higher than garnet from sample CS16_292b. No significant and systematic chemical variation was noticed from border to border among all analyzed crystals (Table S4b) and garnet composition falls in the spessartine field (Figure 7a). Epidote is abundant and shows a composition depleted in Mn^{3+} ranging from epidote end-member to Fe^{3+}-rich clinozoisite (Figure 7b; Table S7a,j) and is responsible for the green coloration within these rocks.

4.5. Mineralogy and Mineral Chemistry of Calc-Silicate-Dominated Horizons

Moving towards the top of the formation, the thickness of the green layers increases and green-red horizons up to 80 cm thick are visible (Figure 3c,d). These beds are intercalated within the main foliation of the kyanite-bearing paragneisses/mica schists at proximity of carbonate/dolomitic rocks. They typically show calc-silicate assemblages more or less enriched in Mg- and Mn-bearing minerals with typical mineralogical zonation showing tremolitic hornblende, diopside, quartz, anorthite/bytownite, clinozoisite/epidote, grossular, titanite and pink clinozoisite from an outer band adjacent to host mica schists towards the center of the layer (Figure 3c,d) [20,24].

The dark green external part of the calc-silicate is dominated by Mn-bearing varieties of diopside (up to 1.44 wt. % MnO) and hornblende (up to 2.75 wt. % MnO) (Table S8a,c,d). The progressive change in color from pale green to red (samples KT01-02, Th01-03) is mainly due to variation in epidote composition, whose EPMA analyses are presented in Table S7b,e–g, and reported in the Mn–Al–Fe ternary plot of Figure 7b.

Green epidote is reported in green layers parallel to rock foliation as in sample CS16_297 previously described and at the outer margins of the calc-silicate layers along with hornblende, anorthite/bytownite (Table S3d) and titanite (Table S9b) (Figure 6c,d). The composition evolves from epidote to clinozoisite end-members (Figure 7b). It may contain up to 1.04 wt. % MnO. LREE were not analyzed for epidote mineral group; their presence may mark the low total of some of the analyses.

Mn-poor clinozoisite exhibits subhedral to euhedral crystals. The pale pink crystals are usually found in the intermediate zone of the calc-silicate layers mostly associated with quartz, anorthite, titanite, diopside and hornblende. Available microprobe data revealed Fe_2O_3 content between 1.36 and 6.11 wt. % (sample KT02) and between 1.54 and 11.06 wt. % (samples Th01 and Th03), and minor Mn_2O_3 content below 0.67 wt. % (Table S7b,e,g).

Near gem-quality Mn-poor clinozoisite, with the same optical and chemical properties as clinothulite variety described by Bocchio et al. [64], is locally observed in the inner part of the calc-silicate layers, intergrown with Mn-grossular, titanite and quartz, both in the matrix of the rock and in fissures cross-cutting the foliation. The crystals occur in deep pink to red colors and form translucent subhedral to euhedral crystals up to 10 cm (Figures 3e and 6e; Table S7b,f). EPMA data revealed low contents of Fe_2O_3 (<3.15 wt. % in sample KT02 and <2.90 wt. %, in sample Th02) and Mn_2O_3 (<0.72 wt. % in sample KT02 and <0.21 wt. %, in sample Th02), however high enough to be responsible for the pink-red coloration.

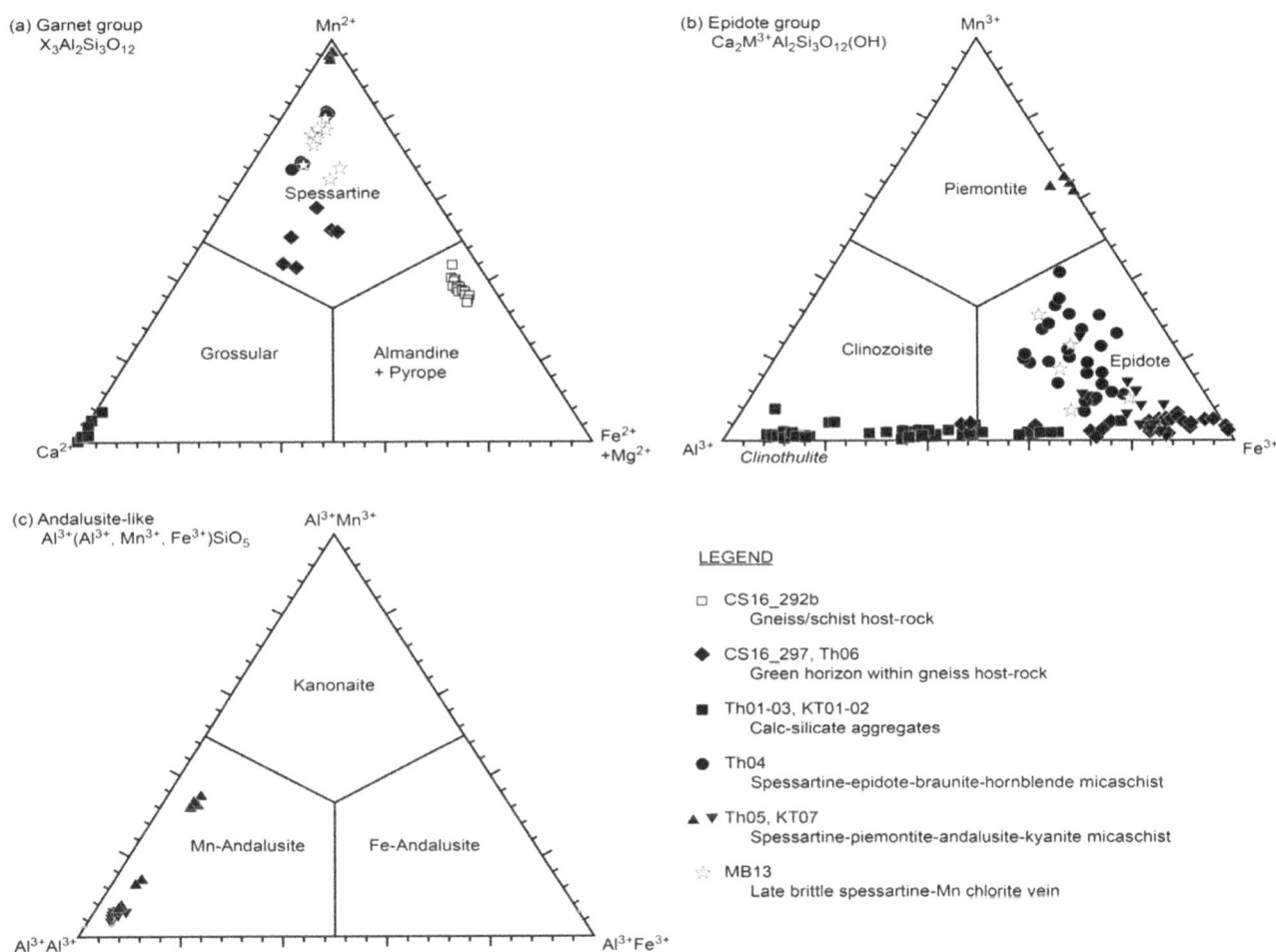

Figure 7. Ternary plots showing compositions of the garnet, epidote and andalusite group minerals. (**a**) Garnet composition considering spessartine, grossular and almandine + pyrope end-members. Garnet composition is $X_3Al_2Si_3O_{12}$ with the X site occupied by divalent cations, Mn^{2+} for spessartine, Ca^{2+} for grossular, Fe^{2+} for almandine and Mg^{2+} for pyrope, e.g., [65]. (**b**) Epidote group minerals, $Ca_2M^{3+}Al_2Si_3O_{12}(OH)$, classified according to one M^{3+} cation substitution with Al in octahedral coordination with $M^{3+} = Al^{3+}$, Mn^{3+} or Fe^{3+}, corresponding to the clinozoisite, piemontite and epidote end-members, respectively [66,67]. Composition analyses for deep pink to red clinozoisite (clinothulite) are shown. (**c**) Andalusite-like minerals, $Al^{3+}(Al^{3+}Mn^{3+}Fe^{3+})SiO_5$, classified according to octahedral Al substitution with Mn^{3+} or Fe^{3+} [2,68].

In the calc-silicate lenses, pure grossular to Mn-bearing grossular form euhedral yellowish crystals of near gem-quality in close association with pink to red colored Mn-bearing clinozoisite and quartz (Figures 3e and 6e). MnO content ranges from 0.08 to 3.60 wt. % (Table S4c,e). All data are plotted in the ternary Mn–Ca–Fe + Mg diagram (Figure 7a).

4.6. Mineralogy and Mineral Chemistry of Mn-rich Schist Layers and Quartz Lenses

Near the top of the Trikorfo area, the formation is dominated by schist layers (samples Th04, Th05 and KT07) and intercalations of quartz lenses (sample KT08) particularly enriched in Mn-bearing minerals as shown by the assemblage of piemontite and Mn-rich epidote together with Mn-rich andalusite and spessartine (Figures 3f, 4d and 5f). Although EPMA analyses were not entirely satisfying (MnO between 27.99 and 45.31 wt. % and maximum total of all analyzed elements of 51.80 wt. %), probably reflecting the heterogeneity and variable Mn oxidation state of the minerals, Mn-oxides could be identified (Table S10). In both lithologies, piemontite and Mn-rich epidote form dark pink/purple-colored prismatic crystals with typical piemontite pleochroism. Microprobe analyses revealed piemontite in the andalusite-bearing mica schists with Mn_2O_3 content up to 12.69 wt. % and Mn-rich epidote in association within the spessartine–braunite–quartz aggregates with Mn_2O_3 in the range of 0.13 to 6.59 wt. % (Figure 7b, Table S7c,h,i). Phlogopite has X_{Mg} of 0.96–0.97 (Table S2d). Plagioclase is oligoclase/andesine with a composition between An_{22} and An_{34} (Table S3c).

Garnet is also marked by enrichment in Mn (Figure 7a). Translucent, orange, euhedral crystals of spessartine up to 1 cm are found (i) in intercalated spessartine–epidote–braunite–hornblende mica schists, and (ii) in spessartine–piemontite–andalusite–kyanite mica schists. Within the spessartine–epidote–braunite–hornblende mica schists, spessartine commonly forms large euhedral crystals with numerous inclusions of epidote and is surrounded by epidote and magnesio–hornblende. Spessartine from the spessartine–piemontite–andalusite–kyanite mica schists is also associated with muscovite, phlogopite, anorthite and hematite. Electron microprobe analyses of spessartine indicate X_{Mn} in the range 64.74–97.19 (Table S4e).

A late brittle vein cross-cutting the Mn-rich schist layers was sampled (MB13; Figure 5). Up to 1 cm near gem-quality euhedral crystals of spessartine (X_{Mn} of 81.23; Figure 7a; Table S4d) are observed together with quartz, Mn-rich epidote (up to 6.67 wt. % Mn_2O_3; Figure 7b; Table S7d), Mn-bearing clinochlore (up to 1.64 wt. % MnO; Figure A1; Table S11d), Mn-bearing hornblende (up to 1.99 wt. % MnO; Table S8b) and almost pure hematite (Table S6c).

Accessory phases include monazite (relatively abundant in discrete layers), titanite and zircon. Tourmaline is very abundant in quartz lenses within mica schists, where it occurs as black crystals (schorl) up to 10 cm and associated with muscovite and chlorite (clinochlore). However not analyzed, olive green-colored tourmaline (crystals up to 3 cm) is observed. Tourmaline is also present in fissures intergrown with clinochlore, titanite and quartz. Finally, up to 10 cm tremolite white crystals were documented in sheared marbles [20].

4.7. Mineralogy and Mineral Chemistry of Kyanite–Quartz Veins

4.7.1. Different Colors of Kyanite Crystals

Among the many spectacular minerals observed at the Trikorfo area, kyanite, found at the contact between quartzo-feldspathic coarse-grained veins and the paragneiss/mica schist rocks, towards the top of the series, is of particular interest. The crystals show a general orientation parallel to the regional foliation but may also be unoriented (Figure 4a). The up to 20 cm crystals are slightly bent as a result of weak plastic deformation. Over a small area of a few km^2, besides the Fe-rich syn-metamorphic kyanite already described in the metamorphic paragenesis (Table S12a), we report at least four dominant types of kyanite based on their color: (i) dark blue-black (Figure 4a, Table S12b), (ii) zoned crystals with dark blue cores and transparent-green rims (Figure 4b,c), (iii) zoned crystals with dark blue cores and yellow rims (Figure 4d) and (iv) orange crystals generally associated with green Mn-rich andalusite (Figure 4e; Table S12c). Most of the crystals are inappropriate for faceting as marketable gemstones due to common fractures, but are at least proper for cabochon-shaped material. In addition, with regard to their overall aesthetic aspect, mainly large size and vivid color, these kyanite crystals are referred to as being of near gem-quality.

4.7.2. Bulk Element Concentration of Individual Kyanite Crystals

When observed down their <100> faces, most kyanite crystals show blue-green-colorless bands parallel to the *c*-axis. Representative kyanite samples have been selected for bulk element concentration analyses (zoned kyanite THA16_1, _2, _3, yellow THA08, pale orange THA12 and deep orange THA13; Figure 8). The elements are classified as oxides and trace elements according to their content in the host rock sample CS16_292b (* symbol in Figures 9a, 10 and 11).

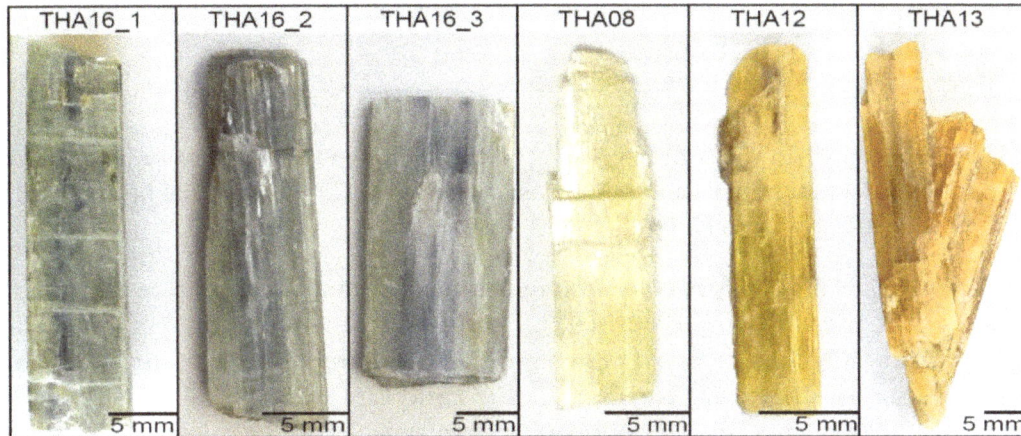

Figure 8. Photographs of individual kyanite crystals prior to bulk rock analyses.

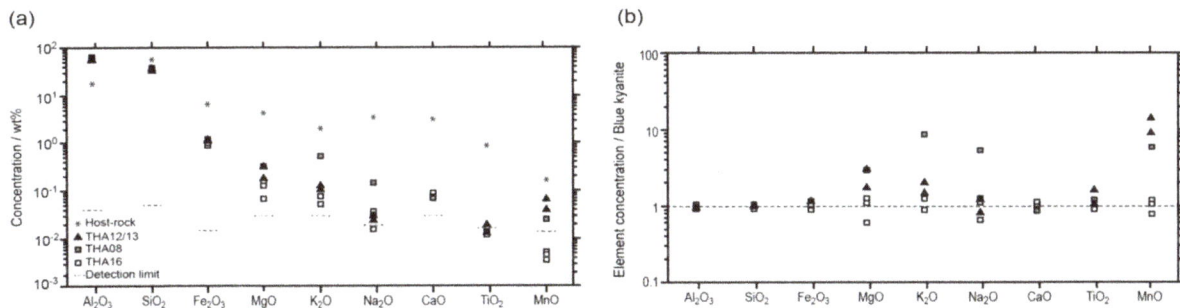

Figure 9. Distribution of measured oxides in CS16_292b host-rock and in blue-green (THA16), yellow (THA08) and orange (THA12 and THA13) kyanite samples. (**a**) Raw concentration. (**b**) Concentration normalized to the average value of the three blue-green THA16 samples.

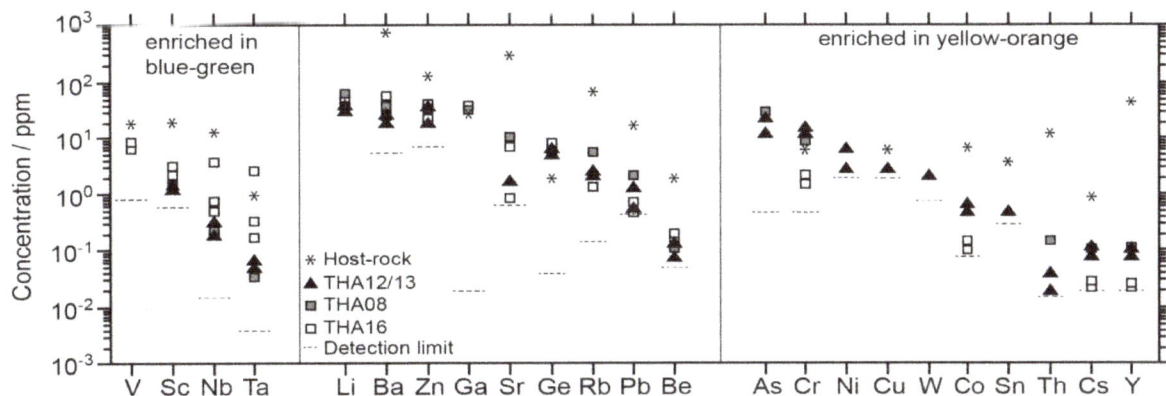

Figure 10. Trace element composition of host-rock CS16_292b and kyanite samples ordered by decreasing concentration. The left part shows elements which are preferentially found in blue-green THA16 kyanite samples, the central part where there is no significant difference among the samples and the right part reports the elements with significant enrichment in yellow THA08 and orange THA12/13 kyanite samples.

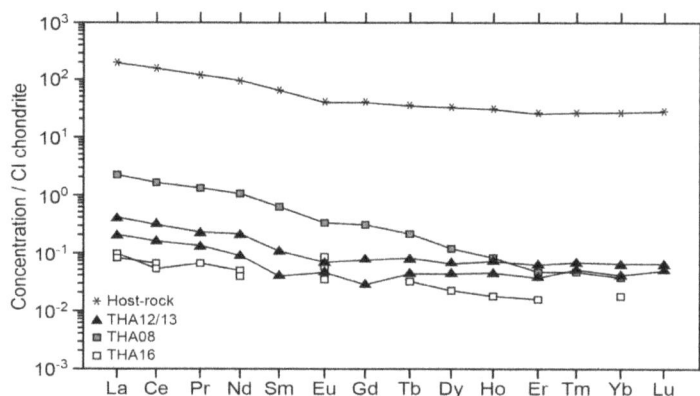

Figure 11. REE spider diagram of blue-green (THA16), yellow (THA08) and orange (THA12 and THA13) kyanite individual crystals and of the host rock sample CS16_292b. The normalizing factors are related to CI chondrite values [69].

The amount of Al_2O_3 and SiO_2 is comparable between all analyzed kyanite samples, higher than 96 wt. % Al_2SiO_5. The measured oxides show a decreasing concentration from Fe_2O_3, MgO, K_2O, Na_2O, CaO, MnO and TiO_2 (Figure 9a). These elements generally show a preferential enrichment in the orange kyanite (Figure 9b), except Na_2O, K_2O, CaO and LOI content (Table S1b), which appear to be correlated but randomly distributed between the samples. This is likely due to alteration phases in microfissures. The total amount of Fe_2O_3, MgO, MnO and TiO_2 increases with coloration from blue-green (1.10–1.21 wt. %), to yellow (1.51 wt. %) and orange (1.42–1.61 wt. %). Yellow and orange samples show markedly higher amounts of MnO (up to one order of magnitude) and MgO (Figure 9b).

Minor elements from the bulk rock analyses are found with amounts lower than 100 ppm in kyanite samples. Ba, Zn and Ga are the most abundant elements. Blue-green kyanite is significantly enriched in V, Nb, Sc and Ta. Yellow and orange samples of kyanite contain predominant amounts of As, Cr, Ni, Cu, W, Co, Sn, Th, Cs and Y. For all other trace element contents, there is no significant difference among the analyzed samples (Figure 10; Table S1c).

The rare earth element concentration measured in each sample was normalized to chondritic abundance (Figure 11). All kyanite samples have comparable chondrite-normalized REE patterns as CS16_292b host-rock with a slight decrease until Gd and then relatively flat towards heavy REE. As for all other elements, REE are less abundant in blue-green samples than in yellow and orange kyanite. The yellow (THA08) sample is characterized by a remarkable enrichment in light REE compared to other samples. This is to be correlated with the higher content of this sample with the elements attributed to alteration phases in microfissures such as K, Na, Li, Sr and Ba.

4.7.3. EPMA X-ray Maps and Element Distribution

For all the samples, Al and Si variations are difficult to interpret. Both elements appear relatively homogeneous throughout the crystals and their maximum/minimum concentration seems to be slightly anti-correlated, likely due to a problem of surface homogeneity. A decrease of Al is noticed on the edges of THA16 marking alteration or re-equilibration with minerals in the surrounding rock. The element Mn also does not show any obvious variation in the different analyzed areas. The only point of interest regarding Mn is that, accordingly with bulk rock analyses, the concentration is lower in blue THA16 (<0.10 wt. % MnO) than in the yellow THA08 (<0.18 wt. % MnO) and orange THA13 (<0.16 wt. % MnO) kyanite samples.

Fe concentration however shows very contrasting behavior. In kyanite samples THA16 (<1.27 wt. % FeO) and THA08 (<1.59 wt. % FeO), the variations are straight and parallel to the c-axis. At this scale, the higher concentration seems to be correlated with the clearer parts of the kyanite crystals. In sample THA13 (<1.56 wt. % FeO), Fe distribution seems to be oblique regarding the orientation of the c-axis and could be related to plastic deformation of the analyzed crystal (Figure 12).

(a) THA16

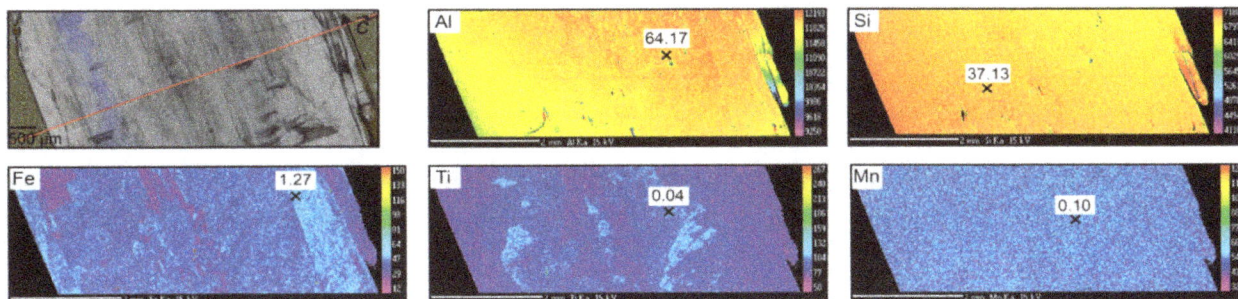

(b) THA08

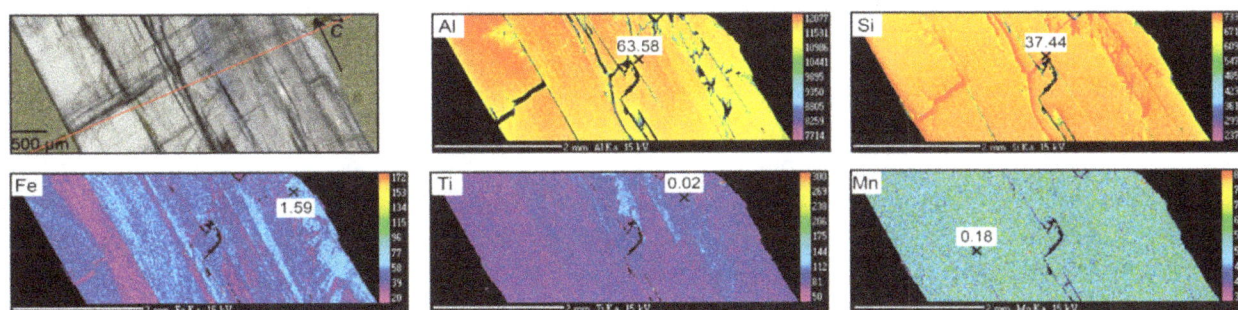

(c) THA13

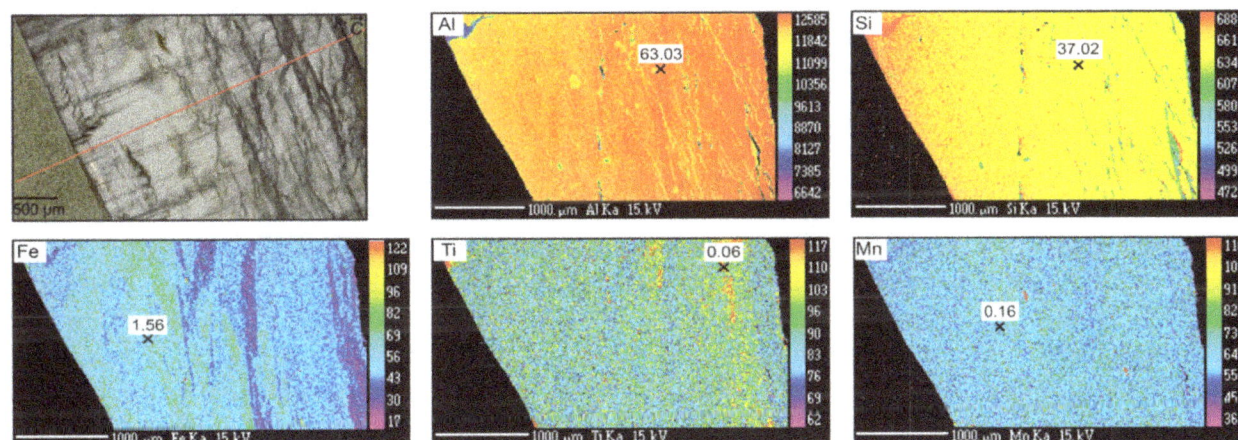

Figure 12. Electron probe microanalyses (EPMA) semi-quantitative X-ray maps of Al, Si, Fe, Ti and Mn on the <100> faces of blue-green THA16 (**a**), yellow THA08 (**b**) and orange THA13 (**c**) kyanite samples, with color scales expressed in counts, not calculated to concentration. The red line corresponds to the EPMA quantitative profile analyses (Table S12d–f). The maximum values (in oxide wt. %) measured along the profile for a given element is reported on each map.

Ti distribution is discontinuous and does not seem to be related with crystallographic orientation. Three marked areas of higher concentration are highlighted in THA16 (<0.05 wt. % TiO_2) and seem to correspond to the blue area of the crystal, at least for the left one. It is also the case to the right of the crystal, but the blue coloration is barely visible. The direct correspondence between blue coloration and Ti concentration is however difficult to establish firmly as most of the blue area is below the sample surface. In sample THA08 (<0.02 wt. % TiO_2) also, the irregular distribution of Ti reveals enrichment at the central right part of the crystal where a thin blue zonation is observed. On THA13 (<0.06 wt. % TiO_2), Ti distribution follows that of Fe, though anti-correlated.

4.7.4. Orange Kyanite and Green Andalusite Assemblages

This atypical polymorphic association is observed towards the top of the formation (i) in kyanite–phlogopite–piemontite–quartz–plagioclase ± muscovite ± spessartine ± braunite mica schists, and (ii) in quartz-feldspar-muscovite lenses. Mn-rich andalusite forms dark green-colored subhedral to euhedral crystals up to 7 cm. In both cases, kyanite and andalusite seem to have grown at equilibrium (Figure 6f). Green andalusite is commonly zoned likely due to twinning, and exhibits deformation lamellae [29]). EPMA data reveal elevated Mn_2O_3 contents, up to 15.98 wt. % (Figure 7c; Table S13a,b).

4.7.5. LA-ICPMS Profiles

The limits of detection of LA-ICMPS are significantly lower than those of EPMA and many elements can be analyzed sequentially during the same analytical run. Edge to edge sections were conducted at the same position, or approximately, as that for the EPMA profiles on the three kyanite samples (Figure 13). Al and Si variations serve as internal standards to provide quantitative elemental evolution through kyanite and are not reported on the figures. All analyses of Ca, Cu, Ag and Au were below the limits of detection. For each sample, the elements are reported according to three different groups with (i) the dominating elements (Fe, Mn, Ti, Mg and Zn) on top, (ii) the elements generally enriched in alteration minerals (K, Na, Li, Sr and Ba) in the middle, and (iii) the other elements, mainly transition metals, only found as traces in the host-rock (Cr, Pb and V) on the lower part.

Kyanite sample THA08 shows large variations in the concentration of the elements K, Na, Li, Sr and Ba. Moreover, it can be noticed that these elements are mainly concentrated either on the sides of the crystals or along fractures inside the crystals, which are common in THA08. Accordingly, these elements are interpreted as the result of alteration or late equilibration with host-rock minerals (feldspars and micas like muscovite and phlogopite).

In all samples, Fe is by far the most common element besides Al and Si. Local concentration decreases are observed where the amount of alteration elements increases, as is clearly visible on sample THA08. Mn and Ti are in the same order of magnitude in THA16 and THA08 but with Ti slightly higher in the blue-green kyanite. Mn concentration does not correlate with any color change, whereas that of Ti shows a relatively good correspondence with the blue color bands, even if they are below the sample surface. This is especially well expressed in THA08 where Ti > Mn at the intersection with the blue band. It is also visible on THA16, where the three Ti highs seen in the EPMA mapping are distinguishable, and correspond to a blue color on the left. However, it must be kept in mind that blue color is mostly below the crystal surfaces. Also, it must be noted that the laser ablation itself down-cuts through the sample (ablation spot of 32 µm) and hence what was actually analyzed may largely be below the surface. In sample THA13, Mn dominates largely over Ti and is one order of magnitude higher as in samples THA16 and THA08. The Mg concentration variation seems to partly correlate with edges and fractures and can be interpreted at least in part due to equilibration with host-rock phlogopite. Along the profiles, Zn is equally present and its concentration evolution is marked by rapid variations up to one order of magnitude. In some parts of the kyanite crystals, Zn is almost as abundant as Mn, Ti and Mg. These variations are not related to any characteristic feature of the crystals. The other elements do not show any obvious correlations. V is found above the limit of detection (1 ppm) only in sample THA16. No systematic evolution of Cr could be observed, because of concentration too close to the limit of detection (15 ppm). Pb shows the same evolution as the elements attributed to alteration and post-crystallization re-equilibration, with higher concentrations at the edges of the crystals and within the fractures.

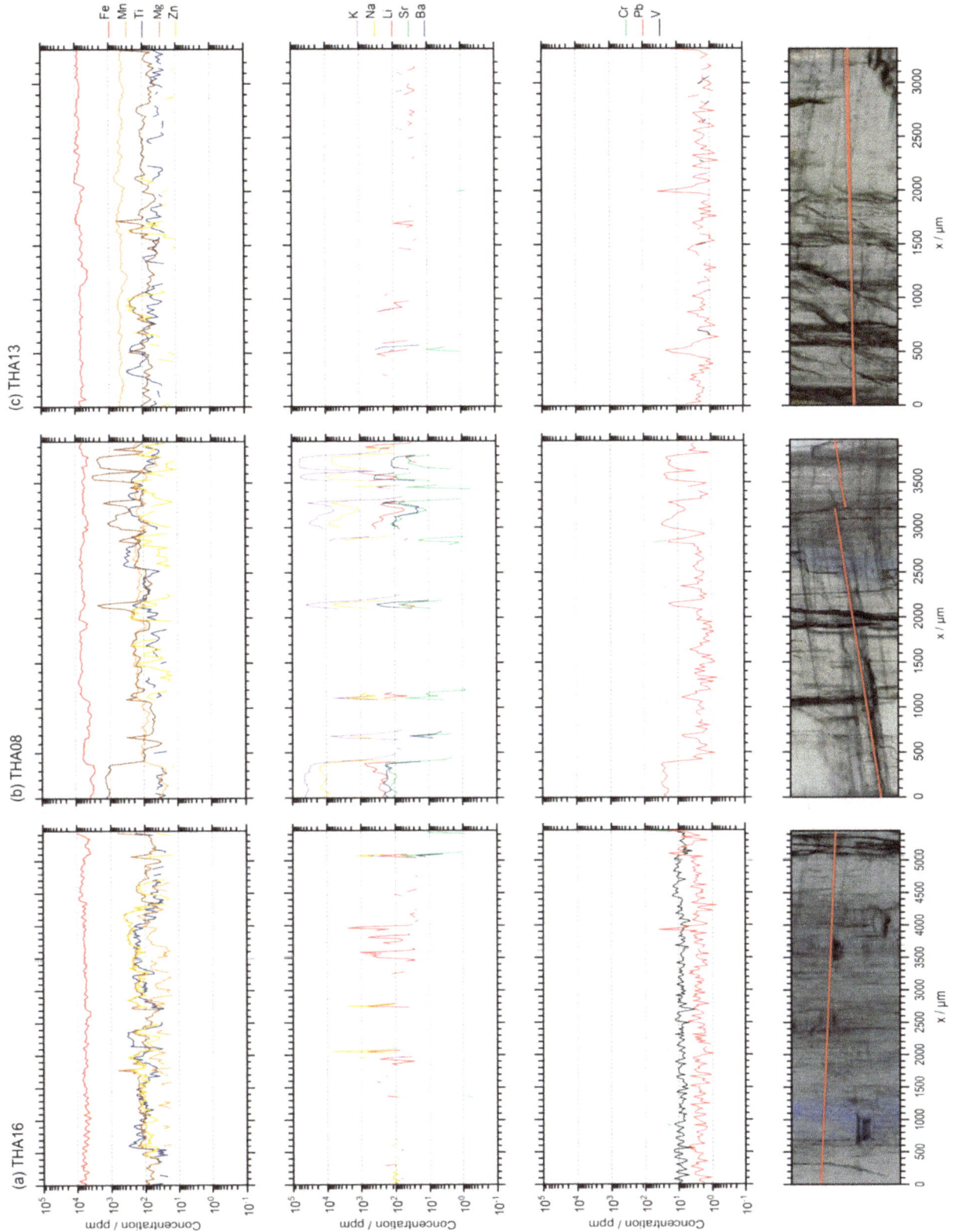

Figure 13. Quantitative Laser Ablation-Inductively Coupled Plasma Mass Spectroscopy (LA-ICPMS) analyses (in ppm) on the <100> faces of blue-green THA16 (**a**), yellow THA08 (**b**) and orange THA13 (**c**) kyanite samples from border to border of the crystals. The location of the profiles (red line) is approximately at the same position as that of EPMA analyses. The element concentration lines might be discontinuous due to values below detection limit.

4.7.6. UV-Near-Infrared Spectroscopy

All spectra (blue, yellow and orange kyanite samples) show an absorption continuum rising from infrared or mid-visible to UV, making absorption range from 315 nm to below (Figure 14). They also all show sharp and weak absorption peaks at 370, 382, 432, 446 nm. Spectra of blue samples are dominated by large absorption bands at 520 and 620 nm that, combined with the previous spectral features, generate a transmission window around 480 nm, in the blue sample. Spectra of orange and yellow kyanite samples are similar. In addition to the common spectral features described initially, both show a band at about 470 nm and a weaker band at 890 nm. These two bands are weaker in the yellow than in the orange kyanite sample. As a consequence, transmission increases from the green (550 nm) toward the infrared, hence the orange and yellow colors of the samples. The allocation of these spectral features to specific ions or defects is proposed in the Discussion section.

4.8. Phase Diagram Calculation of the Paragneiss

The PT pseudo-section is presented in Figure 15 for the PT range of ca. 550–750 °C and ca. 5.5–10.5 kbars. Calculation results show that the assemblage Grt–St–Bt–Pl–Ky–Qtz–Ti–Hem formed between 630–710 °C and 7.8–10.4 kbars, in the upper amphibolite facies. Calculations in the different systems, ranging from NCKFMASH to MnNCKFMASHT(O), allow for discussing the effects of MnO, TiO_2 and Fe_2O_3. The addition of MnO results in an expansion of the garnet stability field with the garnet-in line shifting to lower temperatures and pressures. Because of the occurrence of Ti-bearing hematite in the mica schists, the role of oxidation on phase relations was investigated through a T-XFe_2O_3 phase diagram section. The addition of TiO_2 and Fe_2O_3 controls the field stability of Fe^{2+}–Fe^{3+}–Ti oxide phases. The minimum XFe_2O_3 value to stabilize the Ti-bearing hematite instead of rutile is 0.1 (at constant P = 8.5 kbars). It appears that the stable assemblage garnet (Grt)–staurolite (St)–biotite (Bt)–plagioclase (Pl)–kyanite (Ky)–quartz (Qtz) with Ti–hematite (Ti–Hem) is stable between ca. 630 and 710 °C for a median XFe_2O_3 ($Fe_2O_3/(Fe_2O_3 + FeO)$) of 0.5 (Figure A2). In consequence the phase diagram section was computed with this XFe_2O_3 molar ratio. Considering the P-T range used for this pseudosection (550–750 °C; 5.5–10.5 kbars), hematite-rich ilmenite with ca. 50–70% of hematite end-member is stabilized in each stability field of ilmenite and, thus, labelled as "Ti–Hem" in Figure 15. For the specific PT field of the described paragenesis (ca. 670 °C and 9.1 kbars), the calculated ilmenite in the ilmenite–hematite solid solution is around 33%, consistent with the 10–35 % measured by EPMA (Table S6a).

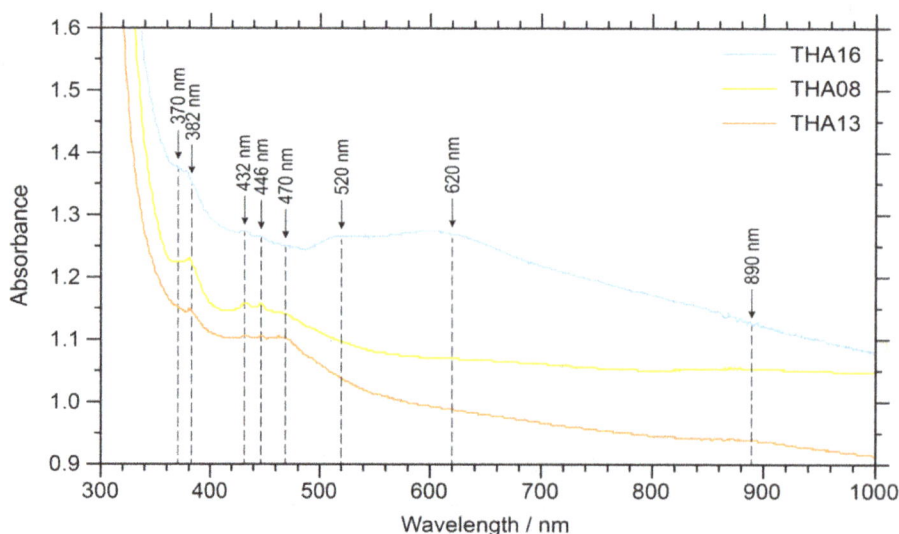

Figure 14. UV near-infrared spectra of kyanite of various colors. Spectra are shifted and re-scaled vertically for clarity. Absorbance values are those of sample THA08.

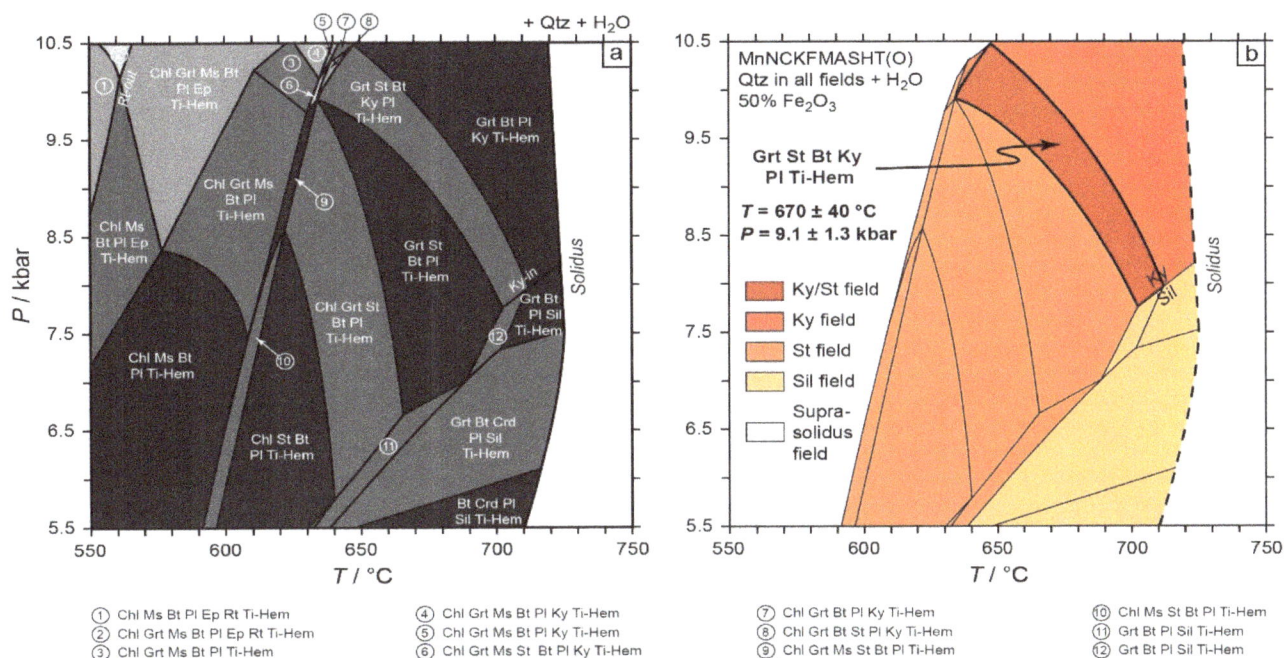

Figure 15. PT pseudo-section calculated for sample CS16_292b. (**a**) Pseudo-section showing labelled metamorphic assemblage fields. Assemblage fields are shaded according to the number of degrees of freedom, with higher-variance assemblages represented by darker shading. Ky-in and Rt-out lines are shown. (**b**) Highlight of the stability fields of kyanite, staurolite and sillimanite index minerals. The interpreted peak metamorphic assemblage field is shaded in red and labelled with bold text. Ky-Sil univariant line is shown. The right white domain corresponds to the supra-solidus field. Dashed line represents the solidus. Mineral abbreviation: biotite (Bt), chlorite (Chl), cordierite (Crd), epidote (Ep), garnet (Grt), kyanite (Ky), muscovite (Ms), plagioclase (Pl), quartz (Qtz), rutile (Rt), sillimanite (Sil), staurolite (St) and Ti-hematite (Ti-Hem).

Chlorite thermometry [70] made on retrogressed phlogopite of the paragneiss CS16_292b and schist sample KT03 indicates temperatures in the range 398–334 °C (Table S2c for phlogopite and Table S11a,b for chlorite composition). The values given by the clinochlore analyzed within sample piemontite–andalusite–muscovite–kyanite schist KT07 are in the range 280–326 °C (Table S11c), while the values from Mn-bearing clinochlore from the late brittle vein MB13 are from 233 to 315 °C (Table S11d).

5. Discussion

The Trikorfo area on Thassos Island represents a unique mineralogical locality where near gem-quality Mn-rich minerals are described. A detailed structural/mineralogical study permits to describe metamorphic parageneses and local metasomatic assemblages. The dominant formation is made of paragneisses/mica schists containing a metamorphic paragenesis, including kyanite. Towards the top of the formation, different units are described parallel to the metamorphic foliation, with some of them bearing large and colored minerals: (i) calc-silicate dominated horizons, (ii) Mn-rich schist layers and quartz lenses, (iii) kyanite-bearing quartz-feldspar coarse-grained veins, and (iv) late brittle veins cross-cutting the metamorphic foliation. The discussion is divided into three parts where (i) the near gem-quality minerals are described and the elemental substitutions leading to their color variation are discussed, (ii) the distinction between metamorphic parageneses and local metasomatic assemblages is made, (iii) a *PT*-deformation exhumation path for the conditions of formation of the different units is proposed. Kyanite is found within the metamorphic paragenesis and as near gem-quality in quartz-feldspar coarse-grained veins; it is therefore the guidance mineral of the study and the discussion.

5.1. Colored Minerals and Cation Substitution in the Trikorfo Area, Thassos Island

Dispersed metal ions in silicate and calc-silicate crystal structures lead to color variation of particular interest for mineral collectors. Silicate and calc-silicate minerals of the Trikorfo area are large and vividly colored, close to gem-quality. Moreover, thanks to trivalent cation substitutions (Mn^{3+} and Fe^{3+}) with Al, the area hosts an outstanding variety of colors within the minerals. This chapter aims to depict the main substitutions observed within kyanite, andalusite, epidote and garnet minerals.

The common natural color of kyanite (Al_2SiO_5) is blue but it can also be green, yellow, orange, white, black, grey or colorless as a function of the nature of elemental substitutions [4]. Moreover, several alternating colors can be found along bands parallel to the *c*-axis in a single kyanite crystal, marking the presence of trace elements in the crystal structure. Kyanite has two Al sites in six-coordination and one Si site in four-coordination. The occupying site and bonding state of transition elements determine the color and physical properties of the crystals [71]. In the Trikorfo area, besides syn-metamorphic crystals, kyanite (zoned blue to green/yellow-transparent, yellow and orange) is also observed in quartz-feldspar coarse-grained veins or in quartz lenses. Bulk chemical composition analyses reveal that each type of color kyanite has its own signature with a characteristic distribution of measured oxides and trace elements in the minerals (Figures 9 and 10). V and Cr distribution is markedly different between zoned blue (up to ~10 ppm and ~2 ppm, respectively) and orange kyanite (below detection limit and up to ~15 ppm, respectively), suggesting a local control from the host-rock composition. However, the low concentrations point to a similar felsic origin [4] that should be attributed to interaction with the main paragneiss unit of the area. Orange kyanite localities are very rare on Earth. The Loliondo deposit (Tanzania) is characterized by elevated MnO content up to 0.23 wt. %. Polarized optical absorption spectra show that the orange coloration is governed by crystal field d-d transitions of Mn^{3+} replacing Al in six-fold coordinated triclinic sites of the kyanite structure [9]. Even if orange kyanite from the Trikorfo area is richer in MnO than the yellow and blue zoned crystals, their MnO content (0.07 wt. % for the bulk sample analysis and locally up to 0.16 wt. % for EPMA analyses) remain far below that of orange kyanite from Tanzania. This is consistent with an orange color less intense in kyanite from Trikorfo than from Loliondo. All kyanite samples show Zn in a significant amount but without any evident correlation with crystal structure or color. The total Cr content is higher in the yellow/orange samples than in the blue kyanite (Figure 10). However, LA-ICPMS sections do not show conclusive information regarding the distribution of Cr due to concentration close to the limit of detection of 15 ppm (Figure 13). Moreover, kyanite colored by Cr^{3+} shows absorption spectra dominated by two broad absorption bands centered at 405–420 nm and 595–625 nm [72,73]. No such absorptions are observed in any of our samples, making Cr^{3+} an unrealistic chromophore in this case. This is consistent with Cr content very low in our samples (typically 1 to 20 ppm; Figure 10; Table S1c) compared to blue kyanite colored by dispersed Cr^{3+} (typically Cr_2O_3 > 1 wt. %; [73]). Hereafter, only Mn^{3+}, Fe^{3+} and Fe^{2+}–Ti^{4+} substitutions with Al^{3+} are thus discussed. The kyanite crystals analyzed by UV-near infrared spectroscopy show weak absorption bands at 370, 381, 432 and 446 nm due to isolated Fe^{3+} [7]. The large absorption bands at 520 and 620 nm, responsible for the blue color of THA16, are due to Fe^{2+}–Ti^{4+} inter-valence charge transfer (IVCT; [74–76]). No spectral features due to the Fe^{2+}-Fe^{3+} IVCT were found. In the spectra of yellow (THA08) and orange (THA13) kyanite samples, the band at 470 nm, that importantly contributes to color, is consistent with the main band attributed to isolated Mn^{3+} [9]. Hence, the origin of the yellow to orange color in our samples is attributed to isolated Mn^{3+}, with some minor contribution of isolated Fe^{3+}. The contribution of the weak 890 nm band remains unattributed.

EPMA X-ray maps and LA-ICPMS profiles show some relation between the blue zones and the Ti content of the zoned kyanite crystals THA16 and THA08 (Figure 12a,b and Figure 13a,b). Interestingly, orange kyanite (THA13) contains more TiO_2 than blue, zoned crystals. This shows that the presence of Ti alone is not sufficient to generate a blue color: Ti^{4+} atoms need to be crystallographically linked to Fe^{2+} atoms for the charge-transfer phenomenon to occur, and hence the blue color to appear. If for any reason, the charge transfer does not occur (titanium atoms may be isolated, or iron ions may be trivalent), the blue color due to IVCT does not appear. A similar behavior is documented in other colored minerals such as corundum: some yellow sapphires sometimes contain more titanium than blue sapphires colored by Fe^{2+}–Ti^{4+} IVCT, e.g., [77]. Again, in this case, titanium concentration alone is not correlated with the blue color.

Orange kyanite is associated with green andalusite in quartz lenses (Figure 3f). Mn^{3+} might substitute for Al^{3+} in andalusite mineral formula, resulting in deep green, Mn-rich andalusite (Al, $Mn)_2SiO_5$ [10,78]. This andalusite variety is named viridine in reason of its brilliant green color [12]. Viridine occurrences (Germany, Belgium, Japan, Australia, Brazil and Tanzania) usually provide small or altered crystals [16,78]. Large viridine crystals of the Trikorfo area reach up to ten centimeters large, are vivid-to-deep green and unaltered. Their occurrence is promising for exploration of gem-quality, marketable viridine in the area. Compared to kyanite and sillimanite, andalusite incorporates maximum contents of Mn^{3+} due to its crystal structure [2]. Experimental works quantified that the Mn_2SiO_5 end-member cannot be higher than 6 mol% for kyanite and 1.7 mol % for sillimanite, but can reach 22 mol % for andalusite [2,79]. In the present study, Mn_2O_3 can reach up to 15.98 wt. %, yielding 17.5 mol % $Mn_2Si_2O_5$, close to the maximum possibly accepted by andalusite.

In the Trikorfo area, epidote group minerals are represented by green epidote, pink to red Mn-rich epidote and Mn-rich clinozoisite (variety clinothulite), purple piemontite, and allanite, which is only reported at the thin-section scale by Voudouris et al. [20]. Epidote, clinozoisite and piemontite represent the Fe-, Al- and Mn-rich end-members respectively of the epidote mineral family. Solid solution involves substitution of trivalent cations (Al^{3+}, Fe^{3+} or Mn^{3+}) in three non-equivalent octahedral sites in the basic formula for epidote $Ca_2M^{3+}Al_2Si_3O_{12}(OH)$ [66,67]. True piemontite (up to 12.69 wt. % Mn_2O_3) is restricted to Mn-rich andalusite bearing samples, indicating that these mica schist layers are locally particularly enriched in manganese. Mn-bearing epidote (up to 7.76 wt. % Mn_2O_3) is usually associated with spessartine-Mn oxides-quartz assemblages. The Mn-poor clinozoisite/epidote and Mn-poor clinozoisite (clinothulite) from the Trikorfo area contain up to 0.72 wt. % Mn_2O_3 which is responsible for their pink to red color. In accordance to Bonazzi and Menchetti [80] and Franz and Liebscher [67] the coloration in pink epidote/clinozoisite is due to Mn^{3+} and not Mn^{2+} in the mineral structure.

Four types of garnet are recognizable on the basis of their colors: (i) syn-metamorphic millimetric red almandine (± pyrope)-spessartine, (ii) plurimillimetric Ca–Fe-rich yellow to orange spessartine within the green horizons, (iii) centimetric almost pure pale yellowish brown grossular associated with epidote and clinozoisite in calc-silicate aggregates, and (iv) centimetric deep amber orange almost pure spessartine associated with Mn-rich andalusite, muscovite and orange kyanite. The colors of these garnet crystals directly reflect their main dominant cation (Ca^{2+}, Fe^{2+} or Mn^{2+}). Mn-rich garnet is found within the metamorphic assemblage and the composition to almost pure spessartine is only very localized to certain mica schist layers in association with orange kyanite and green andalusite, and to brittle veins cross-cutting Mn-rich horizons close to the top of the Trikorfo area.

5.2. Metamorphic Parageneses versus Metasomatic Assemblages

The Trikorfo area is dominated by garnet–kyanite–biotite–hematite–plagioclase ± staurolite ± sillimanite paragneisses/mica schists (sample CS16_292b), with intercalations of metacarbonate units near the base and top of the formation and local calc-silicate layers and Mn-rich mica schists, also close to the top of the formation. Field and mineralogical observations suggest that the epidote-grossular-bearing calc-silicate layers represent local inhomogeneities in the bedding of the sedimentary protoliths and/or compositional changes that have been produced by localized metasomatic processes during regional metamorphism, as suggested for similar layers at Therapio, Evros area [81]. In this sense, the mineralogy of the calc-silicate layers could have developed during prograde metamorphism of a Mn-rich, calcareous pelitic protolith, followed by vein formation and local metasomatic reactions during retrograde metamorphism accompanying the exhumation of Thassos Island during the Oligocene-Miocene.

Alternatively, the skarn-similar mineralogy of the calc-silicate layers (e.g., grossular, diopside, hornblende and epidote) could have been formed by fluids released by granitoids during contact metamorphism with the carbonate (calcite and dolomite) rocks found at the top of the Trikorfo area. With the lack of radiometric data, the exact age (i.e. Hercynian?) of the orthogneiss at Trikorfo is not known but is likely older than alpine deformation. Although not exposed on Thassos Island, Oligo-Miocene magmatism is described nearby in the Rhodope area with the intrusion of syntectonic plutons 30 km to the north of Thassos Island, e.g., [52–54]. Melfos and Voudouris [27] considered that widespread gold mineralization in the area is related to this Miocene magmatic activity. Tourmaline at Trikorfo also occurs as a retrograde mineral, resulting in tourmalinite and quartz-tourmaline veins cross-cutting the metamorphic foliation. According to van Hinsberg et al. [82] metasomatic introduction of boron on the retrograde path is most commonly associated with the intrusion of late granites in orogenic belts.

The Trikorfo area shows the superposition of late secondary minerals over the metamorphic paragenesis. These silicates and calc-silcates, including a large variety of garnet and epidote composition, kyanite, andalusite, diopside, hornblende, Ti-hematite, magnetite and tourmaline indicate post-metamorphic metasomatic reactions where hot, chemically active fluids that altered the metamorphic assemblage. Furthermore Mn-rich mineralogical assemblages are found. This type of assemblage is relatively rare on Earth because it requires highly oxidized environments [15–18]. The large diversity of mineral compositions and assemblages observed at the scale of the Trikorfo area points to a local control on the mineralogy and fO_2 conditions during post-metamorphic metasomatic reactions.

For example, five different garnet-epidote assemblages are described with increasing oxidation state as indicated by the amount of Mn_2O_3 in the epidote: (i) Mn-poor grossular (up to 3.60 wt.% MnO) and Mn-poor epidote to clinozoisite (up to 1.01 wt. % Mn_2O_3) within the calc-silicate aggregates, (ii) Mn-poor spessartine (up to 24.91 wt. % MnO) associated with Mn-poor epidote (up to 1.16 wt. % Mn_2O_3) in the green horizons within gneiss host-rock, (iii) Mn-rich spessartine (up to 36.95 wt. % MnO) with Mn-rich epidote (up to 7.76 wt. % Mn_2O_3) within the spessartine–epidote–braunite–hornblende schist, (iv) almost pure spessartine (up to 42.87 wt. % MnO) associated with piemontite (up to 12.69 wt. % Mn_2O_3), Mn-rich andalusite (up to 15.98 wt. % Mn_2O_3) and orange Mn-bearing kyanite in piemontite–andalusite–muscovite–kyanite mica schists, and (v) spessartine-clinochlore-epidote (up to 6.67 wt. % Mn_2O_3). In concordance with the experimental observations of Keskinen and Liou [18], garnet which coexists with piemontite is uniformly more spessartine-rich.

Garnet and epidote are minerals showing large Fe–Al–Mn solid solutions sensitive to PT and fO_2 environmental conditions. In the case of epidote composition, oxygen fugacity is more important than temperature [18]. Due to extensive solid solution in terms of Mn/Fe and (Mn + Fe)/Al, Mn-rich garnet and epidote may coexist with element ratio of the different phases sensitive to temperature and degree of oxidation. With decreasing fO_2, piemontite becomes poorer in Mn and garnet and epidote minerals might tend to become more aluminous with increasing temperature [18].

Spessartine can be stable over a wide range of fO_2 and PT conditions from greenschist to amphibolite-granulite facies and is evidence of elevated amounts of Mn in the rock [15,83,84]. Natural occurrences of piemontite have shown evidence for metamorphism at high oxygen fugacity [18]. A highly oxidizing environment is shown by the presence of Ti-hematite, Fe^{3+}- and Mn^{3+}-rich epidote, Mn^{3+}-rich andalusite, Fe^{3+}–Mn^{3+}-rich kyanite, and so on. The intercalation of decimeter-thick layers of epidote-spessartine-Mn-rich andalusite in less oxidized levels with epidote-spessartine(almandine) assemblages implies (i) significant fO_2 gradients at restricted scale, and (ii) internally controlled fO_2 resulting from oxidized protoliths containing minerals such as Mn-oxides capable of buffering fO_2 to high pressure levels during metamorphism [17,18]. Preservation of the fO_2 gradients supports evidence for a reduced mobility of oxygen throughout high-pressure metamorphism [17] or the localized circulation of highly oxidized fluids during metasomatism. The strong variability of fO_2 dependent mineral assemblages, at all scales, within the Trikorfo area thus indicates a strong local buffering of fO_2 during post-metamorphic metasomatic reactions. As for fO_2, the mineralogical assemblages also suggest a local control in Mn concentration. Indeed, Mn-rich layers, which are mostly found at the top of the series, are scarce and never exceed a few centimeters thick.

Orange kyanite was found coexisting together with Mn-rich andalusite in quartz-feldspar coarse-grained veins and as post-deformation assemblages in mica schists towards the top of the unit in the Trikorfo area. Mn-rich andalusite may also have formed during decompression in veins within host rocks containing kyanite, e.g., [85]. However, petrographic observations suggest growth at equilibrium between kyanite and andalusite. Co-existing Al_2SiO_5 polymorphs occur in various metamorphic rocks and may form either during regional metamorphism or a combination of regional and contact metamorphism [86,87]. When kyanite, sillimanite and andalusite are free of chemical impurities, their equilibrium coexistence is invariant. However, these three polymorphs show extensive solid solution and can incorporate significant amounts of Fe^{3+} and Mn^{3+}, substituting for Al, in highly oxidized rocks [16,88,89]. This results in an increase of the variance of the system and the possible coexistence of kyanite, andalusite and sillimanite in mutual equilibrium over a measurable range of PT conditions [90]. The presence of Mn-bearing andalusite, and especially the association with braunite, quartz and spessartine, indicates high fO_2 [15].

Structural and mineralogical observations thus point to a two-stage evolution of the Trikorfo area, where metamorphic units were affected locally by metasomatic events during exhumation. Fluid circulation leads to fluid-rock interactions and crystallization of large, vividly-colored minerals. The chemistry of the metasomatic mineralogical assemblages is controlled by the chemistry of the protolith more or less enriched in manganese and by local fO_2 buffer.

5.3. PT-Deformation Conditions of Metamorphic Equilibrium and Metasomatic Reactions

The Rhodope domain exposes a large variety of HP-rocks of the Variscan continental crust and Mesozoic sediments that were subjected to Alpine subduction (Early Cretaceous-Eocene) and subsequent exhumation (Early Oligocene-Miocene) [91]. The recorded conditions can be as high as >19 kbars and 750–800 °C with local partially amphibolitized kyanite eclogites, e.g., [37,91].

Thassos Island, belonging to the South Rhodope metamorphic core complex, displays Oligocene to Miocene metamorphic and exhumation history [36,40]. In response to crustal thickening during Alpine tectonics, the geothermobarometric analyses of amphibole gneisses suggest metamorphic conditions of approximately 620 °C/4.7 kbars for the lower unit and 580 °C/2.4 kbars for the intermediate unit [47,48]. On the basis of the observation of a garnet–kyanite–sillimanite assemblage, Dimitriadis [24] estimated the conditions achieved by the mica schists of the Trikorfo area around 5.5 ± 1.5 kbars and 600 ± 50 °C. Brun and Sokoutis [40] reported an unreferenced personal communication of Kostopoulos (2004) stating that the garnet amphibolites yielded temperatures ranging from 545 to 660 °C, as described in Dimitriadis [24], but at significantly higher pressures from 6.5 to 9.5 kbars. On the basis of the garnet–staurolite–kyanite–plagioclase–quartz–phlogopite–Ti–hematite assemblage, our thermodynamic modeling indicates metamorphic conditions at Trikorfo of ca. 9.1 ± 1.3 kbars and 670 ± 40 °C, at significantly higher P and T conditions than previously estimated. The field given in Figure 15 is the only one where kyanite and staurolite are stable together. Instead of staurolite, Dimitriadis [24] reported the presence of sillimanite growing at the expense of kyanite within the metamorphic gneiss. The conditions of the Trikorfo formation should rather stand in the high-temperature/low-pressure part of the domain indicated by our stable metamorphic assemblage, with sillimanite post-dating kyanite. The MnO content of the calculated equilibrated garnet composition is a bit higher than that measured in sample CS16_292b. It is probably due to the fact that the solid solution model used in our thermodynamic calculation is not fully appropriate. The addition of MnO in the system has a small effect on the garnet mode but a great effect on the garnet stability field. Indeed, in agreement with White et al. [92], the incorporation of MnO in our phase diagram calculations (in the MnNCKFMASHT(O) system) leads to a shift to lower PT conditions of the garnet-in line, compared to a system without MnO. The PT conditions obtained for the peak metamorphic assemblage recorded by the kyanite-bearing paragneisses are in good agreement with the geological context and therefore we are confident with the calculated phase diagram, even if the modeled garnet composition is not strictly the same than the measured one. Indeed, this MP-HT metamorphism recorded by paragneisses from the intermediate unit is comparable to those estimated at the Thassos scale. This amphibolite metamorphism, which is associated with the continuous development of the extensive low-angle S_n foliation since ductile to brittle deformation, could mark the beginning of the nappe stack exhumation.

Kyanite and kyanite–andalusite assemblages are found at the interface between metamorphic host rock and quartzo–feldspathic veins. The crystals are generally weakly deformed and aligned parallel to regional lineation. This indicates a local reaction where a SiO_2-rich fluid interacts with the Al_2O_3-rich host-rock [93], as indicated by their Cr/V ratio [4], to form either kyanite or kyanite-andalusite as a function of the *PT* conditions and the variance of the system linked to available MnO [90]. This reaction occurs at ductile conditions while the rock still records stretching lineation. The temperatures derived from the Ti-in kyanite thermometer from Müller et al. [4] are in the range 670–790 °C coherent with the Perple_X modeling. This would then indicate that metasomatic kyanite form at or close to the metamorphic peak conditions. However, EPMA and LA-ICPMS data (Figures 12 and 13) show that Ti distribution is variable within the crystals and these temperatures should be considered with caution. Except for some of the kyanite and andalusite crystals, the other metasomatic silicate and calc-silicate minerals of Thassos are usually not oriented. Dimitriadis [24] suggested retrogression under static conditions as evidenced by undeformed matted fibrolite and unoriented growth of chlorite. However, continuous deformation during the main kinematic event is evidenced by coherent stretching and mineral lineation and brittle vein opening to temperatures as low as 230 °C, as indicated by clinochlore growth in late brittle veins. The presence of non-oriented minerals should be the result of crystal growth from hydrothermal fluids during metasomatic reactions. The presence of calc-silicate minerals (grossular, diopside, hornblende, epidote), especially located to the top of the series where the paragneisses/mica schists are in contact with marble units, implies decarbonation during metasomatic reactions.

6. Conclusions

The garnet–kyanite–biotite–hematite–plagioclase ± staurolite ± sillimanite paragneiss of the Trikorfo area (Thassos Island, Rhodope Massif, northern Greece) recorded metamorphic conditions of ca. 670 ± 40 °C and ca. 9.1 ± 1.3 kbars. Structural, petrographic and mineralogical observations indicate that localized metasomatic reactions occurred during the exhumation of the HP unit locally containing layers enriched in Mn with a strong local fO_2 buffering. Metasomatism is likely the result of Miocene intrusion of granitoids, which are related to ore deposits on Thassos Island. Metasomatic reactions first occurred under ductile conditions in an extensive context close to the kyanite-andalusite stability curve and continued until purely brittle conditions as indicated by the presence of late veins cross-cutting the metamorphic foliation. The result is a mineralogical site that could be regarded as a unique geotope with uncommon (Mn-rich) silicate and calc-silicate mineral assemblages similar to what is observed in skarn deposits where hot active hydrothermal fluids interact with carbonate rocks.

Crystals of kyanite, green andalusite, garnet (grossular and spessartine) and red epidote–clinozoisite are large (up to several centimeters), show vivid colors, and are suitable for cabochon-shaped gemstones. As such, the Trikorfo locality can be regarded as a promising area for the exploration of true, facetable gemstones. Their genesis due to metasomatic reactions also underlines the important role of metasomatism for gemstone formation in general, as previously noted in the literature [94–96].

Our combined petrographic, geochemical and spectroscopic observations permit us to demonstrate that the orange color of kyanite in the Trikorfo area is mainly due to Mn^{3+} substitution. Although the Mn content is significantly lower than that in the Loliondo (Tanzania) deposit, Thassos Island (Rhodope, Greece) can now be classified as the second locality worldwide where Mn-rich orange kyanite is reported.

Author Contributions: Conceptualization, A.T.; Data curation, A.T.; Funding acquisition, A.T. and P.V.; Investigation, A.T., P.V., A.E., C.S., K.T., M.B., B.R., C.M., I.G., M.E. and C.P.; Software, A.E.; Writing—original draft, A.T. and P.V.

Acknowledgments: The authors warmly acknowledge A. Flammang, J. Moine and O. Rouer (GeoRessources lab., Univ. Lorraine, France) and P. Stutz and S. Heidrich (Institute of Mineralogy and Petrology, University of Hamburg, Germany) for the quality of sample preparation and EPMA analyses. We are grateful to R. Mosser-Ruck (GeoRessources lab., Univ. Lorraine, France) for fruitful discussions. Three anonymous journal reviewers are thanked for their helpful and constructive comments.

Appendix A

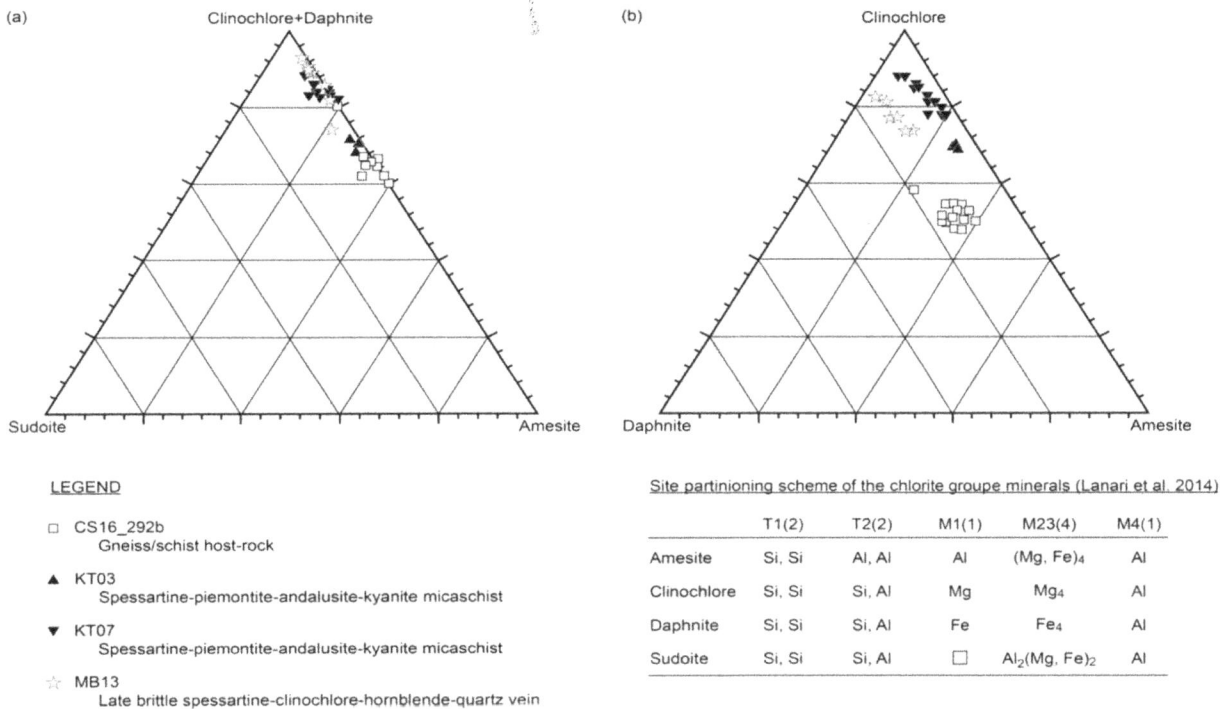

LEGEND

☐ CS16_292b
   Gneiss/schist host-rock

▲ KT03
   Spessartine-piemontite-andalusite-kyanite micaschist

▼ KT07
   Spessartine-piemontite-andalusite-kyanite micaschist

☆ MB13
   Late brittle spessartine-clinochlore-hornblende-quartz vein

Site partinioning scheme of the chlorite groupe minerals (Lanari et al. 2014)

	T1(2)	T2(2)	M1(1)	M23(4)	M4(1)
Amesite	Si, Si	Al, Al	Al	$(Mg, Fe)_4$	Al
Clinochlore	Si, Si	Si, Al	Mg	Mg_4	Al
Daphnite	Si, Si	Si, Al	Fe	Fe_4	Al
Sudoite	Si, Si	Si, Al	☐	$Al_2(Mg, Fe)_2$	Al

Figure A1. Composition of chlorite group minerals analyzed by EPMA in (**a**) Clinochlore + Daphnite–Sudoite–Amesite and (**b**) Clinochlore–Daphnite–Amesite ternary plots. The mineral names are according the site partitioning scheme used in Lanari et al. [97].

Appendix B

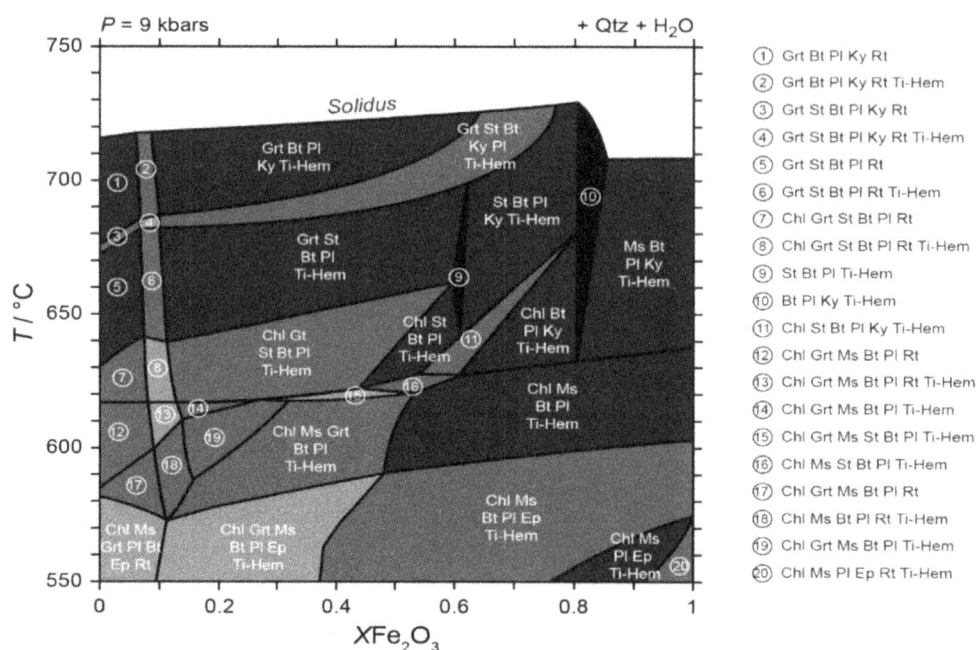

Figure A2. T-XFe$_2$O$_3$ pseudosection calculated for 9 kbars and 550–750 °C for sample CS16_292b. Assemblage fields are shaded according to the number of degrees of freedom, with higher-variance assemblages represented by darker shading. Mineral abbreviation: biotite (Bt), chlorite (Chl), epidote (Ep), garnet (Grt), kyanite (Ky), muscovite (Ms), plagioclase (Pl), quartz (Qtz), rutile (Rt), staurolite (St) and Ti-hematite (Ti-Hem).

References

1. Abs-Wurmbach, I.; Langer, K. Synthetic Mn^{3+}-kyanite and viridine, (Al$_{2-x}$Mn$_x^{3+}$)SiO$_5$, in the system Al$_2$O$_3$-MnO-MnO$_2$-SiO$_2$. *Contrib. Mineral. Petrol.* **1975**, *49*, 21–38. [CrossRef]

2. Abs-Wurmbach, I.; Langer, K.; Seifert, F.; Tillmanns, E. The crystal chemistry of (Mn^{3+}, Fe^{3+})-substituted andalusites (viridines and kanonaite), (Al$_{1-x-y}$Mn$_x^{3+}$Fe$_y^{3+}$)$_2$(O|SiO$_4$): Crystal structure refinements, Mössbauer, and polarized optical absorption spectra. *Z. Für Krist.* **1981**, *155*, 81–113.

3. Pearson, G.R.; Shaw, D.M. Trace elements in kyanite, sillimanite and andalusite. *Am. Mineral.* **1960**, *45*, 808–817.

4. Müller, A.; van den Kerkhof, A.M.; Selbekk, R.S.; Broekmans, M.A.T.M. Trace element composition and cathodoluminescence of kyanite and its petrogenetic implications. *Contrib. Mineral. Petrol.* **2016**, *171*, 70. [CrossRef]

5. Faye, G.H.; Nickel, E.H. On the origin of colour and pleochroism of kyanite. *Can. Mineral.* **1969**, *10*, 35–46.

6. Meinhold, K.D.; Frisch, T. Manganese-silicate-bearing metamorphic rocks from central Tanzania. *Schweiz. Mineral. Petrogr. Mitt.* **1970**, *50*, 493–507.

7. Chadwick, K.M.; Rossman, G.R. Orange kyanite from Tanzania. *Gems Gemol.* **2009**, *45*, 146–147.

8. Gaft, M.; Nagli, L.; Panczer, G.; Rossman, G.R.; Reisfeld, R. Laser-induced time-resolved luminescence of orange kyanite Al$_2$SiO$_5$. *Opt. Mater.* **2011**, *33*, 1476–1480. [CrossRef]

9. Wildner, M.; Beran, A.; Koller, F. Spectroscopic characterisation and crystal field calculations of varicoloured kyanites from Loliondo, Tanzania. *Mineral. Petrol.* **2013**, *107*, 289–310. [CrossRef]

10. Hålenius, U. A spectroscopic investigation of manganian andalusite. *Can. Mineral.* **1978**, *16*, 567–575.

11. Smith, G.; Hålenius, U.; Langer, K. Low temperature spectral studies of Mn^{3+}-bearing andalusite and epidote type minerals in the range 30,000–5000 cm^{-1}. *Phys. Chem. Miner.* **1982**, *8*, 136–142. [CrossRef]

12. Klemm, G. Uber Viridin, eine Abart des Andalusites. *Not. Ver Erdk Darmstadt* **1911**, *32*, 4–13.

13. Burns Roger, G. *Mineralogical Applications of Crystal Field Theory*, 2nd ed.; Cambridge University Press: Cambridge, UK, 1970; ISBN 0-521-43077-1.

14. Manning, P.G. The optical absorption spectra of the garnets almandine-pyrope, pyrope and spessartine and some structural interpretations of mineralogical significance. *Can. Mineral.* **1967**, *9*, 237–251.

15. Abs-Wurmbach, I.; Peters, T. The Mn-Al-Si-O system: An experimental study of phase relations applied to parageneses in manganese-rich ores and rocks. *Eur. J. Mineral.* **1999**, *11*, 45–68. [CrossRef]

16. Jöns, N.; Schenk, V. Petrology of whiteschists and associated rocks at Mautia Hill (Tanzania): Fluid infiltration during high-grade metamorphism. *J. Petrol.* **2004**, *45*, 1959–1981. [CrossRef]

17. Reinecke, T. Phase relationships of sursassite and other Mn-silicates in highly oxidized low-grade, high-pressure metamorphic rocks from Evvia and Andros islands, Greece. *Contrib. Mineral. Petrol.* **1986**, *94*, 110–126. [CrossRef]

18. Keskinen, M.; Liou, J.G. Stability relations of Mn–Fe–Al piemontite. *J. Metamorph. Geol.* **1987**, *5*, 495–507. [CrossRef]

19. Voudouris, P.; Graham, I.; Melfos, V.; Sutherland, L.; Zaw, K. Gemstones in Greece: Mineralogy and Crystallizing Environment. In Proceedings of the 34th IGC Conference, Brisbane, Australia, 5–10 August 2012.

20. Voudouris, P.; Graham, I.; Mavrogonatos, K.; Su, S.; Papavasiliou, K.; Farmaki, M.-V.; Panagiotidis, P. Mn-andalusite, spessartine, Mn-grossular, piemontite and Mn-zoisite/clinozoisite from Trikorfo, Thassos island, Greece. *Bull. Geol. Soc. Greece* **2016**, *50*, 2068–2078. [CrossRef]

21. Klemme, S.; Berndt, J.; Mavrogonatos, C.; Flemetakis, S.; Baziotis, I.; Voudouris, P.; Xydous, S. On the color and genesis of prase (green quartz) and Amethyst from the island of Serifos, Cyclades, Greece. *Minerals* **2018**, *8*, 487. [CrossRef]

22. Ottens, B.; Voudouris, P. *Griechenland: Mineralien, Fundorte, Lagerstätten*; Christian Weise Verlag: München, Germany, 2018.

23. Voudouris, P.; Melfos, V.; Mavrogonatos, C.; Tarantola, A.; Götze, J.; Alfieris, D.; Maneta, V.; Psimis, I. Amethyst Occurrences in Tertiary Volcanic Rocks of Greece: Mineralogical, Fluid Inclusion and Oxygen Isotope Constraints on Their Genesis. *Minerals* **2018**, *8*, 324. [CrossRef]

24. Dimitriadis, E. Sillimanite grade metamorphism in Thasos island, Rhodope massif, Greece and its regional significance. *Geol. Rhodopica* **1989**, *1*, 190–201.

25. Voudouris, P.; Constantinidou, S.; Kati, M.; Mavrogonatos, C.; Kanellopoulos, C.; Volioti, E. Genesis of alpinotype fissure minerals from Thasos island, Northern Greece—Mineralogy, mineral chemistry and crystallizing environment. *Bull. Geol. Soc. Greece* **2013**, *47*, 468–476. [CrossRef]

26. Vason, M.; Martin, S. Metamorphosed iron-manganese deposits from the island of Thassos (Western Rhodope region, northern Greece). *Ofioliti* **1993**, *18*, 181–186.

27. Melfos, V.; Voudouris, P. Fluid evolution in Tertiary magmatic-hydrothermal ore systems at the Rhodope metallogenic province, NE Greece. A review. *Geol. Croat.* **2016**, *69*, 157–167. [CrossRef]

28. Zachos, S. Geological Map of Greece, Thassos Sheet. 1982. Available online: https://www.worldcat.org/title/geological-map-of-greece-thassos-sheet-phyllo-thasos-institogto-geologikon-kai-metalletikoi-ereunon-i-geologiki-chartografioi-egiye-apo-to-s-zachos/oclc/492645828 (accessed on 24 April 2019).

29. Su, S.; Graham, I.; Voudouris, P.; Mavrogonatos, K.; Farmaki, M.V.; Panagiotidis, P. *Viridine, Piemontite and Epidote Group Minerals from Thassos Island, Northern Greece*; AGI: Alexandria, VA, USA, 2016; T37.P2; p. 1715.

30. Voudouris, P. Minerals of Eastern Macedonia and Western Thrace: Geological framework and environment of formation. *Bull. Geol. Soc. Greece* **2005**, *37*, 62–77.

31. Voudouris, P.; Melfos, V.; Katerinopoulos, A. Precious stones in Greece: Mineralogy and geological environment of formation. In Proceedings of the Understanding the Genesis of Ore Deposits to Meet the Demand of the 21st Century, Moscow, Russia, 21–24 August 2006; p. 6.

32. Le Pichon, X.; Bergerat, F.; Roulet, M.J. Plate kinematics and tectonics leading to the Alpine belt formation: A new analysis. In *Processes in Continental Lithospheric Deformation*; Clark, S.P., Burchfiel, B.C., Suppe, J., Eds.; Geological Society of America: Boulder, CO, USA, 1988; Volume 218, pp. 111–131.

33. Dewey, J.F.; Helman, M.L.; Knott, S.D.; Turco, E.; Hutton, D.H.W. Kinematics of the western Mediterranean. *Geol. Soc. Lond. Spec. Publ.* **1989**, *45*, 265–283. [CrossRef]

34. Bonneau, M. La Tectonique de L'arc égéen Externe et du Domaine Cycladique. Ph.D. Thesis, Université Pierre et Marie Curie, Paris, France, 1991; p. 443.

35. van Hinsbergen, D.J.J.; Hafkenscheid, E.; Spakman, W.; Meulenkamp, J.E.; Wortel, R. Nappe stacking resulting from subduction of oceanic and continental lithosphere below Greece. *Geology* **2005**, *33*, 325–328. [CrossRef]

36. Wawrzenitz, N.; Krohe, A. Exhumation and doming of the Thasos metamorphic core complex (S Rhodope, Greece): Structural and geochronological constraints. *Tectonophysics* **1998**, *285*, 301–332. [CrossRef]

37. Liati, A.; Seidel, E. Metamorphic evolution and geochemistry of kyanite eclogites in central Rhodope, northern Greece. *Contrib. Mineral. Petrol.* **1996**, *123*, 293–307. [CrossRef]

38. Liati, A.; Gebauer, D. Constraining the prograde and retrograde P-T-t path of Eocene HP rocks by SHRIMP dating of different zircon domains: Inferred rates of heating, burial, cooling and exhumation for central Rhodope, northern Greece. *Contrib. Mineral. Petrol.* **1999**, *135*, 340–354. [CrossRef]

39. Liati, A. Identification of repeated Alpine (ultra) high-pressure metamorphic events by U–Pb SHRIMP geochronology and REE geochemistry of zircon: The Rhodope zone of Northern Greece. *Contrib. Mineral. Petrol.* **2005**, *150*, 608–630. [CrossRef]

40. Brun, J.-P.; Sokoutis, D. Kinematics of the Southern Rhodope Core Complex (North Greece). *Int. J. Earth Sci.* **2007**, *96*, 1079–1099. [CrossRef]

41. Moulas, E.; Kostopoulos, D.; Connolly, J.A.D.; Burg, J.-P. P-T estimates and timing of the sapphirine-bearing metamorphic overprint in kyanite eclogites from Central Rhodope, northern Greece. *Petrology* **2013**, *21*, 507–521. [CrossRef]

42. Dürr, S.; Altherr, R.; Keller, J.; Okrusch, M.; Seidel, E. The Median Aegean Crystal-line Belt: Stratigraphy, Structure, Metamorphism, Magmatism. *IUCG Sci. Rep.* **1978**, *38*, 455–477.

43. Jacobshagen, V.; Dürr, S.; Kockel, F.; Kopp, K.O.; Kowalczyk, G.; Berckhemer, H. Structure and Geodynamic Evolution of the Aegean Region. In *Alps, Apennines, Hellenides*; Cloos, H., Roeder, D., Schmidt, K., Eds.; IUCG: Stuttgart, Germany, 1978; pp. 537–564.

44. Lister, G.S.; Banga, G.; Feenstra, A. Metamorphic core complexes of Cordilleran type in the Cyclades, Aegean Sea, Greece. *Geology* **1984**, *12*, 221–225. [CrossRef]

45. Melfos, V.; Voudouris, P. Cenozoic metallogeny of Greece and potential for precious, critical and rare metals exploration. *Ore Geol. Rev.* **2017**, *89*, 1030–1057. [CrossRef]

46. Liati, A.; Fanning, C.M. Eclogites and Country rock orthogneisses representing upper Permian Gabbros in Hercynian Granitoids, Rhodope, Greece: Geochronological Constraints. *Mitt. Österr. Mineral. Ges.* **2005**, *150*, 88.

47. Schulz, B. Syntectonic heating and loading-deduced from microstructures and mineral chemistry in micaschists and amphibolites of the Pangeon complex (Thassos island, Northern Greece). *Neues Jahrb. Für Geol. Paläontol. Abh.* **1992**, *184*, 181–201.

48. Bestmann, M.; Kunze, K.; Matthews, A. Evolution of a calcite marble shear zone complex on Thassos Island, Greece: Microstructural and textural fabrics and their kinematic significance. *J. Struct. Geol.* **2000**, *22*, 1789–1807. [CrossRef]

49. Dinter, D.A. Late Cenozoic extension of the Alpine collisional orogen, northeastern Greece: Origin of the north Aegean basin. *Gsa Bull.* **1998**, *110*, 1208–1230. [CrossRef]

50. Kounov, A.; Wüthrich, E.; Seward, D.; Burg, J.-P.; Stockli, D. Low-temperature constraints on the Cenozoic thermal evolution of the Southern Rhodope Core Complex (Northern Greece). *Int. J. Earth Sci.* **2015**, *104*, 1337–1352. [CrossRef]

51. Peterek, A.; Polte, M.; Wölfl, C.; Bestmann, M.; Lemtis, O. Zur jungtertiären geologischen Entwicklung im SW der Insel Thassos (S-Rhodope, Nordgriechenland). *Erlanger Geol. Abh.* **1994**, *124*, 29–59.

52. Dinter, D.A.; Royden, L. Late Cenozoic extension in northeastern Greece: Strymon Valley detachment system and Rhodope metamorphic core complex. *Geology* **1993**, *21*, 45–48. [CrossRef]

53. Sokoutis, D.; Brun, J.P.; Driessche, J.V.D.; Pavlides, S. A major Oligo-Miocene detachment in southern Rhodope controlling north Aegean extension. *J. Geol. Soc.* **1993**, *150*, 243–246. [CrossRef]

54. Kilias, A.A.; Mountrakis, D.M. Tertiary extension of the Rhodope massif associated with granite emplacement (Northern Greece). *Acta Vulcanol.* **1998**, *10*, 331–337.

55. Carignan, J.; Hild, P.; Mevelle, G.; Morel, J.; Yeghicheyan, D. Routine Analyses of Trace Elements in Geological Samples using Flow Injection and Low Pressure On-Line Liquid Chromatography Coupled to ICP-MS: A Study of Geochemical Reference Materials BR, DR-N, UB-N, AN-G and GH. *Geostand. Newsl.* **2001**, *25*, 187–198. [CrossRef]

56. Pouchou, J.-L.; Pichoir, F. *Quantitative Analysis of Homogeneous or Stratified Microvolumes Applying the Model "PAP."* In *Electron Probe Quantitation*; Heinrich, K.F.J., Newbury, D.E., Eds.; Springer US: Boston, MA, USA, 1991; pp. 31–75. ISBN 978-1-4899-2617-3.

57.   Jochum, K.P.; Weis, U.; Stoll, B.; Kuzmin, D.; Yang, Q.; Raczek, I.; Jacob, D.E.; Stracke, A.; Birbaum, K.; Frick, D.A.; et al. Determination of Reference Values for NIST SRM 610–617 Glasses Following ISO Guidelines. *Geostand. Geoanalytical Res.* **2011**, *35*, 397–429. [CrossRef]

58.   Paton, C.; Hellstrom, J.; Paul, B.; Woodhead, J.; Hergt, J. Iolite: Freeware for the visualisation and processing of mass spectrometric data. *J. Anal. At. Spectrom.* **2011**, *26*, 2508–2518. [CrossRef]

59.   Connolly, J.A.D. Computation of phase equilibria by linear programming: A tool for geodynamic modeling and its application to subduction zone decarbonation. *Earth Planet. Sci. Lett.* **2005**, *236*, 524–541. [CrossRef]

60.   Holland, T.J.B.; Powell, R. An improved and extended internally consistent thermodynamic dataset for phases of petrological interest, involving a new equation of state for solids. *J. Metamorph. Geol.* **2011**, *29*, 333–383. [CrossRef]

61.   White, R.W.; Powell, R.; Holland, T.J.B.; Worley, B.A. The effect of TiO_2 and Fe_2O_3 on metapelitic assemblages at greenschist and amphibolite facies conditions: Mineral equilibria calculations in the system $K_2O–FeO–MgO–Al_2O_3–SiO_2–H_2O–TiO_2–Fe_2O_3$. *J. Metamorph. Geol.* **2000**, *18*, 497–511. [CrossRef]

62.   White, R.W.; Powell, R.; Holland, T.J.B.; Johnson, T.E.; Green, E.C.R. New mineral activity–composition relations for thermodynamic calculations in metapelitic systems. *J. Metamorph. Geol.* **2014**, *32*, 261–286. [CrossRef]

63.   Newton, R.C.; Charlu, T.V.; Kleppa, O.J. Thermochemistry of the high structural state plagioclases. *Geochim. Cosmochim. Acta* **1980**, *44*, 933–941. [CrossRef]

64.   Bocchio, R.; Diella, V.; Adamo, I.; Marinoni, N. Mineralogical characterization of the gem-variety pink clinozoisite from Val Malenco, Central Alps, Italy. *Rend. Lincei* **2017**, *28*, 549–557. [CrossRef]

65.   Geller, S. Crystal chemistry of garnets. *Z. Krist.* **1967**, *125*, 1–45. [CrossRef]

66.   Armbruster, T.; Bonazzi, P.; Akasaka, M.; Bermanec, V.; Chopin, C.; Gieré, R.; Heuss-Assbichler, S.; Liebscher, A.; Menchetti, S.; Pan, Y.; et al. Recommended nomenclature of epidote-group minerals. *Eur. J. Mineral.* **2006**, *18*, 551–567. [CrossRef]

67.   Franz, G.; Liebscher, A. Physical and Chemical Properties of the Epidote Minerals: An introduction. *Rev. Mineral. Geochem.* **2004**, *56*, 1–81. [CrossRef]

68.   Schreyer, W.; Bernhardt, H.-J.; Fransolet, A.-M.; Armbruster, T. End-member ferrian kanonaite: An andalusite phase with one Al fully replaced by $(Mn, Fe)^{3+}$ in a quartz vein from the Ardennes mountains, Belgium, and its origin. *Contrib. Mineral. Petrol.* **2004**, *147*, 276–287. [CrossRef]

69.   Sun, S.S.; McDonough, W.F. Chemical and isotopic systematics of oceanic basalts: Implications for mantle composition and processes. *Geol. Soc. Lond. Spec. Publ.* **1989**, *42*, 313–345. [CrossRef]

70.   Cathelineau, M. Cation site occupancy in chlorites and illites as a function of temperature. *Clay Miner.* **1988**, *23*, 471–485. [CrossRef]

71.   Furukawa, Y.; Yoshiasa, A.; Arima, H.; Okube, M.; Murai, K.; Nishiyama, T. Local Structure of Transition Elements (V, Cr, Mn, Fe and Zn) in Al_2SiO_5 Polymorphs. *Aip Conf. Proc.* **2007**, *882*, 235–237.

72.   Rossman, G.R. Optical spectroscopy. In *Spectroscopic Methods in Mineralogy and Geology*; Hawthorne, F.C., Ed.; De Gruyter: Vienna, Austria, 1988; Volume 18, pp. 205–254.

73.   Platonov, A.N.; Tarashchan, A.N.; Langer, K.; Andrut, M.; Partzsch, G.; Matsyuk, S.S. Electronic absorption and luminescence spectroscopic studies of kyanite single crystals: Differentiation between excitation of FeTi charge transfer and Cr^{3+} dd transitions. *Phys. Chem. Miner.* **1998**, *25*, 203–212. [CrossRef]

74.   White, E.W.; White, W.B. Electron Microprobe and Optical Absorption Study of Colored Kyanites. *Science* **1967**, *158*, 915–917. [CrossRef] [PubMed]

75.   Rost, F.; Simon, E. Zur Geochemie und Färbung des Cyanits. *Neues Jahrb. Mineral.-Mon.* **1972**, *9*, 383–395.

76.   Parkin, K.M.; Loeffler, B.M.; Burns, R.G. Mössbauer spectra of kyanite, aquamarine, and cordierite showing intervalence charge transfer. *Phys. Chem. Miner.* **1977**, *1*, 301–311. [CrossRef]

77.   Bonizzoni, L.; Galli, A.; Spinolo, G.; Palanza, V. EDXRF quantitative analysis of chromophore chemical elements in corundum samples. *Anal. Bioanal. Chem.* **2009**, *395*, 2021–2027. [CrossRef]

78.   Novák, M.; Škoda, R. Mn^{3+}-rich andalusite to kanonaite and their breakdown products from metamanganolite at Kojetice near Třebíč, the Moldanubian Zone, Czech Republic. *J. Geosci.* **2007**, *52*, 161–167. [CrossRef]

79.   Abs-Wurmbach, I.; Langer, K.; Schreyer, W. The Influence of Mn^{3+} on the Stability Relations of the Al_2SiO_5 Polymorphs with Special Emphasis on Manganian Andalusites (Viridines), $(Al_{1-x}Mn_x^{3+})_2(O/SiO_4)$: An Experimental Investigation. *J. Petrol.* **1983**, *24*, 48–75. [CrossRef]

80. Bonazzi, P.; Menchetti, S. Manganese in Monoclinic Members of the Epidote Group: Piemontite and Related Minerals | Reviews in Mineralogy and Geochemistry. *Rev. Min. Geochem.* **2004**, *56*, 495–551. [CrossRef]

81. Kassoli-Fournaraki, A.; Michailidis, K. Chemical composition of tourmaline in quartz veins from Nea Roda and Thasos areas in Macedonia, northern Greece. *Can. Mineral.* **1994**, *32*, 607–615.

82. van Hinsberg, V.J.; Henry, D.J.; Dutrow, B.L. Tourmaline as a Petrologic Forensic Mineral: A Unique Recorder of Its Geologic Past. *Elements* **2011**, *7*, 327–332. [CrossRef]

83. Symmes, G.H.; Ferry, J.M. The effect of whole-rock MnO content on the stability of garnet in pelitic schists during metamorphism. *J. Metamorph. Geol.* **1992**, *10*, 221–237. [CrossRef]

84. Geiger, C.A.; Armbruster, T. $Mn_3Al_2Si_3O_{12}$ spessartine and $Ca_3Al_2Si_3O_{12}$ grossular garnet: Structural dynamic and thermodynamic properties. *Am. Mineral.* **1997**, *82*, 740–747. [CrossRef]

85. Kerrick, D.M. *The Al_2SiO_5 Polymorphs*; Reviews in Mineralogy; Mineralogical Society of America: Washington, DC, USA, 1990; ISBN 978-0-939950-27-0.

86. Larson, T.E.; Sharp, Z.D. Stable isotope constraints on the Al_2SiO_5 'triple-point' rocks from the Proterozoic Priest pluton contact aureole, New Mexico, USA. *J. Metamorph. Geol.* **2003**, *21*, 785–798. [CrossRef]

87. Sepahi, A.A.; Whitney, D.L.; Baharifar, A.A. Petrogenesis of andalusite–kyanite–sillimanite veins and host rocks, Sanandaj-Sirjan metamorphic belt, Hamadan, Iran. *J. Metamorph. Geol.* **2004**, *22*, 119–134. [CrossRef]

88. Abraham, K.; Schreyer, W. Minerals of the viridine hornfels from Darmstadt, Germany. *Contrib. Mineral. Petrol.* **1975**, *49*, 1–20. [CrossRef]

89. Kramm, U. Kanonaite-rich viridines from the Venn-Stavelot Massif, Belgian Ardennes. *Contrib. Mineral. Petrol.* **1979**, *69*, 387–395. [CrossRef]

90. Grambling, J.A.; Williams, M.L. The Effects of Fe^{3+} and Mn^{3+} on Aluminum Silicate Phase Relations in North-Central New Mexico, U.S.A. *J. Petrol.* **1985**, *26*, 324–354. [CrossRef]

91. Mposkos, E.; Krohe, A. Petrological and structural evolution of continental high pressure (HP) metamorphic rocks in the Alpine Rhodope Domain (N. Greece). In *Proceedings, Third International Conference on the Geology of the Eastern Mediterranean*; Panayides, I., Xenophontos, C., Malpas, J., Eds.; Geological Survey Department: Nicosia, Cyprus, 2000; pp. 221–232.

92. White, R.W.; Powell, R.; Johnson, T.E. The effect of Mn on mineral stability in metapelites revisited: New a–x relations for manganese-bearing minerals. *J. Metamorph. Geol.* **2014**, *32*, 809–828. [CrossRef]

93. El Mahi, B.; Zahraoui, M.; Hoepffner, C.; Boushaba, A.; Meunier, A.; Beaufort, D. Kyanite-quartz synmetamorphic veins: Indicators of post-orogenic thinning and metamorphism (Western Meseta, Morocco). *Pangea* **2000**, *33/34*, 27–47.

94. Simonet, C.; Fritsch, E.; Lasnier, B. A classification of gem corundum deposits aimed towards gem exploration. *Ore Geol. Rev.* **2008**, *34*, 127–133. [CrossRef]

95. Groat, L.A.; Laurs, B.M. Gem Formation, Production, and Exploration: Why Gem Deposits Are Rare and What is Being Done to Find Them. *Elements* **2009**, *5*, 153–158. [CrossRef]

96. Yakymchuk, C.; Szilas, K. Corundum formation by metasomatic reactions in Archean metapelite, SW Greenland: Exploration vectors for ruby deposits within high-grade greenstone belts. *Geosci. Front.* **2018**, *9*, 727–749. [CrossRef]

97. Lanari, P.; Wagner, T.; Vidal, O. A thermodynamic model for di-trioctahedral chlorite from experimental and natural data in the system $MgO–FeO–Al_2O_3–SiO_2–H_2O$: Applications to P–T sections and geothermometry. *Contrib. Mineral. Petrol.* **2014**, *167*, 968. [CrossRef]

9

Emerald Deposits: A Review and Enhanced Classification

Gaston Giuliani [1,*], **Lee A. Groat** [2], **Dan Marshall** [3], **Anthony E. Fallick** [4] **and Yannick Branquet** [5]

[1] Université Paul Sabatier, GET/IRD et Université de Lorraine, CRPG/CNRS, 15 rue Notre-Dame des Pauvres, BP 20, 54501 Vandœuvre cedex, France

[2] Department of Earth, Ocean and Atmospheric Sciences, University of British Columbia, Vancouver, BC V6T 1Z4, Canada; groat@mail.ubc.ca

[3] Department of Earth Sciences, Simon Fraser University, Burnaby, BC V5A 1S6, Canada; marshall@sfu.ca

[4] Isotope Geosciences Unit, S.U.E.R.C., Rankine Avenue, East Kilbride, Glasgow G75 0QF, Scotland, UK; fallickt@gmail.com

[5] Institut des Sciences de la Terre d'Orléans (ISTO), UMR 7327-CNRS/Université d'Orléans/BRGM, 45071 Orléans, France; yannick.branquet@univ-orleans.fr

* Correspondence: giuliani@crpg.cnrs-nancy.fr

Abstract: Although emerald deposits are relatively rare, they can be formed in several different, but specific geologic settings and the classification systems and models currently used to describe emerald precipitation and predict its occurrence are too restrictive, leading to confusion as to the exact mode of formation for some emerald deposits. Generally speaking, emerald is beryl with sufficient concentrations of the chromophores, chromium and vanadium, to result in green and sometimes bluish green or yellowish green crystals. The limiting factor in the formation of emerald is geological conditions resulting in an environment rich in both beryllium and chromium or vanadium. Historically, emerald deposits have been classified into three broad types. The first and most abundant deposit type, in terms of production, is the desilicated pegmatite related type that formed via the interaction of metasomatic fluids with beryllium-rich pegmatites, or similar granitic bodies, that intruded into chromium- or vanadium-rich rocks, such as ultramafic and volcanic rocks, or shales derived from those rocks. A second deposit type, accounting for most of the emerald of gem quality, is the sedimentary type, which generally involves the interaction, along faults and fractures, of upper level crustal brines rich in Be from evaporite interaction with shales and other Cr- and/or V-bearing sedimentary rocks. The third, and comparatively most rare, deposit type is the metamorphic-metasomatic deposit. In this deposit model, deeper crustal fluids circulate along faults or shear zones and interact with metamorphosed shales, carbonates, and ultramafic rocks, and Be and Cr ($\pm$V) may either be transported to the deposition site via the fluids or already be present in the host metamorphic rocks intersected by the faults or shear zones. All three emerald deposit models require some level of tectonic activity and often continued tectonic activity can result in the metamorphism of an existing sedimentary or magmatic type deposit. In the extreme, at deeper crustal levels, high-grade metamorphism can result in the partial melting of metamorphic rocks, blurring the distinction between metamorphic and magmatic deposit types. In the present paper, we propose an enhanced classification for emerald deposits based on the geological environment, i.e., magmatic or metamorphic; host-rocks type, i.e., mafic-ultramafic rocks, sedimentary rocks, and granitoids; degree of metamorphism; styles of minerlization, i.e., veins, pods, metasomatites, shear zone; type of fluids and their temperature, pressure, composition. The new classification accounts for multi-stage formation of the deposits and ages of formation, as well as probable remobilization of previous beryllium mineralization, such as pegmatite intrusions in mafic-ultramafic rocks. Such new considerations use the concept of genetic models based on studies employing chemical, geochemical, radiogenic, and stable isotope, and fluid and solid inclusion fingerprints. The emerald occurrences and deposits are classified into two main types: (Type I) Tectonic magmatic-related with sub-types hosted in: (IA) Mafic-ultramafic rocks (Brazil, Zambia, Russia, and others); (IB) Sedimentary rocks (China, Canada, Norway, Kazakhstan, Australia); (IC) Granitic rocks (Nigeria). (Type II) Tectonic metamorphic-related with sub-types hosted in: (IIA) Mafic-ultramafic rocks (Brazil, Austria); (IIB) Sedimentary rocks-black shale (Colombia, Canada, USA); (IIC) Metamorphic rocks (China, Afghanistan, USA); (IID) Metamorphosed and remobilized either type I deposits or hidden granitic intrusion-related (Austria, Egypt, Australia, Pakistan), and some unclassified deposits.

Keywords: emerald deposits; classification; typology; metamorphism; magmatism; sedimentary; alkaline metasomatism; fluids; stable and radiogenic isotopes; genetic models; exploration

1. Introduction

Emerald is the green gem variety of the mineral beryl, which has the ideal formula of $Be_3Al_2SiO_{18}$. It is considered one of the so-called precious gems and in general the most valuable after diamond and ruby. The color of emerald is of greater importance than its clarity and brilliance for its valuation on the colored gem market. In the Munsell color chart, emerald exhibits a green color palette that is the consequence of peculiarities of its formation in contrasting environments (Figure 1). The pricing of emerald is unique in terms of the color and weight in carats. In 2000, an exceptional 10.11 ct Colombian cut gem was sold for US$1,149,850 [1]. In October 2017, Gemfield's auction of Zambian emeralds generated revenues of US$21.5 million and companies placed bids with an average value of $66.21 per carat [2]. This auction included the 6100 ct *Insufu* rough emerald extracted from the Kagem mine in 2015. Lastly, on the 2th of October, the remarkable emerald called *Inkalamu* ("the lion elephant") was extracted from the same mine [3]. Other giant crystals have been discovered in Colombia, such as *el Monstro* (16,020 ct) and *Emilia* (7025 ct), both from the Gachalá mines. In 2017, a large piece of biotite schist with several emerald crystals was discovered in the Carnaíba mine, Brazil; the specimen weighed 341 kg with 1.7 million ct of emerald, of which 180,000 ct were of gem quality. The specimen has been valued at approximately US$309 million [4].

The present article assesses the state of our knowledge of emerald then and now, through several questions regarding their locations on the planet; their crystal chemistry; pressure-temperature conditions of crystallization; the source of the constituent elements, i.e., beryllium (Be), chromium (Cr,) and vanadium (V); and their age of formation; and also proposes a new classification scheme.

Exploration beyond the 21st century may require a comprehensive data base of the typology of emerald deposits to understand why some emerald occurrences are economic in terms of quantity and quality and most are not. These future efforts will improve exploration guidelines in the field, including plate tectonics and its consequences in terms of modeling our landscape through time and within the Wilson cycle of continents.

Figure 1. Emeralds' worldwide photographs: (**a**) Emerald crystals on quartz and adularia, Panjshir Valley, Afghanistan, 6.6 × 4.4 cm. Specimen: Fine Art Mineral. Photograph: Louis-Dominique Bayle, le Règne Minéral; (**b**) emerald on pyrite, Chivor, Colombia, 3.9 × 2.6 cm. Collection MulitAxes. Photograph: Louis-Dominique Bayle, le Règne Minéral; (**c**) Emerald in quartz, Kagem mine, Zambia, longest crystal: 7.9 × 1.2 × 1.2 cm. The Collector's Edge. Photograph: Louis-Dominique Bayle, le Règne Minéral; (**d**) emerald in quartz vein and minor potassic feldspar, Dyakou, China, longest crystal: 1.5 cm. Specimen DYKO6-zh. Photograph: Dan Marshall; (**e**) emerald in plagioclase, Carnaíba, Brazil, longest crystal: 6 cm. Photograph: Gaston Giuliani.

2. Worldwide Emerald Deposits

Emerald is rare, but it is found on all five continents (Figure 2). Colombia, Brazil, Zambia, Russia, Zimbabwe, Madagascar, Pakistan, and Afghanistan (Figure 3) are the largest producers of emerald [5]. Emerald deposits occur mainly in the Precambrian series in Eastern Brazil, Eastern Africa, South Africa, Madagascar, India, and Australia, and younger volcano-sedimentary series or ophiolites in Bulgaria, Canada, China, India, Pakistan, Russia, and Spain. Colombian emerald deposits, which produce most of the world's high-quality emeralds, are unique in that they are located in sedimentary rocks, i.e., the Lower Cretaceous black shales (BS) of the Eastern Cordillera basin. Other deposits are hosted in Alpine-type veins, also called Alpine-type clefts. Emerald is found in veins and cavities in the European Alps (Binntal) as well as in the United States (Hiddenite). Nigerian emeralds are unique and are located in pegmatitic pods.

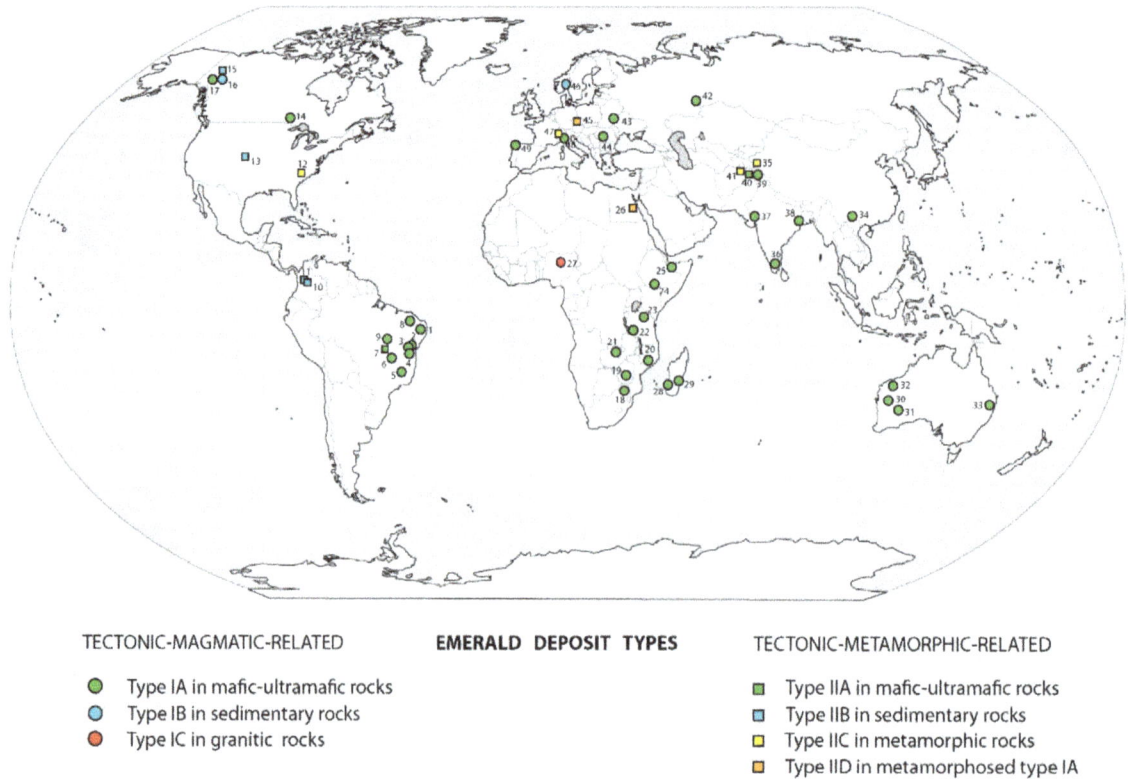

Figure 2. The location of emerald deposits and occurrences worldwide reported following their geological types, i.e., tectonic magmatic-related (type I) and tectonic metamorphic-related (type II): *Brazil*: 1. Fazenda Bonfim; 2. Socotó; 3. Carnaíba; 4. Anagé, Brumado; 5. Piteiras, Belmont mine, Capoierana, Santana dos Ferros; 6. Pirenópolis, Itaberai; 7. Santa Terezinha de Goiás; 8. Tauá, Coqui; 9. Monte Santo. *Colombia*: 10. Eastern emerald zone (Gachalá, Chivor, Macanal); 11. Western emerald zone (Yacopí, Muzo, Coscuez, Maripi, Cunas, La Pita, La Marina, Peñas Blancas). *United States*: 12. Hiddenite; 13. Uinta. *Canada*: 14. Dryden; 15. Mountain River; 16: Lened; 17. Tsa da Gliza. *South Africa*: 18. Gravelotte. *Zimbabwe*: 19. Sandawana, Masvingo, Filibusi. *Mozambique*: 20. Morrua. *Zambia*: 21. Kafubu, Musakashi. *Tanzania*: 22. Sumbawanga; 23. Manyara. *Ethiopia:* 24. Kenticha (Halo-Shakiso). *Somalia*: 25. Boorama. *Egypt*: 26. Gebels Sikaït, Zabara, Wadi Umm Kabu. Nigeria: 27. Kaduna. *Madagascar*: 28. Ianapera; 29. Mananjary. *Australia*: 30. Poona; 31. Menzies; 32. Wodgina; 33. Emmaville, Torrington. *China:* 34. Dyaku; 35. Davdar. *India:* 36. Sankari Taluka; 37. Rajasthan (Bubani, Rajgarh, Kaliguman); 38: Gubaranda (Orissa state). *Pakistan*: 39: Khaltaro; 40. Swat valley. *Afghanistan*: 41. Panjshir valley. Russia: 42. Urals (Malyshevo). *Ukraine*: 43. Wolodarsk. *Bulgaria*: 44. Rila. *Austria*: 45. Habachtal. *Norway*: 46. Eidswoll. *Switzerland*: 47. Binntal. *Italia*: 48. Val Vigezzo. *Spain*: 49. Franqueira.

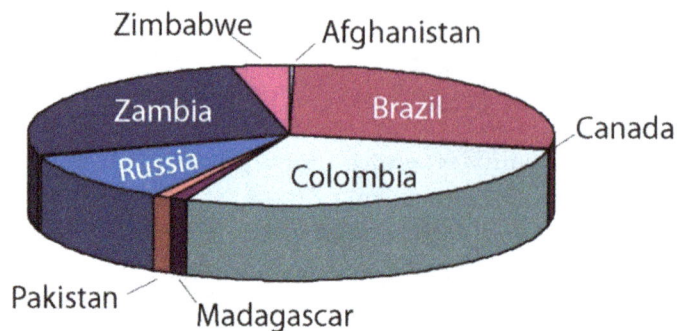

Figure 3. Emerald production worldwide in 2005.

3. Crystal Chemistry of Emerald

Beryl is hexagonal and crystallizes in point group $6/m\,2/m\,2/m$ and space group $P6/m2/c2/c$. It is a cyclosilicate mineral. The crystal structure, as shown in Figure 4, is characterized by six-membered rings of silica tetrahedra lying in planes parallel to (0001). The Al or Y site is surrounded by six O atoms in octahedral coordination, and both the Be and silica (Si) sites are surrounded by four O atoms in tetrahedral coordination. The SiO_4 tetrahedra polymerize to form six-membered rings parallel to (001); stacking of the rings results in large channels parallel to c. The channels are not uniform in diameter; in fact, they consist of cavities with a diameter of approximately 5.1 Å separated by bottlenecks with a diameter of approximately 2.8 Å. The channels can be occupied by alkali ions (such as Na^+, Li^+, K^+, Rb^+, Cs^+, etc.) whose presence is required to balance reductions in positive charges when cation substitutions occur in the structure. Neutral H_2O and CO_2 molecules [6] and noble gases, such as argon, helium [7], xenon, and neon [8], are generally also present in variable amounts in the channels.

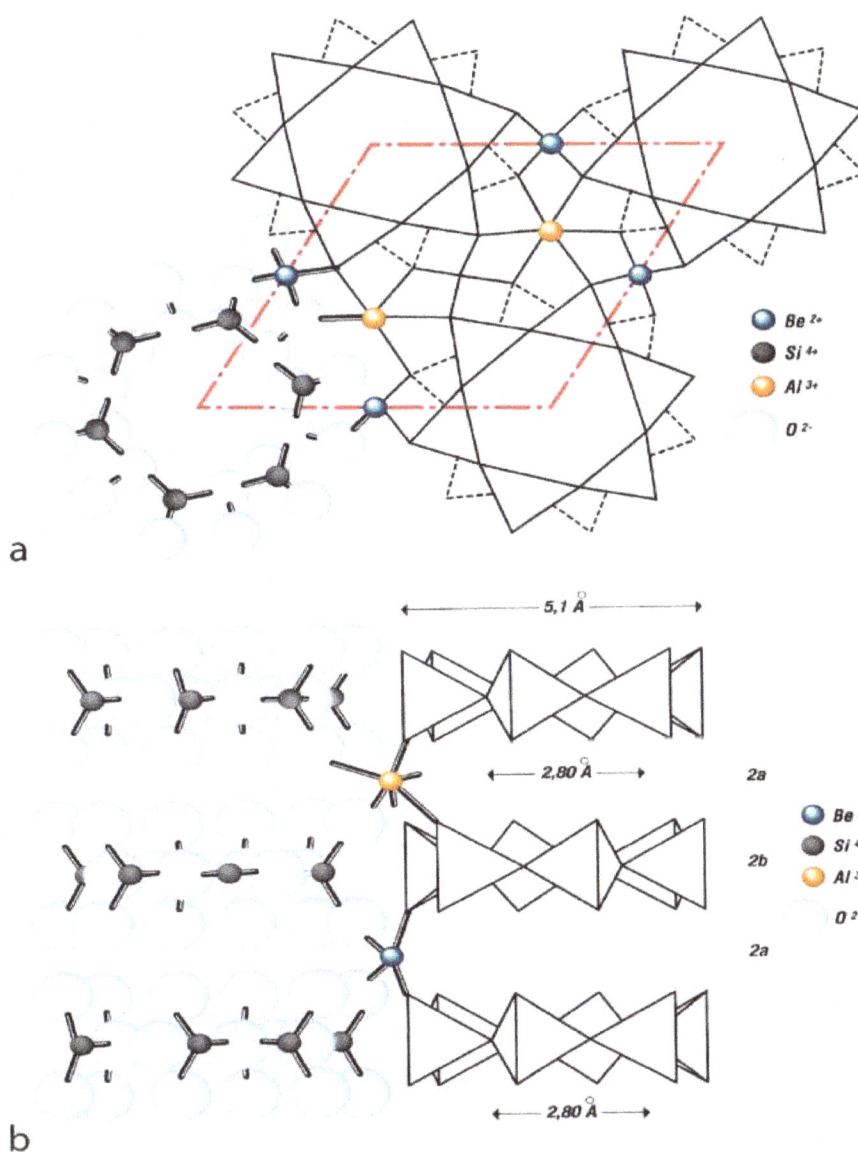

Figure 4. Structure of beryl [9] in: (**a**) Apical view: Hexagonal silicate rings stacked parallel to the c axis (normal to the drawing) are held together by Al^{3+} (octahedral site) and Be^{2+} (tetrahedral site). The radii of the ions are respected in the drawing; (**b**) lateral view perpendicular to the c axis showing the hexagonal silicate rings and the bottleneck (2b site) and the open cage (2a site) structure.

Emerald was defined by [10] as "the yellowish green, green or bluish green beryl which reveals distinct Cr and/or V absorption bands in the red and blue violet ranges of their absorption spectra". Quantitatively, Cr and V substitutions range between 25 [6] and 34,000 ppm [11] for Cr and from 34 ppm [12] to 10,000 ppm [13] for V. Emerald crystals typically exhibit a prismatic habit (Figure 5) characterized by eight faces and their corresponding growth sectors: Six $\{10\bar{1}0\}$ first order prismatic faces and two pinacoidal $\{0001\}$ faces. Small additional $\{10\bar{1}2\}$ and $\{11\bar{2}2\}$faces can also be present.

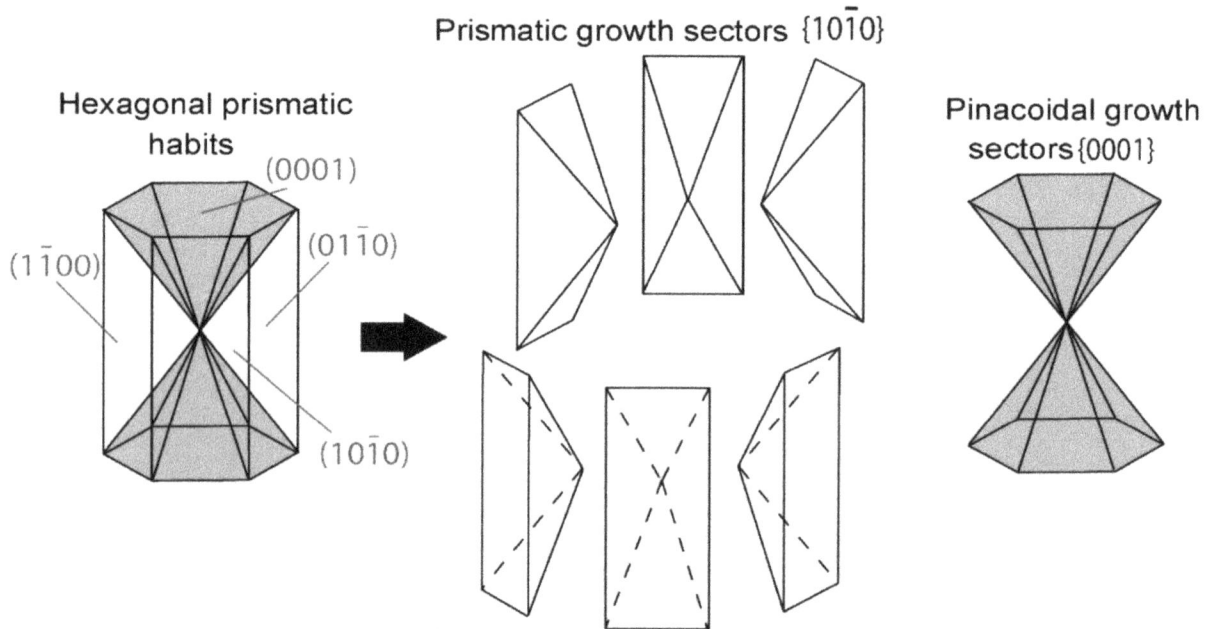

Figure 5. Habits of emerald crystals, characterized by eight main faces and their corresponding growth sectors: Six $\{10\bar{1}0\}$ first order prismatic faces and two pinacoidal $\{0001\}$ faces.

Representative emerald compositions from the literature are listed in Table 1. Most substitutions occur at the Y site. Figure 6 shows Al versus the sum of other Y-site cations for 499 emerald compositions from the literature; as expected, they show an inverse relationship. Figure 7 shows a slight deviation from a 1:1 correlation between Mg + Mn + Fe and the sum of monovalent cations. This graph suggests that, to achieve charge balance, the substitution of divalent cations for Al at the Y site is coupled with the substitution of a monovalent cation for a vacancy at a channel site. There are two sites in the channels; these are referred to as the $2a$ (at 0,0,0.25) and $2b$ (at 0,0,0) positions. Artioli et al. [14] suggested that in alkali- and water-rich beryls, H_2O molecules and the larger alkali atoms (Cs, Rb, K) occupy the $2a$ sites and Na atoms occupy the smaller $2b$ positions, but in alkali- and water-poor beryl, both Na atoms and H_2O molecules occur at the $2a$ site and the $2b$ site is empty. The amount of water in beryl can be difficult to determine, but Giuliani et al. [15] derived the following equation from existing experimental data for emerald:

$$H_2O \text{ (in wt.\%)} = [0.84958 \times Na_2O \text{ (in wt.\%)}] + 0.8373. \tag{1}$$

This equation has been improved [16] and is best defined by the relationship:

$$H_2O = 0.5401 \ln Na_2O + 2.1867a. \tag{2}$$

Table 1. Representative composition with the structural formulas of emerald samples from the recent literature (in wt.%) [1].

Elementss	1	2	3	4	5	6	7	8	9	10	11	12	13	14	15	16
SiO_2	64.50	64.94	63.58	65.98	67.13	66.74	66.30	66.04	64.71	64.75	65.57	65.57	66.49	66.28	67.13	64.39
TiO_2	0.00	0.00	n.d.	0.01	0.00	0.00	0.01	0.01	n.d.	n.d.	0.04	0.01	0.00	0.06	0.01	0.01
Al_2O_3	11.54	10.61	17.36	16.81	18.77	18.67	18.60	18.17	17.23	14.83	17.24	14.51	18.29	18.10	16.52	14.48
Sc_2O_3	n.d.	n.d.	n.d.	0.06	0.01	0.00	0.05	0.01	0.03	n.d.	n.d.	n.d.	n.d.	n.d.	n.d.	0.02
V_2O_3	0.15	0.10	0.02	0.17	0.05	0.01	0.10	0.01	0.23	0.03	0.07	0.00	0.17	0.05	0.00	0.03
Cr_2O_3	1.83	3.39	0.25	0.35	0.20	0.26	0.02	0.05	0.01	0.57	0.07	0.09	0.05	0.24	0.37	0.17
Fe_2O_3	1.34	1.27	n.d.	n.d.	n.d.	n.d.	n.d.	n.d.	n.d.	n.d.	n.d.	n.d.	n.d.	n.d.	n.d.	n.d.
La_2O_3	0.06	0.12	n.d.	n.d.	n.d.	n.d.	n.d.	n.d.	n.d.	n.d.	n.d.	n.d.	n.d.	n.d.	n.d.	n.d.
Ce_2O_3	0.02	0.00	n.d.	0.02	n.d.	n.d.	n.d.	n.d.	n.d.	n.d.	n.d.	n.d.	n.d.	n.d.	n.d.	n.d.
BeO	13.40	13.45	12.85	13.65	13.99	13.91	13.82	13.73	13.54	13.50	13.47	13.59	13.83	13.78	13.80	13.37
MgO	3.49	3.36	0.58	0.95	0.04	0.03	0.06	0.15	0.88	2.24	0.87	2.63	0.05	0.16	0.87	2.38
CaO	0.21	0.17	n.d.	0.00	0.00	0.00	0.01	0.01	0.02	0.02	0.00	0.06	0.00	0.00	0.00	0.04
MnO	0.10	0.00	n.d.	0.03	0.00	0.01	0.01	0.00	n.d.	0.01	0.00	0.01	0.00	0.00	0.01	0.01
FeO	n.d.	n.d.	0.37	0.17	0.21	0.19	0.12	0.15	0.26	0.65	0.33	0.96	0.71	0.40	0.14	0.63
Li_2O	n.d.	n.d.	0.28	0.03	n.d.	n.d.	n.d.	n.d.	n.d.	n.d.	0.11	0.02	0.01	0.01	0.01	n.d.
Na_2O	0.83	0.89	1.13	0.62	0.05	0.06	0.08	0.27	0.82	1.76	0.82	1.84	0.04	0.10	0.70	1.86
K_2O	2.04	1.88	0.05	0.03	0.01	0.02	0.00	0.01	0.02	0.03	0.00	0.24	0.02	0.05	0.00	0.04
Rb_2O	n.d.	n.d.	n.d.	n.d.	n.d.	n.d.	n.d.	0.00	n.d.	0.14	n.d.	n.d.	n.d.	n.d.	n.d.	n.d.
Cs_2O	0.02	0.02	0.13	n.d.	n.d.	n.d.	0.01	0.11	0.10	0.03	n.d.	n.d.	n.d.	n.d.	n.d.	n.d.
H_2O	2.09	2.12	2.25	1.93	0.57	0.67	0.82	1.48	2.08	2.49	2.08	2.52	0.45	0.94	1.99	2.52
Total	101.62	102.32	98.85	100.81	101.04	100.57	100.01	100.20	99.94	101.05	100.67	102.05	100.11	100.17	101.55	99.96
apfu																
Si^{4+}	6.012	6.030	5.963	6.015	5.992	5.991	5.991	6.005	5.970	5.991	5.998	6.012	5.998	6.001	6.066	6.013
Ti^{4+}	0.000	0.000	0.000	0.001	0.000	0.000	0.001	0.001	0.000	0.000	0.003	0.001	0.000	0.004	0.001	0.001
Al^{3+}	1.268	1.161	1.919	1.806	1.975	1.975	1.981	1.947	1.873	1.617	1.858	1.568	1.945	1.931	1.759	1.594
Sc^{3+}	0.000	0.000	0.000	0.005	0.001	0.000	0.004	0.001	0.002	0.000	0.000	0.000	0.000	0.000	0.000	0.002
V^{3+}	0.011	0.007	0.002	0.012	0.004	0.001	0.007	0.001	0.017	0.002	0.005	0.000	0.012	0.004	0.000	0.002
Cr^{3+}	0.135	0.249	0.019	0.025	0.014	0.018	0.001	0.004	0.001	0.042	0.005	0.007	0.004	0.017	0.026	0.013
Fe^{3+}	0.094	0.089	0.000	0.000	0.000	0.000	0.000	0.000	0.000	0.000	0.000	0.000	0.000	0.000	0.000	0.000
La^{3+}	0.002	0.004	0.000	0.000	0.000	0.000	0.000	0.000	0.000	0.000	0.000	0.000	0.000	0.000	0.000	0.000
Ce^{3+}	0.001	0.000	0.000	0.001	0.000	0.000	0.000	0.000	0.000	0.000	0.000	0.000	0.000	0.000	0.000	0.000

Table 1. *Cont.*

Elementss	1	2	3	4	5	6	7	8	9	10	11	12	13	14	15	16
Be^{2+}	3.000	3.000	2.894	2.989	3.000	3.000	3.000	3.000	3.000	3.000	2.960	2.993	2.996	2.996	2.996	3.000
Mg^{2+}	0.485	0.465	0.081	0.129	0.005	0.004	0.008	0.020	0.121	0.309	0.119	0.359	0.007	0.022	0.117	0.331
Ca^{2+}	0.021	0.017	0.000	0.000	0.000	0.000	0.001	0.001	0.002	0.002	0.000	0.006	0.000	0.000	0.000	0.004
Mn^{2+}	0.008	0.000	0.000	0.002	0.000	0.001	0.001	0.000	0.000	0.001	0.000	0.001	0.000	0.000	0.001	0.001
Fe^{2+}	0.000	0.000	0.029	0.013	0.016	0.014	0.009	0.011	0.020	0.050	0.025	0.074	0.054	0.030	0.011	0.049
Li^{+}	0.000	0.000	0.106	0.011	0.000	0.000	0.000	0.000	0.000	0.000	0.040	0.007	0.004	0.004	0.004	0.000
Na^{+}	0.150	0.160	0.205	0.110	0.009	0.010	0.014	0.048	0.147	0.316	0.145	0.327	0.007	0.018	0.123	0.337
K^{+}	0.243	0.223	0.006	0.003	0.001	0.002	0.000	0.001	0.002	0.004	0.000	0.028	0.002	0.006	0.000	0.005
Rb^{+}	0.000	0.000	0.000	0.000	0.000	0.000	0.000	0.000	0.000	0.008	0.000	0.000	0.000	0.000	0.000	0.000
Cs^{+}	0.001	0.001	0.005	0.000	0.000	0.000	0.000	0.004	0.004	0.001	0.000	0.000	0.000	0.000	0.000	0.000

[1] Compositions renormalized on the basis of 3 (Be + Li) and 18 O per formula unit. $H_2O = 0.5401 \ln Na_2O + 2.1867$ [16]. **1**. Ianapera, Madagascar. Ultramafic host, Type-3 core [11]. **2**. Ianapera, Madagascar. Ultramafic host, Type-3 rim [11]. **3**. Taylor 2, Canada. Average of 51 analyses with Li = (Na + K + Cs) - (Mg + Fe), which assumes all Fe as Fe^{2+} [17]. **4**. Davdar, China. Average of 48 analyses; Ce_2O_3 by neutron activation, LiO 0.03 wt.% by fusion with Na_2O_2 [18]. **5**. Emmaville-Torrington, Australia. "Line 1", average of 27 analyses [19]. **6**. Emmaville-Torrington, Australia. "Line 2", average of 31 analyses [19]. **7**. Byrud, Norway. Average of 38 analyses [20]. **8**. Poona, Australia. Average of 37 analyses [16]. **9**. Lened, Canada. Average of 88 analyses [21]. **10**. Fazenda Bonfim, Brazil. Average of approximately 130 analyses [22]. **11**. Malyshevsk, Russia. Average of five analyses [23]. **12**. Alto Ligonha, Mozambique. Average of three analyses [23]. **13**. Jos, Nigeria. Average of five analyses [23]. **14**. Sumbawanga, Tanzania. Average of five analyses [23]. **15**. Muzo, Colombia. Average of three analyses [23]. **16**. Ethiopia [this study, average of 32 analyses].

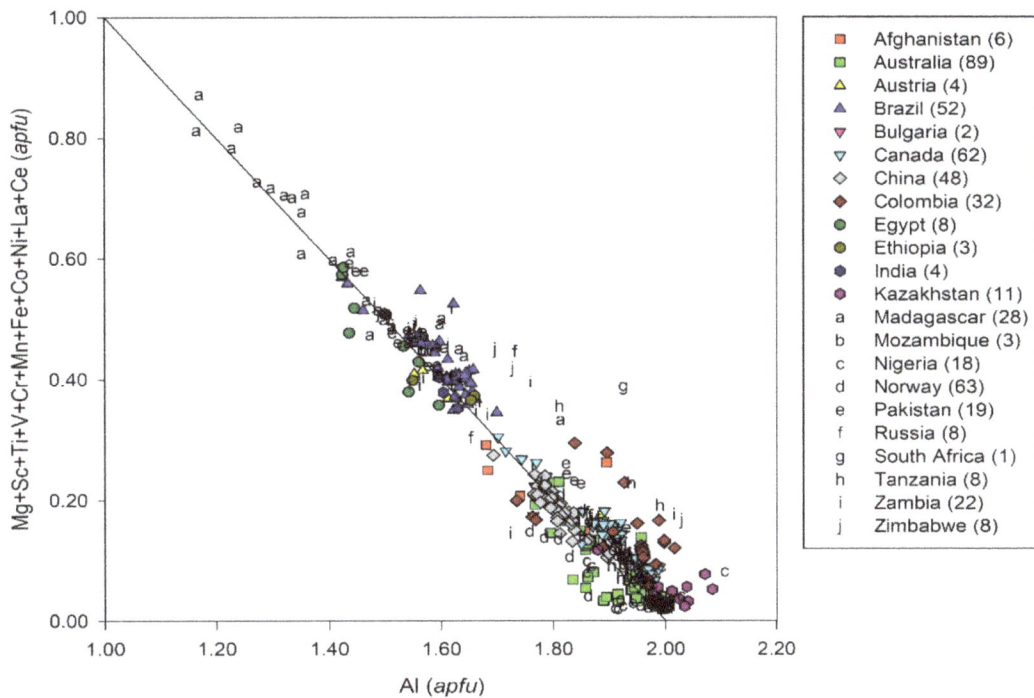

Figure 6. Al versus the sum of other *Y*-site cations, in atoms per formula unit, for 499 emerald analyses from the literature. The number of analyses per country is given in parentheses in the legend. Sources of data: [11–15], [16] (average of 37 analyses), [17–20], [21] (average of 88 analyses), [22] (average of approximately 130 analyses), [23–43], [44] (Kazakhstan values are averages of 11 analyses), [45] (average of 10 analyses), [46], [47] (two averages of five analyses each), [48–51], [52] (average of 55 analyses), and this study.

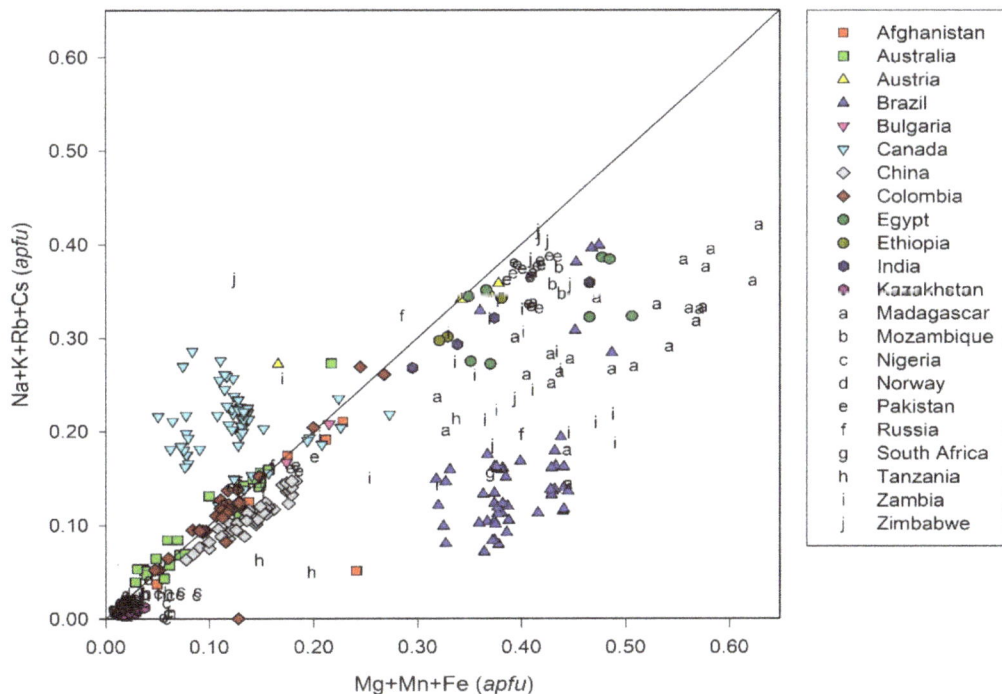

Figure 7. Mg + Mn + Fe vs. monovalent channel-site cations, in atoms per formula unit, for analyses from the literature. Sources of data are the same as in Figure 6.

Unfortunately, it is difficult to obtain accurate analyses of Be, Li, and ferric-ferrous ratios in beryl. Both Be and Li are too light to measure with the electron microprobe, which cannot distinguish between

Fe^{3+} and Fe^{2+}. Thus, most published analyses of beryl are renormalized on the basis of 18 O and 3 Be atoms per formula unit and Fe is generally reported as exclusively ferric or ferrous. Points that lie to the right of the 1:1 line in Figure 6 indicate that some of the Fe in a given emerald is present as Fe^{3+}. Likewise, points that lie above the line could suggest the presence of Li^+, which may substitute for Be^{2+} at the Be site. Charge balance is maintained by adding a monovalent cation to a channel site. Beryllium can be analyzed with LA-ICP-MS (Laser Ablation–Inductively Coupled Plasma–Mass Spectrometry), SIMS (Secondary Ionisation Mass Spectrometry, or "ion microprobe" [23]), or by using wet chemical techniques. However, there is so much Be in the structure that the accuracy of such analyses is suspect. Lithium may be analyzed by the same techniques as Be, but because the concentrations are much lower, the accuracy would presumably be better. The amount of Li can also be estimated in Fe-free beryl from the number of monovalent cations at the channel sites.

The main substituents for Al at the Y site are plotted as oxides in Figure 8a. Magnesium is the main substituent in emeralds from most localities. The elements responsible for most of the variation in color in emerald crystals are plotted as oxides in Figure 8b. In most cases, the Cr_2O_3 content is much greater than that of V_2O_3; the main exceptions are for samples from the Lened occurrence in Canada [21,47], the Davdar occurrence in China [18], the Muzo mine in Colombia [1], the Mohmand district in Pakistan [1], and Eidswoll in Norway [13]. The accuracy of data obtained by LA-ICP-MS is primarily dependent on the standards used for the analysis. Currently, the NIST 610 and 612 glasses are used for calibration standards, but it is likely that emerald standards are necessary to obtain accurate data for trace elements. Beryl has a wide stability field; the lower limit in the presence of water is between 200 °C and 350 °C, depending on the pressure and coexisting minerals [53]. However, the high-temperature stability and melting relationships remain unclear, partly because beryl may contain significant amounts of H_2O at the channel sites, and water has a significant effect on stability [52]. The effect of other channel constituents, such as Na, may be similar. Although thermodynamic data exist for beryl, the lack of experimental data for anything more complex than the BASH (BeO-Al_2O_3-SiO_2-H_2O) system can be a barrier to understanding the formation of natural occurrences, such as those where Be-bearing minerals occur in metamorphic rocks [52].

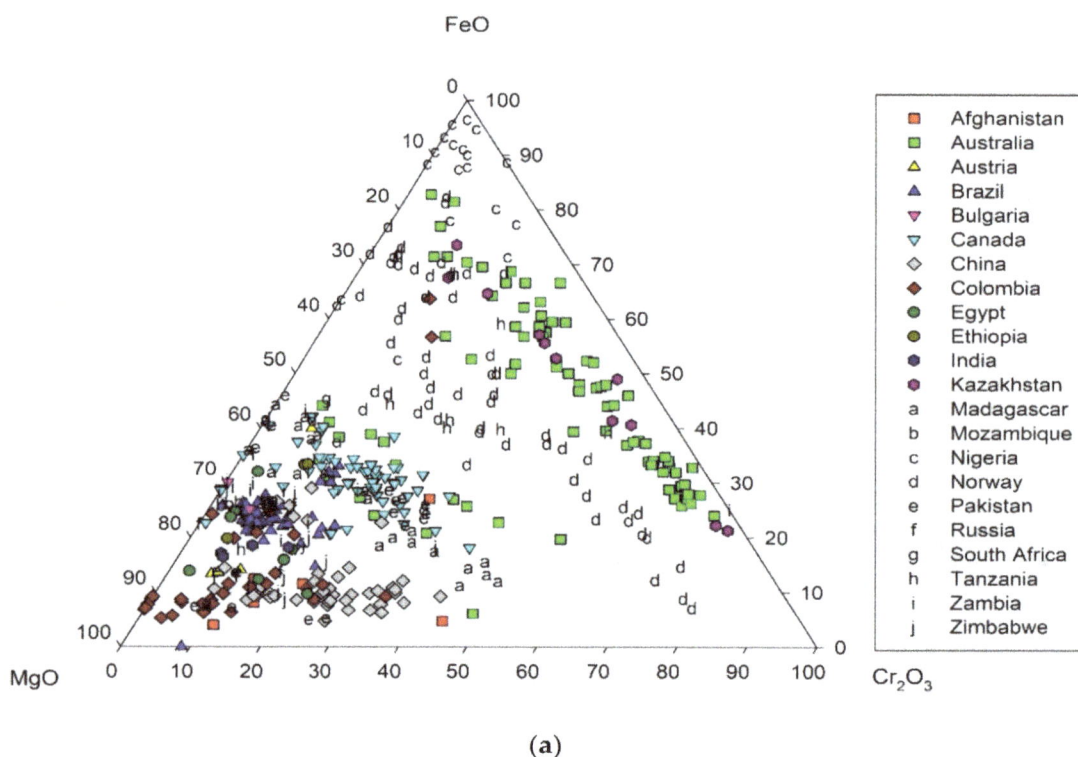

(a)

Figure 8. *Cont.*

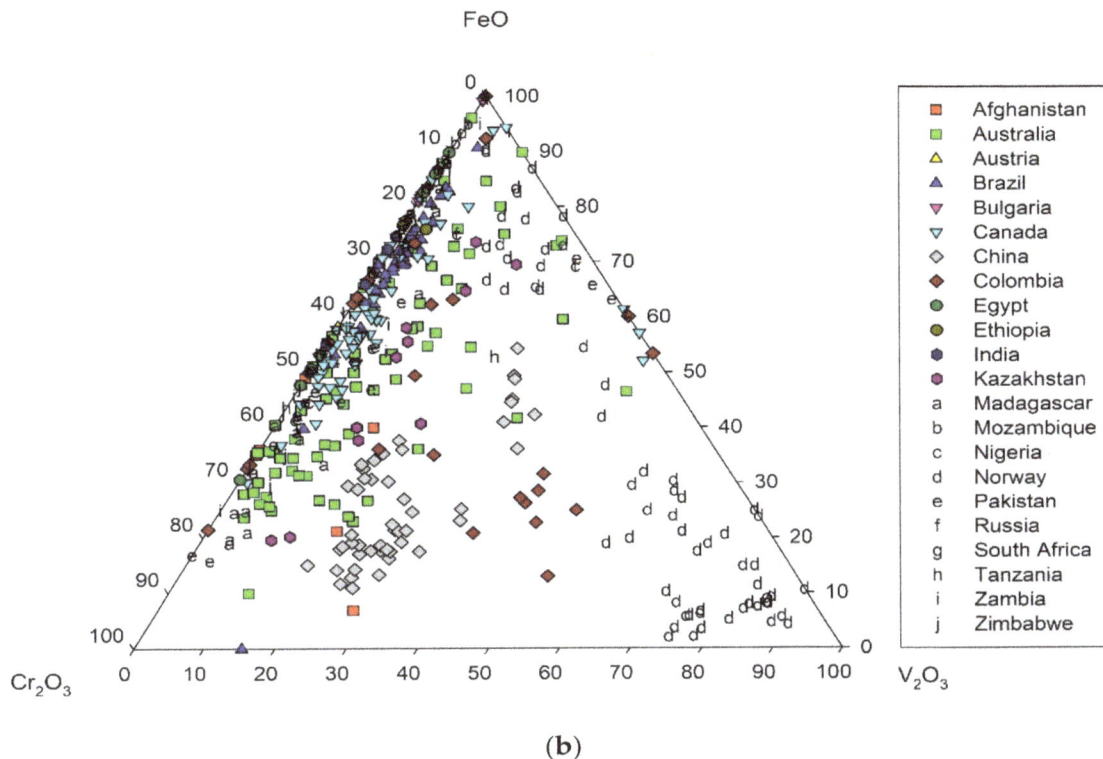

(b)

Figure 8. (**a**) Plot of emerald compositions in terms of FeO-MgO-Cr$_2$O$_3$ (wt.%). Data from the literature (with all Fe as FeO). Sources of data are the same as in Figure 6. The diagram is after [39]. (**b**) Plot of emerald compositions in terms of FeO-Cr$_2$O$_3$-V$_2$O$_3$ (wt.%). Data from the literature (with all Fe as FeO). Sources of data are the same as in Figure 6. The diagram is after [39].

4. Sources of Be, Cr, and V: The Formation of Emerald

The sources of Be, Cr, and V of emerald. The Cr-V-bearing beryl is rare because (i) its constituent metals have opposite affinities and behavior in the continental and oceanic lithospheres, and (ii) there is only 2 ppm of Be in the upper continental crust. Beryllium is a lithophile element, which has a strong affinity for oxygen producing beryl and chrysoberyl (BeAlO$_4$) at higher temperatures and at medium to low temperatures joins with silica to form silicate minerals, such as beryl, bertrandite [Be$_4$Si$_2$O$_7$(OH)$_2$], and/or Be-bearing micas. These minerals have a relatively low-density (ρ_{beryl} = 2.76) and are concentrated in the crust (ρ_{crust} = 2.7). Chromium and V are high-density transition metals that generally concentrate in the core and the mantle of the earth. Although Cr and V show both lithophile and siderophile affinities, they tend to form solid solutions with iron (Fe) to form, e.g., chromite (FeCr$_2$O$_4$) at high temperatures. Progressive changes in tectonic processes since the Archean preserved metamorphosed mafic-ultramafic rocks (M-UMR) and ancient mantle-related terranes in the upper continental crust. Consequently, Cr and V are more common than Be in the upper continental crust, with concentrations of ~100 ppm each.

Beryllium is present in crustal granites and associated dyke swarms of pegmatites, aplites, and quartz veins. The felsic rocks (FELSR), with more than 70% silica (SiO$_2$), are intrusive into the continental lithosphere. The highest Be concentration occurs in granites and rhyolites (~5 ppm). The two-mica granites have concentrations between 5 and 10 ppm, while specialized granites have more than 200 ppm. Beryllium is carried by Be-minerals, but also by feldspar and micas. Chromium and V are more highly concentrated in M-UMR of the oceanic lithosphere that contain less than 50% SiO$_2$ (Figure 9a). These rocks, which are generally termed peridotites, are often metamorphosed into serpentinites containing up to 14 wt.% H$_2$O (Figure 9a,b), or into talc-chlorite-carbonate schists. Chromium is dominantly sourced from chromite and V is sourced from V-bearing magnetite or coulsonite (Fe^{2+}V$^{3+}_2$O$_4$). Nevertheless, Cr-V and Be can be present in sedimentary rocks and are highly concentrated in shale (Figure 9c). Indeed, the weathering and erosion of M-UMR and FELSR on the continents delivers to the sea fine-grained clastic sediments composed of mud with Cr-V-Be-bearing grains and organic matter, as well as tiny fragments of quartz and calcite.

Figure 9. Potential geological sources of chromium (Cr) and vanadium (V) necessary for emerald formation in mafic-ultramafic rocks and black shales: (**a**) A lherzolite (Lrz) that is transformed into serpentinite (Srp), region of Montemaggiore on the Corsican cape, France; (**b**) Photomicrograph (plane polarized light) of an antigorite vein (V2) cross cutting lizardite veins (V1) and olivine (Ol). Fine-grained magnetite (Mag) is present in the centre of V1; V2 boundaries contain minute secondary olivine grains. Photo: B. Debret; (**c**) Colombian black shale crosscut by pyrite (Py) -bearing calcite (Ca) veinlets from the Coscuez mine. Photograph: G. Giuliani; (**d**) Pie diagram representation of the concentration (in %) and relative distribution of beryllium (Be) between the different phases extracted from a black shale containing 4 ppm of ^{9}Be, Coscuez mine [54]. OM: Organic matter; [Fe,Mn(O,OH)] = oxy-hydroxides of iron and manganese.

In anoxic and reducing environments, the un-oxidized organic matter imparts a dark colour to the black shale (BS). The Be contents of BS for Colombian deposits vary between 2 and 6 ppm, while Cr-V concentrations range between 100 and 1000 ppm. At the Coscuez emerald deposit (Figure 9d), the Be content of the BS is around 5 ppm. Beryllium mobility is principally associated with the breakdown of iron-manganese oxy-hydroxide phases. The amount of Be that can be mobilized (~0.7 ppm) may represent up to 18 wt.% of the total Be contained in the BS [54].

Emerald formation requires Be and Cr-V inter-reservoir circulation of fluids to mobilize these elements. Metasomatism and fluid/rock interaction is the main mechanism for element mobilization in sedimentary (black shales) or granitic rocks for Be and metamorphosed-mafic-ultramafic rocks for Cr and V. These mineralizing fluids are trapped by emerald within the large channels parallel to the *c* axis (molecular components) or in fluid inclusions (several constituents in different systems, i.e., H_2O-$NaCl$-CO_2-($\pm H_2S$)-($\pm N_2$)-($\pm CH_4$)-(K, Mg, Be, F, B, Li, SO_4, P, Cs)) within primary growth planes and secondary trails of fractures. Fluid inclusions are important fingerprints for emerald, and microthermometry, Raman spectrometry, and mass spectrometry for O-H isotope signatures make it possible to determine the nature and origin of the fluids [16,55].

5. Classification of Emerald Deposits

5.1. Genetic Classifications

The genetic classification schemes developed for emerald deposits in the 21st century were reviewed by [1]. The classification schemes are ambiguous and not particularly useful when it comes to understanding the mechanisms and conditions that led to the formation of an emerald deposit [52]. There is no ideal combination of mineralogical, chemical, geochemical, and physical parameters or combinations of these data with the age of the formation of the deposits and O-isotope composition of emerald [56].

Dereppe et al. [57] used artificial neural networks (ANN) to classify emerald deposits based on 450 electron microprobe analyses of emeralds from around the world. They defined five categories of deposits with "bad scores". These important misclassifications affected essentially the shear zone and thrust-related deposits in mafic–ultramafic rocks and oceanic suture rocks, such as in Santa Terezinha de Goiás, Brazil, and either Panjsher valley (Afghanistan) or Swat valley (Pakistan).

Schwarz and Giuliani [58,59] recognized two main types of emerald deposit: Those related to granitic intrusions (type I) and those where mineralization is mainly controlled by tectonic structures, such as a fault or a shear zone (type II). Most emerald deposits fall into the first category and are subdivided based on the presence or absence of biotite schist at the contact. Type II deposits are subdivided into schist without pegmatite and black shale with veins and breccia. However, a number of emerald deposits of type I have been influenced by syntectonic events (e.g., Carnaíba, Brazil; Poona, Australia; Sandawana, Zimbabwe) or remain unclassified, such as the Gravelotte (Leydsdorp) deposit in South Africa [58].

Schwarz et al. [60] classified emerald deposits based upon their appearance in the field following several sketched geological profiles drawn by G. Grundmann: (i) Pegmatites without phlogopite schist; (ii) pegmatite and greisen with phlogopite schist; (iii) schist without pegmatite with (iiia) phlogopite schists, (iiib) carbonate-talc schist and quartz lens, (iiic) phlogopite schist and carbonate-talc schist; and (iv) black shale with breccia and veins.

Barton and Young [53] divided emerald deposits into those with a direct igneous connection and those where such a connection was indirect or absent. Further subdivisions were done based on the chemistry of the magma and/or the nature of the emerald-bearing metasomatic rocks (greisen and vein-like, skarn type, biotite schist, and vein-related). However, a number of emerald deposits cannot be unambiguously classified using this scheme (Kaduna in Nigeria, Swat valley in Pakistan and Eastern desert deposits in Egypt).

Schwarz and Klemm [61] used LA-ICP-MS to obtain approximately 2600 spot analyses of 40 major and trace elements from ca. 650 emerald samples from 21 different occurrences worldwide. The classification of the deposits was "non-genetic descriptive", but was used instead as a geographic fingerprint. Nowadays, the gemological laboratories routinely use LA-ICP-MS for deciphering the geographic origin of individual stones [12,62–65], which is important for provenance and international trade certification. New standards and protocols for emerald analysis are being applied in gem testing laboratories for geographic and geological applications [66].

Aurisicchio et al. [23] obtained major and trace element data for emerald from several world deposits and reported important modifications regarding the origin of Be in some type II deposits (in the classification proposed by Schwarz and Giuliani [58], i.e., type I granitic intrusions- and type II tectonic-related emerald deposits).

5.2. A Revised Classification for Emerald Deposits

We start by asking why a reclassification is desireable. Emerald deposits are relatively rare and form in a limited number of geological settings. Existing classification systems or models used to describe emerald formation are too restrictive and attempts at reclassification have proved inadequate to date.

Emerald is a medium to high temperature by-product of the Earth system and is hosted by diverse rock types of different ages. The first task of a field geologist is to identify the emerald-bearing rocks and to define the geological environment as magmatic, sedimentary, or metamorphic. Then, each emerald occurrence worldwide is linked to a major geological and tectonic regime related to the movement of lithospheric plates. Regional stresses in the lithosphere generate deformation subsystems with characteristic geometry, chemistry, and geology. The systems are either (i) closed (diabatic) systems with an exchange of heat with the surrounding rock and with very limited or absent fluid-rock interaction, or (ii) open systems with the migration of material and fluids along shear-zones and faults, resulting in the mobilization and re-precipitation of elements and tectonic melange zones. The tectonic systems are then characterized as compressive or extensional sub-systems with variable amounts of vertical and strike-slip movement encompassing large areas within orogens. Consequently, a more inclusive classification system for emerald deposits should consider tectonic, magmatic, and metamorphic-related types, geological environment, magmatic/metamorphic/sedimentary host rocks, and metasomatic conditions.

So, what would an enhanced classification look like? Giuliani et al. [55] classified emerald deposits into three broad types based on worldwide production in 2005 (see Figure 3): (i) The magmatic-metasomatic type (Ma) accounting for about 65% of the production; and (ii) the sedimentary-metasomatic (Se) and metamorphic metasomatic (Me) types, for 28% and 7%, respectively.

Marshall et al. [16] proposed an enhanced classification for emerald deposits based on the Me, Ma, and Se models, but also including the temperature of formation. They examined the possibility, at deeper crustal levels and high grade metamorphism, of the possible remobilization of previous beryl or emerald occurrences and partial melting of metamorphic rocks blurring the distinction between Me and Ma types.

In this work, we propose classifying emerald occurrences into two main types (Table 2):

(Type I) Tectonic magmatic-related with sub-types hosted in:

(IA)   Mafic-ultramafic rocks (Brazil, Zambia, Russia, and others);
(IB)   Sedimentary rocks (China, Canada, Norway, Kazakhstan Australia);
(IC)   Granitic rocks (Nigeria).

(Type II) Tectonic metamorphic-related with sub-types hosted in:

(IIA)   M-UMR (Brazil, Austria);
(IIB)   Sedimentary rocks-black shale (Colombia, Canada, USA);
(IIC)   Metamorphic rocks (China, Afghanistan, USA);
(IID)   Metamorphosed type I deposits or hidden-granitic intrusion-related (Austria, Egypt, Australia, Pakistan), and some unclassified deposits.

6. Different Types of Emerald Deposits

We will now examine the main emerald deposits worldwide using the above classification.

Table 2. New typological classification of the emerald occurrences and deposits worldwide.

Type of Deposit	Tectonic-Metamorphic-Related					Tectonic-Magmatic-Related		
Geological Environment	SEDIMENTARY	Metamorphic				Granitic		
Metamorphic Conditions	Anchizone to Greenschist facies	Greenschist to granulite facies				Greenschist to granulite facies		
Host-rocks	Sedimentary Rocks	Metamorphic rocks				Mafic-UltraMafic Rocks (M-UMR)	Sedimentary Rocks (SR)	Granitoids
Type	TYPE IIB — carbonate platform sediments	TYPE IIC — Metamorphism of SR	TYPE IIC — Migmatites	TYPE IIA — Metamorphism of M-UMR	TYPE IID — Metamorphosed Type IA, Mixed IA and IIA in M-UMR, and unknown	TYPE IA	TYPE IB	TYPE IC
						pegmatite- aplite- quartz- greisen veins, pods, metasomatites		
Mineralization Style	veins and/or metasomatites	veins	veins	shear zone	shear zone, metasomatites, veins, boudins, fault	veins and/or metasomatites, skarns	veins and/or metasomatites, skarns	pods
Origin of the Fluid	Metasomatic-hydrothermal	Hydrothermal	Hydrothermal	Metamorphic-metasomatic	(Magmatic-Metasomatic) with a metamorphic remobilization	Metasomatic-hydrothermal	Metasomatic-hydrothermal	Metasomatic-hydrothermal
Deposits	Colombia (Eastern and western emerald zones) Canada (Mountain River) USA (Uinta (?): question about the presence of emerald)	China (Davdar) Afghanistan (Panjsher)	USA (Heddenite)	Austria (Habachtal) Brazil (Itaberai, Santa Terezinha de Goiás) Pakistan (Swat-Mingora–Gujar-Kili, Barang)	Austria (Habachtal) ?: Probably metamorphic remobilization of type IA deposit; Brazil (Santa Terezinha de Goias) ?: Probably related to hidden granitic intrusive cut by thrust and emerald-bearing shear-zone; Pakistan (Swat-Mingora)? Probably related to undeformed hidden granitic intrusives; Australia (Poona): Probably metamorphic remobilization of Tye IA deposit; Egypt (Djebel Sikait, Zabara, Umm Kabu): Probably metamorphic remobilization of Type IA deposit; Zambia (Musakashi): Unknown genesis, vein style, fluid inclusion indicates affinities with Types IIB and IIC	Brazil (Carnaíba, Socotó, Itabira, Fazenda Bonfim, Pirenópolis, etc.) Canada (Tsa da Gliza, Taylor 2) Bulgaria (Rila) Urals (Malysheva, etc.) Pakistan (Khaltaro) Afghanistan (Tawak) India (Rajhastan, Gubaranda) South Africa (Gravelotte) Zambia (Kafubu, etc.) Tanzania (Manyara, Sumbawanga) Mozambique (Rio Maria, etc.) Australia (Menzies, Wodgina, etc.) Ethiopia (Kenticha) Madagascar (Ianapera, Mananjary) Zimbabwe (Sandawana, Masvingo, Filibusi) Somalia (Boorama) Ukraine (Wolodarsk)	Norway (Eidsvoll) China (Dyakou) Canada (Lened) Australia (Emmavile, Torrington) Kazakstan (Delgebetey)	Nigeria (Kaduna)

6.1. Tectonic Magmatic-Related (Type I)

Type I deposits are found in all five continents (Figure 2). The main geological environment is characterized by the presence of aluminous to peraluminous granitoids formed in continental collision domains. The collision increases the continental lithosphere thickness, resulting in increased pressure and temperature, and producing a zone of higher grade of metamorphism and partial melting of rocks, forming felsic magmas. These continental collisions have occurred at different geological times throughout the Wilson cycles of the supercontinents, with the formation of emerald deposits (Figure 10; [67]) during the following orogenies: Eburnean or Transamazonian (2.0 Ga (giga-annum)), Pan-African/Brasiliano (490–520 Ma (mega-annum)), Hercynian (300 Ma), Uralian (299–251 Ma), Yenshan (125–110 Ma), and Himalayan (25–9 Ma). In contrast, the emerald deposit at Kaduna in Nigeria is associated with Mesozoic (213–141 Ma) ring complexes with peralkaline granites formed in a volcanic to subvolcanic continental environment [68]. In the case of the Byrud mine in Norway, the emerald is related to Permo-Triassic intrusions associated with the evolution of the Oslo Paleorift. This rift was characterized by a succession of volcanic rocks and the emplacement of batholiths and the intrusion of syenitic dykes and sills [69].

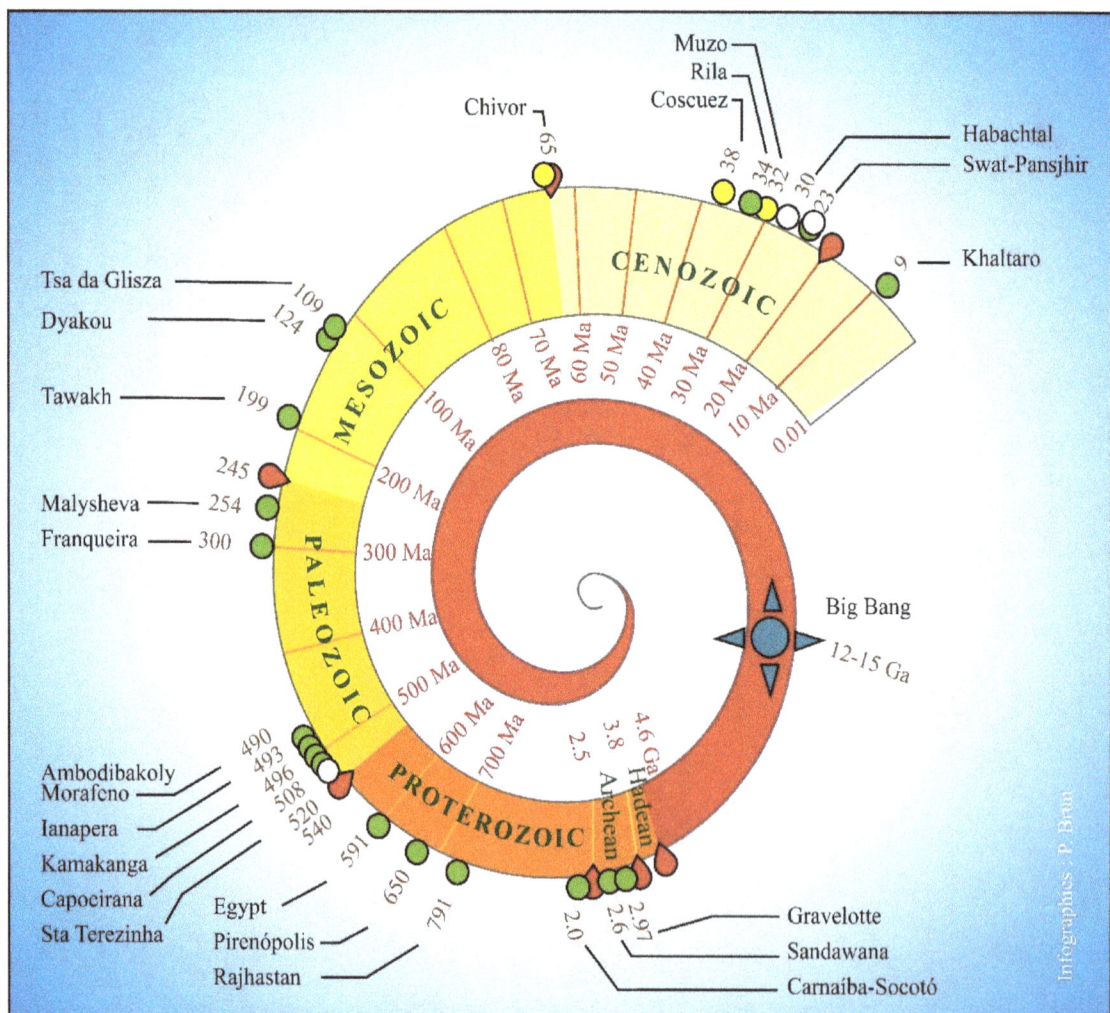

Figure 10. The spiral time of emerald. The oldest emerald formed during Archean times (2.97 Ga) in South Africa (Gravelotte deposit) and the youngest during Cenozoic times (9 Ma) in Pakistan (Kaltharo deposit). Nomenclature of the circles: Red = limit of geological era, green circle = emerald deposits related to the Tectonic magmatic-related types (types IA, IB, and IC); yellow and white circles = Tectonic metamorphic-related types (types IIA, IIB, IIC, and IID) with deposits hosted either in mafic-ultramafic rocks (white circle) or in sedimentary rocks (yellow circle).

Type I emerald deposits, in terms of quantity and gem quality, remain the world's most important emerald deposits. Three sub-types are distinguished according to the emerald host-rocks.

6.1.1. Sub-Type IA: Tectonic Magmatic-Related Emerald Deposits Hosted in M-UMR (Brazil, Zambia, Russia, and Others; See Table 2 and Figure 2)

These deposits produce high volumes of emerald. Why?

There are numerous Be- and/or Be-Ta-Nb-Sn-W-bearing granites of different ages accompanied by subordinate amounts of M-UMR in the continental crust.

There are large volumes of metamorphosed M-UMR in the Precambrian series of the continental crust, such as the recently discovered southern Ethiopian emerald deposit (Halo-Shakiso) hosted in the Archean and Proterozoic volcano-sedimentary series. It is associated with pegmatites and quartz veins similar to the proximal Ta-Nb-Be-bearing pegmatites from the nearby Kenticha mine.

At the regional scale and in several emerald mining districts, there are extended and continuous zones of mineralization related to granites, pegmatites, and quartz veins, as in the Kafubu mining area in Zambia where the mining licences extend for approximately 15 km of the strike length [50,70]. The development of modern mining on a large scale, as done by the company, Gemfields Group Ltd., through underground and huge open-cast mining, allows for the production of large quantities of high-quality commercial-grade gems.

Another example is the mining district of Itabira-Nova Era in Minas Gerais, located in Quadrilátero Ferrifero, Brazil (Figure 11). In 1978, the future deposit of Itabira was discovered on a private property called the "Itabira farm" [71]. After three years of development, the mine was mechanized and named the Belmont mine [72]. In 1988, emeralds were discovered at Capoierana near Nova Era, 10 km southeast of the Belmont mine [73]. In 1998, the deposit of Piteiras was officially discovered. The emerald deposit is located 15 km southeast of Itabira city, between the Belmont and Capoierana mines. The mineralized zone extends for over 5 km and the mines are located on two thrust zones (Brasiliano orogenesis) juxtaposing ultramafic rocks and highly deformed granites called "Borrachudos". At the regional scale, an extended emerald mining district was defined based on the geological continuity of the thrusts [74]; this district accounts for the majority of the Brazilian production.

In this type of deposit, granitic magmas have intruded M-UMR within volcano-sedimentary sequences or greenstone belts (Figure 12) and fluids expelled from the granite circulated within the thermal aureole, sometimes mixing with metamorphic fluids, and reacted with the pegmatite or quartz veins and M-UMR, forming emerald. The deposits are associated with vast amounts of fluid-rock interactions producing intense K-Na-Mg metasomatism of M-UMR and granite. This genetic pathway is the most common ([1,16,22,75,76]) and is characterized by emeralds contained in magnesium-rich micaceous rocks known as phlogopite schists or phlogopitites, or "glimmerite" in the Russian literature, "black wall zones" in [77], or "sludite" in Brazil. Generally, the pegmatite and the M-UMR experienced intense fluid infiltration and metasomatism where: (i) The M-UMR is converted into the emerald-bearing micaceous-host rock; (ii) the granitic rock itself is transformed into a feldspar-rich rock called plagioclasite comprised of albite-andesine feldspar. Quartz is dissolved and the pegmatite becomes a desilicated pegmatite (Figure 13). The metasomatic processes are highly variable and dependent upon P-T conditions, the extent of fluid circulation, and the timing between granitic intrusions and regional deformation.

At Carnaíba, the vein-like metasomatic rocks, in which phlogopite is by far the most abundant mineral, have a longitudinal extension that may extend several hundreds of meters, whereas their thickness does not exceed a few metres. They display a clear zoning and are organized, in many places in the M-UMR, symetrically around intrusive aplo-pegmatite dikelets, which channel the fluids [78].

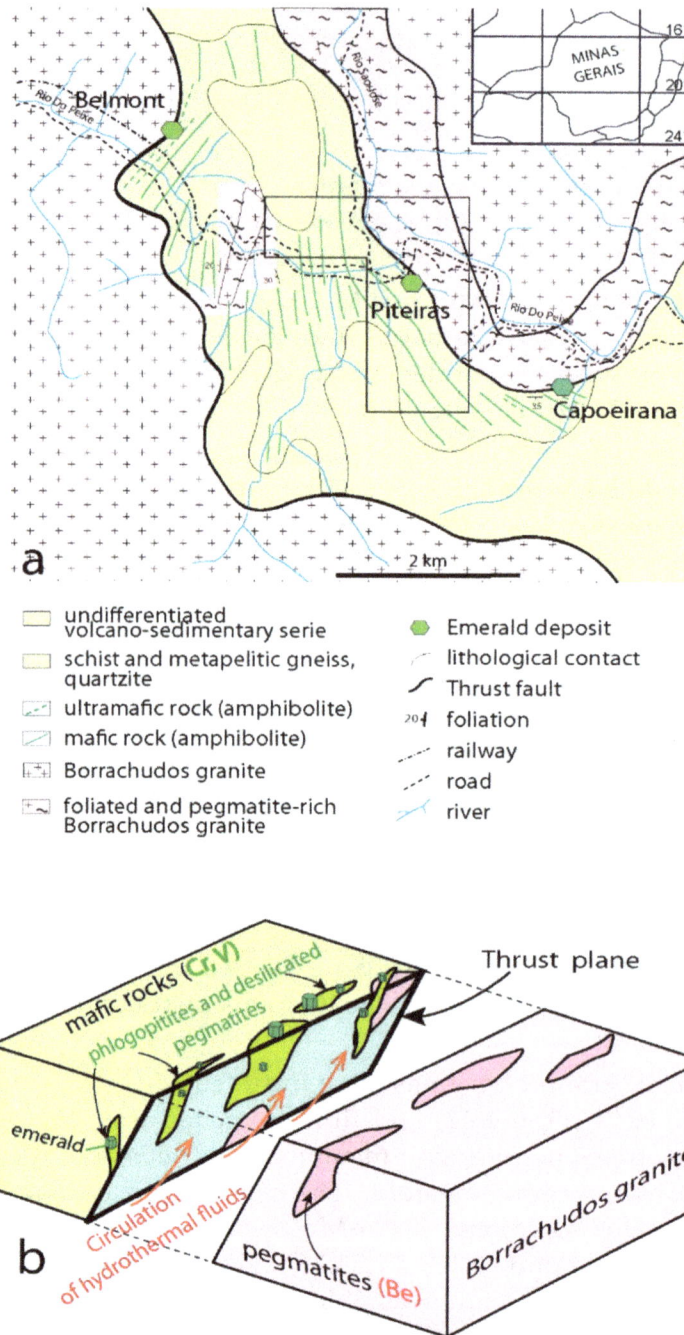

Figure 11. The emerald mining district of Itabira-Nova Era in Minas Gerais, Brazil, with the mines of Belmont, Piteiras, and Capoierana: (**a**) Geological map of the extended regional mining district. The mining concession of Piteiras is bounded by the polygon area plotted in the figure. Three deposits are currently mined. All occur in the same geological context. Ultramafic bodies (chromium—Cr, vanadium—V, and magnesium-rich-bearing rocks) are in contact with the deformed granites of Borrachudos (beryllium—Be, aluminium-silica-rich pegmatites), modified after [73]; (**b**) schematic diagram representing the probable formation of the Piteiras-Capoeirana-Belmont emerald deposits. During the Brasiliano orogenesis (508 Ma), the hydrothermal fluid (red arrows) circulated along the thrust planes, altering granites and associated Be-bearing pegmatites and Cr-V-bearing mafic-ultramafics. The hydrothermal fluids interacted with both rocks and were enriched in all the elements necessary for emerald crystallization. Emerald crystallized in plagioclasite (desilicated pegmatite) and phlogopitite.

quartz vein

desilicated pegmatite and
biotite schist

biotite schist
with emerald and
plagioclase

peridotite and serpentinite

quartz-tourmaline vein

Figure 12. Idealized model for the Tectonic magmatic-related emerald type. The model is based on the emplacement in the crust of a granitic massif, with its pegmatite and aplite dikes and their tourmaline- (Tr) or beryl- (Brl) bearing quartz veins, intruding mafic (metabasalt) and/or ultramafic rocks (metaperidotite, serpentinite). The fluid circulations from the granite into the surrounding rocks and granitic dykes (arrows), preferentially along the contacts between the pegmatite or aplite or quartz veins and the regional rocks, transform the mafic rocks into a magnesium-rich biotite schist and the pegmatite into an albite-rich plagioclasite. Emerald (Em) and apatite (Ap) precipitates in the rocks affected by the fluid/rock interaction. It can precipitate in the pegmatite, aplite, plagioclasite, and quartz veins and their adjacent phlogopite schist zones.

Figure 13. Desilicated emerald-bearing pegmatites: (**a**) Desilicated pegmatites crosscutting metabasites (mb) from the Kafubu mine (Zambia). Photograph: Dietmar Schwarz; (**b**) desilicated pegmatite from the Carnaíba mine, Brazil. At the contact of a pegmatite (pg) and a serpentinite, the metasomatic fluid dissolved quartz, mobilized Cr and V from the chromite-bearing serpentinite, and Be from the beryl-bearing pegmatite to form emerald (Em). The fluid transformed the serpentinite into a phlogopitite (ph) and the pegmatite in an albite-rich rock (ab). The three arrows indicate the limit of the dissolution of quartz (q) from the pegmatite. Photograph: Gaston Giuliani.

These metasomatic rocks, called exo-F-phlogopitites (1.3 to 4 wt.% F), formed a metasomatic column that consists of seven zones, from the central desilicated granitic vein (zone 7) to the enclosing serpentinite (Figure 14):

Zone 6: Coarse-grained F-phlogopite + apatite + emerald + quartz;
Zone 5: Fine-grained F-phlogopite + apatite + emerald;
Zone 4: F-phlogopite + spinel (chromite + magnetite);
Zone 3: F-phlogopite + spinel + amphibole (actinolite + tremolite);
Zone 2: F-phlogopite + spinel + ampbibole + talc; and
Zone 1: spinel + amphibole (or dolomite) + talc + serpentine + chlorite.

From the inner to the outer zones, (i) phlogopite composition evolves with continuous or discontinuous decrease in the Al and Fe contents with an increase in the Si and Mg contents and K/Al ratio; (ii) amphibole evolves from actinolite to tremolite; and (iii) spinel from Al-chromite to Cr-magnetite.

The aplopegmatite dikelets were transformed into plagioclasites (with disseminated phlogopite and little hornblende) with irregular commonly centimeter-thick phlogopite rims, embayments, and veinlets called endo-phlogopitites by analogy with the endo-skarns [78].

The strong chemical gradients in the different zones of the metasomatic column are characterized by a change in the composition of phlogopite (Figure 15; [78,79]). From zone 6 to zone 2, the Al and Fe contents decrease while Si and Mg increase. This evolution is discontinuous and two abrupt changes are observed, one at the front of zones 6/5, where the evolution corresponds to an increase in Si and Mg and a decrease in the Al, Fe, and Na contents and the other within zone 4. Such pattern of evolution are coherent with infiltration metasomatism. The potassium content remains constant in zones 6 and 5, suddenly decreases within zone 4, and remains constant up to zone 2.

Over the whole column, the F content of phlogopite is in the range of 1.3 to 4 wt.% and the highest values are observed in the phlogopite of the outer zone that has the highest Mg/(Fe + Mg) ratios, in agreement with the so-called "Fe-F avoidance effect" [80].

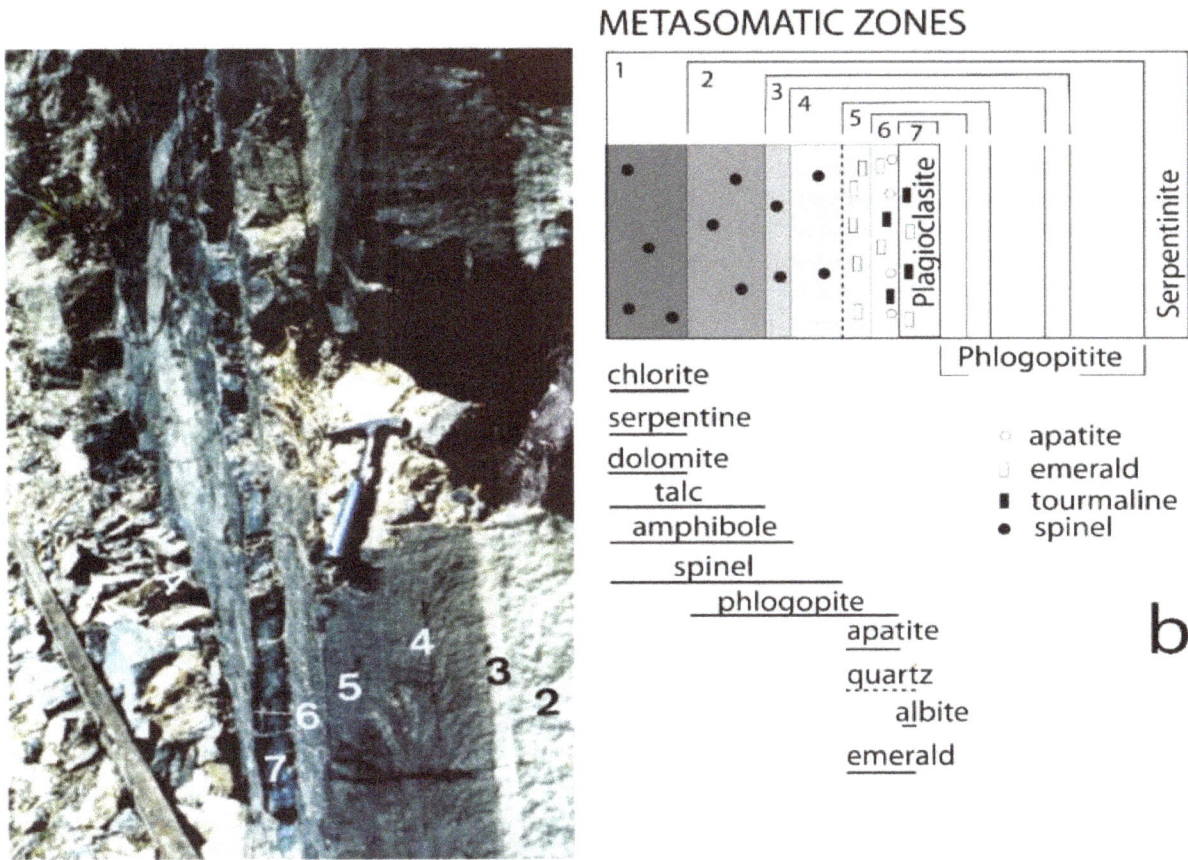

a   METASOMATIC ZONING AROUND A PEGMATITE VEIN

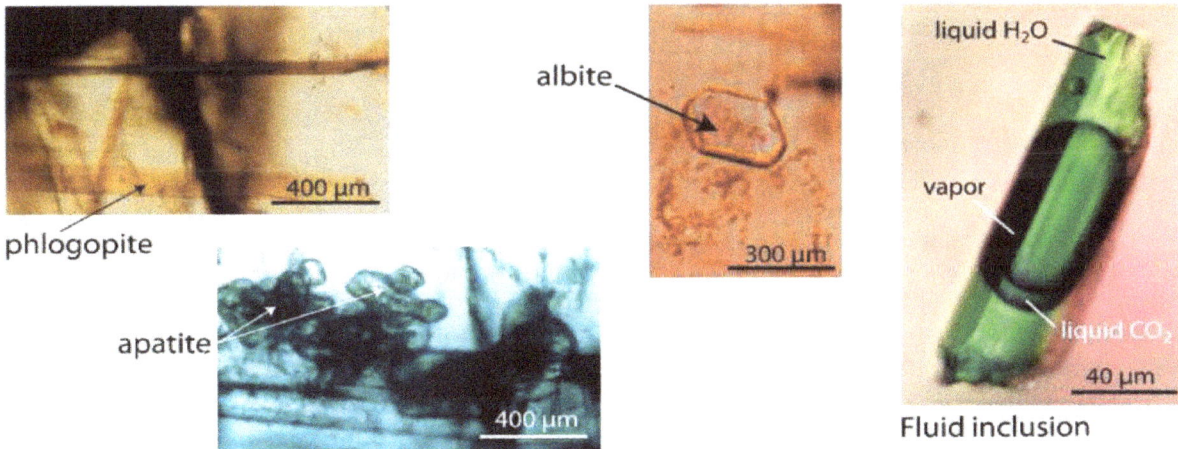

C   INCLUSIONS IN EMERALD FROM ZONES 6 and 7

Figure 14. The metasomatism related to fluid circulation between aplopegmatite dikes and mafic-ultramafic rocks: (**a**) Metasomatic column formed by different mineralogical zones (zones 7 to 2) developed around a central pegmatite vein crosscuting serpentinites, Braúlia prospecting pit, Carnaíba; (**b**) vein-like metasomatic rocks and their mineralogical composition. The pegmatite is transformed into plagioclasite (zone 7) and the metasomatic rocks consist of six zones, from the desilicated pegmatite to the enclosing serpentinite (zones 6 to 2). Emerald is found in the plagioclasite and in zones 6 (coarse-grained phlogopitite) and 5 (fine-grained phlogopitite). Zone 1 is the protolith formed by the serpentinite (not seen on the photograph); (**c**) solids and fluid inclusion in emeralds from zones 7 and 6 (Photographs: Dietmar Schwarz and Gaston Giuliani).

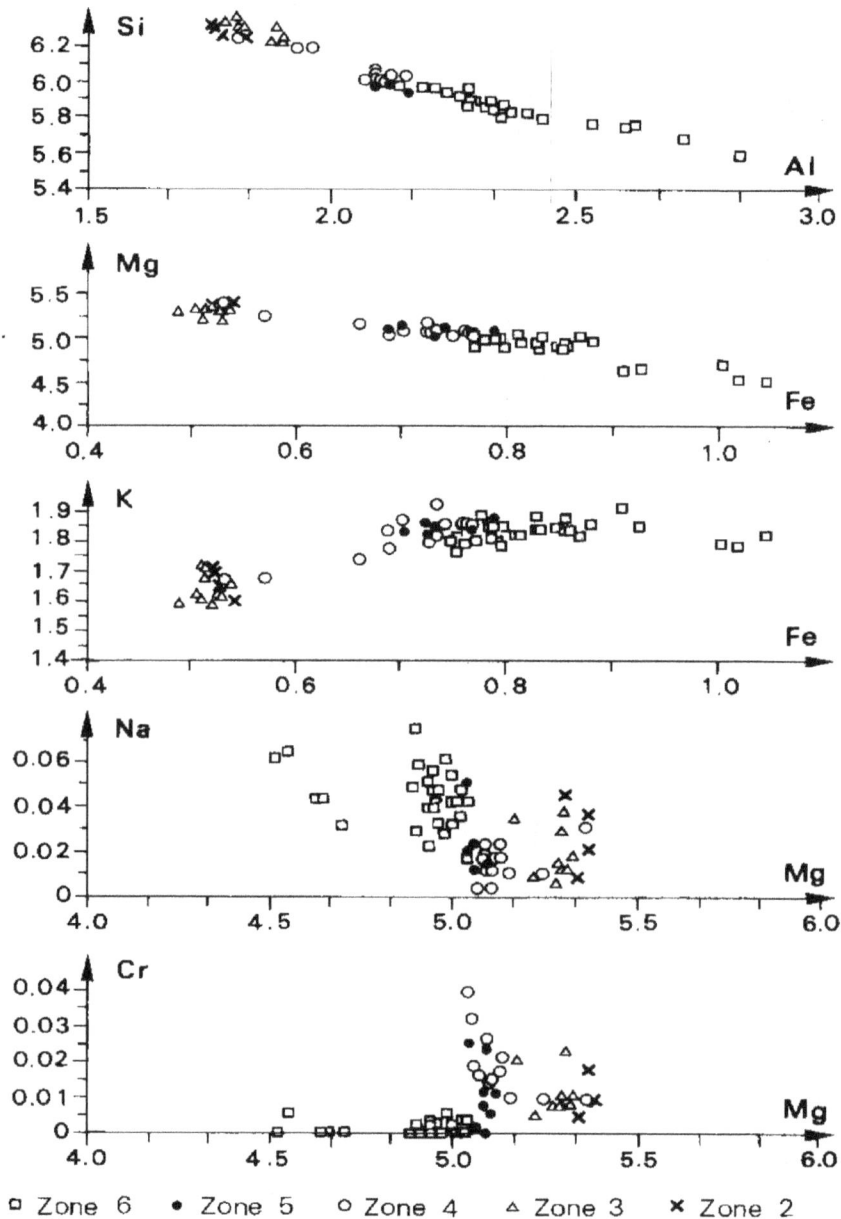

Figure 15. Evolution of the phlogopite composition in the phlogopitites of the different metasomatic zones (zones 6 to 2, see Figure 14 for the repartition of the different zones from the plagioclasite to the serpentinite) at Braúlia prospecting pit, Carnaíba mine, Brazil. Composition calculated in atom per formula (electron microprobe data from [79]). From the inner to the outer zones, phlogopite composition evolves with a continuous or discontinuous decrease in the Al and Fe contents and an increase of the Si and Mg contents and K/Al ratio, corresponding to changes in the petrographic habitus of the phlogopite.

Chromium contents are usually very low (<0.05 wt.%) in zones 6 and 5 and present a peak (up to 0.7 wt.%) at the front of zones 5/4, at the place where the first crystals of chromite are observed. In the outer zones 4 to 2, the Cr contents of phlogopite are nearly constant (0.25 wt.%). Chromium dispersed in the original rock can be concentrated at the front advancing outwards, like a "sweep effect" [81]. The almost uniform Cr content of phlogopite in the outer zones, 4 to 2, may result from the establishment of a exchange equilibrium between mica and chromite, Cr behaving as an inert component.

Mass transfer calculations show that the infiltration metasomatism in serpentinites is accompanied by important transfers of material, concerning mainly Al, K, F, Si, Mg, Ca, and H_2O (Figure 16). The strong enrichment in K, Rb, Li, Be, Nb, P, and F suggests that the metasomatic fluids have a magmatic origin, since some of these elements (including Rb, Nb, and Be) behave incompatibly during magmatic differentiation.

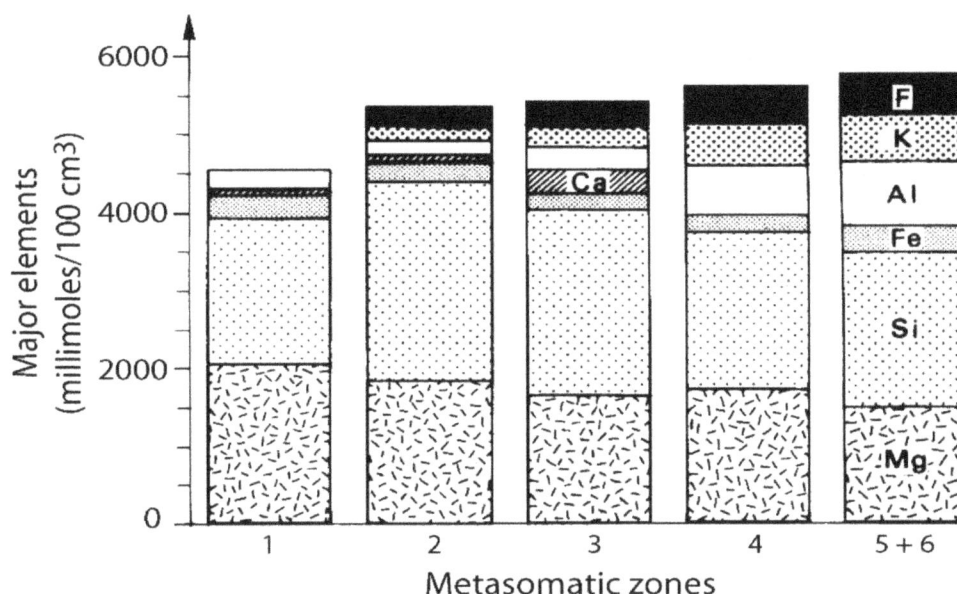

Figure 16. Mass balance calculations realized on the metasomatic column at Bode prospecting pit, Carnaíba mine, Bahia, Brazil (modified from [79]). Chemical mass balance was realized on major elements' molar composition (in millimoles/100 cm^3) of the different metasomatic zones (1 to 5 + 6). It indicates the supply of many elements, such as K, Si, Al, and F, over the whole column, and Fe, in the inner most zones of 6 and 5, the leaching of Mg and Ca from the outer to the inner zones.

The association of phlogopitite–plagioclasite corresponds to a "bimetasomatic system" in the sense given by [82]. The hydrothermal solutions that developed the potassic metasomatism in the serpentinites (phlogopitites) are alkaline, undersaturated with respect to microcline and quartz of the intrusive aplopegmatite, and albitizing.

The occurrence of emeralds is considered to be due to the efficiency of the metasomatic trap rather than a significant magmatic pre-enrichment in Be. The occurrence of strong chemical gradients in the zones of preferential circulation of the solutions (zones 6 and 5 of the exo-phlogopitites, plagioclasites, and endo-phlogopitites) constitutes highly favorable conditions for the beryl formation.

Elevated temperature and alumni and silica activities are the main factors for emerald formation in the IA sub-type (Figure 17). At high temperatures (T > 600 °C) and medium to high silica activity, beryl, phenakite (Be_2SiO_4), and chrysoberyl ($BeAl_2O_4$) are stable [53]. Emerald occurs in quartz veins, such as at the Kafubu deposits in Zambia [50,70], or greisens (an association of quartz and white micas), such as at the deposits of Delbegetey in Kazakhstan [51] and Khaltaro in Pakistan [83]. At lower silicate activities, albite is always stable and emerald is replaced by alexandrite (Cr-chrysoberyl) + phenakite. This is the case for emerald-bearing desilicated pegmatites in M-UMR, such as in Russia and Brazil. At higher Al_2O_3 activities, corundum can crystallize, such as at the Poona emerald deposit in Australia [84].

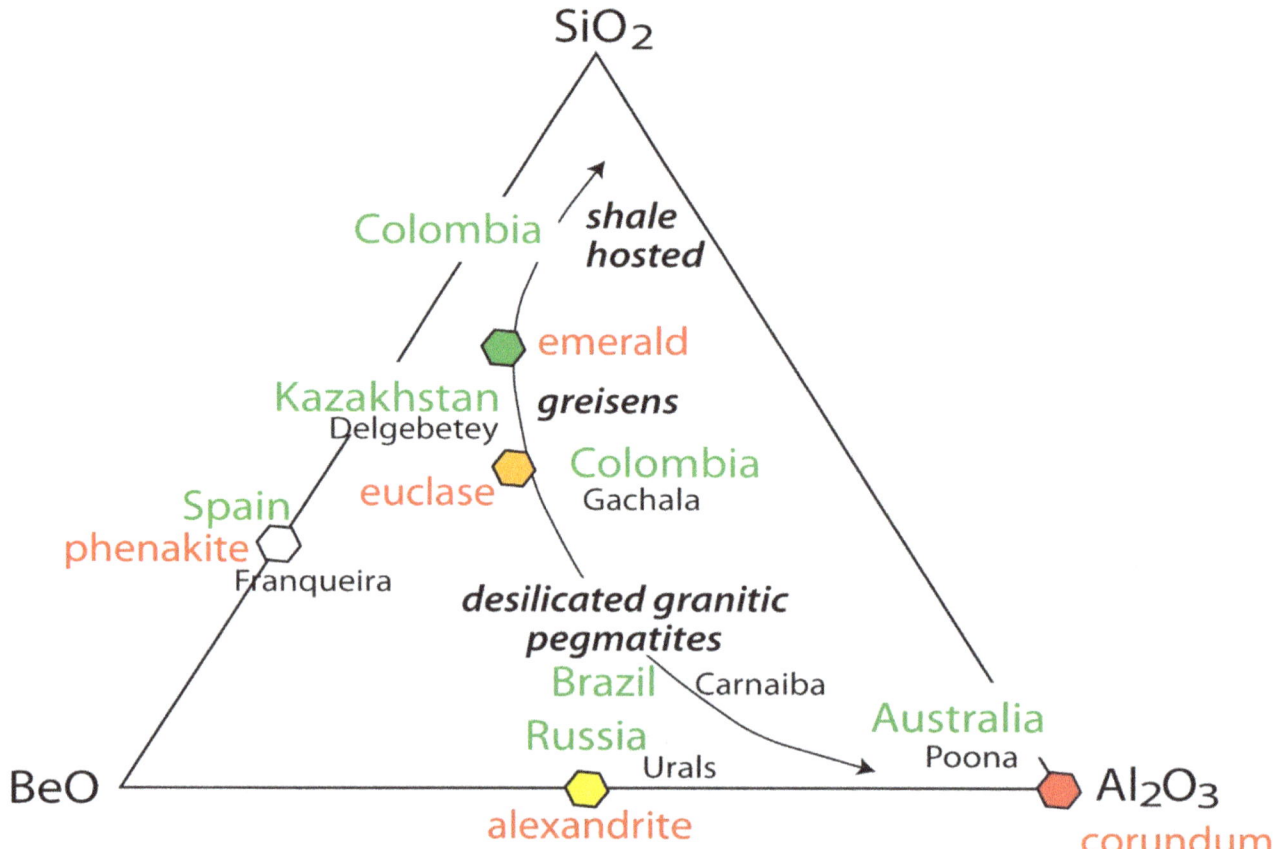

Figure 17. The ternary diagram of BeO-Al$_2$O$_3$-SiO$_2$ visualises the major types of emerald deposits as a function of their typical mineral assemblages, such as alexandrite, euclase, phenakite, and corundum ('BASH' system modified from [53]). Follow the discussion in the text.

The classic model of these granitic-metasomatic deposits presents much variability in terms of the geometry, chemical composition of the rocks, geological contacts, etc.:

1. At the Tsa da Gliza deposit in Canada, Be-bearing magmatic fluids from the neighboring granite reacted with the Cr- and V-bearing M-UMR [46]. Emeralds formed in aplite, pegmatite, and quartz veins surrounded by biotite schists. There is a continuum between the crystallization of granitic rocks, fluid rock-interaction, and emerald formation (Figure 18).

2. In the Kafubu area in Zambia (Figure 1c), emerald is found predominantly in metamorphosed M-UMR with phlogopite schist or in quartz-tourmaline veins adjacent to pegmatites.

3. At the Sandawana deposit in Zimbabwe, pegmatites intruded the M-UMR, but they are folded and fluid circulation in shear zones formed phlogopite schist. The fluid-rock interaction is coeval with the regional deformation [52]. Such phenomena are also found in the Carnaíba deposits in Brazil, where the dissolution of quartz from pegmatite is common [76].

4. At the Ianapera deposit in Madagascar, two coeval emerald deposits coexist (Figure 19; [11]): (1) A proximal one, formed at the contact between pegmatites and UMR; and (2) a distal one, hosted in biotite schist in fractures developed in mafic rocks with widespread fluid circulation affecting all the geological formations. Similarly, the Trecho Novo and Trecho Velho deposits at Carnaíba, Brazil [78] are also distal in nature.

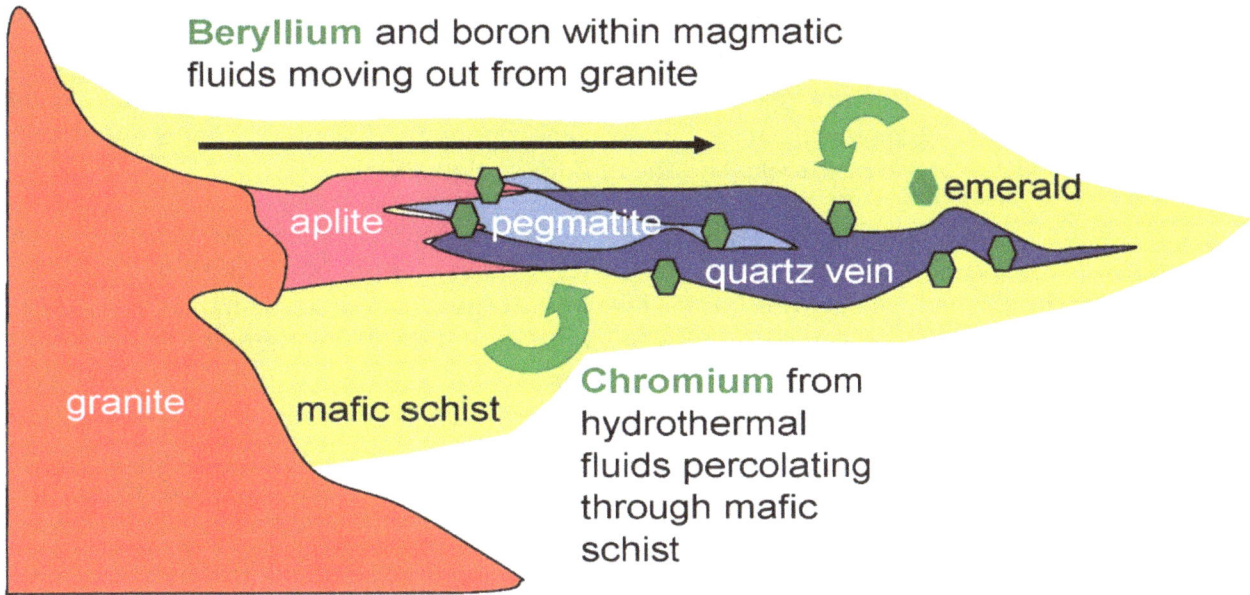

Figure 18. Genetic model proposed for the emerald mineralization at Tsa da Glisza, Yukon Territory, Canada.

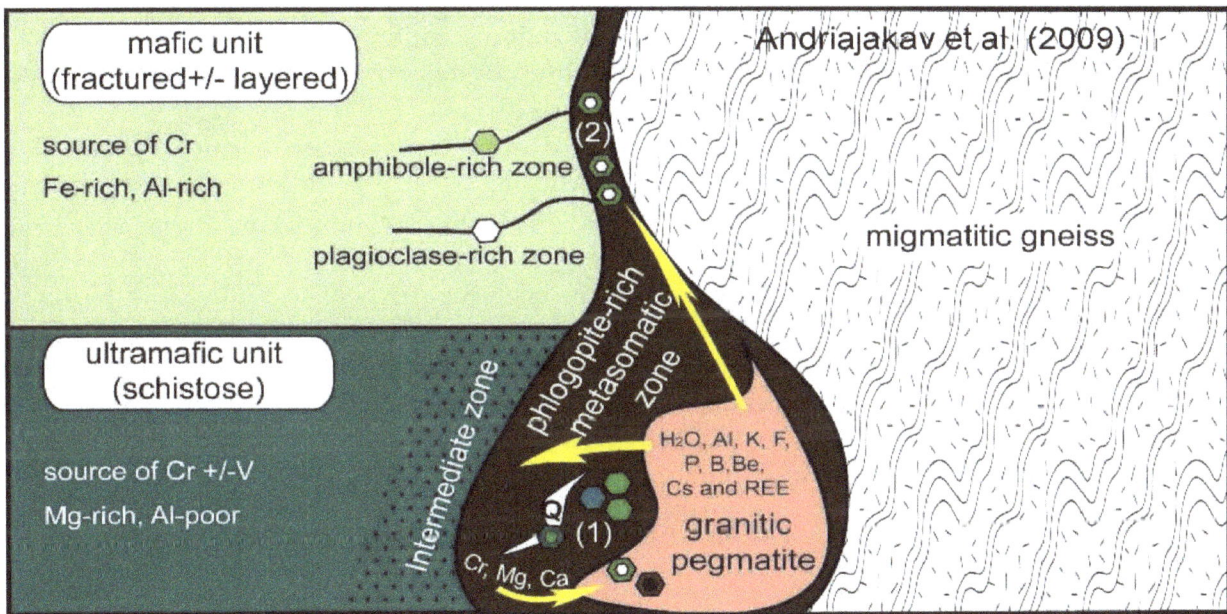

Figure 19. Schematic representation of the two styles of mineralization evidenced for the Ianapera emerald deposit, Madagascar: (**1**) Proximal mineralization occurs at the contact of pegmatite veins with ultramafic units. Emerald mineralization is hosted in metasomatic phlogopitite and desilicated pegmatites or quartz-tourmaline veins, at the contact between migmatitic gneiss and garnet amphibolite; (**2**) distal style is formed by phlogopite veins crosscutting mafic rocks. In addition to phlogopite, these veins contain Mg-amphibole, apatite, and dolomite, and minor quartz, calcite, zircon, plagioclase, and chlorite.

6.1.2. Sub-Type IB: Tectonic Magmatic-Related Emerald Deposits Hosted in (Meta)-Sedimentary Rocks (China, Canada, Kazakhstan, Norway, Australia; See Table 2)

These emerald occurrences are associated with granites that intrude sedimentary or meta-sedimentary lithologies. These deposits are generally sub-economic and several variants are described, depending on the nature of the host-rocks and the mineralization styles.

1. At Dyakou (China), the Lower Cretaceous porphyritic granite intruded biotite granofels, quartzite, gneiss, and plagioclase amphibolite of Lower and Upper Neoproterozoic formations. The intrusion formed skarns and dyke swarms of quartz veins, which are crosscut by pegmatites [85]. Emerald in quartz veins is less abundant than in the pegmatites, but of higher quality (Figure 1b). Some pegmatites show a local zoning with an outer zone enriched in K-feldspar and an inner zone of emerald and quartz.

2. At the Lened V-rich emerald occurrence (Canada), Be and other incompatible elements (i.e., W, Sn, Li, B, and F) in the emerald, vein minerals, and surrounding skarn were derived during the terminal stages of crystallization of the proximal Lened pluton [21,47]. Decarbonation during pyroxene-garnet skarn formation in the host carbonate rocks probably caused local overpressuring and fracturing that allowed ingress of magma-derived fluids and formation of quartz-calcite-beryl-scheelite-tourmaline-pyrite veins. The vein fluid was largely igneous in origin, but the dominant emerald chromophore V was mobilized by metasomatism of V-rich sedimentary rocks (avg. 2000 ppm V) that underlie the emerald occurrence [21].

3. At Delbegetey (Kazakhstan), the emerald mineralization is confined to the granite that hornfelsed carboniferous sandstones. Emerald is found in muscovite greisen formed in the wall-rocks of muscovite-tourmaline-fluorite-bearing quartz veins [51].

4. At Eidswoll (Norway), the emerald Byrud Gård mine is related to Permo-Triassic alkaline intrusions. The V-bearing emerald occurs in Middle Triassic pegmatite veins that intruded Cambrian Alum shales and quartz syenite sills [20]. Vanadium and Cr were probably leached from the alum shales by the mineralizing fluids [13].

5. At Emmaville-Torrington (Australia), emerald is located in pegmatite, aplite, and quartz veins associated with the Mole granite. The granite intrudes a Permian metasedimentary sequence consisting of meta-siltstones, slates, and quartzites [19]. The emerald-bearing pegmatite veins contain quartz, topaz, K-feldspar, and mica. Emerald is embedded in cavities and surrounded by dickite in the quartz-topaz veins. In the quartz lodes, emerald is associated with Sn-W-F minerals. At the Heffernan's Wolfram mine, emerald occurs with wolframite in vugs in the pegmatites [86].

6.1.3. Sub-Type IC: Tectonic Magmatic-Related Emerald Deposits Hosted in Peralkaline Granites (Nigeria)

These emerald occurrences are located in the Jurassic younger granite ring complexes of the anorogenic magmatism of Nigeria [68]. These granites crystallized in a volcanic-subvolcanic environment. They are generally peralkaline with perthitic K-feldspar, sodic amphiboles, and alkaline pyroxene. The roof zone of the intrusions is characterized by disseminated tin, tungsten, niobium, tantalum, and zinc mineralization related to sodic or potassic metasomatism, sheeted quartz vein systems, pegmatite pods and veinlets, replacement bodies, and fissure-filling veins (Figure 20). The emerald mineralization is located in sporadic pegmatitic pods with quartz and feldspar, as well as topaz and gem quality aquamarine. The source of chromium for the emerald is debated, but it could be the consequence of the mode of emplacement of the granites, which involved mechanisms of underground cauldron subsidence. The caldera produced ignimbrite, rhyolite, and thin Cr-bearing basic flows, which collapsed during doming or swelling and intrusion of the younger granites. Assimilation of the previous basic volcanic rocks or local fluid interaction between the pegmatites and the basic flows could have enriched the metasomatic fluids in Cr and Fe, resulting in the formation of emerald-aquamarine beryls.

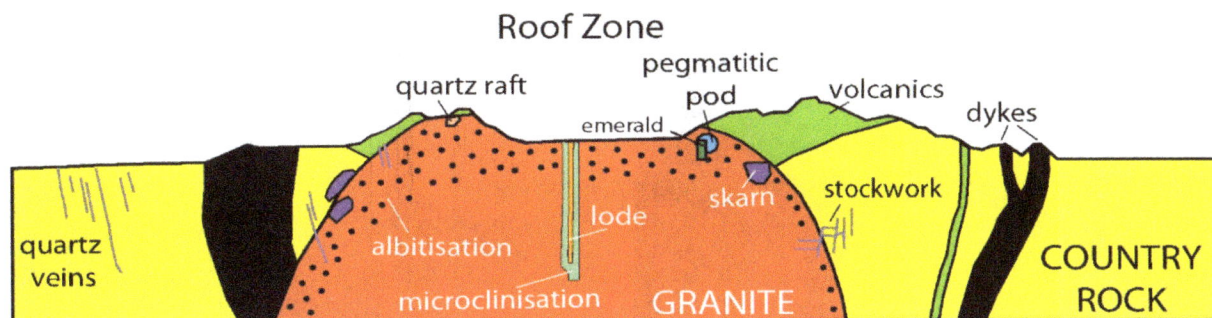

Figure 20. Ring complexes of Nigeria and emerald mineralization. Idealized cross-section showing structural setting and styles of mineralization.

6.2. Tectonic Metamorphic-Related (Type II)

Type II deposits are found in metamorphic environments with facies varying from high anchizone metamorphic illite crystallinity (Colombian black shales) to medium pressure conditions where the most exposed metamorphic rocks belong to the greenschist (ex: Swat talc-carbonate schist, Pakistan), amphibolite (ex: Habachtal metamorphic series, Austria), and up to granulite facies, with local partial melting of the rocks (migmatites of Hiddenite, USA). The majority of these deposits are characterized by the absence of granitic intrusions, and the deposits are linked to the circulation of fluids and metasomatism in thrusts, shear zones, or vein systems. Four sub-types are proposed in the new classification of emerald deposits (IIA, IIB, IIC, and IID).

6.2.1. Sub-Type IIA: Tectonic Metamorphic-Related Emerald Deposits Hosted in M-UMR (Brazil, Austria, Pakistan; See Table 1)

The Brazilian emerald deposits are partly associated with shear-zones cross-cutting M-UMR. The absence of a magmatic influence in their formation is constrained by field, geochemical, fluid inclusion, and stable isotope studies [15,76]. This is the case for the Santa Terezinha de Goiás deposit and the occurrence of Itaberaí. The economic importance of this deposit type is waning due to smaller emeralds, emerald dissemination, and higher artisanal mining cost and today represents only 4% of the total production of Brazil. The main Brazilian production is from type I deposits in Minas Gerais (74%) and Bahia (22%) [87]. Nevertheless, type IIA must be considered when prospecting in the Archean and Precambrian volcano-sedimentary series or greenstone belts.

1. The Santa Terezinha de Goiás deposit, located in central Goiás, produced 155 tons of emerald between 1981, date of the discovery, and 1988 [88]. The emerald grade was between 50 and 800 g/t. The infiltration of hydrothermal fluids is controlled by tectonic structures, such as the thrust and shear zones (Figure 21a). Pegmatite veins are absent and the mineralization is stratiform (Figure 21b). Emerald is disseminated within phlogopitites and phlogopitized carbonate-talc schists of the metavolcanic sedimentary sequence of Santa Terezinha [88–90]. Talc-schists provide the main sites for thrusting and the formation of sheath folds [91]. Emerald-rich zones are commonly found in the cores of sheath folds and along the foliation (Figure 21c). Two types of ore can be distinguished [88]: (i) A carbonate-rich ore composed of dolomite, talc, phlogopite, quartz, chlorite, tremolite, spinel, pyrite, and emerald; (ii) a phlogopite-rich ore composed of phlogopite, quartz, carbonates, chlorite, talc, pyrite, and emerald.

The distal São José two-micas granite, located 5 km from the emerald deposit, is a syntectonic foliated granite, which underwent a polyphase ductile deformation coeval to that observed in the emerald deposit [92]. C and S structures in the granite indicate shear deformation along a typical frontal thrust ramp and the granite overthrusted the Santa Terezinha sequence where the emerald deposit is located. D'el-Rey Silva and Barros Neto [92] suggested that the granite most probably was the source of Be for the formation of emerald in the Santa Terezinha de Goiás deposit.

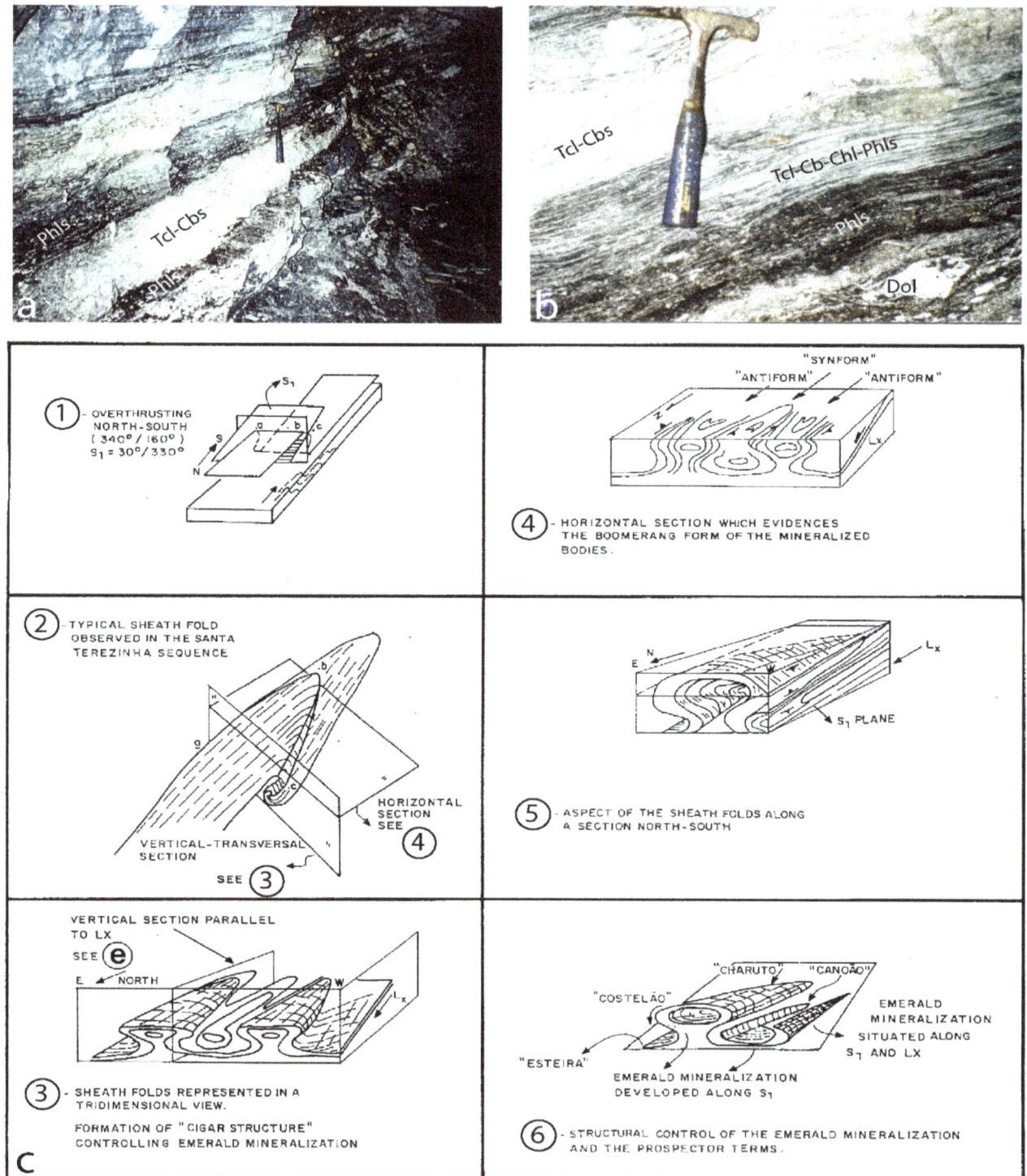

Figure 21. The Santa Terezinha de Goiás (Campos Verdes) emerald deposit, Goiás, Brazil: (**a**) Carbonated and phlogopite-rich emerald ores. The phlogopite schists (phls) underline the foliation of the talc-carbonate schists (Tcl-Cbs); (**b**) carbonates lenses (Dol) are observed within the talc-carbonate schists (Tcl-Cbs). The phlogopitisation affects the carbonated-talc-schists and the talc-carbonate-chlorite schist (Tcl-Cb-Chls) showing the "bed-by-bed" fluid injection along foliation planes; (**c**) structural evolution of the Santa Terezinha volcano-sedimentary sequence and the controls of the emerald mineralization on the basis of the structural study done on the Trecho Novo 167 underground mine (EMSA company property). Photographs: Gaston Giuliani.

2. The Habachtal deposit in the Austrian Alps has been studied in detail [77,93–97]. This alpine deposit is located in a contact zone, which overthrusts the volcano-sedimentary series of Habachtal (Habach Formation) on the ortho-augengneisses (central gneisses). The Paleozoic Habach formation is composed of a series of amphibolites, acid metavolcanics tranformed in muscovite schists, and black pelites with interlayered serpentinites and talc series. Two metamorphic events, one occurring before the Alpine event (P < 3 kb and T < 450 °C) and one occuring during the Alpine event (4.5 < P < 6 kb and 500 < T < 550 °C) were superimposed. The mineralized "blackwall zone", the equivalent of a phlogopitite, is a tectonic or shear zone 100 m wide, formed from UMR (serpentinites) pinched between orthogneisses and amphibolites. Emerald is disseminated in the "blackwall zone" phlogopitites, talc-actinolite, and chlorite schists. The metasomatic process involves fluid percolation that extracted Be from the muscovite schists (average Be content = 36 ppm) and Cr from the serpentinites (Cr content = 304 ppm) to facilitate the crystallization of emerald (Figure 22). Fluid inclusions trapped by emerald belong to the H_2O-CO_2-NaCl system [96] with two generations of fluid inclusions: An early generation (XCO_2 < 4 vol.%) and a late one (XCO_2 up to 11 vol.%). Emerald-metasomatic fluids were related to hydration phenomena due to the alpine metamorphism [96].

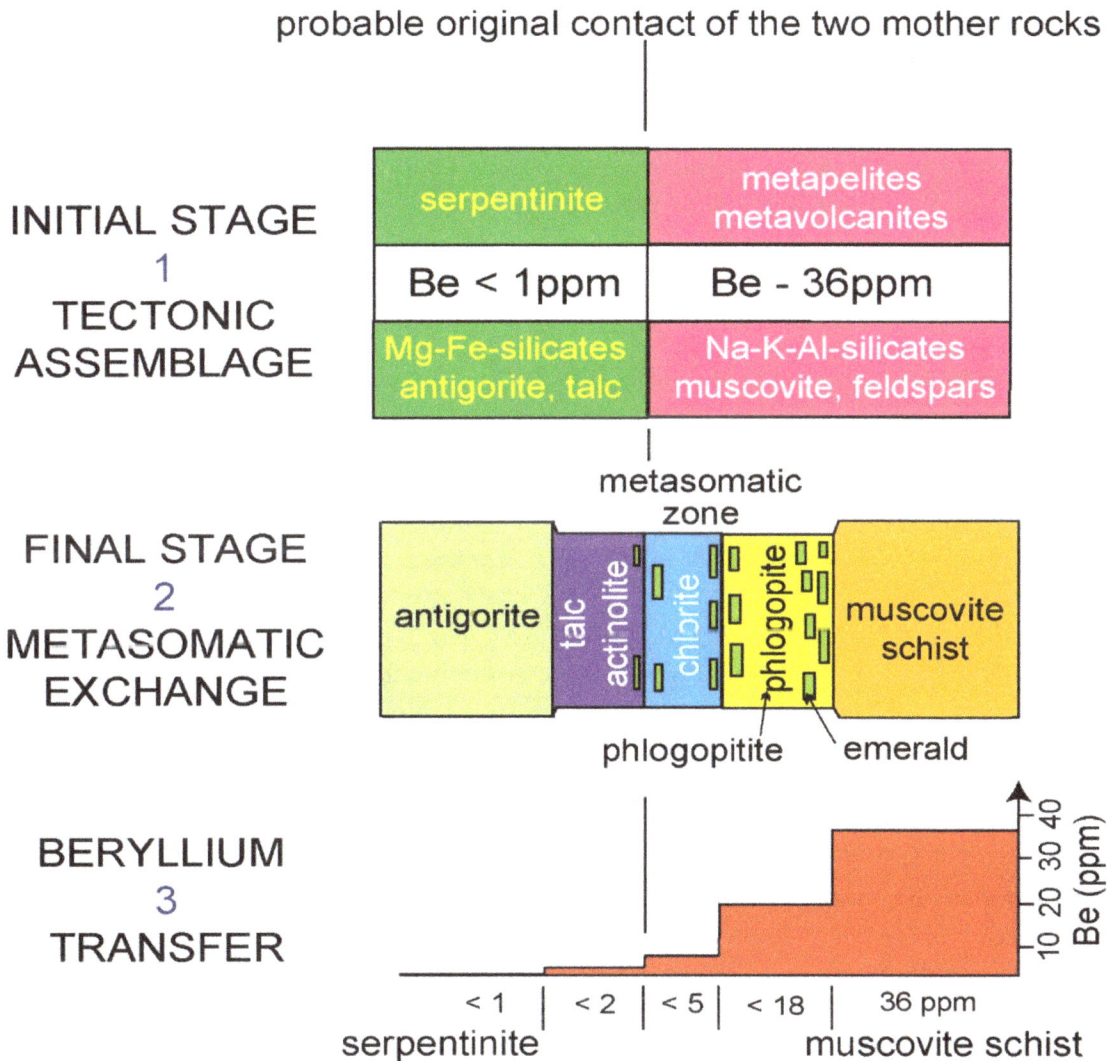

Figure 22. Model of formation of the Habachtal emerald deposit: (1) Initial stage showing the different lithologies with their respective Be contents; (2) final stage after regional metamorphism showing the final metasomatic rocks assemblages; (3) beryllium liberated from the muscovite schists is incorporated into the emerald crystals.

3.   The Swat-Mingora–Gujar Kili–Barang deposits are controlled by the Main Mantle Thrust [86,98–100]. The suture zone that marks the collision of the Indo-Pakistan plate with the Kohistan arc sequence is composed of a number of fault bounded rock melanges (blueschist, greenschist, and ophiolitic melanges).

The ophiolitic melange, which hosts the Pakistani emerald deposits, is composed mainly of altered ultramafic rocks with local cumulate, pillow lavas, and metasediments. Emerald occurs within hydrothermally altered serpentinites that show metasomatic zoning [100]: An outer zone composed of talc-magnesite ± chlorite ± micas; an intermediate zone consisting of talc-magnesite with dolomite veins; and an inner zone with dolomite-magnesite-talc schists and quartz-dolomite ± tourmaline ± fuschite veins. Emerald occurs disseminated in the inner and intermediate zones within or spatially associated with quartz-carbonate veins. Dilles et al. [100] obtained a $^{40}Ar/^{39}Ar$ Oligocene age of 23.7 ± 0.1 Ma for a fuchsite-quartz vein in the Swat emerald deposit.

Isotopic study of magnesite from the outer zone showed that the mineral association resulted from early metamorphic fluids [101], whereas the inner and intermediate zones resulted from the infiltration of hydrothermal fluids, which carried Si, Be, B, K, and Ca. Chromium came from the dissolution of chromite crystals in the serpentinites. Arif et al. [102] analysed Cr, Be, B, and other trace element contents in the ophiolitic rocks of the Indus suture zone in Swat. They showed that the Cr present in the Cr-bearing silicates (emerald, Cr-tourmaline, and Cr-muscovite) was derived from the original protolith. Beryllium and boron enrichments were found only in M-UMR affected by fractures and fluid circulation. In addition, analyses of small granitic dykes cutting granitic gneisses showed extreme B and Be enrichment. In consequence, Arif et al. [102] argued that the Be and B are sourced from a probable hidden leucogranite in depth.

6.2.2. Sub-Type IIB: Tectonic Metamorphic-Related Emerald Deposits Hosted in Sedimentary Rocks: Black Shales (Colombia, Canada, USA; Table 2)

Colombian emerald deposits are unique. Why?

The Colombian emerald deposits are located on both sides of the Eastern Cordillera sedimentary basin, with an eastern zone comprising the mining districts of Gachalá, Chivor, and Macanal, and a western zone, including the mines of La Glorieta-Yacopi, Muzo, Coscuez, La Pita, Cunas, and Peñas Blancas (Figure 23). Their distribution along both sides of the basin ensures abundant production of high-quality emerald. These deposits are unique because there is no connection with granites in their formation [103,104].

The emerald mineralization is hosted in Lower Cretaceous (135–116 Ma) sedimentary rocks composed of a thick succession of sandstone, limestone, black shale, and evaporites. The salt and sulphate rocks were necessary for the formation of emerald. These intercalations of evaporites are found both in the Guavio (Figure 24) and Rosablanca formations, in the eastern and western emerald zones, respectively. The evaporites are responsible for the high salinity of the basinal brines (~40 wt.% equivalent NaCl) and the circulation of H_2O-NaCl-CO_2-(Ca-K-Mg-Fe-Li-SO_4) fluids trapped by emerald during its growth ([105]; Figure 25).

The tectonic-sedimentary evolution of the Eastern Cordillera basin is unique (Figure 26). The formation of the emerald deposits is related to changes in the acceleration and convergence of the Nazca and South American plates that took place at: (1) 65 Ma, forming the emerald deposits on the eastern side of the basin [106]; and at (2) 38–32 Ma, generating the Muzo and Coscuez deposits on the western side [103]. Formation at different ages and conditions resulted in two drastically different styles of mineralization in the eastern and western emerald zones [107,108].

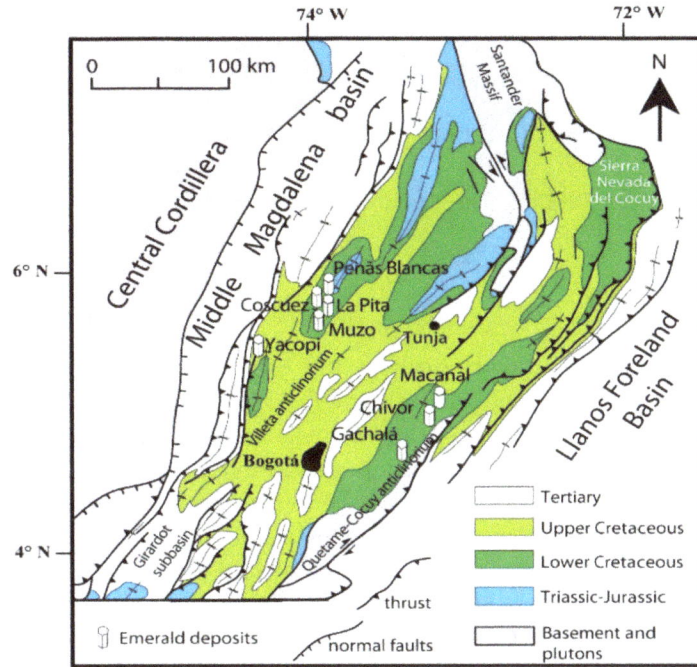

Figure 23. Simplified geological map of the basin from the Eastern Colombian Cordillera. The emerald deposits are hosted by Lower Cretaceous sedimentary rocks forming two mineralized zones located, respectively, on the eastern and western borders of the basin. On the western border, with the mining districts of La Glorieta-Yacopi, Muzo, Coscuez, La Pita, Cunas, and Peñas Blancas, and on the eastern border are Gachalá, Chivor, and Macanal.

Figure 24. Evaporites in the the Chivor mining area: (**a**) The gypsum deposits present decametric lenses of white gypsum/dolostone alternations hosted in a black matrix made of crushed and dismembered black shales similar to the main evaporitic regional breccia level in emerald deposits; (**b**) Dolomicritic beds are in alternation with nodulated beds of gypsum. Photographs: Yannick Branquet.

In the eastern emerald deposits, the Chivor mining district presents extensional structures extending from a brecciated regional evaporitic level (Figure 27), which acted as a local, gravity-driven detachment [107,108]. The brecciated rock unit in the Chivor area, which is in excess of 10 km long and 10 m thick (Figure 28a,b), is stratabound, i.e., parallel to the sedimentary strata, and dominantly composed of hydrothermal breccia (Figure 28c) made up of fragments of the hanging wall (carbonated carbon-rich BS, limestone, and whitish albitite (albitized black shale)) cemented by carbonates and pyrite (Figure 28d).

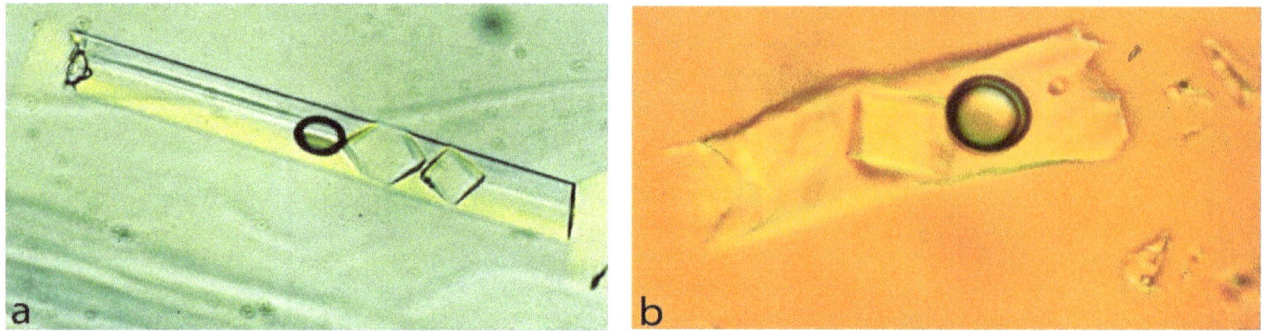

Figure 25. Highly saline basinal fluids trapped by fluid inclusions in Colombian emerald is strong evidence for the evaporitic origin of the mineralizing fluid: (**a**) Emerald from Oriente mine in Chivor (Eastern emerald zone). Primary multi-phase fluid inclusion trapped in an emerald. The cavity, 180 μm long, contains from the right to the left, two cubes of sodium chloride (halite), a rounded gas bubble, and two minute crystals of calcite all of which are wetted by a salty water occupying 75 vol.% of the cavity; (**b**) emerald from Coscuez (western emerald zone). The cavity, 40 μm long, contains 75 vol.% of salt water solution, 10 vol.% of gas corresponding to the vapour bubble, 15 vol.% of cubic halite crystal (NaCl), and a rounded crystal of carbonate (on the right of the cavity). The vapour phase is rimmed by liquid carbon dioxide. Photographs: Hervé Conge.

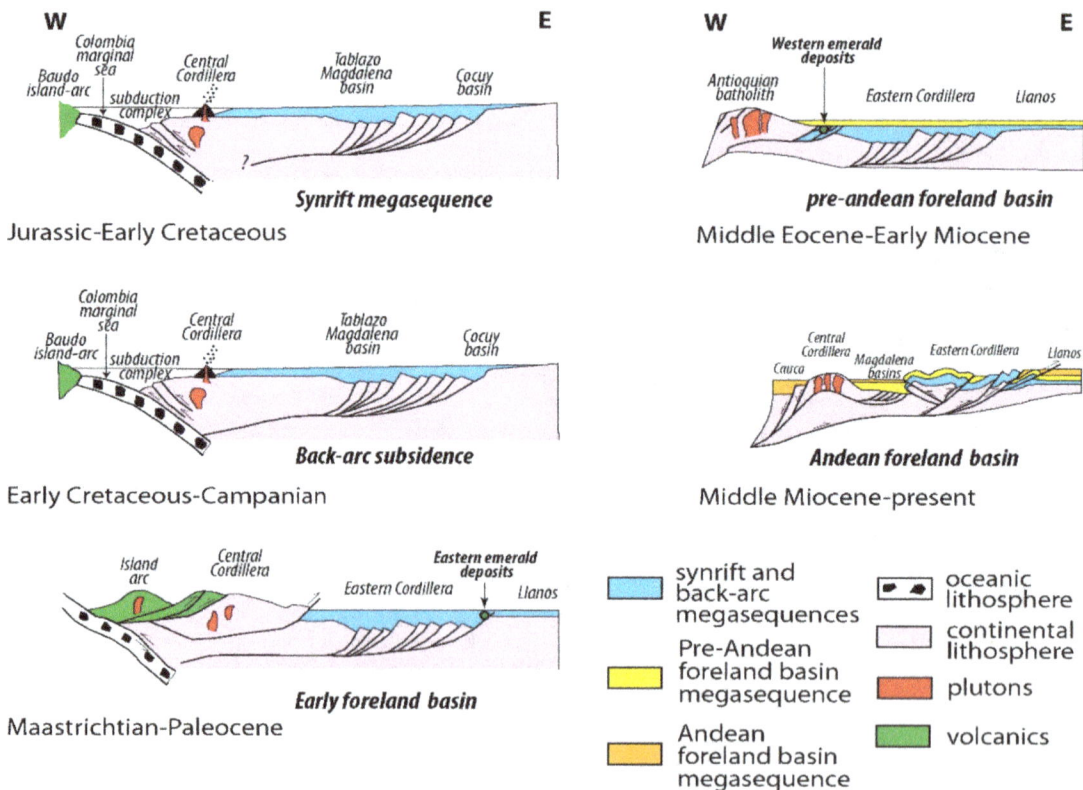

Figure 26. Basin development and tectonic history of the Llanos and Eastern Cordillera basins, and Middle Magdalena Valley in Colombia (modified from [109]). Four episodes of deformation have been recognized in the Tertiary of central Colombia: (**1**) Late Cretaceous–early Paleocene, with the formation at 65 Ma of the emerald deposits from the eastern zone; (**2**) middle Eocene with the creation of folds and thrusts in the Middle Magalena valley and western border of the Eastern Cordillera basin, and with the formation, between 38–32 Ma, of the emerald deposits from the western; (**3**) late Oligocene–early Miocene; and (**4**) late Miocene–Pliocene at 10.5 Ma where the Eastern Cordillera was uplifted and eroded, with the outcropping of the emerald deposits from the eastern and western emerald zones.

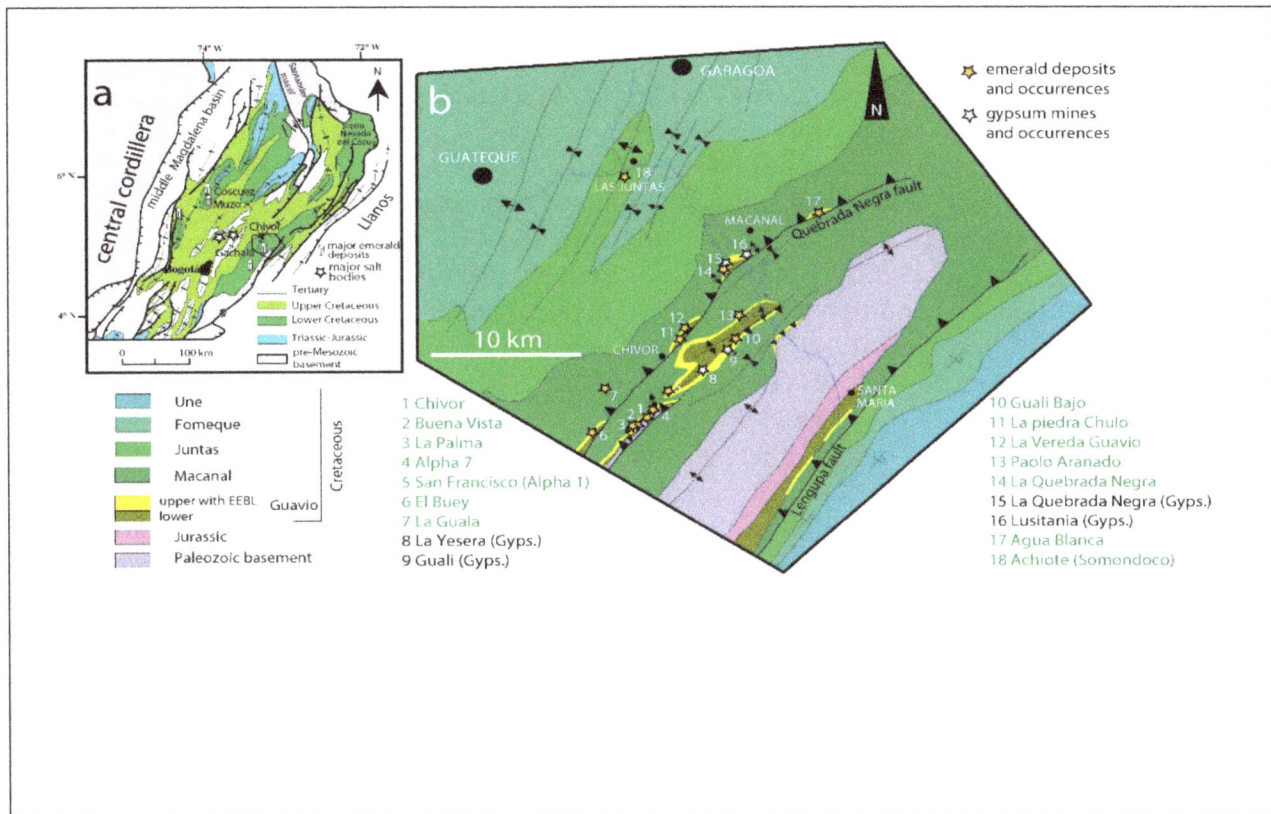

Figure 27. Deposits from the eastern emerald zone in Colombia:(**a**) Simplified geological map of the Eastern Cordillera with the location of the main emerald deposits. Inset is the location of Figure 27b; (**b**) geological map of the Chivor area. All emerald and gypsum deposits and occurrences are hosted within the Berriasian Upper Guavio Formation.

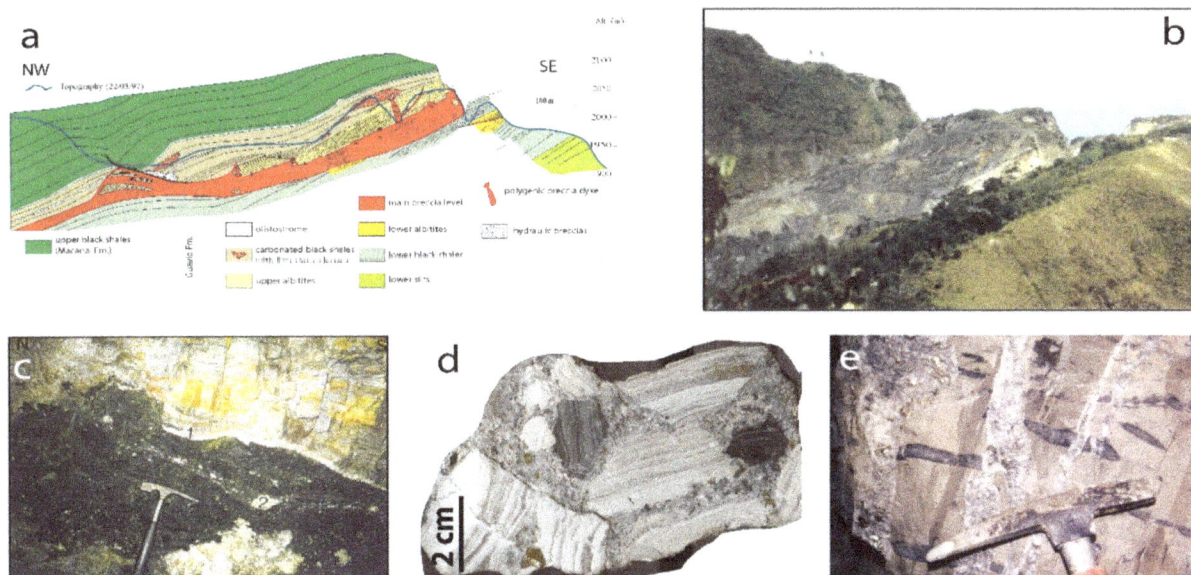

Figure 28. The Chivor mining area: (**a**) Geological cross-section through the Chivor emerald deposits; (**b**) south-eastern field view of the cross-section; (**c**) Chivor Klein pit. Upper contact of the main breccia level (in black) with albitites (1). The transport of clasts of albitite (2) within the breccia is marked by tails; (**d**) Oriente deposit. Polygenic breccia formed by clasts of albitite and black shales, cemented by pyrite, carbonates, and albite; (**e**) Oriente deposit. Carbonate-pyrite-emerald-bearing veins crosscutting albitite showing some remnants of black shale. Photographs **b** to **d**: Yannick Branquet; Photograph **e**: Gaston Giuliani.

All the mineralized structures, i.e., sub-vertical veins, extending from the roof of the brecciated level (Figure 28e), are mineralized listric faults that attest of the bulk extensional structures extending from the brecciated regional evaporitic level. At 65 Ma, to a small amount of horizontal stretching a large flow of hydrothermal fluids responsible for emerald deposition occurred through the regional evaporitic level. The observed structures of the breccia are diagnostic of hydraulic fracturing associated with the evaporite solution within the salt-bearing main breccia level. At the Cretaceous–Tertiary boundary, the overburden of the Guavio Formation in the Chivor area was about 5–6 km [110]. At that time, the area was slightly uplifted in an incipient foreland bulge [111]. This was [111] interpreted as extensional structures observed in the emerald deposits, resulting from flexural extension. Following this model, the emerald-related hydrothermal event recorded an abrupt change in the thermal and dynamic conditions of the Eastern Cordillera basin, which triggered regional-scale hot, deep, and over-pressured brines migration [110].

The western emerald deposits, such as Muzo (Figure 29a) and Coscuez (Figure 30), are characterized by compressive fold and thrust structures formed along tear faults. These tectonic structures are synchronous with the circulation of the hydrothermal fluids and emerald deposition. The deposits are hectometre-sized at most and display numerous folds, thrusts, and tear faults [112]. In the Muzo deposit, thrusts are marked by the calcareous BS, which overly siliceous BS (Figure 29b). All the tectonic contacts are marked by cm- to m-thick hydrothermal breccia called "cenicero", i.e., ashtray, by the local miners (Figure 29c,d). These white- or red-coloured breccias outline the thrust planes, which are associated with intense hydraulic fracturing due to overpressured fluids [113]. Multistage brecciation corresponds to successive fault-fluid flow pulses and dilatant sites resulting from shear-fracturing synchronous to the thrust fault propagation.

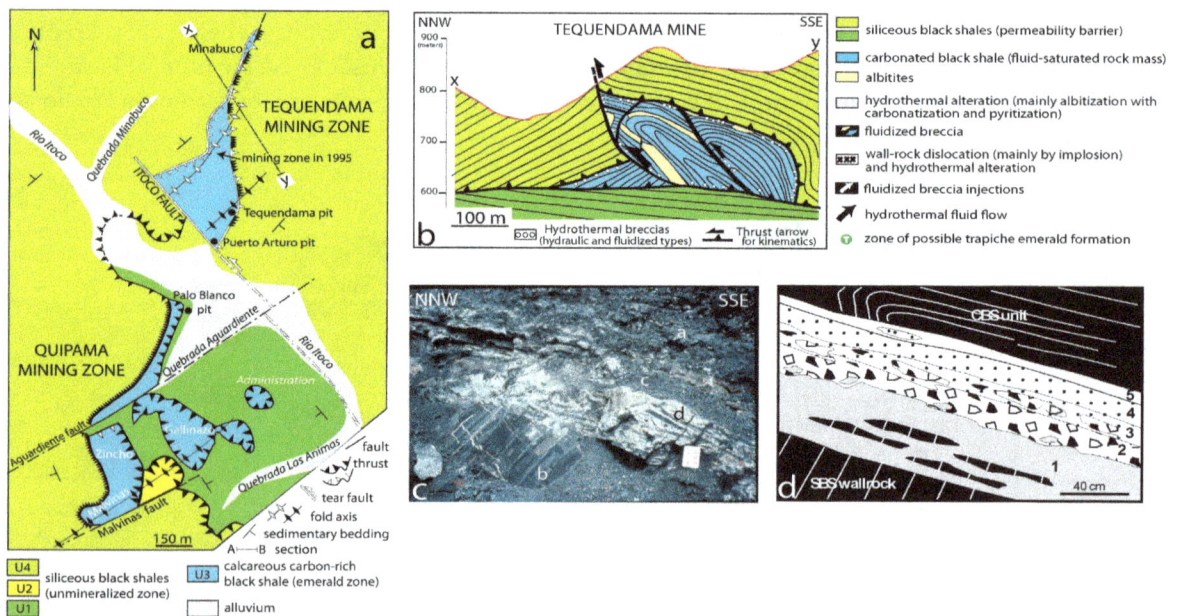

Figure 29. Tectonic style and lithologies of the Muzo emerald mining district, western emerald zone: (a) Geological map produced between 1994 and 1996. U1 through U4 represent the different tectonic units. The units, U1, U2, and U4, are comprised of barren siliceous black shales (cambiado), while U3 is composed of emerald-bearing calcareous carbon-rich black shale; (b) Tequendama mine cross-section along x-y (see Figure 29a). The cross-section shows the different lithologies and tectonic structures described by [112] and [107,113]. The thrusts and tear faults are associated with the presence of breccia, extensional veins, and related potential zones for trapiche emerald in the wall-rock dislocations and hydrothermal alteration [114]; (c) thrust verging NNW at Tequendama mine. The SSE dipping calcareous black shale unit (a) thrusts over the NNW dipping siliceous black shale unit (b). The siliceous beds are truncated at high angles by the thrust plane characterized by a hydrothermal breccia (c) and a thrust-parallel vein (d). Photograph: Gaston Giuliani; (d) texture and mineralogy of a thrust-parallel

and layered breccia called "cenicero" by the miners, outcrops of Puerto Arturo in 1994—95, Tequendama mine [107,113]. Layering of the breccia zoning: 1 = carbonate thrust vein; 2 = cemented breccia with pyrite and calcite crystals; 3 = tectonically reworked breccia with a red melange cemented by pyrite and ankerite; 4 = red layer totally cemented by pyrite and ankerite ("cenicero rojo"); 5 = pure pyrite layer ("cordón piritoso"). CBS = calcareous black shale; SBS = siliceous black shale.

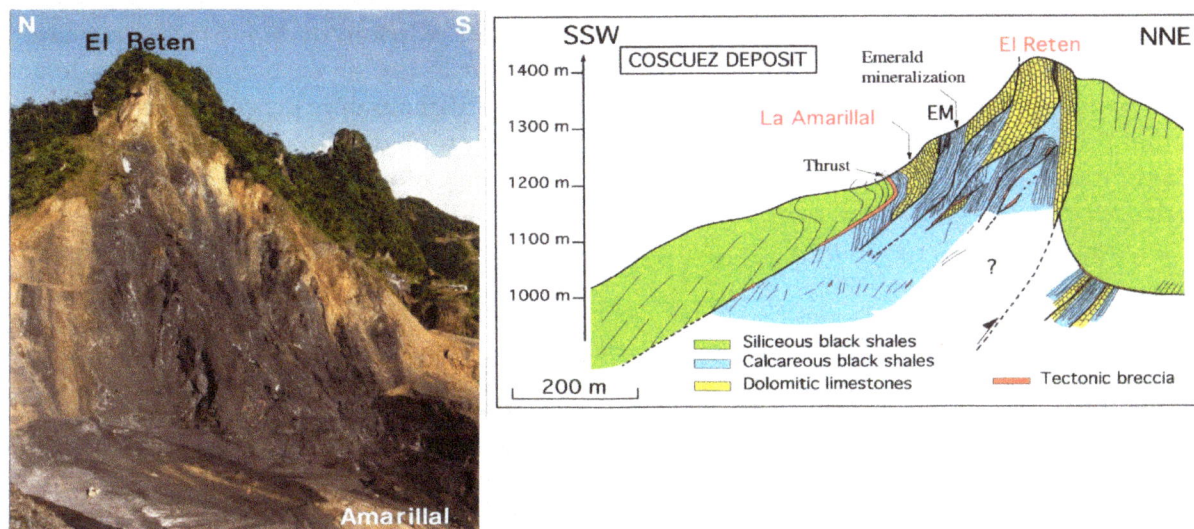

Figure 30. Tectonic style and lithologies of the Coscuez emerald mining district, western emerald zone: (**a**) General view of the Coscuez deposit in 1996. Photograph: Gaston Giuliani; (**b**) geological cross-section of the Coscuez deposit showing the lithostratigraphic column of the formations and the tectonic style marked by thrusts and faults. The emerald mineralization is linked to tectonic hydrothermal breccia zones, faults, thrusts, and stockwork veins [107,113].

The thermal-reduction of sulphate, at 300–330 °C, in the presence of organic matter in the black shale during the formation of the emerald-bearing veins is unique for Be-bearing mineralization [104,115]. Sulphates (SO_4^{2-}) in minerals of evaporitic origin are reduced by the organic matter of the BS to form hydrogen sulphide (H_2S) and hydrogen carbonate bounding (HCO_3^-), which are responsible for the precipitation of pyrite, carbonates, and bitumen in the veins via the following reactions:

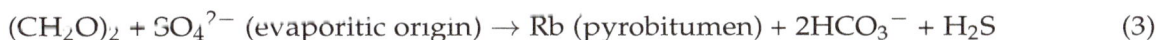

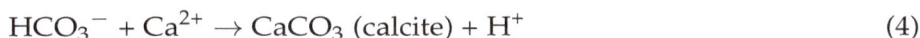

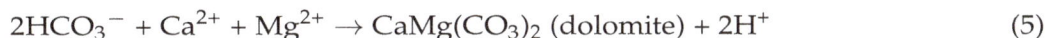

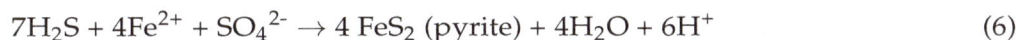

$$(CH_2O)_2 + SO_4^{?-} \text{ (evaporitic origin)} \rightarrow Rb \text{ (pyrobitumen)} + 2HCO_3^- + H_2S \tag{3}$$

$$HCO_3^- + Ca^{2+} \rightarrow CaCO_3 \text{ (calcite)} + H^+ \tag{4}$$

$$2HCO_3^- + Ca^{2+} + Mg^{2+} \rightarrow CaMg(CO_3)_2 \text{ (dolomite)} + 2H^+ \tag{5}$$

$$7H_2S + 4Fe^{2+} + SO_4^{2-} \rightarrow 4 FeS_2 \text{ (pyrite)} + 4H_2O + 6H^+ \tag{6}$$

The consequence of pyrite precipitation is the depletion of iron in the fluid prior to emerald precipitation. This impacted the gemmological features of the emerald as follows: (i) Iron-free absorption spectra with sharp Cr-V bands; (ii) iron-poor chemical fingerprinting; and (iii) optical properties that correlate with the low trace elements content [116].

Although, as noted above, the emerald deposits in Colombia are unique, two minor occurrences that show similar recipes for emerald formation have been reported:

1. An emerald occurrence was described near Mountain River in the northern Canadian Cordillera [117]. These emerald veins are hosted within siliciclastic strata in the hanging wall of the Shale Lake thrust fault. The emerald formed as a result of inorganic thermal-chemical sulphate reduction via the circulation of deep-seated hydrothermal carbonic brines through basinal

siliciclastic, carbonate, and evaporitic rocks (Figure 31). The deep-seated H_2O-$NaCl$-CO_2-N_2 brines, with a salinity up to 24 wt.% equivalent $NaCl$, were driven along deep basement structures and reactivated normal faults related to tectonic activity associated with the development of a back-arc basin during the Late Devonian to Middle Mississippian (385–329 Ma). The Mountain River emerald occurrence thus represents a similar and small-scale variation of the Colombian-type emerald deposit model [117].

2.   Three emeralds were reported in the Uinta Mountains in Utah, USA [118,119]. The discovery was realized in the Neoproterozoic Red Pine shale, which is overlain by Paleozoic carbonate rocks. Based on the study of fibrous calcite hosted by the Mississippian carbonate units, subjacent to the hypothetic emerald-bearing shale, [120] proposed an amagmatic process for the formation of emeralds. The authors combined their chemical and isotope data on calcite, limestone, and the Red Pine shale using a model of formation similar to the Colombian type and involving thermal reduction of sulphates between 100 and 300 °C. Emeralds were not described [120] and one can question their existence in the Uinta Mountains. No one has described these emeralds since the discovery of [118].

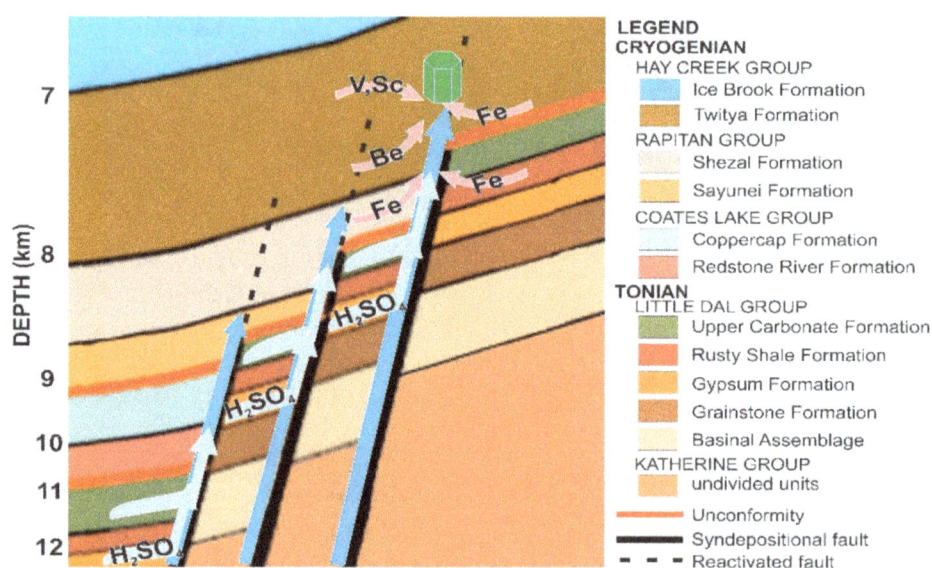

Figure 31. Idealized schematic model for the emerald mineralization of the Mountain River emerald occurrence located in Devonian to Mississippian platform sediments [117]. Emerald resulted from inorganic thermochemical sulphate reduction (light blue arrow) via the circulation of brines along a reactivated normal fault (dark blue arrow). The high-salinity brines result from the dissolution of evaporites lenses in the sediments. Be, V, Sc, and Fe were mobilized from the sedimentary formations (pinkish arrows).

6.2.3. Sub-Type IIC: Tectonic Metamorphic-Related Emerald Deposits Hosted in Metamorphic Rocks Other than M-UMR and Black Shales (Afghanistan, China, USA; Table 2)

This sub-type includes emerald-bearing quartz vein and veinlet deposits located in medium pressure metamorphic rocks from the greenschist to granulite facies:

1.   The Panjsher emerald deposits in Afghanistan (Figure 1a) are located in the Herat-Panjsher suture zone along the Panjsher Valley. The suture zone, which marks the collision of the Indo-Pakistan plate with the Kohistan arc sequence, contains a number of faults, such as the Herat-Panjsher strike-slip fault, which was mainly active during the Oligocene-Miocene [121]. The emerald deposits lie southeast of the Herat-Panjsher Fault in the Khendj, Saifitchir, and Dest-e-rewat Valleys. The deposits are hosted in the Proterozoic metamorphic basement formed by migmatite, gneiss, schist, marble, and amphibolite. The basement is overlain to

the northwest by a Paleozoic metasedimentary sequence crosscut by Triassic granodiorite [86]. During the Oligocene, the Proterozoic rocks of the Panjsher valley were affected by the intrusion of granitoids [86] 20 km north and south of the emerald mining district [122]. The emerald deposits are hosted by metamorphic schists that have been affected by intense fracturing, fluid circulation, and hydrothermal alteration, resulting in intense albitization and muscovite-tourmaline replacements [122]. Emerald is found in vugs and quartz veins associated with muscovite, tourmaline, albite, pyrite, rutile, dolomite, and Cl-apatite [122]. Ar-Ar dating on a muscovite from the emerald-bearing quartz veins at the Khendj mine gave an Oligocene age of 23 ± 1 Ma [123]. At the moment, the sources of Cr and Be remain unclear.

2. The Davdar emerald deposit is located in the western part of Xinjiang Province, China. The deposit is formed by emerald-bearing quartz-carbonate veins associated with a major northwest-southeast trending fault zone [18]. The deposit is hosted by lower Permian meta-sedimentary rocks, including sandstone, dolomitic limestone, siltstone, and shale, which have been metamorphosed at upper greenschist conditions [124] to produce metasedimentary host rocks, which include quartzite, marble, schist, and phyllite, prior to the emplacement of the emerald-bearing veins. Basaltic dykes of an unknown age, which are up to 10 m wide and crop out along strike lengths of up to 200 m, are the only intrusive igneous rocks known in the area. The dykes are emplaced along the northwest-southeast fault zone; no visible contacts are exposed between the dykes and the emerald-bearing veins. The emerald-bearing veins, which are up to about 20 cm wide, contain epidote, K-feldspar, tourmaline group minerals, carbonates, and iron oxides. Quartz and emerald crystals up to a few centimeters long are found in the veins. Alteration haloes up to a few cm wide occur around the veins. In the sandstone and dolomitic limestone, the alteration halo is barely visible, but it is conspicuous as a bleached white halo in the phyllite. The alteration halo is generally enriched in fine-grained silica with variable amounts of quartz, biotite, muscovite, feldspar, carbonate, and tourmaline. It is representative of a retrograde metamorphic assemblage typical of greenschist facies minerals (epidote, plagioclase, potassic feldspar, quartz, biotite, and chlorite). Emerald typically occurs in the quartz veins and not in the host rocks or the alteration haloes.

3. Hiddenite emerald was discovered in North Carolina northeast of the community of Hiddenite in 1875. Since then, a number of notable samples have been discovered, primarily from the Rist and North American Gem mines [125]. Over 3500 carats of emerald were extracted from the latter in the 1980s, including the 858 ct (uncut) "Empress Caroline" crystal [126]. At the Rist property the emeralds occur in quartz veins and open cavities (50% of the veins) that occupy NE-trending sub-vertical fractures in folded metamorphic rocks [126,127]. The hiddenite area is underlain by Precambrian migmatitic schists, gneisses, and interlayered calc-silicate rocks, metamorphosed in the upper amphibolite facies. The area is locally intruded by the Rocky face leucogranite. The quartz veins range in size from 2 to 100 cm wide, 30 cm to 7 m long, and 10 cm to 5 m high. Most of the veins are not interconnected and represent tensional gash fractures that sharply crosscut the prominent metamorphic fabric of the host rocks, suggesting that they formed during late or post metamorphic brittle-ductile deformation [126]. Wise and Anderson [127] identified four cavity assemblages: (1) An emerald-bearing assemblage composed of albite, beryl, calcite, dolomite, siderite, muscovite, cryptocrystalline quartz, rutile, and sulfides with clays; (2) a Cr-spodumene-bearing assemblage, which includes calcite, muscovite, and quartz. The green Cr-bearing spodumene, locally referred to as "hiddenite", occurs in only minor amounts; (3) a calcite assemblage dominated by calcite and quartz; and (4) an amethyst assemblage characterized by amethystine quartz, calcite, muscovite, and chabazite. Emerald and spodumene rarely occur together in the same vein or cavity. Within the emerald-bearing cavities, beryl crystals up to 20 cm in length are closely associated with dolomite, muscovite, and quartz. The crystals are typically color-zoned with a pale green to colorless core and an emerald-green rim. Speer [126] described the veins and reported that the emeralds occur as free-standing crystals attached to

cavity walls and as individual collapsed fragments. The collapsed crystals that have fallen from walls of the cavities show cementation phenomena, while the attached crystals exhibit dissolution, re-growth, and over-growth. Bleached wall-rock alteration halos up to 9 cm wide and rich in silica and chlorite are commonly peripheral to veins and crystal cavities. Wise and Anderson [127] pointed out that the emerald and Cr-spodumene mineralization in quartz veins and cavities is similar to what is seen in alpine-type fissures. In the absence of a pegmatitic or granitic body, the source of Be and Li remains in question; the source of Cr and V is also uncertain, given that M-UMR are unknown in the area. Speer [126] specified that the veins originated as hydrothermal filling of tensional sites during the waning ductile/brittle stages of metamorphism. Apparently, the geological setting and genesis of the Hiddenite emerald occurrences are unique.

6.2.4. Sub-Type IID: Tectonic Metamorphosed or Remobilized Type IA Deposits, Tectonic Hidden Granitic Intrusion-Related Emerald Deposits, and Some Unclassified Deposits (Egypt, Australia, perhaps also Brazil, Austria, Pakistan, Zambia; Table 2)

This sub-type includes deposits probably genetically linked to hidden granitic intrusions (Swat valley) and those where metamorphism has blurred the distinction between metamorphic and magmatic origin (Habachtal, Eastern desert in Egypt). This sub-type permits reclassification and debate on the genesis of several deposits, including the following located in M-UMR:

1.   Those for which the genesis is considered to be the consequence of regional metamorphism but with multi-stage emerald formation (Eastern desert of Egypt).
2.   The sub-type IIA (Santa Terezinha de Goiás, Habachtal, and Swat Valley) where metamorphic-metasomatic deep crustal fluids circulated along faults or shear zones and interacted with M-UMR with apparently no magmatic intrusion.
3.   The mineralization stages for the Poona deposit where emerald and ruby are associated.

It is also useful for considering deposits in meta-sedimentary rocks for which insufficient geological knowledge renders the genesis and classification uncertain (Panjsher Valley and Davdar deposits) and for the Hiddenite occurrences where the source of Be and origin of the mineralizing fluids are unknown.

The Egyptian emerald occurrences of Gebels Zabara, Wadi Umm Kabu, and Sikait occur in a N-W trending band circa 45 km long in the Nugrus thrust [128–130]. The deposits are located in a volcano-sedimentary sequence featuring an ophiolitic tectonic melange composed of metamorphosed M-UMR overlying biotite orthogneiss. Syntectonic intrusions of leucogranites and pegmatites occurred along the ductile shear-zone [43]. The study of [130] described three beryl-emerald generations that crystallized during magmatic, post-magmatic hydrothermal, and regional-metamorphic events. The genetic succession was based on chemical and microstructural studies. The original colorless Cr-poor beryl and phenakite of pegmatite origin has been partly replaced by the formation of Cr-rich beryl (emerald) through K-Mg metasomatism. At the Gebel Sikait, [130] described the occurrence of emerald in phlogopitites (i) at the contact between meta-pegmatite, meta-pelite, and meta-greisen veinlets of up to 10 cm in thickness and (ii) in folded quartz layers. At Gebel Zabara, the emerald is either in phlogopitites and talc-carbonate-chlorite-actinolite schists present in the serpentinite bodies, or in quartz veins. At Gebel Umm Kabo, emerald is within phlogopitites in contact with small lenses of quartz. The dating of phlogopite by K-Ar returned ages of 520 to 580 Ma [131] and by Rb/Sr returned ages of 591 ± 5.4 Ma, confirming the Panafrican orogen.

Based on microtextures, [128,129] suggested that emerald formation occurred during low-grade regional metamorphism. The emerald formation was controlled in detail by the local availability of Be present in the beryl-bearing meta-pegmatite and quartz veins, Cr in the meta-M-UMR, and the metamorphic fluids, all in the context of the late Pan-African tectonic-thermal event. This genetic model has been challenged [43], who pointed to the intrusions of syntectonic leucogranites with the presence of greisens and beryl-emerald-bearing pegmatites and quartz veins along the shear zones. Fluid inclusion studies have shown the presence of H_2O-$NaCl$-CO_2-CH_4 fluids with a salinity between 8 to 22 wt.% equivalent NaCl and a temperature of homogenization between 260 and 390 °C [43]. The oxygen isotope data for emeralds were consistent with both magmatic and metamorphic origins for the source of the mineralizing fluid [132]. Grundmann and Morteani [130] confirmed the existing $\delta^{18}O$ values with new isotopic data in the range of 9.9 to 10.7‰. They concluded that the complex interplay of magmatic and regional magmatic events during the genesis of the emeralds makes it impossible to relate their genesis to a particular event. The Pan-African regional metamorphic model is consistent with the remobilization of syntectonic Cr-poor beryl quartz veins and beryl-phenakite-bearing pegmatites.

The Egyptian emerald occurrence is a good example of the proposed sub-type IID emerald deposit: Magmatism, deformation, and remobilization by metamorphic-metasomatic fluids of the mixed Be and Cr reservoirs. The formation of emerald occurred during a regional tectonic event with the syntectonic intrusion of leucogranites and the injection of pegmatites, thrust and shear zone deformation accompanied by fluid circulation (with probable mixing of magmatic and metamorphic fluids), and reaction with rocks of different composition. The remobilization of Be, Cr, and V occurred at the same time through continuing regional tectonic activity. The oxygen isotopic composition of emerald with $\delta^{18}O$ values around 10‰ is similar to the oxygen magmatic signatures found for other worldwide type IA emerald deposits [132].

With the proposed sub-type IID, the genesis of the Santa Terezinha de Goiás, Habachtal, Swat Valley, and Poona deposits can be discussed and finally classified using the chemistry of the emerald and O-H stable isotopes.

At the Santa Terezinha de Goiás deposit, the ductile-fragile deformation coeval with the mineralization was strongly assisted by fluids migrating along shear planes under lithostatic fluid pressure at 500 °C [15]. The O-H isotopic composition of phlogopite and emerald is consistent with both magmatic (evolved crustal granites) and metamorphic fluids. This hypothesis was also proposed [133] based on fluid inclusion studies showing the mixing between carbonic and aqueous fluids. Nevertheless, considering the absence of granites and pegmatites in the underground mine, which reaches depths of up to 400 m [134], the low beryllium concentration in the Santa Terezinha volcano-sedimentary series (Be < 2 ppm), the lack of tourmaline in the metasomatic rocks, the control of the mineralization by shear zone structures, and the CO_2-H_2O-NaCl-($\pm N_2$) composition of the fluids, a metamorphic origin was preferred for the parental fluids of the emeralds [15,76]. The metamorphic hypothesis involves some input of Be-bearing metamorphic fluids released at the greenschist-amphibolite transition (T = 400–500 °C) or fluids generated at higher grades of metamorphism and channeled along transcrustal structures at the brittle-ductile transition. Such specific features are similar to those found for gold deposits in the vicinity of the emerald deposit in the Goiás metallogenetic province. The mineralizing fluids are channeled along lineaments and second order structures where CO_2 unmixing, wall-rock interaction, and concomitant ore precipitation are promoted during temperature and pressure fluctuations.

Chemical analyses of Santa Terezinha de Goiás emeralds led [65] and [23] to question the proposed metamorphic origin [76]. The emeralds have the highest Cs contents ever reported for emerald, with values between 907 and 980 ppm, and with Li contents between 142 and 155 ppm. The high content of Cs supports another hypothesis that magmatic fluids mixed with metamorphic fluids [23]. Following this genetic scheme, despite the absence of pegmatites and granites up to a 400 m depth in the mine and along the metamorphic strike, the influence of the magmatic fluids is evidenced by the chemistry of emerald. In such a ductile shear zone environment, the intrusion of felsic granitoids into the volcano-sedimentary sequences along thrusts is common [135] and could have happened at the Santa Terezinha deposit. This new chemical result confirms the hypothesis proposed by D'el-Rey Silva and Barros Neto [92] that the probable source of Be for emerald was the intrusion of syntectonic two-mica granite at São José do Alegre, located 5 km to the southwest of the emerald deposit. Whole rock Sm/Nd data from the granite and the volcano-sedimentary sequence at Santa Terezinha yielded ages of 510 ± 110 Ma (n = 6) and 556 ± 77 Ma, respectively [92,136]. $^{40}Ar/^{39}Ar$ ages on phlogopite grains from the emerald-bearing phlogopitites yielded ages of 550 ± 4 and 522 ± 1 Ma, respectively [137]. These ages show considerable overlap in a large window, which characterizes the Brasiliano orogenesis [138].

At the Habachtal deposit, the emerald-bearing phlogopitites called the "blackwall zone" are located at the tectonic contact between the orthogneisses and amphibolites of the Habach group. The metasomatic "blackwall zone" is formed in sheared melange zones surrounding tectonic lenses of the serpentinite-talc series at the contact with the tourmaline-garnet-mica-bearing metapelitic unit of the Habach Group.

Detailed textural studies on emerald-tourmaline and plagioclase porphyroblasts [77,93,95] recorded three metamorphic episodes and crystallization for these minerals. Deformation enhanced fluid circulation and metasomatic reaction and produced the emerald-tourmaline-bearing phlogopitites. Fluid inclusions in the emerald show similar characteristics to those in syn-metamorphic Alpine fissures in the Habach Formation. Grundmann and Morteani [77] proposed the genesis of emerald through syntectonic growth during regional metamorphism.

Trumbull et al. [97] used B isotopes of coexisting tourmaline in the metapelites and phlogopitites. The $\delta^{11}B$ isotope values suggest that two separate fluids were channelled and partially mixed in the shear zone during the formation of the metasomatic rocks. A regional metamorphic fluid carried isotopically light B as observed in the metapelite ($-14 < \delta^{11}B < -10‰$) and a fluid derived from the serpentinite association carried isotopically heavier B ($-9 < \delta^{11}B < -5‰$) typical for Middle Oceanic Ridge Basalt or an altered oceanic crust.

Grundmann and Morteani [77] pointed out that the source of Be was either the Be-rich garnet-mica schist series or the biotite-plagioclase gneisses (Be up to 36 ppm). Zwaan [52] was critical of this interpretation and warned that, in cases where pegmatitic sources of Be are not apparent, one must proceed with caution since fluids can travel far from granites and pegmatites. He pointed out that pegmatites do occur in the Zentral gneiss and the Habachtal emeralds contain up to 370 ppm of Cs [139], which suggest a pegmatitic source. However, the Cs data produced [23] are very different ($79 < Cs < 157$ ppm). The possibility of metamorphism of a pre-existing Be-bearing felsic rock occurrence cannot be excluded and must be considered when constructing a model in a medium- to high-grade metamorphic regime [16]. The question is similar to what has been reported for the origin of tungsten for the Felbertal scheelite deposit [139], which is located in the same metamorphic series as the Habachtal emeralds and is now considered to represent metamorphic remobilization of a Be-W-enriched Variscan granite [140]. This hypothesis is likely correct for emerald because (1) Hercynian aquamarine-bearing pegmatites were found in the Habach series [141] and (2) scheelite disseminations with chalcopyrite and molybdenite in the banded gneiss series of the Habachtal emerald deposit are drawn in the lithologic cross-section presented [77]. Following this hypothesis, the source of Be is magmatic.

The Swat-Mingora–Gujar Kili–Barang emerald deposits are thrust controlled and there is no magmatic or pegmatite intrusions visible in the field [86,101]. Emerald is either disseminated in carbonate-talc-fuchsite-tourmaline-quartz schists or in quartz veins and a network of fractures in magnesite rocks. The mean oxygen isotopic composition of emerald is remarkably uniform at $\delta^{18}O = 15.6 \pm 0.4‰$. The mean hydrogen isotopic composition of the channel waters is $\delta D = -42.2 \pm 6.6‰$ and that of the fluid calculated from hydrous minerals, such as tourmaline and fuchsite, is $\delta D = -47 \pm 7.1‰$. These O-H isotope data are consistent with both metamorphic and magmatic origins [101]. However, a magmatic origin is favored because the measured δD values of fuchsite and tourmaline are comparable to those found for muscovite and tourmaline from granites, such as the Makaland granitoid, exposed 45 km to the southwest of Mingora. The mineralization was probably caused by modified ^{18}O-enriched hydrothermal solutions derived from an S-type granitic magma [101].

The magmatic model proposed for Pakistani emerald deposits can be constrained by the $^{40}Ar/^{39}Ar$ ages obtained on the different rocks: 83.5 ± 2 Ma for the Shangla blue-schist melange, 22.8 ± 2.2 Ma for the tourmaline-beryl-fluorite-bearing Makaland Granite [142], and 23.7 ± 0.1 Ma for a fuschite mica from a quartz vein in the Swat emerald deposit [100]. However, the chemical data for Swat emerald presented [23] show low Cs contents (61 to 74 ppm), which are unusual for granite-related emeralds.

Finally, the genesis of the Swat emerald deposits can result from both metamorphic and magmatic contributions, with up to now a magmatic source not identified in the emerald mining districts.

At the Poona deposit, three styles of emerald mineralization have been identified in M-UMR from the Precambrian series of the northern Murchinson Domain [143]: (a) Emerald in phlogopitites formed at the contact of beryl-granite- muscovite-bearing pegmatites and M-UMR; (b) emerald with ruby-sapphire, topaz, and alexandrite in banded fluorite-margarite-beryl-bearing banded greisens in phlogopitites; and (c) quartz-margarite-topaz-bearing veins in phlogopitites.

A multi-stage mineralizing episode was proposed [143] of: First, the intrusion of granites (probably between 2724–2690 Ma) and circulation of fluids in the M-UMR produced greisens and quartz veins with topaz, beryl, quartz, and muscovite. The first episode was affected by regional metamorphism of greenschist to lower amphibolite facies. Metasomatic reactions occurred at the borders of the greisen zones with the formation of ruby, alexandrite, and emerald. The third episode corresponded to the retrograde phase of metamorphism (probably between 2710–2660 Ma), where corundum and alexandrite were partially or totally replaced by margarite, muscovite, and/or emerald.

Fluid inclusions in emerald combined with the O-H isotope compositions of both lattice and channel fluids of emerald confirmed multiple origins yielding both igneous and metamorphic signatures [16]. This emerald-ruby association is unique worldwide and merits more petrologic and geochemical studies before proposing a coherent genetic scheme.

The genesis of the Panjsher Valley [98] and Davdar [18] emerald deposits is not well understood due to difficulties with access and poor exposure. The similar geographic and geologic environments indicate that these two deposits may share a similar genetic model. Both deposits are hosted in layered meta-sedimentary rocks, with metamorphic facies ranging up to lower amphibolite. Emerald occurs in both veins and host rocks. The veins are predominantly composed of quartz and carbonate, with minor amounts of albitic plagioclase, phlogopite, tourmaline, scheelite, and pyrite. The host rocks vary from shale to carbonate and are thought to be Paleozoic. Proximal to the veins, hydrothermal alteration is dominated by quartz and calcite with lesser amounts of albite, phlogopite, tourmaline, and pyrite. In both localities, there are mafic to felsic intrusions as stocks, dykes, or sills. However, no clear relationship between the intrusive rocks and emerald mineralization has been established. The high-salinity fluids [18,56,144] and meta-sedimentary host rocks combined with the lack of observed igneous association could also be compatible with a tectonic metamorphic-related (type IIB) formational model, but more field work needs to be carried out on these deposits to map the local geology to prove or disprove an igneous link.

Emeralds from the Musakashi deposit in Zambia are of high quality with a bluish color very different from that observed for emerald originating from the Kafubu mining district [63]. Discovered in 2002, the deposit is located ~150 km west of the Kafubu mines. The geology of the deposits is unknown, however, small fragments of emerald were discovered in eluvium adjacent to quartz veins [50]. The internal features, chemical composition, solids, and three-phase fluid inclusions are quite different from those of Kafubu emeralds. Saeseasaw et al. [63] examined three-phase fluid inclusions (halite + vapor + liquid) in emerald from Musakashi, Panjsher Valley, Davdar, and Colombia. They conclude that Musakasi fluid inclusions look like those from Colombia. Nevertheless, the photographs show that they are polyphase and contain a cube of halite and a rounded salt, which is probably sylvite (KCl). Such multiphase inclusions with both halite and sylvite are found in both Panjsher and Davdar emeralds [62]. In addition, the chemistry of these emeralds overlaps with those from Colombian, Panjsher, and Davdar [63]. The geological setting is not precisely known, but the emeralds are very different from those for type IA deposits as shown by their very low Cs contents (3 < Cs < 10 ppm), which are similar to those found for Colombian and Davdar emeralds [63].

The genesis of the Hiddenite deposits remains obscure in terms of the sources of Be, Cr, and V, and the origin of the mineralizing fluids. Fluid inclusions are characterized as two populations of aqueous-carbonic fluids with two populations, those having high- and those having low-CO_2 contents, which underwent immiscibility between 230 and 290 °C, respectively [145]. The deposits are interpreted as late and low temperature metamorphic hydrothermal alpine-type quartz veins cutting meta-sedimentary and migmatitic biotite gneiss. The low temperature hydrothermal mineralization is confirmed by the association of quartz, carbonates, muscovite, and chabazite. The origin of the fluids must be clarified by the O-H isotopes' compositions of the different minerals associated with emerald, but a metamorphic origin has advanced [126,127].

7. Fluid Inclusions in Emerald

Emerald is one of the best hosts for fluid inclusions. It is commonly idiomorphic, well preserved, zoned, and displays growth zones optically, chemically, and especially via cathodoluminescence [20,85,117]. These growth zones facilitate the identification of primary vs. secondary fluid inclusions. Additionally, primary fluid inclusions often form during emerald precipitation and are elongated parallel to the host's c axis (Figure 32). The determination of fluid inclusion chemistry is generally limited to microthermometry [19,146,147], with more refined analyses performed via bulk leachate analyses or LA-ICP-MS or secondary ion mass spectrometry (SIMS) on quartz-hosted fluid inclusions petrographically determined as synchronous to emerald hosted inclusions [148,149]. In addition to fluid chemistry, fluid inclusions studies have proven most useful in determining the pressures and temperatures of emerald formation [47] and for determining if boiling is responsible for emerald colouration [19].

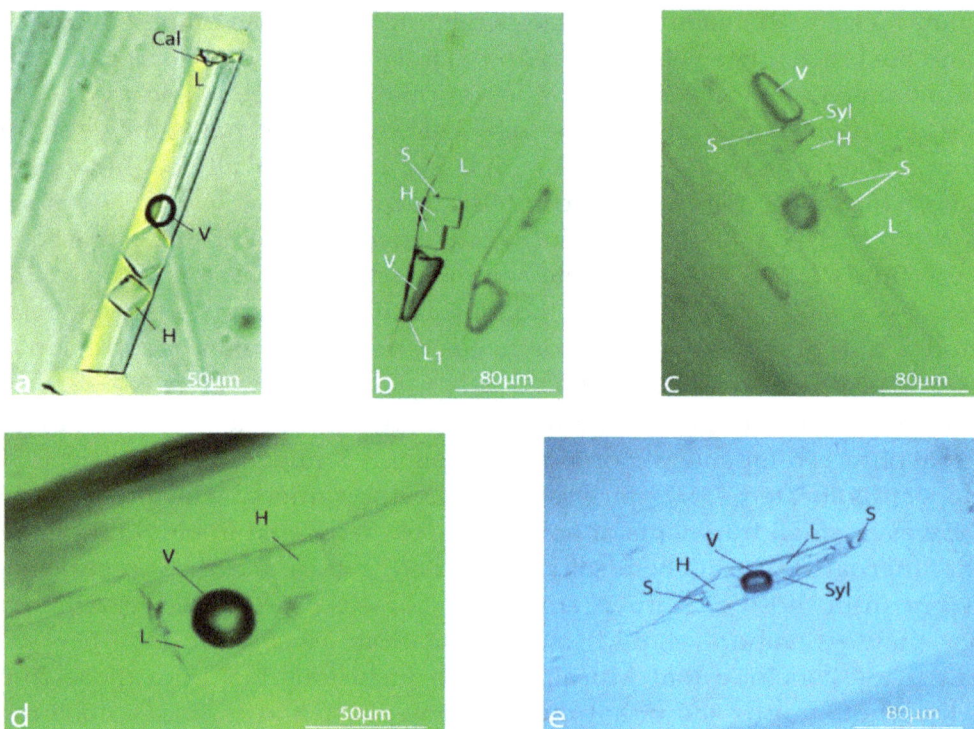

Figure 32. Primary fluid inclusions in emeralds: (**a**) Colombian fluid inclusion presenting a jagged and shredded outline. The cavity contains 75 vol.% of salted water solution (L), 10 vol.% of gas corresponding to the vapour bubble (V), 15 vol.% of halite (NaCl) daughter mineral, and a crystal of carbonate (Ca); (**b**) fluid inclusion in a Colombian emerald showing three cubes of halite (H), the liquid phase (L), the contracted vapour phase (V), a minute black phase (S), and a thin rim of liquid carbon dioxide (L_1) rim visible at the bottom part of the vapour phase; (**c**) multiphase fluid inclusions from Panjsher emerald (Afghanistan). They contain vapour and liquid phases (V + L), a cube of halite (H),

usually a primary sometimes rounded salt of sylvite (Syl), and aggregates of several anisotropic grains (S). The volume and the concentration of NaCl and KCl are different from those observed in Colombia. The overall salinity is estimated to 30 to 33 wt.% eq. NaCl (Vapnik and Moroz, 2001; (**d**) three-phase fluid inclusion in emerald from Nigeria emerald showing halite (H), liquid (L), and vapour (V) phases. Generally, the primary halite-bearing fluid inclusions are associated with coeval monophase or biphase (V + L) fluid inclusions; (**e**) multiphase fluid inclusion from the Davdar emeralds (China). The cavities contain liquid (L) and vapour (V) phases with daughter minerals, like halite (H) and sometimes sylvite (syl), and aggregates or multiple solid inclusions (S). The morphology of the cavities and the infilling looks like those found in emeralds from Afghanistan. Photographs: Gaston Giuliani.

Emerald-hosted fluid inclusions are, in general, aqueous dominant, with a wide range of salinities from dilute to salt saturated (Table 3). Numerous emerald deposits also have gaseous species contained within the fluid inclusions; the gaseous phases are generally dominated by CO_2, but other species, such as CH_4, N_2, and H_2S, have been identified via Raman spectroscopy in a number of emerald studies [47,150,151]. Raman spectroscopy can also be used to identify accidental and daughter inclusions within fluid inclusions [152]. Raman analyses of emerald hosted inclusions often prove challenging, as beryl/emerald is generally fluorescent and thus the inclusion spectrum is lost in the fluorescence from the host. However, different laser wavelengths and confocal Raman spectrometers can be used to limit the effects of host fluorescence; these applications to gases and solid inclusions contained within fluid inclusions were reviewed [153–155].

Bulk leachate analyses of fluid inclusions by thermal decrepitation or crushing are limited by contamination via the emerald host. However, leachate analyses of fluid inclusions hosted in quartz precipitated synchronously with emerald have proven successful for general chemistry and especially for halogen sourcing [105]. A further limitation of the bulk methods is the presence of multiple fluid inclusion generations and careful detailed petrographic studies should be undertaken to determine if specific quartz and emerald-hosted fluid inclusions are amenable to bulk techniques.

Hydrogen isotopes [156] are generally determined from trapped channel fluids, which are analogous to primary fluid inclusions. The advantage of channel fluid extraction is that channel fluids are normally three (or more) orders of magnitude more abundant than fluids trapped in fluid inclusions. The oxygen isotope signature of fluids responsible for emerald precipitation is generally determined via the measurement of structural oxygen within emerald/beryl or from other synchronously precipitated silicates. The hydrogen isotope signature of the fluids may also be inferred from the analyses of synchronously precipitated hydrous silicates, such as mica or tourmaline [101]. Additionally, the extraction of the channel fluids may also yield a quantitative determination of the weight percent of H_2O present in the emerald and this can be used to complement electron microprobe and LA-ICP-MS analyses to determine emerald chemistry.

Fluid fingerprinting to determine emerald provenance has proven incredibly useful. The pioneering studies [15,76] are consistently used to determine provenance and fluid sources, as done for the Biintal occurrence in Switzerland ([157]; Figure 33). Halogen bulk leachate analyses [105] can test for the presence of fluid interaction with evaporitic source rocks as well as providing more detailed information on the various dissolved salts present in the fluid inclusions. Although not yet routinely used, LA-ICP-MS, and to a lesser extent SIMS, can provide very detailed chemical analyses of formation fluids of emerald deposits, as they provide precise analyses of most of the elements of the periodic table and potentially their isotopes.

Table 3. Fluid inclusion data and oxygen isotope composition of several emerald deposits worldwide following the enhanced classification proposed for emerald deposits.

Type of Environment and Deposit	Tectonic Magmatic-Related Granitic Rocks in M-UM and SR (Type I)	Tectonic Metamorphic-Related	
		(Meta) Sedimentary Rocks (Type IIB)	Metamorphic Rocks (Types IIA- IIC-IID)
Temperature	300-680°C	300-330°C*	350-400°C [*1] 260-550°C [*2]
Pressure	0.5 to 7 kbar	3 to 4 kbar	1.6 kbar [*1D] 4-4 to 5 kbar [*2]
Salinity	2 to 45 wt.% eq. NaCl	40 wt.% eq. NaCl	30 to 33 wt.% eq. NaCl [*1P] 35 to 41 wt.% eq. NaCl [*1D] 2 to 38 wt.% eq.NaCl [*2]
Composition	H_2O-NaCl-($\pm CO_2$)-($\pm N_2$)-($\pm CH_4$)-(K,Be,F,B,Li,P,Cs)	H_2O-NaCl-($\pm CO_2$)-($\pm N_2$)-($\pm$hydrocarbon liquid)-F15(K,Mg,Fe,Li,SO4,Pb,Zn)	H_2O-NaCl-($\pm CO_2$) [*1D] H_2O-NaCl-KCl-$FeCl_2$-($\pm CO_2$) [*1P] H_2O-NaCl-($\pm CO_2$) [*2] CO_2 or CO_2-H_2O-NaCl-($\pm N_2$ ($\pm CH_4$) [*2] or H_2O-CH_4-CO_2-NaCl [*2]
Oxygen isotopes	$6.0 < \delta^{18}O < 15‰$	$16.2 < \delta^{18}O < 24.5‰$	Panjsher: $13.25 < \delta^{18}O < 13.9‰$ Davdar: $14.4 < \delta^{18}O < 15.8‰$ Sta Terezinha: $12.0 < \delta^{18}O < 12.4‰$ Gravelotte: $9.5 < \delta^{18}O < 9.7‰$ Gebel Sikait: $9.8 < \delta^{18}O < 10.7‰$ Habachtal: $6.5 < \delta^{18}O < 7.3‰$
Origin of the fluid	Metasomatic-Hydrothermal	Basinal brines that have dissolved evaporites	Metamorphic-Metasomatic

Salinity in wt.% eq. NaCl = weigth per cent equivalent NaCl; kbar = kilobars; $\delta^{18}O$ = ratio $^{18}O/^{16}O$ in per mil (‰).
Type IIB deposits: * = Colombia. Some Types IIA, IIC, IID: [*1] = deposits of Panjsher ([P]) and Davdar ([D]); [*2] = deposits of Santa Terezinha de Goiás (Brazil), Gravelotte (South Africa), Gebel Sikait (Egypt), Habachtal (Austria)

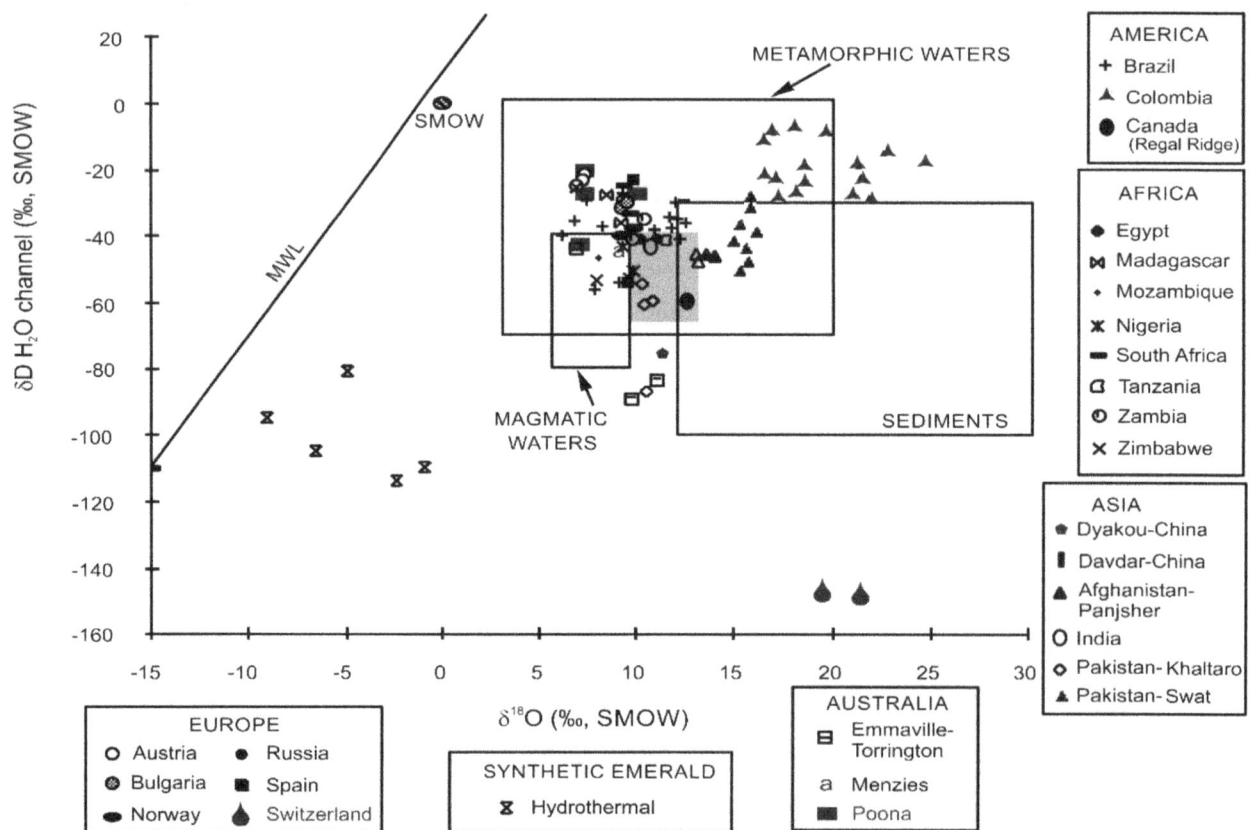

Figure 33. Channel δD H_2O versus $\delta^{18}O$ for emerald worldwide [15,76,157,158]. The isotopic compositional fields are from [159], including the extended (Cornubian) magmatic water box (grey). MWL = Meteoric Water Line, SMOW = standard mean ocean water.

The use of fluid inclusion petrography combined with microthermometry and halogen and cation bulk leachate analyses is useful for the discussion of the source of salts present in fluid inclusions [160]. The example of multi-phase halite-bearing fluid inclusions in emeralds (Figure 32) from Colombia (Eastern and Western emerald zones) and Afghanistan (Panjsher Valley) is very illustrative.

In Colombia, fluids trapped by emerald are commonly three-phase fluid inclusions (Figure 32a) characterized by the presence of a daughter mineral, i.e., halite ($NaCl$). At room temperature, the cavities contain 75 vol.% of salty water, i.e., aqueous brine (liquid H_2O), 10 vol.% of gas corresponding to the vapor bubble (V), and 15 vol.% of halite daughter mineral (H). However, some Colombian emeralds have multiphase fluid inclusions presenting a liquid carbonic phase (CO_2) forming up to 3 vol.% of the total cavity volume (Figures 25b and 32b), minute crystals of calcite (Figure 32a), very rare liquid and gaseous hydrocarbons [38], and sometimes two or three cubes of halite (Figure 32a,b), and sylvite (KCl).

The high Cl/Br ratio of the fluids (between 6300 and 18,900) indicates that the strong salinity of the brines is derived from the dissolution of halite of an evaporitic origin (Figure 34a; [105]). Cation exchanges, especially calcium, with the black shale host rocks are strong when compared to most basinal fluids (Figure 35; [161]) and are due to the relatively high temperature of the parent brines of emerald (T~300 °C). Indeed, these fluids are enriched in Ca (16,000 to 32,000 ppm), base metals (Fe ~5000 to 11,000 ppm; Pb ~125–230 ppm; Zn ~170–360 ppm), lithium (Li~400–4300 ppm), and sulfates (SO_4 ~400–500 ppm). In comparison, they have a composition and Fe/Cl and Cl/Br ratio similar to the fluids of the geothermal system of the Salton Sea in California (Figure 34b; [162]). The K/Na ratios confirm the Na-rich character of the fluids and the strong disequilibrium between K-feldspar and albite, as shown by the huge albitisation of the black shale (Figure 35).

In Afghanistan, primary multiphase halite-sylvite-bearing fluid inclusions (Figure 32c) are common for the Panjsher emeralds [15,86]. The fluids associated with emerald have total dissolved salts (TDS) between 300 and 370 g/L and the trapping temperature of the fluid is about 400 °C [144].

Crush-leach analyses of fluid inclusions indicate that the fluids are Cl-Na-rich and contain sulfates (140 < SO_4 < 4300 ppm) and lithium (170 < Li < 260 ppm), but very low to zero fluorine contents [160]. The K/Na ratio of the fluid inclusion confirms the disequilibrium, at ~400 °C, between K-feldspar and albite that drives the Na-metasomatism of the metamorphic schists and the deposition of albite in the veins (Figure 35). Crushing demonstrates that fluids are dominated by $NaCl$ with Cl/Br ratios much greater than that of seawater (Figure 34a), indicating that the salinity was derived by the dissolution of halite. Thus, the high Cl/Br ratios are consistent with halite dissolution. The I/Cl versus Br/Cl ratios diagram (Figure 34b) also shows that the fluid inclusions have low I contents, which are also typical of brines derived from evaporite dissolution. They are comparable to the Hansonburg and contemporary fluids from the Salton Sea geothermal brines, both of which have dissolved evaporites [163,164].

Although it is not possible to unambiguously classify emerald deposits based solely on their fluid composition, there are some general observations that can prove useful: (1) The presence of salt cubes can limit possible modes of formation, as it is very unlikely that we can generate a salt saturated fluid in a purely metamorphic environment, hence the presence of salt cubes generally implies the input of igneous fluids or an interaction with evaporites and individual $Ca/Na/K$ values may be specific to individual deposits; (2) compressible gases are seen in all types of emerald deposit, but gas ratios may also be used to identify individual deposits; and (3) specific daughter and accidental minerals contained within fluid inclusions cannot be used to identify specific deposit models within our classification system, but may again be used to fingerprint emeralds from specific deposits.

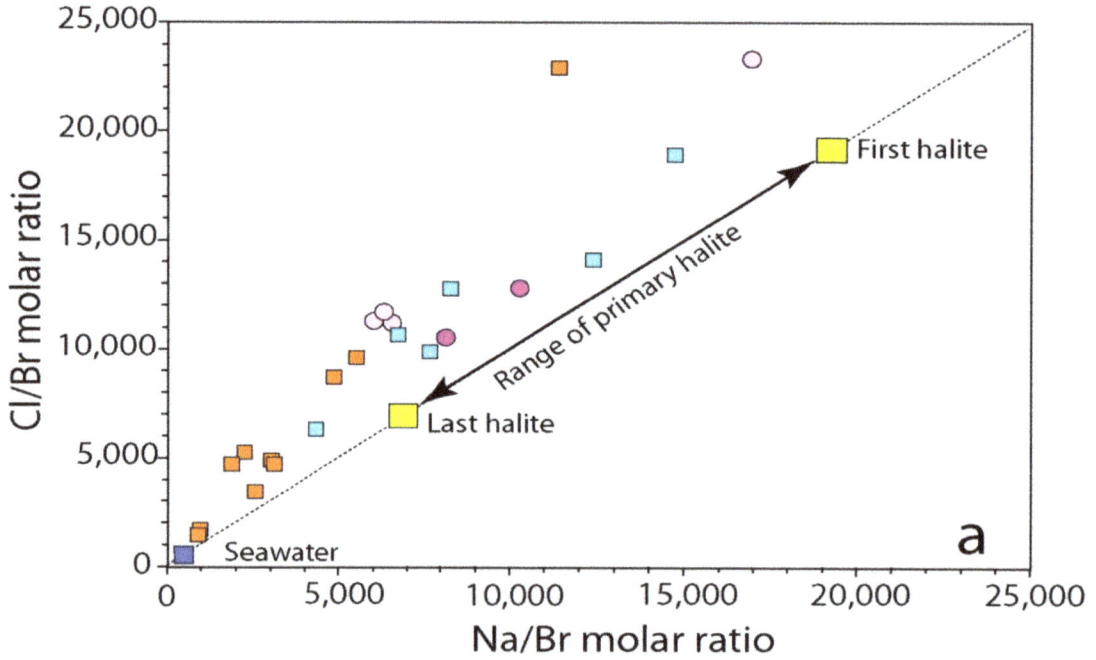

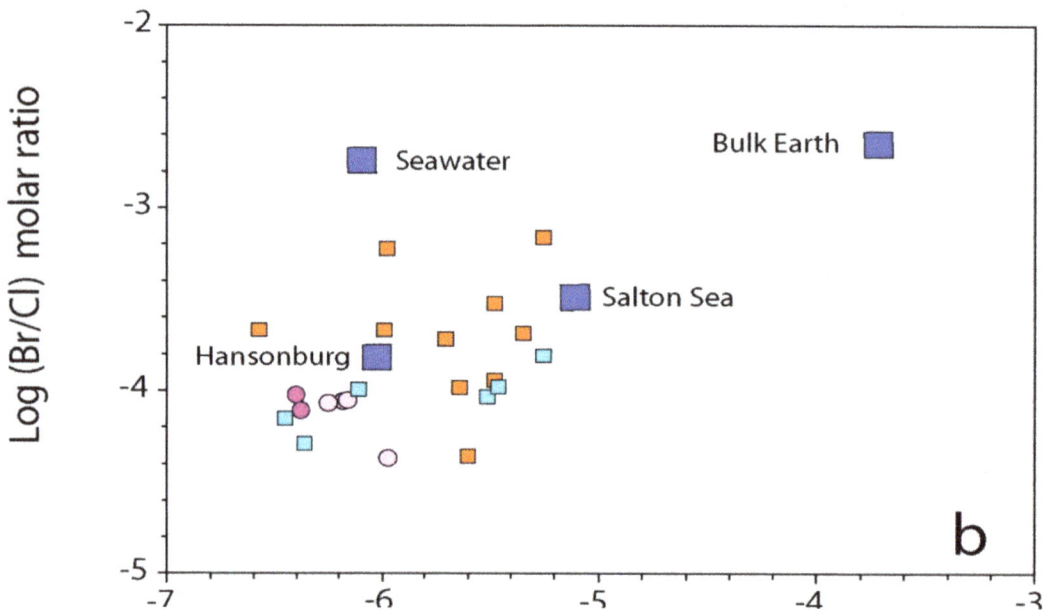

Figure 34. Origin of salinity in the emerald and quartz brines from Colombia and Afghanistan: (**a**) Analyses of the fluid inclusions from both emerald and quartz show a wide range of Na/Br and Cl/Br molar ratios that are much greater than those of primary halite and indicate a substantial loss of Br, typical of recrystallised halite for both emerald deposits; (**b**) log(I/Cl) versus log(Br/Cl) molar ratios of Afghan and Colombian fluid inclusions, which are depleted in both Br and I, indicative of evaporites contribution to the fluids in emerald and quartz. They are compared with the composition of fluids where evaporites are known to be involved, such as for the Salton Sea geothermal brines [163] and Hansonburg [164].

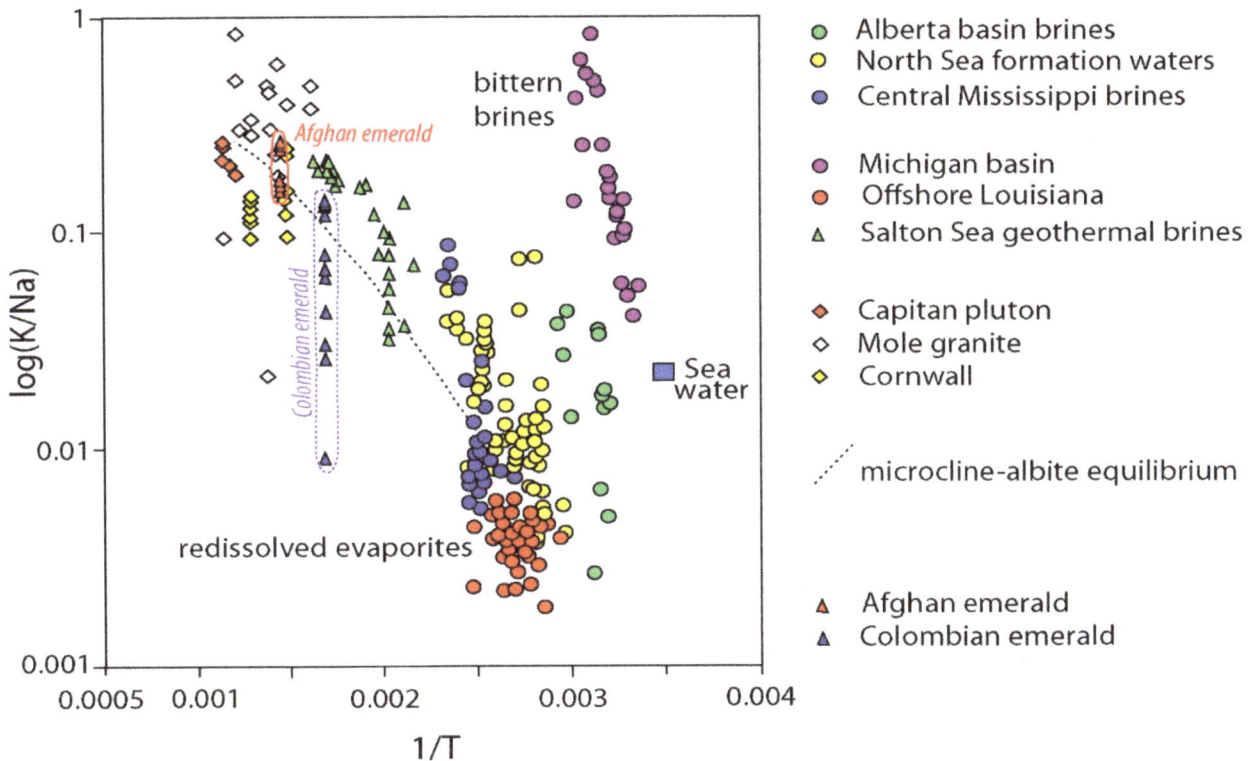

Figure 35. Diagram log (K/Na) molar ratio versus 1/T (°K) showing the evolution of the fluids associated with Colombian and Afghan emeralds relative to crustal fluids, including bittern brines, brines derived by dissolution of evaporites, and magmatic fluids. Sedimentary formation brines deviate significantly from the K-feldspar-albite equilibrium as well as for Afghan and Colombian brines, which are associated with a huge albitisation of their host-rock with the complete consumption of K-feldspar from, respectively, the schist and black shale.

8. Discussion of the Model of Formation of the Emerald Deposits Associated with M-UMR

The tectonic magmatic-related emerald deposits hosted in M-UMR (sub-type IA) occurred during orogenies from the end of the Archean (2.94 Ga) with the Gravelotte deposit in South Africa [165] to the Himalayan (9 Ma) with the Khaltaro deposit in Pakistan [83]. These deposits are related to plate tectonics over ~3.2 billion years with stable convection cells initializing continental drifts and subduction zones, and are related either to continental collisions or continent-to-continent rifting.

The classical model of sub-type IA relates granites with their dyke swarms of aplo-pegmatites and quartz veins in metamorphosed M-UMR, fluid-rock interaction, and infiltrational metasomatism [15,58,166,167]. An alternative genetic model for the formation of these emeralds is a regional tectonic and metamorphic model [77,130,168–170]. This genetic controversy involving sub-type IA deposits shows that (i) they are possibly genetically different; (ii) that alternative models are always dependent on a school of thought, which we absolutely want to apply for all deposits; (iii) these deposits share many common denominators in terms of geological setting, age of formation, nomenclature of rocks, source of the elements, and source of the fluids; and finally, (iv) each deposit belongs to the same family, but with a wide range of genres and uncountable typological varieties that allow us to follow the evolution of a mountain range marked by magmatic events accompanied by metamorphic remobilizations, which sometimes erase the primary geological features of the emerald deposit.

The following discussion examines the possible genetic links existing between the different sub-type IA and sub-types IIA and IID hosted in M-UMR within the geological and dynamic evolution of a continental crust.

8.1. Sub-Type IA Emerald

This is the fruit of the fluid-rock interaction producing metasomatism in both pegmatite and M-UMR, whatever the origin of the fluid. The classical sub-type IA results from the fluid circulation at the contact between these two geochemically contrasting rocks. The most representative deposits described in the literature are those from Carnaíba [76,78], Franqueira [171,172], Khaltaro [83], and Kafubu [70]:

1. The granite emplacement is related to a tectonic event, but the pegmatites are not deformed and metamorphosed. They are clearly intruding the M-UMR of the volcano-sedimentary series, and sometimes roof pendants on the granite as observed in Bode mine at Carnaíba (Figure 36) or at Franqueira;

2. At Carnaíba, the granite and pegmatite intrusions and fluid circulation are coeval [67], and the pegmatite is transformed into plagioclasite with disseminations of phlogopite (endo-plogopitite), and the M-UMR into phlogopitite (exo-phlogopitites). These exo-phlogopitites display clear zonation with a very sharp metasomatic front, i.e., metasomatic columns formed by infiltrational processes [173]. The metasomatic fronts where an additional phase appears in the mineral association correspond to the change of one of the determinant components from mobile to inert: Ca is displaced from the serpentinites to the center of the column for the formation of actinolite-tremolite, but also apatite at the border of the endo-phlogopitite. The occurrence of beryl is restricted to the most aluminous parts of the metasomatic zonation (plagioclasite, endo-phlogopitite, and exo-phlogopitites proximal to the plagioclasite). The inner part of the phlogopitite zonation plays the role of a filter for the Be-bearing fluids and constitutes a very efficient "metasomatic trap" where the mobile behavior of Cr favored the formation of emerald. At Khaltaro in Pakistan, non-deformed pegmatites and quartz veins crosscut amphibolites, which were metasomatized on 20 cm-wide selvages that are symmetrically zoned around the veins [83], as found at Carnaíba (see Figure 14). Mass-balance calculations on the metasomatic column (Figure 37) have shown that (a) in the inner and intermediate metasomatic zones, K, F, H_2O, B, Li, Rb, Cs, Be, Ta, Nb, As, Y, and Sr are gained and Si, Mg, Ca, Fe, Cr, V, and Sc are lost; and (b) in the outer zone, F, Li, Rb, Cs, and As are gained. The oxygen isotope composition of the hydrothermal minerals indicated the circulation of a single fluid of magmatic origin. At Kafubu, the regional metamorphic event pre-dates the emerald formation. The F-B-Li-rich phlogopitites are located at the contact between tourmaline veins and pegmatites with Mg-metabasites. The pegmatites of the Lithium-Cesium-Tantalum family are linked to hidden fertile B-F-Nb-Ta-Li-Cs-rich granite.

3. These deposits sometimes exhibit multi-stage Be-mineralization, as observed at the Carnaíba deposit: A second minor stage of metasomatism affected in some areas the emerald-bearing phlogopitites [15,90]. This stage is related to the intrusion of dyke swarms of quartz-muscovite veinlets with greisenisation of the granites, chloritisation, and muscovitisation of the phlogopitites, general silicification, and muscovitisation of the plagioclasites. This stage involves yellowish to whitish beryl, sometimes with molybdenite, scheelite, and schorlite.

4. These emerald mineralizations are interpreted to be due to the efficiency of the metasomatic trap rather than significant pre-enrichment in Be (5 to 11 ppm of Be in the Carnaíba granite). The occurrence of strong chemical gradients in the zone of preferential circulation of the solutions constitutes highly favorable conditions for the beryl crystallization.

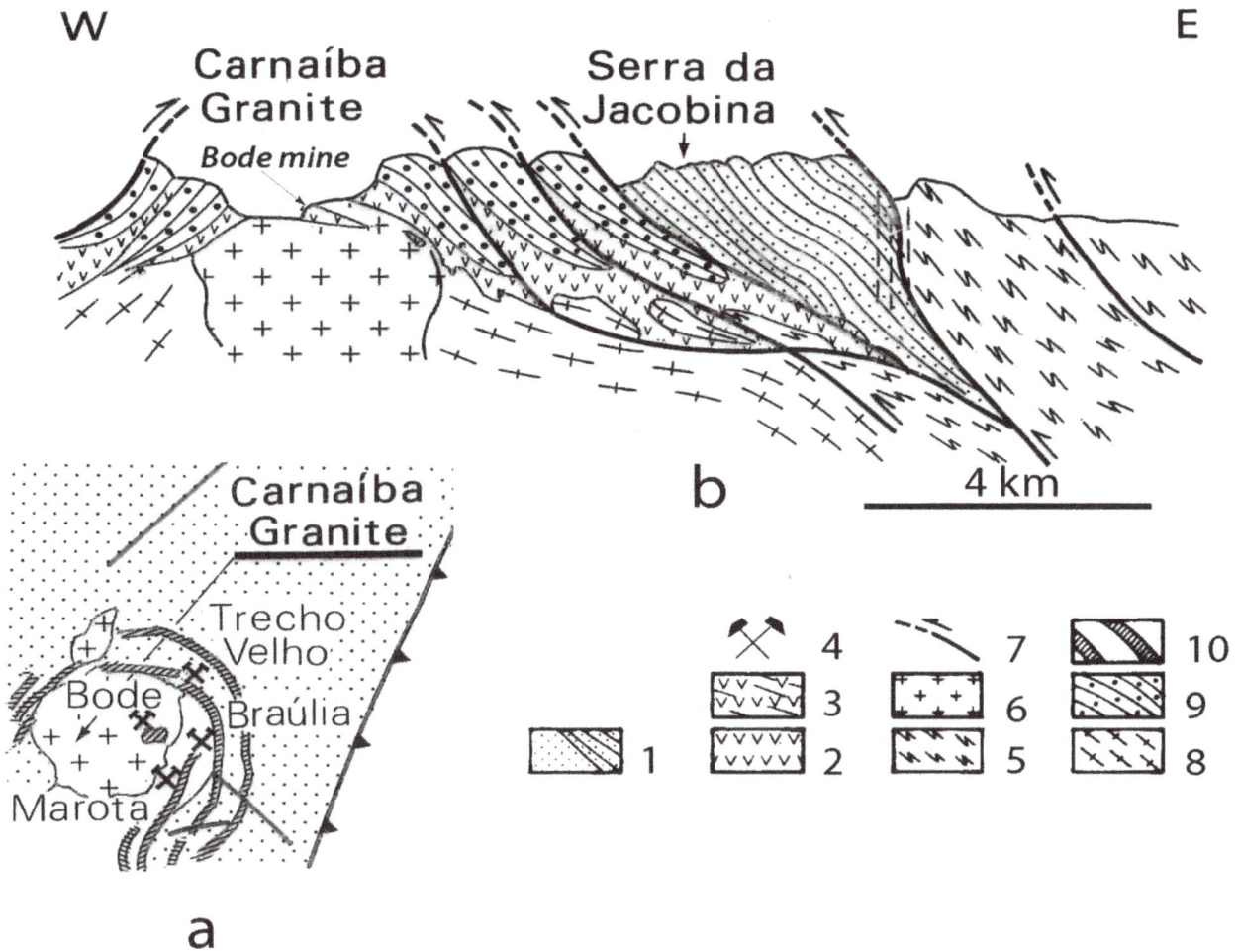

Figure 36. Schematic geological section of the emerald deposits of Carnaíba, Bahia state, Brazil: (a) The granite of Carnaíba and its emerald deposits cited in the present work. The Bode deposit is located in a roof-pendant of serpentinite present at the roof of the granite; (b) The Serra da Jacobina volcano-sedimentary sequence formed by intercalations of quartzite and serpentinite is crosscut by the Carnaíba granite. The pendants of serpentinite are present at the contact and on the roof of the granitic intrusion.

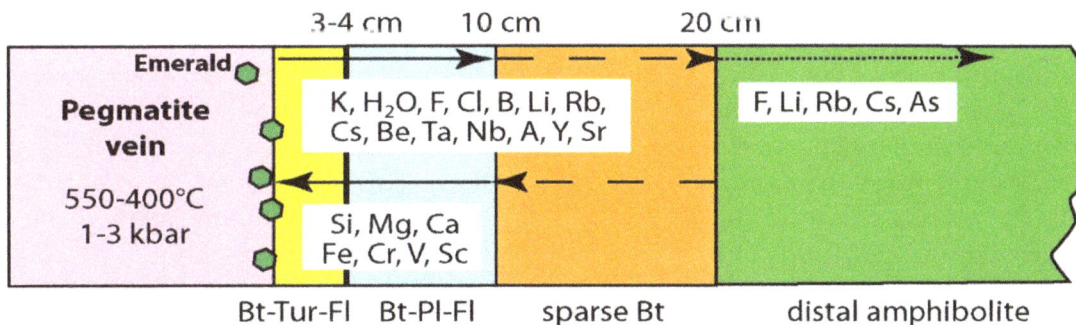

Figure 37. The Khaltaro emerald deposit, Nanga Parbat—Haramosh massif, Pakistan [83]. A symmetrically zoned metasomatic column is formed at the contact of the albitized pegmatites or hydrothermal quartz veins with amphibolite. The mass balance calculation in the metasomatized amphibolite indicates the gains and losses of components (see the transferts in the outer and inner zones). Emerald froms within the vein, near the contact with the altered amphibolite.

8.2. Multi-Stage Formation and Ages of Formation and Remobilization of Type IA Deposits

1. The Precambrian deposits located in the volcano-sedimentary series or greenstone belts are often folded and sheared, but metasomatic processes during emerald formation are generally coeval with the deformation, as in the deposits of Piteiras, Fazenda Bonfim, and Socotó (Brazil); Sumbawanga and Manyara (Tanzania); Kafubu (Zambia); and recently in the Gubaranda area from Eastern India [174].

2. The deposit at Sandawana in Zimbabwe [52,175] presents multi-stage formation; [169] advanced that the classical model of sub-Type IA cannot be applied. The Cs-Nb-Ta-bearing pegmatite veins that intruded UMR suffered the classical desilication with the formation of plagioclasite. During folding, shearing, and regional metamorphism, after the albitisation of the pegmatites, a reactive F-P-Be-Li-rich fluid of pegmatitic origin circulated in a shear zone, affecting the albitites and reacting with the UMR, to form emerald-bearing phlogopitites (Figure 38). Two generations of emerald are found: (i) Fine-grained crystals at the contact between albitite and phlogopitite and (ii) euhedral gem crystals formed later in phlogopitites either away from the albitite or in low-pressure zones next to the albitites. In that case, the albitites acted as incompetent levels, folded and sheared, and forming traps for euhedral emerald.

    The syntectonic pegmatites yielded an age of 2640 ± 40 Ma by U/Pb dating on monazite and an age of 2600 ± 100 Ma by the Pb-Pb method on microlite [176]. The $^{40}Ar/^{39}Ar$ dating on phlogopite and actinolite of the phlogopitites yielded a very disturbed age spectra and variable total gas ages between 2225 and 2447 Ma, with relative plateau ages of 1903 Ma for the phlogopite and of 1936 Ma for the amphibole [52]. Two ages were proposed for the formation of emerald: (i) An Archean age at 2640 Ma, which is the age of the intrusion of the pegmatites, or (ii) a Proterozoic age at around 2000 Ma, which corresponds to a major tectono-metamorphic episode that affected the Limpopo belt formed around the Zimbabwe craton. Zwaan [52] opted for the first hypothesis, considering that the deformation at circa 2000 Ma modified the isotopic argon clock of mica and amphibole, but these integrated Ar-ages between 2200 and 2500 Ma correlate with the Archean thermal event. The complexity of dating rocks that suffered deformational events and remobilization of material illustrates the complexity of classifying ore deposits. The Sandawana deposit belongs to sub-type IA and its genetic link with a magmatic source is obvious in terms of chemical elements, but it could be re-classified as sub-type IID if the age of the emerald is considered to be younger than 2400 Ma.

3. The possible genesis of sub-type IIA deposits, such as those of Swat Valley, Santa Terezinha de Goiás, and Habachtal [168], was discussed previously. The presence of meta-pegmatites can be suspected based on either chemical data or O-H isotope composition of emerald and associated minerals, but has not been found up to now due to the tectonic regime (thrust and shear zone) and the level of observation. These deposits are classified as sub-type IIC based on the geological environment and are considered to be metamorphic with probable mixing of magmatic fluids (high Cs content for the Santa Terezinha emeralds). This hypothesis is strengthened by the Na/Li vs. Cs/Ga chemical diagram presented by Schwarz [65]. Figure 39 shows that these emerald deposits are grouped in one chemical field very different from those of Colombia, Russia, and Nigeria. They are characterized by a high Cs/Ga ratio, indicating appreciable to high amounts of Cs (magmatic source), and a high Na/Li ratio. The high Na content of this emerald is correlated with a high mean H_2O content in the channels, as, determined for Santa Terezinha de Goiás (2.9 wt.%, n = 5), Habachtal (3.1 wt.%, n = 3), and Swat Valley (3.4 wt.%, n = 1) [15,16]. This is not just a coincidence, but is probably a genetic proxy Be-Cs source for emerald in these three deposits, i.e., magmatic sources with huge fluid circulation and metasomatism in a metamorphic environment.

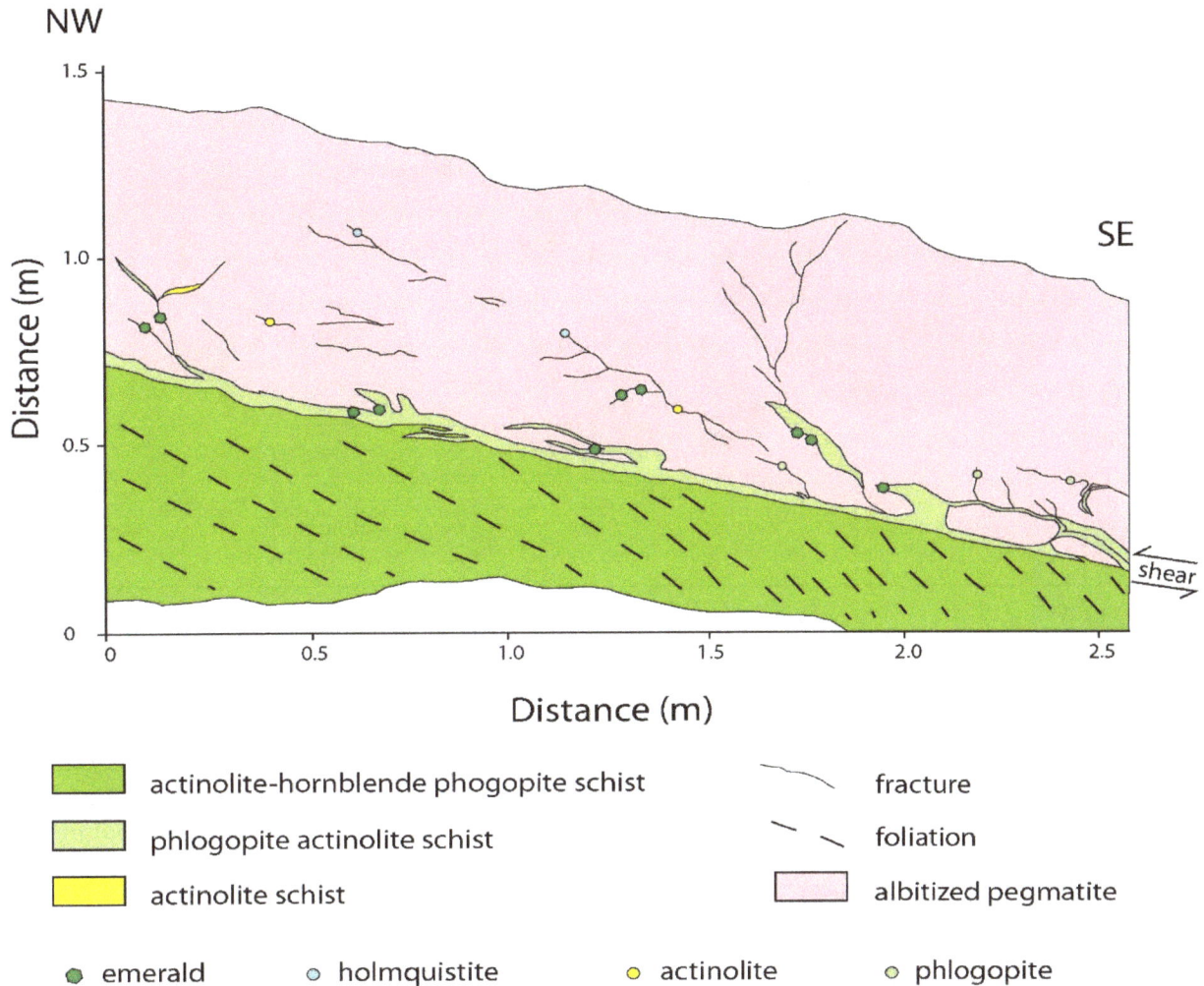

Figure 38. Cross-section of the mineralization at the Zeus underground mine, 200 ft. level, 26/28 stope, Sandawana deposit, Zimbabwe [167]. Fluids infiltrated along the schistosity of the actinolite-hornblende-phlogopite schist, and induced a metasomatic reaction with emerald formation at the foot wall of the pegmatite.

The Habachtal deposit is a complex deposit in terms of the genetic model and the previous discussion about the possible remobilization of Be-W enriched Hercynian pegmatites (possible sub-type IID deposit) by the regional metamorphism [168] opens the debate on the age of the formation of this deposit. The genesis of the emerald is metasomatic, but bound to the regional metamorphism of alpine age [77]. The K-Ar age obtained on phlogopite from the phlogopitites is 22 Ma, while the tracks of fission on apatite yielded ages of 9 Ma. The K-Ar age on muscovite from the muscovite schist is 27 Ma. The Rb/Sr dating realized on the zones of growth of garnets from the Schieferhülle formation, situated structurally above the Habach formation, indicated ages of crystallization between 62.0 and 30.2 Ma [177]. This age around 30 Ma is in agreement with the dates found for the end of the growth of garnet in the central Alps [178]. So, the best estimation established for the growth of the Habachtal emeralds would be situated around 30 Ma [179].

4.   The deposits of the Eastern desert in Egpyt and Poona in Australia are good examples of sub-type IID, where sub-type IA deposits were remobilized by regional metamorphism with deformation and remobilization of older rocks, following the genetic model proposed by Grundmann and Morteani [77].

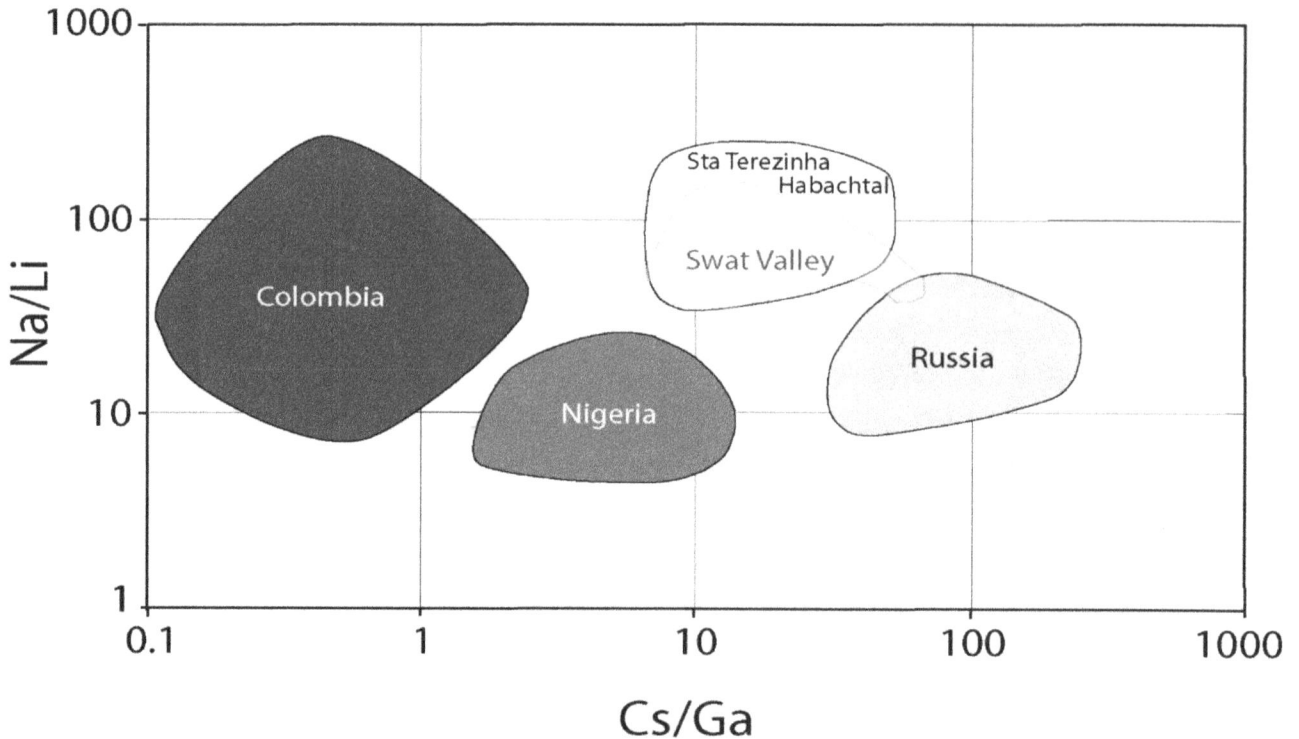

Figure 39. Cs/Ga versus Na/Li diagram of emeralds from Colombia, Nigeria, Russia, Pakistan, Austria, and Swat valley presented by Schwarz [65]. The diagram indicates that the chemical ratios of emerald from Santa Terezinha de Goiás (Brazil), Habachtal (Austria), and Swat valley (Pakistan) plot in the same population field. They are characterized by moderate to high Cs/Ga and very high Na/Li ratios. Discrepancies are observed for the concentrations of Cs and other elements when compared with the analysis presented by Aurisicchio et al. [23].

9. Exploration Now and in the Future

The majority of development of exploration methods for emerald has been for sub-type IIB deposits in Colombia. In Colombia, repeated washing of sediments is an effective and ubiquitous technique employed by local emerald-seeking "guaqueros" to reveal emerald in mineralized drainages. However, as pointed out by Lake [180], it is not effective to directly recover beryl by density separation because its specific gravity (~2.7) is similar to most rock-forming minerals. Geochemistry has also proven to be useful in Colombia. Escobar [181] studied the geology and geochemistry of the Gachalá area and found that Na enrichment and depletion of Li, K, Be, and Mo in the host rocks were very good indicators for locating mineralized areas. Beus [182] presented the results of a United Nations-sponsored geochemical survey of the streams draining emerald deposits in the Chivor and Muzo areas. The spatial distribution of areas with emerald mineralization was linked, on a regional scale, to intersections of the NNE- and NW-trending fault zones. The black shale units in those tectonic blocks that contain emerald mineralization were found to be enriched in CO_2, Ca, Mg, Mn, and Na, and depleted in K, Si, and Al [182]. The results of this study were tested with a stream sediment sampling program in the Muzo area. The results showed that samples collected from emerald-bearing tectonic blocks had anomalously low K/Na ratios. Beus [182] also suggested using a "composite" geochemical ratio that takes into account the albitization and leaching phenomena: $M = Na^3/(K \times Li \times Mo)$ (with Na and K in wt.% and Li and Mo in ppm) where M is an expression of the degree of metasomatic alteration. Other combinations, such as K, Li, Co or K, Li, Ba could be tried to determine the most contrasting value of M [182]. Subsequently, it was discovered that the Na content of the sediments was the best indicator of the mineralized zones in the drainage basins. Several new emerald occurrences were discovered by U.N. teams using the results of this study [183].

More recently, Ringsrud [183] reported that Colombian geologists were analyzing soil samples collected from altered tectonic blocks for Li, Na, and Pb to delineate emerald mineralization. Cheilletz et al. [103] showed that the Be content of black shale outside of the leached mineralized areas ranges from 2 to 6 ppm. Beryllium concentrations in the leached areas were found to range from 0.1 to 3.0 ppm [182]. Cheilletz and Giuliani [184] observed that the spatial coexistence of emerald districts and gypsum and anhydrite deposits could be used to prospect for new deposits.

Branquet et al. [107,113] observed that discovering new deposits will necessitate prospecting that is structurally oriented and focused on finding (1) the stratiform brecciated level in the eastern zone and (2) structural traps along regional tear faults in the western zone.

Geophysical techniques, such as induced polarization (IP) and magnetic surveys, have been used at Chivor and Muzo to delineate pyrite mineralization that is abundant and often associated with emerald-bearing veins [185]. Gutiérrez [186] tested several radiometric and magnetometric techniques at Chivor and Macanal deposits with some success.

Escobar [177] applied the criteria for emerald exploration in Colombia to stream sediment geochemical data for the Yukon and Northwest Territories in Canada. The criteria had to be adjusted for a number of factors, including the absence of Be in the Northwest Territories data and elevated Na due to plutonic alkali feldspar weathering into black shale drainages. Escobar [181] also noted that Colombian-type emerald deposits have relatively small footprints, and regional-scale geochemical surveys with data points 5–10 km apart may not be sufficient to show an emerald occurrence. Despite these constraints, Escobar [181] was able to identify several regions of interest.

With respect to other type II deposits, Arif et al. [187] observed that in the Swat valley in Pakistan, the emerald deposits occur in carbonate-altered ultramafic rocks, which also host Cr-rich dravite and "oxy-dravite". They suggested that the presence of high-Cr magnesian tourmaline, particularly in magnesite–talc-altered ultramafic rocks, can represent a criterion for further emerald exploration in the lower Swat region of Pakistan and in other ultramafic-hosted emerald-producing regions worldwide.

Wise [188] suggested that that the morphology of quartz, calcite, pyrite, and rutile crystals may serve as potential exploration guides for the discovery of hidden emerald deposits in the Hiddenite district of North Carolina. Emerald-bearing veins are characterized by: (1) quartz with multiple generations, including fine-grained doubly terminated crystals and very coarse-grained prismatic crystals; (2) calcite with largely rhombohedral habit, commonly accompanied by dolomite and siderite; (3) pyrite crystals dominated by octahedral faces; and (4) rutile that varies from single untwinned crystals to highly reticulated aggregates.

Some work has also been done in northwestern Canada to explore for type I deposits. Murphy et al. [189] plotted potential Be reservoirs and Cr and V reservoirs in the Yukon Territory and suggested that the best place to look is where the two come together. Lewis et al. [190] noted that all beryl occurrences in the Yukon Territory are intrusion-related, but for an intrusion to become enriched enough to reach Be-saturation to form beryl, it must be "ultrafractionated".

Future efforts will move towards the development of more effective exploration guidelines that consider the geological factors responsible for emerald formation. For type I deposits, this could involve exploring continental collision domains and considering the overlaps of Be and Cr-V reservoirs, mineralogical and geochemical anomalies (F, K, Li, P, B) linked to high intensity fluid/rock interaction, and deposit location, e.g., proximal or distal to the granitic intrusive. This last point is an important one as illustrated by the Ianapera deposit in Madagascar [11,135] where emerald occurs in two different, but coeval settings: (1) A proximal one, formed classically at the contact between pegmatites (and quartz veins) and M-UMR; and (2) a distal one formed outside the roof of the granite, with emerald-bearing phlogopitites in fractures and faults. The fluid percolation affected all the geological formations and different M-UMR, resulting in several types of zoned and un-zoned emerald habits with contrasting Cr-V chemistry.

10. Conclusions and Perspectives

The elaboration of a new classification proposed for emerald occurrences and deposits is the first step to building a new scheme for studying Be-mineralization from the field to the laboratory. The economic emerald deposits, in terms of volume and quality, are the Tectonic magmatic-related type hosted in M-UMR (Brazil, Zambia, Russia) or the Tectonic metamorphic-related type hosted in either low- (Colombia) to medium- (Afghanistan, China) temperature metamorphosed sedimentary rocks or medium-temperature metamorphosed M-UMR (Brazil, Pakistan).

The enhanced classification proposed in the present paper, based on objective geological criteria, takes into account the question of genesis of a number of deposits, which has been brought to the fore by research using new analytical facilities in the 21st century: The following topics were discussed: (i) Magmatic intrusives in M-UMR from classical intrusion to multi-stage formation through fluid circulation, metasomatism, and emerald formation synchronous with folding and shearing of regional metamorphic sequences; (ii) the presence of hidden intrusives that are probably the source of Be and Cs for emerald, such as in the Swat valley and Santa Terezinha de Goías deposits; and (ii) metamorphic remobilization of previous Be-magmatic mineralization over several orogenic episodes, such as for the Habachtal and Egyptian emerald occurrences. All these cases reflect the mixing of magmatic and metamorphic fluids.

The present work, which connects geological knowledge of the formation of gem deposits to studies of the properties and features of individual gems, is a step forward in the great challenge of geographic origin determination. The enhanced classification typology for emerald deposits opens a new framework for mineral exploration guidelines.

Author Contributions: G.G., L.A.G., D.M., A.E.F., and Y.B. contributed with their field and laboratory work, and experience to the realization of this description of the state-of-the-art of the geology and genesis of emerald deposits and the proposal of the enhanced classification typology.

Acknowledgments: The authors acknowledge the three reviewers for their constructive comments. We thank very much the guest editors of the special issue "Mineralogy and Geochemistry of Gems", namely Panagiotis Voudouris, Stefanos Karampelas, Vasilios Melfos, and Ian Graham for their invitation to present an extended review on emerald deposits. We acknowledge with thanks the editorial team of "Minerals" for their handling and the final layout. The authors want to thank Dietmar Schwarz, Vincent Pardieu, Victor Carillo, and Jean-Claude Michelou for their discussion and contribution of photographs. Gaston Giuliani wants to thank the Institut de Recherche pour le Développement (IRD, France) and the French CNRS and the Unversities of Lorraine and Paul Sabatier (CRPG/CNRS-Université de Lorraine and GET/IRD-Université Paul Sabatier) for academic and financial support for research on the geology of gems over the past thirty years. The authors want to thanks Mackenzie Parker for editing.

References

1.   Groat, L.; Giuliani, G.; Marshall, D.; Turner, D. Emerald. In *Geology of Gem Deposits*; Raeside, E.R., Ed.; Mineralogical Association of Canada, Short Course Series 44: Tucson, AZ, USA, 2014; pp. 135–174.

2.   In the News. Gemfields' October emerald auction: US$21.5 million in sales. *Incolor* **2018**, *37*, 22.

3.   Schmidt, S. Une des plus grandes émeraudes au monde a été découverte dans une mine en Zambie. Trust My Sicence, Nature. 2018. Available online: https://trustmyscience.com (accessed on 11 February 2019).

4.   Ibbetson, R. Miner finds giant 5,655-carat emeraldin Zambia worth up to £ 2m. *Daily Mail Online*, 30 October 2018.

5.   Yager, Y.G.; Menzie, W.D.; Olson, D.W. *Weight of Production of Emeralds, Rubies, Sapphires, and Tanzanite from 1995 through 2005*; Open File Report 2008–1013; USGS Reston: Reston, VA, USA, 2008; 9p.

6.   Wood, D.L.; Nassau, K. Characterization of beryl and emerald by visible and infrared absorption spectroscopy. *Am. Mineral.* **1968**, *53*, 777–800.

7.   Damon, P.E.; Kulp, J.L. Excess helium and argon in beryl and other minerals. *Am. Mineral.* **1958**, *43*, 433–459.

8.   Giuliani, G.; Marty, B.; Banks, D. Noble gases in fluid inclusions from emeralds: Implications for the origins

of fluids and constraints on fluid-rock interactions. In Proceedings of the 18th Biennial Meeting of European Current Research on Fluid Inclusions, Siena, Italy, 4–5 July 2005. CD-ROM.

9. Charoy, B. Cristallochimie du béryl: L'état des connaissances. In *L'Émeraude*; Giard, D., Giuliani, G., Cheilletz, A., Fritch, E., Gonthier, E., Eds.; AFG-CNRS-ORSTOM Edition: Paris, France, 1998; pp. 47–54.

10. Schwarz, D.; Schmetzer, K. The definition of emerald: The green variety of beryl colored by chromium and/or vanadium. In *Emeralds of the World, (2002) ExtraLapis English No. 2: The Legendary Green Beryl*; Lapis International, LLC: East Hampton, CT, USA, 2002; pp. 74–78.

11. Andrianjakavah, P.R.; Salvi, S.; Béziat, D.; Rakotondrazafy, A.F.M.; Giuliani, G. Proximal and distal styles of pegmatite-related metasomatic emerald mineralization at Ianapera, southern Madagascar. *Miner. Deposita* **2009**, *44*, 817–835. [CrossRef]

12. Zwaan, J.C.; Jacob, D.E.; Häger, T.; Calvacanti Neto, M.T.O.; Kanis, J. Emeralds from the Fazenda Bonfim region, Rio Grande do Norte, Brazil. *Gems Gemol.* **2012**, *48*, 2–17. [CrossRef]

13. Rondeau, B.; Fritsch, E.; Peucat, J.J.; Nordru, F.S.; Groat, L.A. Characterization of emeralds from a historical deposit: Byrud (Eidsvoll), Norway. *Gems Gemol.* **2008**, *44*, 108–122. [CrossRef]

14. Artioli, G.; Rinaldi, R.; Stahl, K.; Zanazzi, P.F. Structure refinements of beryl by single-crystal neutron and X-ray diffraction. *Am. Mineral.* **1993**, *78*, 762–768.

15. Giuliani, G.; France-Lanord, C.; Zimmermann, J.-L.; Cheilletz, A.; Arboleda, C.; Charoy, B.; Coget, P.; Fontan, F.; Giard, D. Composition of fluids, δD of channel H_2O and $\delta^{18}O$ of lattice oxygen in beryls: Genetic implications for Brazilian, Colombian and Afghanistani emerald deposits. *Int. Geol. Rev.* **1997**, *39*, 400–424. [CrossRef]

16. Marshall, D.; Downes, P.J.; Ellis, S.; Greene, R.; Loughrey, L.; Jones, P. Pressure–temperature-fluid constraints for the Poona emerald deposits, Western Australia: Fluid inclusion and stable isotope studies. *Minerals* **2016**, *6*, 130. [CrossRef]

17. Brand, A.A.; Groat, L.A.; Linnen, R.L.; Garland, M.I.; Breaks, F.W.; Giuliani, G. Emerald mineralization associated with the Mavis Lake Pegmatite Group, near Dryden, Ontario. *Can. Mineral.* **2009**, *47*, 315–336. [CrossRef]

18. Marshall, D.; Pardieu, V.; Loughrey, L.; Jones, P.; Xue, G. Conditions for emerald formation at Davdar, China: Fluid inclusion, trace element and stable isotope studies. *Mineral. Mag.* **2012**, *76*, 213–226. [CrossRef]

19. Loughrey, L.; Marshall, D.; Jones, P.; Millsteed, P.; Main, A. Pressure-temperature-fluid constraints for the Emmaville-Torrington emerald deposit, New South Wales, Australia: Fluid inclusion and stable isotope studies. *Cent. Eur. J. Geosci.* **2012**, *4*, 287–299. [CrossRef]

20. Loughrey, L.; Marshall, D.; Ihlen, P.; Jones, P. Boiling as a mechanism for colour zonations observed at the Byrud emerald deposit, Eidsvoll, Norway: Fluid inclusion, stable isotope and Ar-Ar studies. *Geofluids* **2013**, *13*, 542–558. [CrossRef]

21. Lake, D.J.; Groat, L.; Falck, H.; Cempirek, J.; Kontak, D.; Marshall, D.; Giuliani, G.; Fayek, M. Genesis of emerald-bearing quartz veins associated with the Lened W-skarn mineralization, northwest territories, Canada. *Can. Mineral.* **2017**, *55*, 561–593. [CrossRef]

22. Santiago, J.S.; Souza, V.S.; Filgueiras de, B.C.; Jiménez, F.A.C. Emerald from the Fazenda Bonfim deposit, northeastern Brazil: Chemical, fluid inclusions and oxygen isotope data. *Braz. J. Geol.* **2018**, 1–16. [CrossRef]

23. Aurisicchio, C.; Conte, A.M.; Medeghini, L.; Ottolini, L.; De Vito, C. Major and trace element geochemistry of emerald from several deposits: Implications for genetic models and classification schemes. *Ore Geol. Rev.* **2018**, *94*, 351–366. [CrossRef]

24. Kovaloff, P. Geologist's report on Somerset emeralds. *S. Afr. Mining Eng. J.* **1928**, *39*, 101–103.

25. Zambonini, F.; Caglioto, V. Ricerche chimiche sulla rosterite di San Piero in Campo (Isola d'Elba) e sui berilli in generale. *Gazz. Chim. Ital.* **1928**, *58*, 131–152.

26. Leitmeier, H. Das Smaragdvorkommen in Habachtal in Salzburg und seine Mineralien. *Tscher. Miner. Petrog.* **1937**, *49*, 245–368. [CrossRef]

27. Otero Muñoz, G.; Barriga Villalba, A.M. *Esmeraldas de Colombia*; Banco de la Republica: Bogotá, Colombia, 1948.

28. Simpson, E.S. *Minerals of Western Australia*; Government Printer: Perth, Australia, 1948; Volume 1, pp. 195–207.

29. Gübelin, E.J. Emeralds from Sandawana. *J. Gemmol.* **1958**, *6*, 340–354. [CrossRef]

30. Vlasov, K.A.; Kutakova, E.I. *Izumrudnye Kopi*; Moscow Akademiya Nauk SSGR: Moscow, Russia, 1960; 252p.

31. Martin, H.J. Some observations on southern Rhodesian emeralds and chrysoberyls. *Chamb. Mines J.* **1962**, *4*, 34–38.

32. Petrusenko, S.; Arnaudov, V.; Kostov, I. Emerald pegmatite from the Urdini Lakes, Rila Mountains. *Annuaire de l'Université de Sofia Faculté de Géologie et Géographie* **1966**, *59*, 247–268.

33. Beus, A.A.; Mineev, D.A. *Some Geological and Geochemical Features of the Muzo-Coscuez Emerald Zone, Cordillera Oriental, Colombia*; Empresa Colombiana de Minas, Biblioteca: Bogotá, Colombia, 1972; 55p.

34. Hickman, A.C.J. The Miku emerald deposit. *Geol. Surv. Zambia Econ. Rep.* **1972**, *27*, 35.

35. Garstone, J.D. The geological setting and origin of emeralds from Menzies, Western Australia. *J. R. Soc. West. Aust.* **1981**, *64*, 53–64.

36. Hanni, H.A.; Klein, H.H. Ein Smaragdvorkommen in Madagaskar. *Zeitschrift der Deutschen Gemmologischen Gesellschaft* **1982**, *21*, 71–77.

37. Graziani, G.; Gübelin, E.; Lucchesi, S. The genesis of an emerald from the Kitwe District, Zambia. *Neues Jb. Miner. Monat.* **1983**, 175–186.

38. Kozlowski, A.; Metz, P.; Jaramillo, H.A.E. Emeralds from Sodomondoco, Colombia: Chemical composition, fluid inclusion and origin. *Neues Jb. Miner. Abh.* **1988**, *59*, 23–49.

39. Hammarstrom, J.M. Mineral chemistry of emeralds and some minerals from Pakistan and Afghanistan: An electron microprobe study. In *Emeralds of Pakistan*; Kazmi, A.H., Snee, L.W., Eds.; Van Nostrand Reinhold: New York, NY, USA, 1989; pp. 125–150.

40. Ottaway<sc>, T.L. The Geochemistry of the Muzo Emerald Deposit, Colombia. M.Sc. Thesis, University of Toronto, Toronto, ON, Canada, 1991.

41. Schwarz, D. Australian emeralds. *Austral. Gemmol.* **1991**, *17*, 488–497.

42. Schwarz, D.; Kanis, J.; Kinnaird, J. Emerald and green beryl from central Nigeria. *J. Gemmol.* **1996**, *25*, 117–141. [CrossRef]

43. Abdalla, H.M.; Mohamed, F.H. Mineralogical and geochemical investigations of beryl mineralizations, Pan-African belt of Egypt: Genetic and exploration aspects. *J. Afr. Earth Sci.* **1999**, *28*, 581–598. [CrossRef]

44. Gavrilenko, E.V.; Pérez, B.C. Characterisation of emeralds from the Delbegetey deposit, Kazakhstan. In *Mineral Deposits: Processes to Processing*; Stanley, C.J., Rankin, A.H., Bodnar, R.J., Naden, J., Yardley, B.W.D., Criddle, A.J., Hagni, R.D., Gize, A.P., Pasava, J.., Eds.; Balkema: Rotterdam, The Netherlands, 1999; pp. 1097–1100.

45. Alexandrov, P.; Giuliani, G.; Zimmerman, J.L. Mineralogy, age and fluid geochemistry of the Rila Emerald deposits, Bulgaria. *Econ. Geol.* **2001**, *96*, 1469–1476. [CrossRef]

46. Groat, L.A.; Marshall, D.D.; Giuliani, G.; Murphy, D.C.; Piercey, S.J.; Jambor, J.L.; Mortensen, J.K.; Ercit, T.S.; Gault, R.A.; Mattey, D.P.; et al. Mineralogical and geochemical study of the Regal Ridge showing emeralds, southeastern Yukon. *Can. Mineral.* **2002**, *40*, 1313–1338. [CrossRef]

47. Marshall, D.D.; Groat, L.A.; Falck, H.; Douglas, H.L.; Giuliani, G. Fluid inclusions from the Lened emerald occurrence; Northwest Territories, Canada: Implications for Northern Cordilleran Emeralds. *Can. Mineral.* **2004**, *42*, 1523–1539. [CrossRef]

48. Vapnik, Y.; Sabot, B.; Moroz, I. Fluid inclusions in Ianapera emerald, Southern Madagascar. *Int. Geol. Rev.* **2005**, *47*, 647–662. [CrossRef]

49. Vapnik, Y.; Moroz, I.; Eliezri, I. Formation of emeralds at pegmatite-ultramafic contacts based on fluid inclusions in Kianjavato emerald, Mananjary deposits, Madagascar. *Mineral. Mag.* **2006**, *70*, 141–158. [CrossRef]

50. Zwaan, J.C.; Seifert, A.V.; Vrána, S.; Laurs, B.M.; Anckar, B.; Simons, W.B.S.; Falster, A.U.; Lustenhouwer, W.J.; Muhlmeister, S.; Koivula, J.K.; et al. Emeralds from the Kafubu area, Zambia. *Gems Gemol.* **2005**, *41*, 116–148. [CrossRef]

51. Gavrilenko, E.V.; Calvo Pérez, B.; Castroviejo Bolibar, R.; Garcia Del Amo, D. Emeralds from the Delbegetey deposit (Kazakhstan): Mineralogical characteristics and fluid-inclusion study. *Mineral. Mag.* **2006**, *70*, 159–173. [CrossRef]

52. Zwaan, J.C. Gemmology, geology and origin of the Sandawana emerald deposits, Zimbabwe. *Scr. Geol.* **2006**, *131*, 211.

53. Barton, M.D.; Young, S. Nonpegmatitic deposits of beryllium: Mineralogy, geology, phase equilibria and origin. In *Beryllium: Mineralogy, Petrology, and Geochemistry*; Grew, E.S., Ed.; Mineralogical Society of America: Washington, DC, USA, 2002; Volume 50, pp. 591–691.

54. Giuliani, G.; Bourlès, D.; Massot, J.; Siame, L.L. Colombian emerald reserves inferred from leached beryllium of their host black shale. *Explor. Min. Geol.* **1999**, *8*, 109–116.

55. Giuliani, G.; Branquet, Y.; Fallick, A.E.; Groat, L.; Marshall, D. Emerald deposits around the world, their similarities and differences. *InColor* **2015**, *special issue*, 56–69.

56. Sabot, B. Classification des gisements d'émeraude: Apports des études pétrographiques, minéralogiques et géochimiques. Thèse de Doctorat, Université de Nancy, Nancy, France, 2002.

57. Dereppe, J.M.; Moreaux, C.; Chauvaux, B.; Schwarz, D. Classification of emeralds by artificial neural networks. *J. Gemmol.* **2000**, *27*, 93–105. [CrossRef]

58. Schwarz, D.; Giuliani, G. Emerald deposits—A review. *Austral. Gemmol.* **2001**, *21*, 17–23.

59. Schwarz, D.; Giuliani, G.; Grundmann, G.; Glas, M. Die Entstehung der Smaragde, ein vieldisskutiertes Thema. In *Smaragd, der Kostbarste Beryll, der Teuerste Edelstein*; Schwarz, D., Hochlitner, R., Eds.; ExtraLapis: Charleston, SC, USA, 2001; pp. 68–73.

60. Schwarz, D.; Giuliani, G.; Grundmann, G.; Glas, M. The origin of emerald—A controversial topic. *ExtraLapis Engl.* **2002**, *2*, 18–21.

61. Schwarz, D.; Klemm, L. Chemical signature of emerald. *Int. Geol. Congr. Abstr.* **2012**, *34*, 2812.

62. Giuliani, G.; Groat, L.; Ohnenstetter, D.; Fallick, A.E.; Feneyrol, J. The geology of gems and their geographic origin. In *Geology of Gem Deposits*; Raeside, E.R., Ed.; Mineralogical Association of Canada, Short Course Series 44: Tucson, AZ, USA, 2014; pp. 113–134.

63. Saeseasaw, S.; Pardieu, V.; Sangsawong, S. Three-phase inclusions in emerald and their impact on origin determination. *Gems Gemol.* **2014**, *50*, 114–132.

64. Ochoa, C.J.C.; Daza, M.J.H.; Fortaleche, D.; Jiménez, J.F. Progress on the study of parameters related to the origin of Colombian emeralds. *InColor* **2015**, *special issue*, 88–97.

65. Schwarz, D. The geographic origin determination of emeralds. *InColor* **2015**, *special issue*, 98–105.

66. Hainschwang, T.; Notari, F. Standards and protocols for emerald analysis in gem testing laboratories. *InColor* **2015**, *special issue*, 106–114.

67. Giuliani, G. La spirale du temps de l'émeraude. *Règne Minéral* **2011**, *98*, 31–44.

68. Kinnaird, J.A. Hydrothermal alteration and mineralization of the alkaline anorogenic ring complexes of Nigeria. *J. Afr. Earth Sci.* **1985**, *3*, 229–251. [CrossRef]

69. Larsen, B.T.; Olaussen, S.; Sundwoll, B.; Heeremans, M. The Permo-Carboniferous Oslo Rift through six stages and 65 million years. *Episodes* **2008**, *31*, 52–58.

70. Seifert, A.V.; Žaček, V.; Vrána, S.; Pecina, V.; Zachariáš, J.; Zwaan, J.C. Emerald mineralization in the Kafubu area, Zambia. *Bull. Geosci.* **2004**, *79*, 1–40.

71. Bastos, F.M. Emeralds from Itabira, Minas Gerais, Brazil. *Lapid. J.* **1981**, *35*, 1842–1848.

72. Hänni, H.A.; Schwarz, D.; Fischer, M. The emeralds of the Belmont mine, Minas Gerais, Brazil. *J. Gemmol.* **1987**, *20*, 446–456. [CrossRef]

73. de Souza, J.L.; Mendes, J.C.; da Silveira Bello, R.M.; Svisero, D.P.; Valarelli, J.V. Petrographic and microthermometrical studies of emeralds in the "Garimpo" of Capoeirana, Nova Era, Minas Gerais State, Brazil. *Miner. Deposita* **1992**, *27*, 161–168. [CrossRef]

74. Rondeau, B.; Notari, F.; Giuliani, G.; Michelou, J.-C.; Martins, S.; Fritsch, E.; Respinger, A. La mine de Piteiras, Minas Gerais, nouvelle source d'émeraude de belle qualité au Brésil. *Rev. Gemmol. AFG* **2003**, *148*, 9–25.

75. Walton, L. *Exploration Criteria for Colored Gemstone Deposits in Yukon*; Open file 2004-10; Geological Survey of Canada: Ottawa, ON, Canada, 2004; 184p.

76. Giuliani, G.; Cheilletz, A.; Zimmermann, J.-L.; Ribeiro-Althoff, A.M.; France-Lanord, C.; Féraud, G. Les gisements d'émeraude du Brésil: Genèse et typologie. *Chron. Rech. Min. BRGM* **1997**, *526*, 17–60.

77. Grundmann, G.; Morteani, G. Emerald mineralization during regional metamorphism: The Habachtal (Austria) and Leydsdorp (Transvaal, South Africa) deposits. *Econ. Geol.* **1989**, *84*, 1835–1849. [CrossRef]

78. Rudowski, L.; Giuliani, G.; Sabaté, P. Les phlogopitites à émeraude au voisinage des granites de Campo Formoso et Carnaíba (Bahia, Brésil): Un exemple de minéralisation protérozoïque à Be, Wo et W dans les ultrabasiques métasomatisées. *C. R. Acad. Sci.* **1987**, *304*, 1129–1134.

79. Rudowski, L. Pétrologie et géochimie des granites transamazoniens de Campo Formoso et Carnaíba (Bahia, Brésil) et des phlogopitites à émeraude associées. Thèse de Doctorat, Université Paris VI, Paris, France, 1989.

80. Muñoz, J.L. F-OH and Cl-OH exchange in micas with applications to hydrothermal ore deposits. In *Micas*; Bailey, S.W., Ed.; Mineralogical Society of America: Washington, DC, USA, 1984; Volume 13, pp. 469–493.

81. Guy, B. Contribution à l'étude des skarns de Costabonne (Pyrénées orientales, France) et à la théorie de la zonation métasomatique. Ph.D. Thesis, Université de Paris VI, Paris, France, 1979.

82. Korzhinskii, D.S. The theory of systems with perfectly mobile components and processes of mineral formation. *Am. J. Sci.* **1965**, *263*, 193–205. [CrossRef]

83. Laurs, B.M.; Dilles, J.H.; Snee, L.W. Emerald mineralization and metasomatism of amphibolite, Khaltaro granitic pegmatite hydrothermal vein system, Haramosh mountains, northern Pakistan. *Can. Mineral.* **1996**, *34*, 1253–1286.

84. Grundmann, G.; Morteani, G. Ein neues Vorkommen von Smaragd, Alexandrit, Rubin und Saphir in einem Topas-führenden Phlogopit-felds von Poona, Cue District, West Australien. *Zeitschrift der Deutschen Gemmologischen Gesellschaft* **1995**, *44*, 11–31.

85. Xue, G.; Marshall, D.; Zhang, S.; Ullrich, T.D.; Bishop, T.; Groat, L.A.; Thorkelson, D.J.; Giuliani, G.; Fallick, A.E. Conditions for early Cretaceous emerald formation at Dyakou, China: Fluid inclusion, Ar-Ar, and stable isotope studies. *Econ. Geol.* **2010**, *105*, 339–349. [CrossRef]

86. Kazmi, A.H.; Snee, L.W. Geology of the world emerald deposits: A brief review. In *Emeralds of Pakistan: Geology, Gemology and Genesis*; Kazmi, A.H., Snee, L.W., Eds.; Van Nostrand Reinhold Company: New York, NY, USA, 1989; pp. 165–228.

87. Martins, S. Brazilian emeralds. In Proceedings of the Oral Communication, II World Emerald Symposium, Bogotá, Colombia, 12–14 October 2018.

88. Biondi, J.C. Depósitos de esmeralda de Santa Terezinha (GO). *Rev. Bras. Geociências* **1990**, *20*, 7–24. [CrossRef]

89. Gusmão Costa, S.A. de Correlação da seqüência encaixante das esmeraldas de Santa Terezinha de Goiás com os terrenos do tipo greenstone belt de Crixás e tipologia dos depósitos. In Proceedings of the Boletim de Resumos Espandidos XXXVIII Congresso Brasileiro de Geologia Goiânia, Goiás, Brazil, 1986; Volume 2, pp. 597–614.

90. Giuliani, G.; D'El-Rey Silva, L.J.; Couto, P.A. Origin of emerald deposits of Brazil. *Miner. Deposita* **1990**, *25*, 57–64. [CrossRef]

91. D'el-Rey Silva, L.J.H.; Giuliani, G. Controle estrutural da jazida de esmeraldas de Santa Terezinha de Goiás: Implicações na gênese, na têctonica regional e no planajamento de lavra. In Proceedings of the Boletim de Resumos Espandidos XXXV Congresso Brasileiro de Geologia, Belém, PA, Brazil; 1988; Volume 1, pp. 413–427.

92. D'el-Rey Silva, L.J.H.; de Barros Neto, L.S. The Santa Terezinha-Campos Verdes emerald district, central Brazil: Structural and Sm-Nd data to constrain the tectonic evolution of the Neoproterozoic Brasília belt. *J. S. Am. Earth Sci.* **2002**, *15*, 693–708. [CrossRef]

93. Morteani, G.; Grundmann, G. The emerald porphyroblasts in the penninic rocks of the central Tauern Window. *Neues Jb. Miner. Monat.* **1977**, *11*, 509–516.

94. Grundmann, G. Polymetamorphose und Abschätzung der Bildungsb edin-gungen der smaragd-führenden gesteinsserien der leckbachscharte, Habachtal, Österreich. *Fortschr. Mineral.* **1980**, *58*, 39–41.

95. Grundmann, G.; Morteani, G. Die geologie des smaragdvorkommens im Habachtal (Land Salzburg, Österreich). *Archiv. Lagerstättenforsch. Geol. Bundensanst* **1982**, *2*, 71–107.

96. Nwe, Y.Y.; Grundmann, G. Evolution of metamorphic fluids in shear zones: The record from the emeralds of Habachtal, Tauern window, Austria. *Lithos* **1990**, *25*, 281–304. [CrossRef]

97. Trumbull, R.B.; Krienitz, M.-S.; Grundmann, G.; Wiedenbeck, M. Tourmaline geochemistry and $\delta^{11}B$ variations as a guide to fluid-rock interaction in the Habachtal emerald deposit, Tauern window, Austria. *Contrib. Mineral. Petr.* **2009**, *157*, 411–427. [CrossRef]

98. Bowersox, G.; Snee, L.W.; Foord, E.E.; Seal, R.R., II. Emeralds of the Panjsher Valley, Afghanistan. *Gems Gemol.* **1991**, *27*, 26–39. [CrossRef]

99. Hussain, S.S.; Chaudhry, M.N.; Dawood, H. Emerald mineralization of Barang, Bajaur Agency, Pakistan. *J. Gemmol.* **1993**, *23*, 402–408. [CrossRef]

100. Dilles, J.H.; Snee, L.W.; Laurs, B.M. Geology, Ar-Ar age and stable isotopes geochemistry of suture-related emerald mineralization, Swat, Pakistan, Himalayas. In Proceedings of the Geological Society of America, Annual Meeting, Seattle, WA, USA, 1994; Abstracts. Volume 26, p. A-311.

101. Arif, M.; Fallick, A.E.; Moon, C.J. The genesis of emeralds and their host rocks from Swat, northwestern Pakistan: A stable-isotope investigation. *Miner. Deposita* **1996**, *31*, 255–268. [CrossRef]

102. Arif, M.; Henry, D.J.; Moon, C.J. Host rock characteristics and source of chromium and beryllium for emerald mineralization in the ophiolitic rocks of the Indus suture zone in Swat, NW Pakistan. *Ore Geol. Rev.* **2011**, *39*, 1–20. [CrossRef]

103. Cheilletz, A.; Féraud, G.; Giuliani, G.; Rodriguez, C.T. Time-pressure-temperature formation of Colombian emerald: An ^{40}Ar/^{39}Ar laser-probe and fluid inclusion-microthermometry contribution. *Econ. Geol.* **1994**, *89*, 361–380. [CrossRef]

104. Ottaway, T.L.; Wicks, F.J.; Bryndzia, L.T.; Kyser, T.K.; Spooner, E.T.C. Formation of the Muzo hydrothermal emerald deposit in Colombia. *Nature* **1994**, *369*, 552–554. [CrossRef]

105. Banks, D.; Giuliani, G.; Yardley, B.W.D.; Cheilletz, A. Emerald mineralisation in Colombia: Fluid chemistry and the role of brine mixing. *Miner. Deposita* **2000**, *35*, 699–713. [CrossRef]

106. Cheilletz, A.; Giuliani, G.; Branquet, Y.; Laumonier, B.; Sanchez, A.J.M.; Féraud, G.; Arhan, T. Datation K-Ar et ^{40}Ar/^{39}Ar à 65 ± 3 Ma des gisements d'émeraude du district de Chivor-Macanal: Argument en faveur d'une déformation précoce dans la Cordillère Orientale de Colombie. *C. R. Acad. Sci.* **1997**, *324*, 369–377.

107. Branquet, Y.; Cheilletz, A.; Giuliani, G.; Laumonier, B. Fluidized hydrothermal breccia in dilatant faults during thrusting: The Colombian emerald deposits case. In *Fractures, Fluid Flow and Mineralization*; McCaffrey, K.J.W., Lonergan, L., Wilkinson, J.J., Eds.; Geological Society London: London, UK, 1999; Volume 155, pp. 183–195.

108. Branquet, Y.; Giuliani, G.; Cheilletz, A.; Laumonier, B. Colombian emeralds and evaporites: Tectono-stratigraphic significance of a regional emerald-bearing evaporitic breccia level. In Proceedings of the 13th SGA Biennal Meeting, Nancy, France, 24–27 August 2015; Volume 4, pp. 1291–1294.

109. Cooper, M.A.; Addison, F.T.; Alvarez, R.; Coral, M.; Graham, R.; Hayward, A.B.; Howe, S.; Martinez, J.; Naar, J.; Pena, R.; et al. Basin development and tectonic history of the Llanos basin, Eastern Cordillera, and Middle Magdalena valley, Colombia. *AAPG Bull.* **1995**, *79*, 1421–1443.

110. Branquet, Y.; Cheilletz, A.; Cobbold, P.R.; Baby, P.; Laumonier, B.; Giuliani, G. Andean deformation and rift inversion, eastern edge of Cordillera Oriental (Guateque, Medina area), Colombia. *J. S. Am. Earth Sci.* **2002**, *15*, 391–407. [CrossRef]

111. Bayona, G.; Cortès, M.; Jaramillo, C.; Ojeda, G.; Aristizabal, J.J.; Harker, A.R. An integrated analysis of an orogen-sedimentary basin pair: Latest Cretaceous (Cenozoic evolution of the linked Eastern Cordillera orogen and the Llanos foreland basin of Colombia. *Geol. Soc. Am. Bull.* **2008**, *120*, 1171–1197. [CrossRef]

112. Laumonier, B.; Branquet, Y.; Lopès, B.; Cheilletz, A.; Giuliani, G.; Rueda, F. Mise en évidence d'une tectonique compressive Eocène-Oligocène dans l'Ouest de la Cordillère orientale de Colombie, d'après la structure en duplex des gisements d'émeraude de Muzo et de Coscuez. *C. R. Acad. Sci.* **1996**, *323*, 705–712.

113. Branquet, Y.; Laumonier, B.; Cheilletz, A.; Giuliani, G. Emeralds in the Eastern Cordillera of Colombia: Two tectonic settings for one mineralization. *Geology* **1999**, *27*, 597–600. [CrossRef]

114. Pignatelli, I.; Giuliani, G.; Ohnenstetter, D.; Agrosì, G.; Mathieu, S.; Morlot, C.; Branquet, Y. Colombian trapiche emeralds: Recent advances in understanding their formation. *Gems Gemol.* **2015**, *51*, 222–259. [CrossRef]

115. Giuliani, G.; France-Lanord, C.; Cheilletz, A.; Coget, P.; Branquet, Y.; Laumonier, B. Sulfate reduction by organic matter in Colombian emerald deposits: Chemical and stable isotope (C, O, H) evidence. *Econ. Geol.* **2000**, *95*, 1129–1153. [CrossRef]

116. Bosshart, G. Emeralds from Colombia. *J. Gemmol.* **1991**, *22*, 409–425. [CrossRef]

117. Hewton, M.L.; Marshall, D.D.; Ootes, L.; Loughrey, L.E.; Creaser, R.A. Colombian-style emerald mineralization in the northern Canadian Cordillera: Integration into a regional Paleozoic fluid flow regime. *Can. J. Earth Sci.* **2013**, *50*, 857–871. [CrossRef]

118. Keith, J.D.; Thompson, T.J.; Ivers, S. The Uinta emerald and the emerald-bearing potential of the Red Pine shale, Uinta Mountains, Utah. *Geol. Soc. Am. Abstr. Program* **1996**, *28*, 85.

119. Keith, J.D.; Nelson, S.T.; Thompson, T.J.; Dorais, M.J.; Olcott, J.; Duerichen, E.; Constenius, K.N. The genesis of fibrous calcite and shale-hosted emerald in a nonmagmatic hydrothermal system, Uinta mountains, Utah. *Geol. Soc. Am. Abstr. Program* **2002**, *34*, 55.

120. Nelson, S.T.; Keith, J.D.; Constenius, K.N.; Duerichen, E.; Tingey, D.G. Genesis of fibrous calcite and emerald by amagmatic processes in the southwestern Uinta Mountains, Utah. *Rocky Mt. Geol.* **2008**, *43*, 1–21. [CrossRef]

121. Tapponnier, P.; Mattauer, M.; Proust, F.; Cassaigneau, C. Mesozoic ophiolites, sutures, and large-scale tectonic movements in Afghanistan. *Earth Planet. Sci. Lett.* **1981**, *52*, 335–371. [CrossRef]

122. Sabot, B.; Cheilletz, A.; de Donato, P.; Banks, D.; Levresse, G.; Barrès, O. Afghan emeralds face Colombian cousins. *Chron. R. Min. BRGM* **2000**, *541*, 111–114.

123. Sabot, B.; Cheilletz, A.; de Donato, P.; Banks, D.; Levresse, G.; Barrès, O. The Panjsher-Afghanistan emerald deposit: New field and geochemical evidence for Colombian style mineralisation. In Proceedings of the European Union Geoscience XI, Strasbourg, France, 8–13 April 2001; Section OS 06. p. 548.

124. An, Y. The geological characteristics of the emerald deposit in Tashen Kuergan, Xinjiang. *Xinjiang Non-Ferrous Met.* **2009**, *29*, 9–10.

125. Wise, M.A. New finds in North Carolina. *ExtraLapis Engl.* **2002**, *2*, 64–65.

126. Speer, W.E. Hiddenite district, Alexander Co, North Carolina. In Fieldtrip Guidebook. In *Proceedings of the Southeastern Section 57th Annual Meeting*; Charlotte, NC, USA, 10–11 April 2008, The Geological Society of America: Boulder, CO, USA, 2008; 28p.

127. Wise, M.A.; Anderson, A.J. The emerald- and spodumene-bearing quartz veins of the Rist emerald mine, Hiddenite, North Carolina. *Can. Mineral.* **2006**, *44*, 1529–1541. [CrossRef]

128. Grundmann, G.; Morteani, G. "Smaragdminen der Cleopatra": Zabara, Sikait und Umm Kabo in Ägypten. *Lapis* **1993**, *7–8*, 27–39.

129. Grundmann, G.; Morteani, G. Emerald formation during regional metamorphism: The Zabara, Sikeit and Umm Kabo deposits (Eastern Desert, Egypt). In *Geoscientific Research in Northeast Africa*; Schandelmeier, T., Ed.; A.A. Balkema: Rotterdam, The Netherlands, 1993; pp. 495–498.

130. Grundmann, G.; Morteani, G. Multi-stage emerald formation during Pan-African regional metamorphism: The Zabara, Sikait, Umm Kabo deposits, South Eastern desert of Egypt. *J. Afr. Earth Sci.* **2008**, *50*, 168–187. [CrossRef]

131. Surour, A.A.; Takla, M.A.; Omar, S.A. EPR spectra and age determination of beryl from the Eastern Desert of Egypt. *Ann. Geol. Surv. Egypt* **2002**, *25*, 389–400.

132. Giuliani, G.; France-Lanord, C.; Coget, P.; Schwarz, D.; Cheilletz, A.; Branquet, Y.; Giard, D.; Pavel, A.; Martin-Izard, A.; Piat, D.H. Oxygen isotope systematics of emerald—Relevance for its origin and geological significance. *Miner. Deposita* **1998**, *33*, 513–519. [CrossRef]

133. Duarte, L.D.C.; Pulz, G.M.; D'el-Rey Silva, L.J.H.; Ronchi, L.H.; de Brun, T.M.M.; Juchem, P.L. Microtermometria das inclusões fluidas na esmeralda do distrito mineiro de Campos Verdes, Goiás. In *Caracterização e Modelamento de depósitos Minerais*; Ronchi, L.H., Althoff, F.J., Eds.; Eitoras Unisinos: São Leopoldo, RS, Brazil, 2003; pp. 267–292.

134. Olivera, J.A.P.; Ali, S.H. Gemstone mining as a develoment cluster: A study of Brazil's emerald mines. *Resour. Policy* **2011**, *36*, 132–141. [CrossRef]

135. Salvi, S.; Giuliani, G.; Andrianjakavah, P.R.; Moine, B.; Beziat, D.; Fallick, A.E. Fluid inclusion and stable isotope constraints on the formation of the Ianapera emerald deposit, Southern Madagascar. *Can. Mineral.* **2017**, *55*, 619–650. [CrossRef]

136. de Wit, M.J.; Hart, R.A.; Hart, R.J. The Jamestown ophiolite complex, Barberton mountain belt: A section through 3.5 Ga oceanic crust. *J. Afr. Earth Sci.* **1987**, *6*, 681–730. [CrossRef]

137. Biondi, J.C.; Poidevin, J.L. Idade da mineralização e da sequência Santa Terezinha (Goiás, Brasil). In Proceedings of the Boletim de Resumos Espandidos XXXVIII Congresso Brasileiro de Geologia, Camboriú, Brasil, 23–28 October 1994; pp. 302–304.

138. Ribeiro-Althoff, A.M.; Cheilletz, A.; Giuliani, G.; Féraud, G.; Barbosa Camacho, G.; Zimmermann, J.L. Evidences of two periods (2 Ga and 650-500 Ma) of emerald formation in Brazil by K-Ar and ^{40}Ar/^{39}Ar dating. *Int. Geol. Rev.* **1997**, *39*, 924–937. [CrossRef]

139. Calligaro, T.; Dran, J.-C.; Poirot, J.-P.; Querré, G.; Salomon, J.; Zwaan, J.C. PIXE/PIGE characterization of emeralds using an external micro-beam. *Nucl. Instrum. Meth. B* **2000**, *161–163*, 769–774. [CrossRef]

140. Höll, R.; Maucher, A.; Westenberger, H. Synsedimentary diagenetic ore fabrics in the strata- and time-bound scheelite deposits of Kleinarltal and Felbertal in the Eastern Alps. *Miner. Deposita* **1972**, *7*, 217–226. [CrossRef]

141. Raith, J.G.; Montanuiversität Leoben, Austria. Personal communication, 2010.

142. Kozlik, M.; Raith, J.G.; Gerdes, A. U-Pb, Lu-Hf and trace element characteristics of zircon from the Felbertal scheelite deposit (Austria): New constraints on timing and source of W mineralization. *Chem. Geol.* **2016**, *421*, 112–126. [CrossRef]

143. Maluski, H.; Matte, P. Ages of alpine tectonometamorphic events in the northwestern Himalaya (northern Pakistan) by ^{39}Ar/^{40}Ar method. *Tectonophysics* **1984**, *3*, 1–18.

144. Grundmann, G.; Morteani, G. Alexandrite, emerald, sapphire, ruby and topaz in a biotite-phlogopite fels from the Poona, Cue district, Western Australia. *Austral. Gemmol.* **1998**, *20*, 159–167.

145. Vapnik, Y.; Moroz, I. Fluid inclusions in Panjshir emerald (Afghanistan). In Proceedings of the 16th European Current Research on Fluid Inclusions, Faculdade de Ciências do Porto, Porto, Portugal, 17–20 July 2001; Noronha, F., Doria, A., Guedes, A., Eds.; Memoria n° 7: New York, NY, USA; pp. 451–454.

146. Lapointe, M.; Anderson, A.J.; Wise, M. Fluid inclusion constraints on the formation of emerald-bearing quartz veins at the Rist tract, Heddenite, North Carolina. *Atl. Geol. Abstr.* **2004**, *4*, 146.

147. Diamond, L.W. Introduction to gas-bearing aqueous fluid inclusions. In *Fluid Inclusions: Analysis and Interpretation*; Samson, I., Anderson, A., Marshall, D., Eds.; Mineralalogical Association of Canada: Quebec City, QC, Canada, 2003; Volume 32, pp. 101–158.

148. Bodnar, R.J. Introduction to aqueous-electrolyte fluid inclusions. In *Fluid Inclusions: Analysis and Interpretation*; Samson, I., Anderson, A., Marshall, D., Eds.; Mineralogical Association of Canada: Quebec City, QC, Canada, 2003; Volume 32, pp. 81–100.

149. Audétat, A.; Gunter, D.; Heinrich, C. Formation of a magmatic-hydrothermal ore deposit: Insights with LA-ICP-MS analysis of fluid inclusions. *Science* **2012**, *279*, 2091–2094.

150. Diamond, L.W.; Marshall, D.; Jackman, J.; Skippen, G.B. Elemental analysis of individual fluid inclusions in minerals by Secondary Ion Mass Spectrometry (SIMS): Application to cation ratios of fluid inclusions in an Archaean mesothermal gold-quartz vein. *Geochim. Cosmochim. Acta* **1990**, *54*, 545–552. [CrossRef]

151. Rosasco, G.; Roedder, E.; Simmons, J. Laser-excited Raman spectroscopy for non-destructive analysis of individual phases in fluid inclusions in minerals. *Science* **1975**, *190*, 557–560. [CrossRef]

152. Delhaye, M.; Dhamelincourt, P. Raman microprobe and microscope with Laser excitation. *J. Raman Spectrosc.* **1975**, *3*, 33–43. [CrossRef]

153. Dhamelincourt, P.; Schubnel, H. La microsonde moléculaire à laser et son application à la minéralogie et la gemmologie. *Rev. Gemmol. AFG* **1977**, *52*, 11–14.

154. Burruss, R.C. Raman spectroscopy of fluid inclusions. In *Fluid Inclusions: Analysis and Interpretation*; Samson, I., Anderson, A., Marshall, D., Eds.; Mineralogical Association of Canada: Quebec City, QC, Canada, 2003; Volume 32, pp. 279–290.

155. Burke, E.A.J. Raman microspectrometry of fluid inclusions. *Lithos* **2001**, *55*, 139–158. [CrossRef]

156. Taylor, R.; Fallick, A.; Breaks, F. Volatile evolution in Archean rare-element granitic pegmatites: Evidence from the hydrogen-isotopic composition of channel H_2O in beryl. *Can. Mineral.* **1992**, *30*, 877–893.

157. Marshall, D.; Meisser, N.; Ellis, S.; Jones, P.; Bussy, F.; Mumenthaler, T. Formational conditions for the Binntal emerald occurrence, Valais, Switzerland: Fluid inclusion, chemical composition and stable isotope studies. *Can. Mineral.* **2017**, *55*, 725–741. [CrossRef]

158. Groat, L.A.; Giuliani, G.; Marshall, D.D.; Turner, D. Emerald deposits and occurrences: A review. *Ore Geol. Rev.* **2008**, *34*, 87–112. [CrossRef]

159. Sheppard, S.M.F. Characterization and isotopic variations in natural waters. In *Stable Isotopes in High Temperature Geological Processes*; Valley, J.W., Taylor, H.P., O'Neil, J.R., Eds.; Mineralogical Association of America: Chantilly, VA, USA, 1986; Volume 16, pp. 165–183.

160. Giuliani, G.; Dubessy, J.; Ohnenstetter, D.; Banks, D.; Branquet, Y.; Feneyrol, J.; Fallick, A.E.; Martelat, E. The role of evaporites in the formation of gems during metamorphism of carbonate platforms: A review. *Miner. Deposita* **2017**, *53*, 1–20. [CrossRef]

161. Yardley, B.W.D.; Bodnar, R.J. Fluids in the Continental Crust. *Geochem. Perspect.* **2014**, *3*, 1–127. [CrossRef]

162. Yardley, B.W.D.; Cleverley, J.S. *The Role of Metamorphic Fluids in the Formation of Ore Deposits*; Geological Society: London, UK, 2013; pp. 1–393.

163. Williams, A.E.; McKibben, M.A. A brine interface in the Salton Sea Geothermal System, California: Fluid geochemical and isotopic characteristics. *Geochim. Cosmochim. Acta* **1989**, *53*, 1905–1920. [CrossRef]

164. Bohlke, J.K.; Irwin, J.J. Laser microprobe analyses of Cl, Br, I, and K in fluid inclusions: Implications for sources of salinity in some ancient hydrothermal fluids. *Geochim. Cosmochim. Acta* **1992**, *56*, 203–225. [CrossRef]

165. Poujol, M.; Robb, L.J.; Respaut, J.P. Origin of gold and emerald mineralization in the Murchinson greenstone belt, Kaapval craton, South Africa. *Mineral. Mag.* **1998**, *62A*, 1206–1207. [CrossRef]

166. Beus, A.A. *Geochemistry of Beryllium and Genetic Types of Beryllium Deposits*; W.H. Freeman: San Francisco, CA, USA; London, UK, 1966; 401p.

167. Smirnov, V.L. Deposits of beryllium. In *Ore deposits of the U.S.S.R.*; Pitman Publications: London, UK, 1977; Volume 3, pp. 320–371.

168. Franz, G.; Gilg, H.A.; Grundmann, G.; Morteani, G. Metasomatism at a granitic pegmatite-dunite contact in Galicia: The Franqueira occurrence of chrysoberyl (alexandrite), emerald, and phenakite: Discussion. *Can. Mineral.* **1996**, *34*, 1329–1331.

169. Zwaan, J.C.; Touret, J. Emeralds in greenstone belts: The case of Sandawana, Zimbabwe. *Münchener Geologische Hefte* **2000**, *28*, 245–258.

170. Franz, G.; Morteani, G. Be-minerals: Synthesis, stability, and occurrence in metamorphic rocks. In *Beryllium: Mineralogy, Petrology, and Geochemistry*; Grew, E.S., Ed.; Mineralogical Society of America: Washington, DC, USA, 2002; Volume 50, pp. 551–589.

171. Martin-Izard, A.; Paniagua, A.; Moreiras, D. Metasomatism at a granitic pegmatite-dunite contact in Galicia: The Franqueira occurrence of chrysoberyl (alexandrite), emerald and phenakite. *Can. Mineral.* **1995**, *33*, 775–792.

172. Martin-Izard, A.; Paniagua, A.; Moreiras, D.; Acevedo, R.D.; Marcos-Pascual, C. Metasomatism at a granitic pegmatite-dunite contact in Galicia: The Franqueira occurrence of chrysoberyl (alexandrite), emerald, and phenakite: Reply. *Can. Mineral.* **1996**, *34*, 1332–1336.

173. Korzhinskii, D.S. *Theory of Metasomatic Zoning*; Clarendon Press: Oxford, UK, 1970.

174. Sahu, S.S.; Singh, S.; Stapathy, J.S. Lithological and structural controls on the genesis of emerald occurrences and their exploration implications in and around Gubarabanda area, Singhbhum crustal province Eastern India. *J. Geol. Soc. India* **2018**, *92*, 291–297. [CrossRef]

175. Zwaan, J.C.; Kanis, J.; Petsch, J. Update on emeralds from the Sandawana mines, Zimbabwe. *Gems Gemol.* **1997**, *33*, 80–101. [CrossRef]

176. Holmes, A. The oldest dated minerals of the Rhodesian shield. *Nature* **1954**, *173*, 612–614. [CrossRef]

177. Christensen, J.N.; Selverstone, J.; Rosenfeld, J.L.; DePaolo, D.J. Correlation by Rb/Sr geochronology of garnet growth histories from different structural levels within the Tauern Window, Eastern Alps. *Contrib. Mineral. Petr.* **1994**, *118*, 1–12. [CrossRef]

178. Vance, D.; O'nions, R.K. Isotopic chronometry of zone garnets growth kinetics and metamorphic histories. *Earth Planet. Sci. Lett.* **1990**, *114*, 113–129. [CrossRef]

179. Gilg, A.; Technische Universität München, Germany. Personal comunication, 2009.

180. Lake, D.J. Are there Colombian-type emeralds in Canada's Northern Cordillera? Insights from regional silt geochemistry, and the genesis of emerald at Lened, NWT. M.Sc. Thesis, University of British Columbia, Vancouver, BC, Canada, 2017.

181. Escobar, R. Geology and geochemical expression of the Gachalá emerald district, Colombia. *Geol. Soc. Am. Abstr. Program* **1978**, *10*, 397.

182. Beus, A.A. Sodium—A geochemical indicator of emerald mineralization in the Cordillera Oriental, Colombia. *J. Geochem. Explor.* **1979**, *11*, 195–208. [CrossRef]

183. Ringsrud, R. The Coscuez mine: A major source of Colombian emeralds. *Gems Gemol.* **1986**, *22*, 67–79. [CrossRef]

184. Cheilletz, A.; Giuliani, G. The genesis of Colombian emeralds: A restatement. *Miner. Deposita* **1996**, *31*, 359–364. [CrossRef]

185. Rohtert, W.R.; Independant geologist. Personal comunication, 2017.

186. Gutiérrez, L.H.O. Evaluación magnetométrica, radiométrica y geoeléctrica de depósitos esmeraldíferos. *Geofísica Colombiana* **2003,** *7*, 13–18.

187. Arif, M.; Henry, D.J.; Moon, C.J. Cr-bearing tourmaline associated with emerald deposits from Swat, NW Pakistan: Genesis and its exploration significance. *Am. Mineral.* **2010**, *95*, 799–809. [CrossRef]

188. Wise, M.A. Crystal Morphology of quartz, calcite, pyrite and rutile; Potential tool for emerald exploration in the Hiddenite area (North Carolina). *Geol. Soc. Am. Abstr. Programs* **2012**, *44*, 25.

189. Murphy, D.C.; Lipovsky, P.S.; Stuart, A.; Fonseca, A.; Piercey, S.J.; Groat, L. What about those emeralds, eh? Geological setting of emeralds at Regal Ridge (SE Yukon) provides clues to their origin and to other places to explore. *Geol. Assoc. Canada Mineral. Assoc. Canada Abstr. Program* **2002**, *28*, 154.

190. Lewis, L.L.; Hart, C.J.R.; Murphy, D.C. Roll out the beryl. In *Yukon Geoscience Forum*; Publications du Gouvernement du Canada: Yukon Territory, NT, Canada, 2003.

10

Gem-Quality Zircon Megacrysts from Placer Deposits in the Central Highlands, Vietnam—Potential Source and Links to Cenozoic Alkali Basalts

Vuong Bui Thi Sinh [1,*], **Yasuhito Osanai** [2], **Christoph Lenz** [3,4], **Nobuhiko Nakano** [2], **Tatsuro Adachi** [2], **Elena Belousova** [4] **and Ippei Kitano** [2]

[1] Graduate School of Integrated Sciences for Global Society, Kyushu University, 744 Motooka, Nishi-ku, Fukuoka 819-0395, Japan

[2] Division of Earth Sciences, Faculty of Social and Cultural Studies, Kyushu University, 744 Motooka, Nishi-ku, Fukuoka 819-0395, Japan; osanai@scs.kyushu-u.ac.jp (Y.O.); n-nakano@scs.kyushu-u.ac.jp (N.N.); t-adachi@scs.kyushu-u.ac.jp (T.A.); i.kitano@scs.kyushu-u.ac.jp (I.K.)

[3] Institut für Mineralogie und Kristallographie, Universität Wien, 1090 Wien, Austria; christoph.lenz@univie.ac.at

[4] Australian Research Council (ARC) Centre of Excellence for Core to Crust Fluid Systems (CCFS) and National Key Centre for Geochemical Evolution and Metallogeny of Continents (GEMOC), Department of Earth and Planetary Sciences, Macquarie University, Sydney, NSW 2109, Australia; elena.belousova@mq.edu.au

* Correspondence: 3GS17013Y@s.kyushu-u.ac.jp

Abstract: Gem-quality zircon megacrysts occur in placer deposits in the Central Highlands, Vietnam, and have euhedral to anhedral crystal shapes with dimensions of ~3 cm in length. These zircons have primary inclusions of calcite, olivine, and corundum. Secondary quartz, baddeleyite, hematite, and CO_2 fluid inclusions were found in close proximity to cracks and tubular channels. LA-ICP-MS U-Pb ages of analyzed zircon samples yielded two age populations of ca. 1.0 Ma and ca. 6.5 Ma, that were consistent with the ages of alkali basalt eruptions in the Central Highlands at Buon Ma Thuot (5.80–1.67 Ma), Pleiku (4.30–0.80 Ma), and Xuan Loc (0.83–0.44 Ma). The zircon geochemical signatures and primary inclusions suggested a genesis from carbonatite-dominant melts as a result of partial melting of a metasomatized lithospheric mantle source, but not from the host alkali basalt. Chondrite-normalized rare earth element patterns showed a pronounced positive Ce, but negligible Eu anomalies. Detailed hyperspectral Dy^{3+} photoluminescence images of zircon megacrysts revealed resorption and re-growth processes.

Keywords: zircon megacrysts; placer deposits; rare earth elements (REE); carbonatite-dominant melts; Central Highlands; Vietnam; hyperspectral photoluminescence imaging; LA-ICP-MS

1. Introduction

Zircon ($ZrSiO_4$, tetragonal, $I4_1/amd$) is an accessory mineral in most types of igneous and metamorphic rocks [1]. Zircon megacrysts are often found in placer deposits derived from intraplate basaltic fields as xenocrysts or xenolith debris in alkali basaltic rocks [2]. Alkali basalt fields in South-East Asia commonly contain mantle xenoliths that include garnet lherzolites, spinel lherzolites, and harzburgite, as well as mantle- and/or crust-derived megacrysts of pyroxene, olivine, plagioclase, garnet, zircon, and corundum [3]. Therefore, zircon placers that are related to basaltic magmatism are often associated with megacrysts of other important gem materials (e.g., sapphire, garnet, and spinel), as reported from various localities, including Australia, Vietnam, Cambodia, Thailand, China, Myanmar, Sri Lanka, and Tanzania [4–11]. Zircon material from the Central Highlands, Vietnam, shows brownish-red colors and blue, upon heat-treatment under reducing conditions [12]. This color effect is similar to zircon found in placers in the Ratanakiri district, Cambodia [13]. The Vietnamese zircon deposits are known since the late 1980s, and sporadic mining activities are conducted by local people at numerous small sites. These deposits are exploited by digging to a depth of approximately 1–2 m, then picking the gems by hand after washing the alluvial material. The exploration of placer deposits of gem-quality zircons and other gem megacrysts is of high economic importance in many South-East Asian countries that supply the global gem market [14]. The source characteristics and forming processes of placer zircon play an important role for further exploration of this kind of gem deposits in geologically related regions.

The genesis of zircon megacrysts associated with alkali basalt has been addressed previously, but it is still vigorously debated [2,15–18]. Some authors agree that large zircon and corundum crystals represent xenocrysts that are transported by alkaline-basalt magmas and do not crystallize primarily from them (e.g., Hinton and Upton [2]), while others consider crystallization directly from basaltic magma [19,20]. Various hypotheses have been reported for the origin of zircon megacrysts from various sites in different geological contexts: (1) crystallization from melts derived from a metasomatized upper mantle, as supported by O-isotope studies [15,17,18,21]; (2) formation within a late-stage fractional crystallization process of oceanic island basalt magma [16,19]; and (3) crystallization from a primitive alkaline mafic magma, which later evolved to a less alkaline host magma [20]. According to Cong et al. [10], zircon megacrysts from Cenozoic basalts in northeastern Cambodia crystallized in the mantle during metasomatic events caused by phosphate-rich fluids and/or silicate melts enriched in zirconium. Piilonen et al. [22] recently reported findings of megacrystic zircons as single crystals enclosed in the alkali basalt from Ratanakiri, Cambodia, and suggested a single stage growth in a carbonate-influenced environment.

In the Central Highlands of Vietnam, gem quality zircon is commonly accompanied by corundum (sapphire) in alluvial deposits that are considered eroded from Cenozoic alkali basalt fields in close proximity (e.g., Garnier et al. [17]). In this study, zircon megacrysts from various alluvial deposits in the Central Highlands were systematically studied to investigate their origin and potential links to Cenozoic basalt eruptions. In addition to trace element data and U-Pb geochronological data, photoluminescence (PL) imaging methods were applied to reveal internal growth and potential secondary alteration textures. In this paper, we used mineral abbreviations as given by Whitney and Evans [23].

2. Geological Setting

South-East Asia was formed by an amalgamation of several crustal blocks, including South China, Indochina, Siumasu, Inthanon, the West Burma block, and the Trans Vietnam Orogenic Belt (TVOB) (Figure 1A) [24]. The TVOB was first proposed by Osanai et al. [24] as a zone of Permo-Triassic metamorphic rocks in Vietnam, which were formed by continent-continent collision between South China and Indochina blocks. This orogenic belt is characterized by numerous shear zones with strong deformation, such as the Red River shear zone, Song Ma suture zone, Tam Ky-Phuoc Son shear zone, and the Dak To Kan shear zone (Kontum Massif). The northern extension of these shear zones reaches the Yuan Nan Province through Ailaoshan in China. The Central Highlands lies entirely within the Indochina block and are located in the southern part of the TVOB, partly extending into eastern Vietnam and Laos (e.g., Hutchison [25]) (Figure 1A).

Miocene-Pliocene alkaline and subalkaline basalts are exposed over a vast region of Thailand, Laos, and Vietnam [26]. In southern and central Vietnam, this province stretches across an area of approximately 23,000 km^2 with a thickness up to several hundred meters [3] (Figure 1A). The basalt plateau is accompanied by pull-apart structures composed of short extensional rifts bounded by strike-slip faults [27]. At least two eruptive episodes ("early" and "late") have been reported by Hoang et al. [28]. Tholeiitic, and rarely, alkali basaltic flows represent the "early" episode at Dalat (17.60–7.90 Ma), whereas in the "late" episodes, olivine tholeiite, alkali basalt, basanite, and (rarely) nephelinite erupted at Phuoc Long (<8.00–3.40 Ma), Buon Ma Thuot (5.80–1.67 Ma), Pleiku (4.30–0.80 Ma), Xuan Loc (0.83–0.44 Ma), and the Re Island centers (0.80–0 Ma) [27] (Figure 1B). Hoang et al. [28] proposed that tholeiitic basalts are the most common basalt type in the region and build up much more volumetric mass in comparison to the alkali basalts and the rarely occurring nephelinites which erupted from small volcanoes.

Erosion of these Cenozoic (Neogene-Quaternary) basalts formed placer deposits that represent a major source of gem-quality corundum, zircon, olivine, garnet, pyroxene, and plagioclase [4,14,26]. In South-Central Vietnam, gem-quality zircons have been found in alluvial deposits from six provinces, including Kontum, Gia Lai, Dak Lak, Dak Nong, Lam Dong, and Binh Thuan [12] (Figure 1A,B).

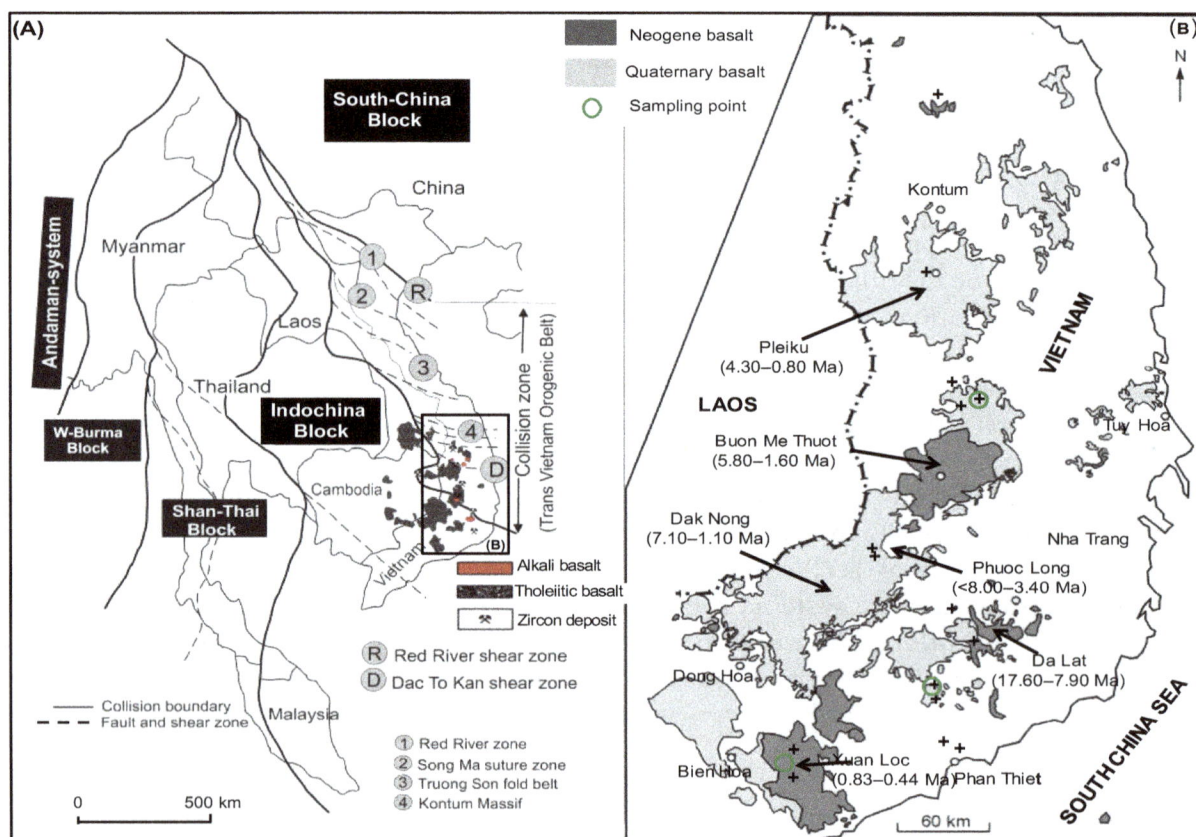

Figure 1. (A) Location of various alluvial zircon deposits in Cambodia and Vietnam within South-East Asia [12,24]. (B) Distribution of Neogene and Quaternary basalts with their K-Ar and Ar-Ar ages (after [4,6]). Reproduced with permission from all authors. Green circles indicate the sampling localities.

3. Sample Description and Methods

In this study, representative samples of zircon megacrysts collected from several alluvial deposits in the Central Highlands, Vietnam, were investigated in more detail. A representative zircon sample (Rata) from the Ratanakiri district, Cambodia was included for comparison. Large, gem-quality zircon specimens have dimensions up to several centimeters (Figure 2). Their color varies from colorless, orange, brownish-orange to dark brown and dark red (Figure 2). Some crystals have anhedral to subhedral shapes with rounded termination, while most grains are euhedral crystals with a typical combination of bipyramid and tetragonal prism (Figure 2B). Occasionally, internal colored zones that comprise an oscillatory change of orange and brown bands are visible to the naked eye (Figure 2B, sample C16).

Selected samples were cut along their long axis for petrographic analyses. Inclusions in zircon samples were observed and identified using an optical microscope attached to a confocal laser Raman spectrometer system (JASCO NSR-3100, JASCO (USA)) at Kyushu University. A 100× objective (with a numerical aperture NA = 0.90) and a green, continuous 532 nm frequency-doubled YAG:Nd laser (with an energy output of approx. 10 mW at the sample surface) was used to perform the Raman spot analyses. Note that potential bias of Raman spectra from the photoluminescence (PL) of Er^{3+} in zircon has been reported when using this laser wavelength [29]. A very minor PL contribution of Er^{3+} was identified in most recorded Raman spectra, but did not hamper the identification of typical Raman bands as obtained from various inclusion phases. Inclusions that were accessible via the polished sample surface were further identified qualitatively using a JEOL JSM-5310S-JED2140 scanning electron microprobe (SEM, JEOL, Ltd., Tokyo, Japan) equipped with an energy-dispersive X-Ray (EDX) detector (Li- doped Si semiconductor) at Kyushu University.

The internal texture of the prepared zircon crystals was investigated using a back-scattered electron (BSE) detector system and a GATAN MiniCL panchromatic cathodoluminescence (CL) detector attached to the above-mentioned SEM system at Kyushu University. In addition, we applied laser-induced photoluminescence spectroscopy using a HORIBA LabRam HR800 Evolution spectrometer equipped with a Peltier-cooled Si detector and a grating of 600 lines per millimeter (Horiba, Ltd., Kyoto, Japan). Confocal PL spot measurements were performed using an Olympus BX80 microscope (manufactured by Olympus Corporation, Tokyo, Japan) with a solid-state continuous laser, operating at 473 nm, to excite the most prominent REE^{3+} photoluminescence emissions in zircon (e.g., Dy^{3+}; Lenz et al. [29]). Hyperspectral images of polished, large zircon single-crystals were produced by measuring a multitude of spots in a point-by-point raster with a step-width of 10–20 μm using a software-controlled Maerzhauser mechanical x-y table. A software-based, automated data treatment procedure for each single spectrum was applied, and the plotted spectral parameters of interest were color-coded. With respect to the large size of investigated zircon crystals, and the comparatively large mapping step-widths, we used a 50× objective (NA = 0.5) and a confocal hole of 200 μm to adjust the lateral spatial resolution to be approximately 8–10 μm. However, we note that a diffraction-limited maximum spatial resolution of ~1 μm (planar) and 2–3 μm in depth may be achieved using the combination of a small confocal hole and a 100× objective.

Zircon U-Pb dating of samples C05, C07, and C16 was carried out by applying laser-ablation inductively coupled plasma mass-spectrometry (LA-ICP-MS) using an Agilent 7500cx quadrupole ICP-MS with a New Wave Research UP-213 YAG:Nd laser at the Kyushu University, that produced a laser-ablation spot of 100 μm. Details of the analytical procedure are presented by Adachi et al. [30]. The zircon standards Temora (417 Ma; [31]) and FC-1 (1099 Ma; [32]) were used for calibration and accuracy checks, respectively. The NIST SRM-611 glass standard was applied to determine the Th/U ratios. Raw data were treated using GLITTER software (Version 4.4.2, Glitter sold through Access Macquarie Ltd., Sydney, Australia) [33] and plotted in concordia diagrams using the Isoplot/Ex 3.7 software (Berkeley Geochronology Center, Berkeley, CA, USA) [34]. Trace element concentrations were obtained from similar zones in close proximity to spots used for U-Pb dating. Details of the analytical procedure are given in Nakano et al. [35] using the theoretical Si concentration in zircon as an internal standard.

Dating of zircon samples DL1, DL2, and DL3 was carried out using an Agilent 7700 quadrupole ICP-MS instrument attached to a Photon Machines Excimer 193 nm laser system at GEMOC, Macquarie University, using a beam diameter of ca. 50 μm with a 5 Hz repetition rate, and energy of around 0.06 J/cm^2 to 8 J/cm^2. Ablation was carried out in He gas to improve the sample transport efficiency, and to provide stable signals with reproducible Pb/U fractionation. Sample analyses were run along with analyses of the GEMOC GJ-1 zircon standard [36]. This standard is slightly discordant, and has a TIMS $^{207}Pb/^{206}Pb$ age of 608.5 Ma [37]. The other well-characterized zircon standard 91500 and Mud Tank were analysed within the run as an independent control on reproducibility and instrument stability. Individual time-resolved data analysis provides isotopically homogeneous segments of the signal to be selected for integration. We corrected the integrated ratios for ablation related fractionation and instrumental mass bias by the calibration of each selected time segment against the identical time segments for the standard zircon analyses. Furthermore, we employed the common-Pb correction procedure described by Andersen [38]. The analyses presented here were corrected assuming recent lead-loss with a common-lead composition corresponding to the present-day average orogenic lead, as given by the second-stage growth curve of Stacey and Kramers [39] for $^{238}U/^{204}Pb = 9.74$. No correction was applied to the analyses that were concordant within a 2σ analytical error in $^{206}Pb/^{238}U$ and $^{207}Pb/^{235}U$, or which had less than 0.2% common lead. Trace element concentrations of samples DL1, DL2, and DL3 were obtained using the same LA-ICP-MS system. Calibration of the relative element sensitivities were done using the NIST-610 standard glass as external calibration. Zircon BCR-2g and GJ-1 reference material were analyzed along the measurement runs as an independent control on reproducibility and instrument stability. Theoretical Zr content of the zircon was used for internal calibration of unknown zircon samples. The precision and accuracy of the NIST-610 analyses were 1–2% for REE, Y, Nb, Hf, Ta, Th, and U at the ppm concentration level, and at 5–10% for Ti (further details see in Belousova et al. [16]).

Figure 2. (A) Typical large, gem-quality, cut zircon samples from the Central Highlands are available on the gem market. **(B)** Selected zircon megacrysts used in this study from the same region. Samples C05, C07, and C16 are from the Buon Ma Thuot and Xuan Loc alkali-basaltic field. Samples DL1, DL2, and DL3 (not shown) are from deposits in the Dak Lak alkali-basaltic field (Figure 1B).

4. Results

4.1. Inclusion Features

Zircon crystals investigated in this study contained various fluid and mineral inclusions throughout the grains, especially along the cracks (Figure 3A–G). Tubular channels, bifurcating cavities, and vesicles in association with fluid inclusions were observed along the 2-dimensional planes (Figure 3A–G). Typical fluid inclusions along the fractures or fissures consisted of two phases, liquid (H_2O) with CO_2 or O_2 bubbles. Mineral inclusions were identified as calcite, hematite, corundum, olivine, baddeleyite, quartz, and feldspar. We identified two different paragenetic types of mineral inclusions: (1) the inclusions that were distributed throughout the grains without fluid inclusions in close proximity; and (2) others that were distributed along the cracks or were associated with tubular channels, bifurcating cavities, and vesicles with fluid inclusions nearby. Calcite inclusions were found in both areas: ovoid and droplet shaped ones sat inside the tubular channels or were associated with fluid inclusions (Figure 3B); whilst square or rectangular shaped calcite inclusions were accumulated along the chains or in groups that did not occur next to the fluid inclusions or within the tubular channels (Figure 3I). We rarely found corundum (Figure 3H) and olivine inclusions. They were of an euhedral and subhedral shape, and appeared in areas where no cracks, fluid inclusions, or tubular channels were present. Quartz, baddeleyite, hematite, and feldspar were found with irregular or ovoid shapes with rounded termination. They were located along the fissures, within tubular channels, cavities or associated with fluid inclusions (Figure 3C–G).

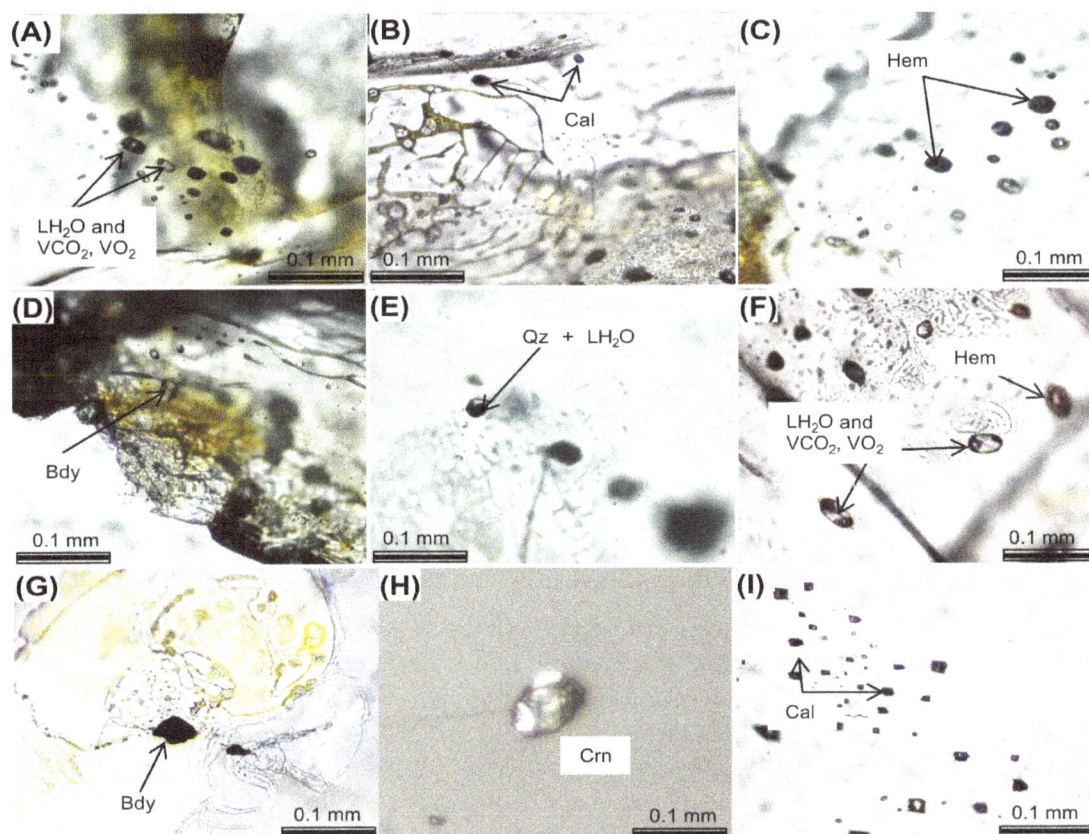

Figure 3. Representative inclusions found in placer zircon megacrysts: (**A**) fluid inclusions (LH_2O: aqueous H_2O VCO_2 and VO_2: vapor CO_2 and O_2, respectively); (**B**) Calcite (Cal); (**C**) hematite (Hem); (**D**) baddeleyite (Bdy) within tubular channels; (**E**) quartz (Qz) within tubular channels; (**F**) hematite and fluid (Hem + Fld) within bifurcating cavities and vesicles; (**G**) baddeleyite (Bdy) in tubular channels, bifurcating cavities, and vesicles; (**H**) corundum (Crn); and (**I**) calcite (cal) inclusion groups and along chains.

4.2. Internal Texture of Zircon Megacrysts

Backscattered-electron images obtained from the analysis of polished sections revealed no internal textural features although high contrast levels were applied (not shown). The latter indicated a homogeneous distribution and/or low concentration levels of trace elements. In contrast, cathodoluminescence images shown in Figure 4, revealed complex internal textures comprised of oscillatory zoning and sector zoning, although overall CL intensities were comparably low, except for the colorless zircon (Zrn-C05, see Figure 4). Anhedral grains were found to be split or broken into sections of former even larger grains (e.g., F03, F04), whereas the euhedral grains were characterized by multiple CL growth bands that retraced multiple distinct growth stages within a single crystal (e.g., C16 in Figure 4). To obtain minor differences in the CL zonation, high intensity contrast-levels were applied. This caused white stripes to be present in the CL images of some samples that did not correlate with the crystal's internal zonation (Sample C01, C07, and F01). They were most prominent along cracks and break-outs of fine material that produced mechanically-induced defects on the sample surface during polishing. Those structural defect centers mostly caused broad and intense CL signals in the UV spectral region, which were detected efficiently using panchromatic CL detectors.

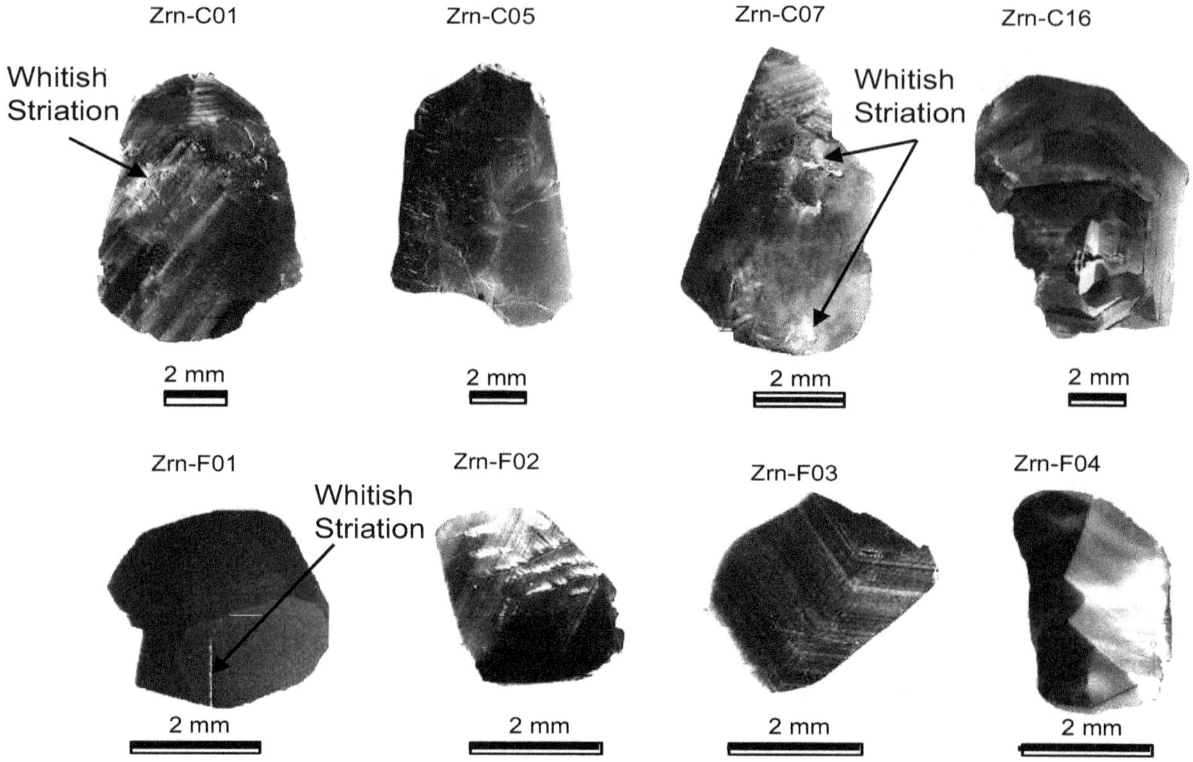

Figure 4. Cathodoluminescence images of the selected zircon samples from the Central Highlands, resemble zircon internal textural features that are typical for zircon of magmatic growth, e.g., oscillatory, sector, and growth zoning. Note, however, that whitish striations (indicated by black arrows) in CL contrasts obtained in the images of samples C01, C07, and F01 are due to mechanically-induced structural defect centers emanating from cracks and break-outs during polishing.

We applied laser-induced PL hyperspectral mapping to the large euhedral crystal grains with the aim of visualizing the internal luminescence distribution patterns as specifically caused by the emissions of REEs (Figure 5). The advantage of applying the latter technique is that the PL of specific REE species may be excited effectively using the appropriate laser sources in the visible spectral range without being obscured by intensive broad-band luminescence features as observed by CL spectroscopy and imaging [29]. Figure 5 shows hyperspectral images of the zircon samples from the Central Highlands (C16, DL2) and from Ratanakiri, Cambodia (Rata). The integrated intensity of the most prominent PL emission of Dy^{3+} ($^4F_{9/2} \to {}^6H_{13/2}$) was used as a plotted spectral parameter (see Figure 6A; and Figure 1 in Lenz et al. [29]). Note that the Dy sub-level bands of Vietnamese and Cambodian (Rata) zircon samples were characterized by exceptionally narrow band-widths (sublevel at 581 nm had a band-width of 11.0 cm^{-1}; see Figure 6A), which indicated a high degree of crystallinity and the absence of structural radiation-damage that might be caused by decay of radioactive U and Th. This was in accordance to findings by Lenz and Nasdala [40] and Zeug et al. [13] who reported very low PL and Raman band widths for zircon from Ratanakiri, Cambodia (Raman ν_3 [SiO$_4$] ~1.8 cm^{-1}). This has been interpreted to be the result of very low α-doses due to low U and Th concentrations in combination with a very young age (ca. 1 Ma).

In this study, we found a systematic linear correlation of the integrated PL intensity of Dy with its concentration as obtained from multiple LA-ICP-MS measurement spots (Figure 6B and Table S1). Furthermore, the latter laser-ablation reference spots of Dy concentrations and their correlation with PL intensity were used as an external calibration for the PL hyperspectral images to infer Dy concentrations from the PL signal in regions of unknown trace-element chemistry (see color-coded scales in Figure 5). Especially in the euhedral crystals (e.g., C16 and DL3), we identified several growth stages that were characterized by abrupt changes in concentrations of trace Dy (REE). Typically, the crystal's cores were found to be enriched in Dy, followed by a decreasing oscillatory REE substitutional budget (see inset of sample C16 in Figure 5). PL patterns further revealed episodical resorption of pre-existent crystals and fast re-growth. Growth zones that were characterized by low REE concentrations cross-cut former zonation patterns and occurred along with large holes and cavities (see arrows in Figure 5, samples C16 and DL3).

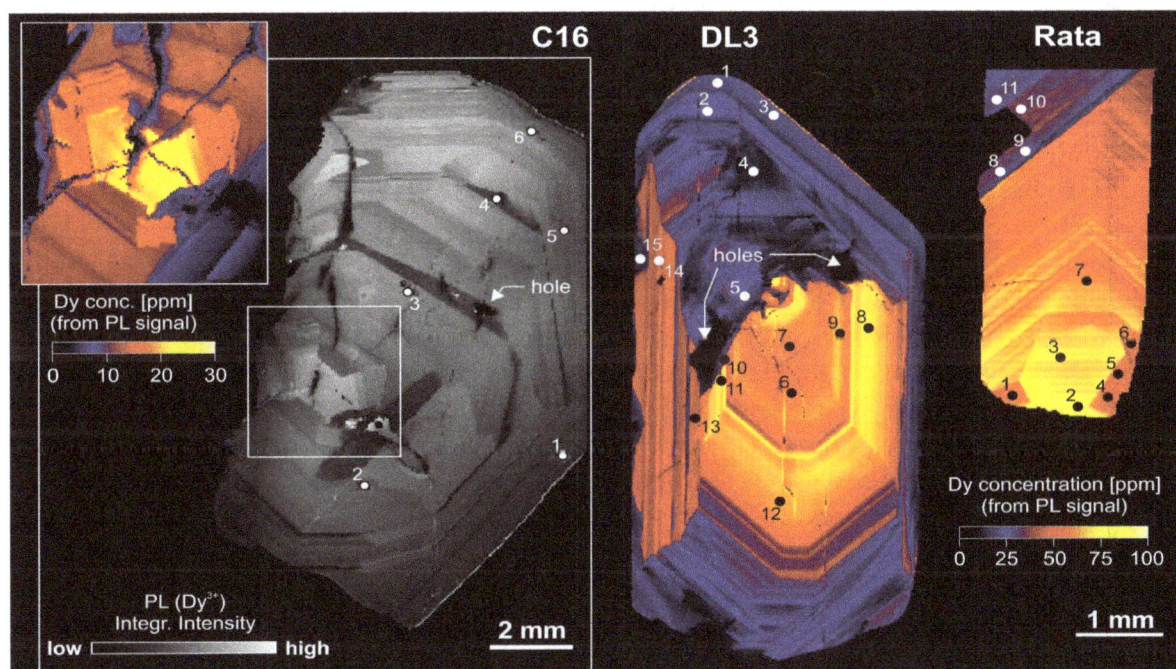

Figure 5. Laser-induced photoluminescence hyperspectral images of zircon samples from the Central Highlands (C16 and DL3) and from Ratanakiri, Cambodia (Rata). The integrated intensity of the Dy^{3+} ($^4F_{9/2} \to {}^6H_{13/2}$) emission (compare Figure 6) is the plotted spectral parameter (grey-scale). Trace-element concentrations of Dy from multiple LA-ICP-MS spots on the grains, were used for external calibration to correlate the Dy concentrations with its PL response. Concentrations given color-coded were back-calculated using the latter correlation (see Figure 6) to extrapolate the Dy distribution in regions of unknown trace-element chemistry. Besides oscillatory and sector zoning, several growth zones may be distinguished that were interpreted to result from resorption and re-growth events, as indicated by the cross-cut zonation patterns, abruptly changing (decreasing) Dy concentrations, and the appearance of large holes and cavities (see arrows).

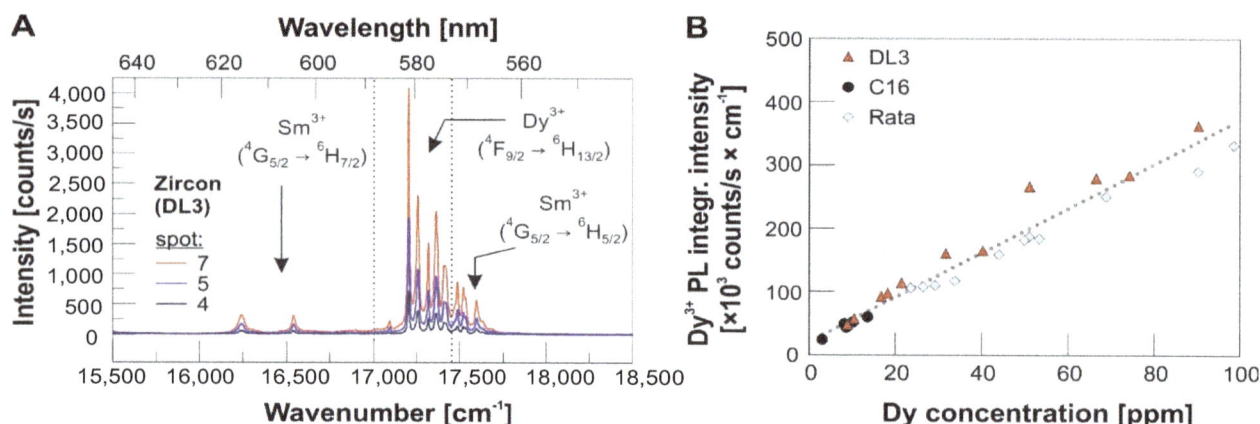

Figure 6. (**A**) Laser-induced PL spectra (laser-excitation wavelength λ_{exc} = 473 nm) obtained from three different measurement spots on the zircon sample DL3 (spots 4, 5, and 7; Figure 5). (**B**) The integrated intensity of the most intensive emission of Dy^{3+} in the spectral range 17,000–17,450 cm^{-1} (see dotted lines in sub-image (**A**)) from multiple spots on samples DL3, C16, and Rata plotted against the Dy trace-element concentration as obtained from LA-ICP-MS (Table S1). Linear regression from the data pairs PL vs. Dy conc. (R^2 = 0.95) was used to calibrate hyperspectral maps presented in Figure 5.

4.3. Geochemical Characteristics

Results from trace-element analyses of Vietnamese zircon samples, in addition to a Cambodian sample from Ratanakiri (Rata) are summarized in Table S1. Representative samples that comprised a colorless (C05), a dark-red (C07), an orange (C16), and three heavily zoned samples (DL1, DL2, DL3), were selected for detailed analyses. All zircon megacrysts had low total Y + REE concentrations typically within a 50–550 ppm range. Somewhat higher concentrations were detected at measurement spots placed in the core region of the heavily zoned samples DL2 and DL3 that were up to 2300 ppm Y + REE (see Table S1). The chondrite-normalized REE concentrations revealed a steep slope from La to Lu (Figure 7). Note that the REE geochemical characteristics of samples from the Central Highlands closely coincided with those of the samples from Ratanakiri (see samples Rata and data from Cong et al. [10]). All samples were characterized by a pronounced positive Ce-anomaly and a lack of Eu-anomaly, which is typical of magmatic zircon crystallized from more evolved magma sources with co-crystallization of feldspars that substitute Eu^{2+} on Ca-sites. The average Eu/Eu* ratio, that reflects the deviation of the measured Eu concentration from a theoretical Eu* concentration—extrapolated from chondrite-normalized Sm and Gd abundance (Eu* = Eu_N / $[Sm_N + Gd_N]^{0.5}$)—was found to be close to 1 (Figure 8A). The obtained REE compositional characteristics excluded crustal derived rocks or granitoids as a potential source of the zircon samples studied, and were more typical of the zircon of kimberlitic and/or carbonatitic origin (Figure 8A). Similar to REEs, the overall U and Th concentrations were found to be comparably low, with typical variability at 10–60 ppm and 3–30 ppm, respectively. Core regions of samples DL2 and 3 were found to be more enriched in U and Th with concentrations up to 500 ppm. Despite the pronounced variation in U and Th concentrations among the different growth zones within the samples, their U/Th ratios showed a clear correlation (Figure 8B). Likewise, the Y and U data pairs in a Y vs. U discrimination diagram scattered appreciably amongst the different growth zones within and among the samples, but coincided well with the compositional fields that were reported for basic, kimberlitic, and carbonatitic rocks (Figure 8C; Belousova et al. [16]). A source rock discrimination based on the Hf composition as proposed by Shnukov et al. [41], is presented in Figure 8D. The samples' Hf composition was found to be very low and shared compositional similarities with zircon from carbonatitic rocks, which had the lowest reported Hf concentrations (<0.7 wt %).

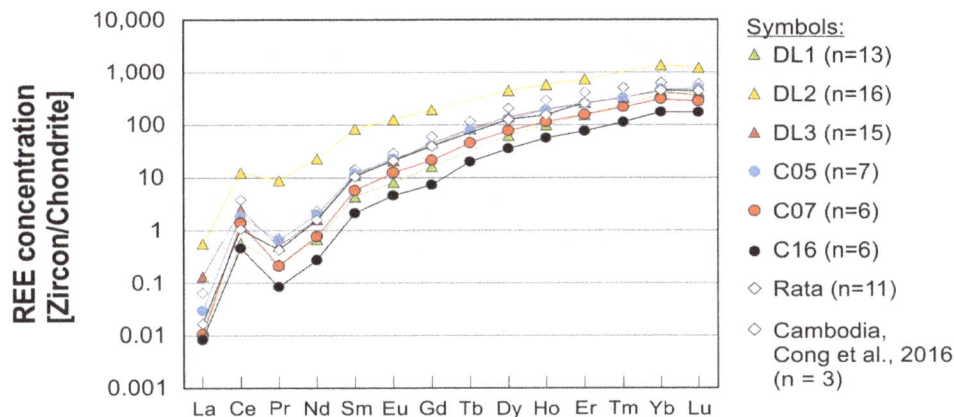

Figure 7. Chondrite-normalized plot of REE concentrations of zircon samples from the Central Highlands compared to samples from Ratanakiri, Cambodia. REE composition of C1 chondrite is described by Sun and McDonough [42]. All samples showed a steep slope rise from La to Lu, towards heavy REEs. A positive pronounced Ce-anomaly, but no Eu-anomaly was detected.

4.4. Geochronology

Results of the measured U-Th-Pb isotopic ratios and calculated ages of samples DL1, DL2, C05, C07, C16 are summarized in Tables S2 and S3 (see Supplementary Materials) and Figure 9. Vietnamese zircon samples from the Central Highlands were found to have Cenozoic, fairly concordant (~50% of discordance in average), and U-Pb ages (Figure 9A). Detailed comparison of the more robust $^{206}Pb/^{238}U$ ages indicated two distinct events of zircon megacrysts formation. Samples DL1, DL2, and C16 had a U-Pb age of around ca. 6.5 Ma, whereas samples C05 and C07 were even younger with an age of ca. 1 Ma. Note that the LA-ICP-MS measurement spots of individual single crystals were from different growth zones among the entire grains, which were clearly discernable in CL imaging, as well as PL hyperspectral mapping, and they had significant differences in trace-element concentrations (results above). We, however, found no systematic difference in U-Pb ages between the growth zones of individual samples within the standard errors.

5. Discussion

Zircon megacrysts from the Central Highlands that were investigated in this study had various colors, ranging from colorless, orange, brownish-orange, and dark brown to dark red. Some were characterized by colored zonation visible to the naked eye. Most of the selected grains showed a euhedral crystal shape with typical combinations of bipyramid and tetragonal prisms. Other grains had a subhedral to anhedral shape with rounded termination (Figure 2B). The latter might result from magmatic resorption on crystal faces and/or erosion during weathering and transportation into the placer. The internal texture of single crystals revealed by CL imaging and PL hyperspectral mapping was characterized by wide to narrow oscillatory and/or sector zoning (Figures 4 and 5). Typical crystal shape and textural features were conclusive hints that the megacrysts had a magmatic origin [1]. In heavily zoned samples (DL1−3, C16), dark-brownish colored zones correlated with the elevated concentrations of REE, U, and Th. On the other hand, a colorless zircon sample (C05, Figure 2B) was found to be more enriched in REEs, Th, and U, than the non-transparent dark-red sample C07 (Figure 2B, Figure 7 and Table S2). Hence, any rigorous interpretation based on color information should be avoided. For example, we found no corroboration for the interpretation that dark-brownish coloration of zircon was related to the presence of structural radiation damage. Multiple zircon samples obtained from placer deposits in Vietnam, including samples from Ratanakiri, Cambodia, had various colors ranging from transparent to non-transparent dark-brown, but all were characterized by very low Raman and PL spectral band-widths that indicated no radiation damage to be present [13].

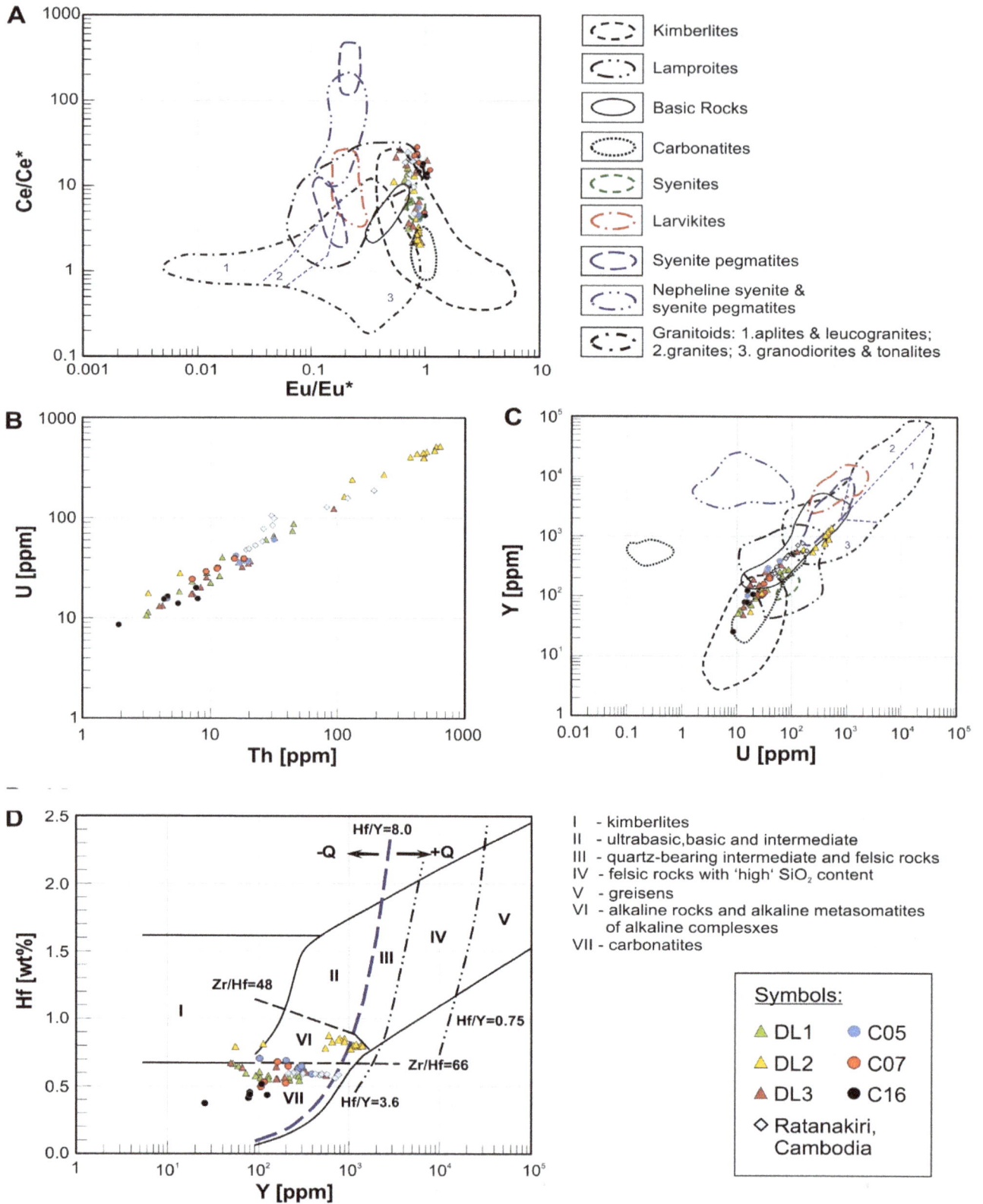

Figure 8. Compositional discrimination diagrams using trace-element concentrations of the studied samples obtained by LA-ICP-MS [16]: Ce/Ce* vs. Eu/Eu* (**A**); U vs. Th (**B**); Y vs. U (**C**); and Hf vs. Y (**D**) [41].

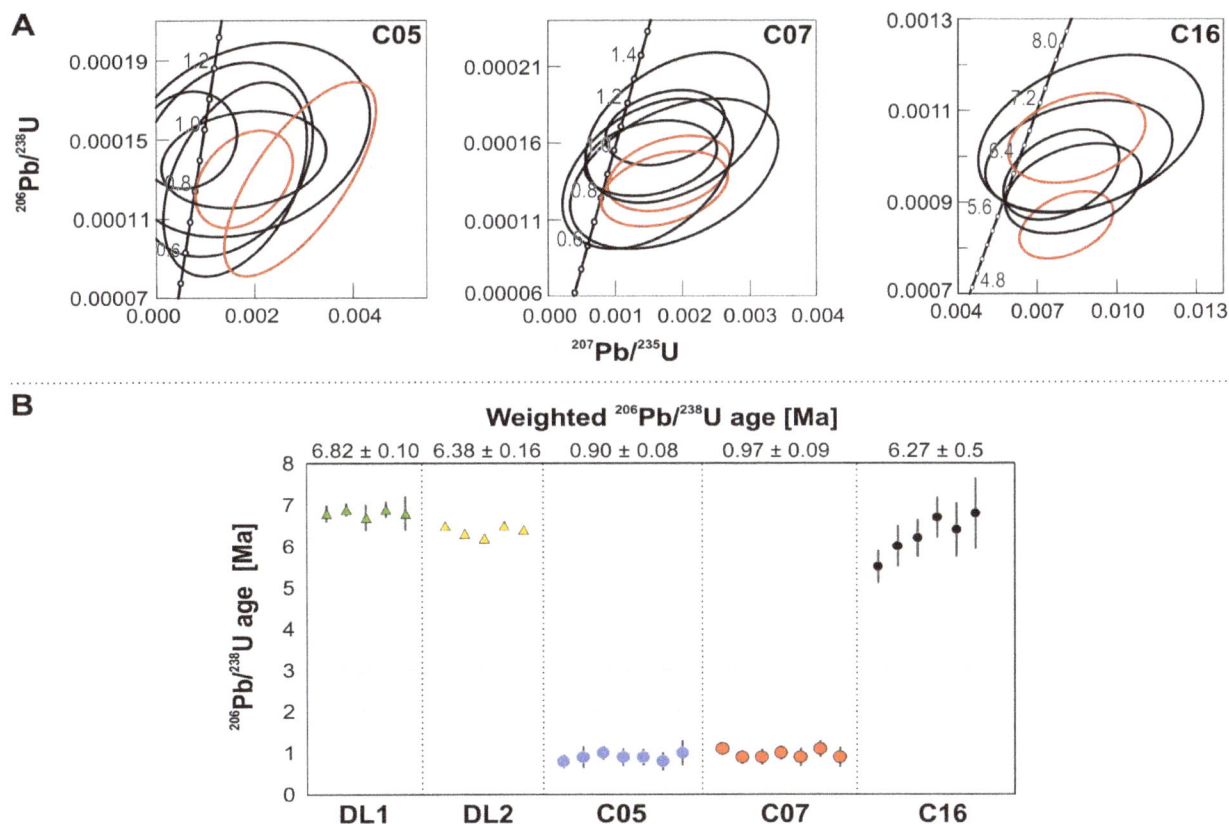

Figure 9. Summary of U-Pb geochronology data obtained by LA-ICP-MS (see details in Tables S2 and S3). (**A**) Representative concordia plots of samples C05, C07, and C16. Ellipses give a 2σ error, and red ellipses correspond to measurement spots placed in the core region of single crystals. (**B**) $^{206}Pb/^{238}U$ ages of the samples analyzed with the mean weighted age given for each sample. Error bars represent 2σ. Note that the ages obtained do not differ systematically amongst the measurement spots placed in the different growth zones within the individual samples. Two distinct sample populations were clearly differentiated (ca. 6.5 Ma and ca. 1 Ma).

We used PL hyperspectral imaging of the Dy^{3+} emission as an effective tool to visualize the distribution of REEs across the zircon samples (Figure 5). The latter technique was more reliable with respect to specifically exciting and detecting the emissions of REEs, because the panchromatic CL imaging is often strongly affected by other luminescence emissions that are caused by defects that are not coupled to the substitution of REEs (see mechanically induced bright CL striations caused by polishing; Figure 4). We further found that PL integrated intensities of Dy^{3+} used for PL hyperspectral imaging, correlated well with the Dy concentrations determined using spot LA-ICP-MS analyses (Figure 6). We, therefore, used PL hyperspectral images to quantitatively visualize Dy concentrations in regions of unknown chemical compositions as extrapolated from Dy^{3+} integrated emission intensities. Note, however, that the correlation of Dy concentrations with their PL response may be strongly hampered in other zircon samples that show a much higher accumulation of structural radiation damage, e.g., Lenz and Nasdala [40] reported that the presence of radiation damage resulted in substantial quenching of the PL intensities.

In this study, we identified multiple growth stages that included the resorption and re-growth of large volumes of the zircon megacrysts (Figure 5, samples C16 and DL3). Outer growth zones irregularly cut the former zones repeatedly and were characterized by lower REE concentrations, which indicated that the growth condition of megacrysts may have changed spontaneously to a more REE undersaturated growth environment that resulted in re-growth of the zones with depleted REE concentrations. Although absolute REE concentrations changed considerably across the different growth zones, their general chondrite-normalized concentration patterns were qualitatively similar.

Chondrite-normalized REE patterns in the core and overgrowth regions had a steep slope from light REE (LREE) to heavy REE (HREE), with pronounced positive Ce-, but with a lack of an Eu-anomaly (Figure 7). The positive Ce anomaly is typically caused by the significant difference between Ce^{4+} and $LREE^{3+}$ to fit into the zircon's structure, as Ce^{4+} has the same charge and similar ionic radius compared to Zr^{4+}. The Eu anomaly of zircon in REE patterns is generally explained by Eu^{2+} fractionation in the plagioclase and/or alkali feldspar that crystallizes before or during zircon formation from the magma [1,43,44]. Therefore, a pronounced negative Eu anomaly is commonly found in zircons from crustal-derived felsic rocks [16,44], whereas no Eu anomaly is found in zircon from feldspar free rocks, such as little to non-fractionated or mantle-derived rocks [45]. Note also that zircons of mantle origin usually have low REE and Y concentrations (zircon in kimberlite from southern Africa with TREE = 5–39 ppm and Y = 11–74 ppm [46]; zircon from Jwaneng kimberlite TREE up to 12 ppm and Y up to 23 ppm [47]), whereas zircons in crustal rocks are more enriched in REEs ranging from 250 ppm to 5000 ppm, with a 1500–2000 ppm average [1,16,48]. Trace-element concentrations of zircon crystals investigated in this study (TREE = 25–309 ppm, Y = 26–392 ppm) shared chemical characteristics comparable to mantle-derived zircon, and had no Eu anomaly (Figure 7). Moreover, the provenance discrimination plots proposed by Belousova et al. [16] and Shnukov et al. [41] revealed that the trace-element composition of Vietnamese zircon megacrysts were consistent with zircon typically found in kimberlite, syenite, carbonatites (Figure 8A–C), or alkaline rocks and alkaline metasomatites of alkaline complexes (Figure 8D). As the observed trace-element chemical characteristics ambiguously meet with those of several discrimination fields, it is hard to define an exact source composition. In fact, zircon samples from carbonatites often fall into various discrimination fields [49,50]. For example, Saava et al. [49] reported that only 45 out of 100 analyzed zircon grains extracted from a carbonatite body meet the trace-element signatures that indicate a carbonatitic source, while all others suggest an alkaline or ultramafic source. This compositional variation has been interpreted to result from the interaction between carbonatitic melt and co-magmatic silicate sources that broaden the chemical variation of carbonatitic zircon. In this study, total REE contents of zircon typically varied in the range of 25 to 309 ppm. These concentrations were slightly higher than those found in the kimberlitic zircon that had typically less than 50 ppm [46,47], but had a much lower concentration than the one from zircon of corundum bearing syenites (up to 3500 ppm [2]). Low concentrations of U and Th (Figure 8B), and low Nb and Ta concentrations are also typical of the zircon that originates from very little, to non-fractionated, Si-poor melts [16]. Trace-element chemical signatures of the zircon megacrysts of south-central Vietnam were found to be very comparable to those of zircon from Ratanakiri, Northeast Cambodia (Figure 7 in Cong et al. [10]; Figure 7 in this study) and other places such as New Zealand [11], East Australia [18], eastern and northeast China [5,8]. Results of a detailed study by Cong et al. [10] and Piilonen et al. [22] using $\delta^{18}O$ and $^{176}Hf/^{177}Hf$ isotopes clearly demonstrated that zircon from Ratanakiri, Cambodia, had a mantle origin and may be derived from metasomatized, partially melted lithospheric mantle material (such as peridotite, harzburgite), with strong carbonatitic geochemical fingerprint. In fact, experimental studies by Foley et al. [51], have demonstrated that mantle metasomatism, as induced by the presence of CO_2 and H_2O, lowers the solidus temperature at upper mantle conditions and promotes partial melting of the peridotites that results in melts with carbonatitic compositions at low degree, and carbonated silicate melts with higher degrees of melting. Early carbonate-dominant melts were characterized by a low Hf compatibility, that is consistent with low Hf concentrations obtained from zircon in this study (Figure 8D) [5,8,10,11]. With increasing degrees of partial melting, Hf is more compatible in the carbonated silicate melt [51].

Owing to the close geographical, geochemical, and geochronological relation of Cambodian zircon from Ratanakiri with material from the Central Highlands in Vietnam, studied here, we considered a very similar genetic origin for the latter. This was further supported by the presence of primary inclusions. Euhedral inclusions of calcite and corundum, and subhedral olivine are present in the zircon host, that is free of fluid inclusions, tubular channels, which shows no indication of visible cracks or alteration (Figure 3). These findings were in accordance to those of Le Bas [52], who interpreted the presence of abundant primary carbonate and olivine inclusions in the zircon megacrysts to be indicative of the growth in a silica-undersaturated, carbonatite-like melt. The experimental work of Baldwin [53] demonstrates that corundum and olivine likely crystallize from carbonatitic melts during early phases. However, no carbonatite bodies were reported from the studied area in the Central Highlands, Vietnam. In fact, the only known occurrence of carbonatites in Vietnam has been found in South Nam Xe, Northwestern Vietnam. Those rift-related carbonatites have reported ages of 28–44 Ma (biotite K-Ar ages) and 30–32 Ma U-Th-Pb isochron ages [54,55], and hence, may not be considered as a potential source. Instead, we found U-Pb zircon ages that were consistent with periods of basaltic volcanic activities in southern Vietnam. Two distinct zircon populations with U-Pb ages of ca. 6.5 Ma and ca. 1.0 Ma were identified. Although zircon megacrysts may be potentially derived from carbonatitic, metasomatized mantle material and do not primarily relate to alkali basalts, the zircon ages obtained were consistent with bi-episodal eruptive events at Buon Ma Thuot, Dak Lak (5.8–1.67 Ma), Pleiku (4.3–0.8 Ma), and the younger basalt eruptions exposed at Xuan Loc (0.83–0.44 Ma) (Figure 1B) [4]. Zircon megacrysts from North-East Cambodia not only occur geographically close to the megacrysts found in the present study, but also share a similar U-Pb age connected to the younger period of volcanic activity in the region (0.98 ± 0.04 Ma, Cong et al. [10]; 0.88 ± 0.22 Ma to 1.56 ± 0.21 Ma, Piilonen et al. [22]). Obtained zircon U-Pb ages were just slightly older than the eruption events that were dated based on the K-Ar ages for the whole rock [4]. This might be due to the difference in closure temperatures for U-Pb ages of zircon and K-Ar ages from the basalt, that indicated rapid extraction of zircon from its formation source, and the comparably fast uplift by the alkali-basalt eruption events [17,22]. Note, however, that the rim zones of resorption and re-growth (visualized by PL images in Figure 5) gave the very same ages as the zircon core regions within statistical errors. We excluded a potential alteration process to explain this texture. Late-stage, fluid-driven dissolution-reprecipitation typically results in geochronological resetting [56], and geochemical signatures typically indicate a progressive REE fractionation, e.g., a more pronounced negative Eu-anomaly. As both criteria are not verified for the overgrowth zones found in the zircon megacrysts, in this study, we assumed that multiple resorption and re-growth processes took place in the deep sub-crustal "carbonatitic" source magma chambers at temperatures higher than U-Pb closure (>900 °C; Cherniak et al. [57]). Multiple, late-Cenozoic, relatively fast uplifts of basaltic melts induced by tectonic weakening along shear-zones in the tectonically active region, may have either caught up with xenolithic zircon megacrysts from and/or mixed with pre-existent carbonatite-dominated melts generated by CO_2 and/or H_2O metasomatization of the upper mantle material. Inclusions of hematite, baddeleyite, quartz, and feldspar, as well as some ovoid or droplet carbonate inclusions along fractures or along channels, that are associated with fluid inclusions of liquid H_2O and vapor bubbles of CO_2, are considered to be of secondary origin (Figure 3). These phases are well known to occur upon zircon alteration. Particularly, the formation of baddeleyite due to the metasomatic alteration of zircon induced by Ca-bearing fluids has been proven by the experimental works of Lewerentz [58]. Therefore, the occurrences of these phases along fluid inclusions in close proximity to fractures, within tubular channels and/or vesicles are interpreted to result from late stage circulation of oxidative fluids in the ascending alkali-basaltic magma. Finally, weathering processes eroded zircon megacrysts from the alkali-basalt hosts in the Central Highlands and resulted in the deposition and enrichment of the nearby alluvial zircon placers found today.

6. Conclusions

Crystal morphology and internal texture of gem-quality, placer zircon megacrysts of the Central Highlands, Vietnam, are indicative of magmatic origin. The geochemical signatures and primary inclusions of calcite, olivine, and corundum indicate that zircon megacrysts might have crystallized from a carbonatite-dominant melt caused by the low-grade partial melting of metasomatized lithospheric mantle. Zircon resorption and re-growth textural features, observed with PL imaging, indicate an extended residence time in the sub-crustal magma chambers at variable REE-saturation levels, but at temperatures higher than closure of the U-Pb system (>900 °C). Zircon megacrysts were subsequently incorporated into ascending alkali basalts as xenocrysts. Two distinct populations with U-Pb ages at ca. 6.5 Ma and ca. 1.0 Ma that correlate with the eruption of alkali basalt fields in the Central Highlands, were identified. Recent reports of the direct findings of megacrystic zircon in the geographically and genetically related basalts of Ratanakiri, Cambodia [22] strongly supports this close relationship of placer zircon with Cenozoic alkali basalts as a potential host. The basalt magma and/or late stage circulation of carbonate-rich fluids must have resulted in the entrapment of further secondary inclusions like CO_2-H_2O fluids, baddeleyite, quartz, hematite, and feldspar along the cracks and tubular channels.

Author Contributions: Concept of study, V.B.T.S.; methodology data acquisition and treatment, V.B.T.S., C.L., N.N., T.A. and E.B.; formal investigation, V.B.T.S., Y.O, C.L., N.N., and I.K.; compilation and writing of the manuscript, V.B.T.S.; review and editing of manuscript, C.L., Y.O., N.N., I.K., T.A., and E.B.

Acknowledgments: The authors thank Le Thi Thu Huong from Senckenberg museum in Frankfurt, Germany, and Nguyen Thi Minh Thuyet from Hanoi University of Science, Vietnam for their help during fieldwork and sample collection in Vietnam. Special thanks to the local zircon miners in Central Highlands, Vietnam for providing insights into their mining operations. We are indebted to three anonymous reviewers for their critical comments on the manuscript.

References

1. Hoskin, P.W.O.; Schaltegger, J.M. The composition of zircon and igneous and metamorphic petrogenesis. *Rev. Mineral. Geochem.* **2003**, *53*, 27–62. [CrossRef]

2. Hinton, R.W.; Upton, B.G.J. The chemistry of zircon: Variation within and between large crystals from syenite and alkali basalt xenoliths. *Geochim. Cosmochim. Acta* **1991**, *55*, 3287–3302. [CrossRef]

3. Hoang, N.; Han, N.X. Petrochemistry of Quaternary basalts of Xuan Loc area (South Vietnam). In *Geology of Cambodia, Laos, Vietnam*; Geological Survey of Vietnam: Hanoi, Vietnam, 1990; Volume 2, pp. 77–88.

4. Hoang, N.; Flower, M. Petrogenesis of Cenozoic basalts from Vietnam: Implication for origins of a 'Diffuse igneous province'. *J. Petrol.* **1998**, *39*, 369–395. [CrossRef]

5. Qiu, Z.; Yang, J.; Yang, Y.; Yang, S.; Li, C.; Wang, Y.; Lin, W.; Yang, X. Trace element and Hafnium isotopes of Cenozoic basalt-related zircon megacrysts at Muling, Heilongjiang province. Northeast China. *Acta Pet. Sin.* **2007**, *23*, 481–492.

6. Izokh, A.E.; Smirnov, S.Z.; Egorova, V.V.; Anh, T.T.; Kovyazin, S.V.; Phuong, N.T.; Kalinina, V.V. The condition of formation of sapphire and zircon in the areas of alkali-basaltoid volcanism in Central Vietnam. *Russ. Geol. Geophys.* **2010**, *51*, 719–733. [CrossRef]

7. Shigley, J.E.; Laurs, B.M.; Janse, A.J.A.; Elen, S.; Dirlam, D.M. Gem localities of the 2000s. *Gems Gemol.* **2010,** *46*, 188–216. [CrossRef]

8. Yu, Y.; Xu, X.; Chen, X. Genesis of zircon megacrysts in Cenozoic alkali basalts and the heterogeneity of subcontinetal lithospheirc mantle, eastern China. *Mineral. Petrol.* **2010**, *1000*, 75–94. [CrossRef]

9. Chen, T.; Ai, H.; Yang, M.; Zheng, S.; Liu, Y. Brownish red zircon from Muling, China. *Gems Gemol.* **2011**, *47,* 36–41. [CrossRef]

10. Cong, F.; Li, S.Q.; Lin, F.C.; Shi, M.F.; Zhu, H.P.; Siebel, W.; Chen, F. Origin of Zircon Megacrysts from Cenozoic Basalts in Northeastern Cambodia: Evidence from U-Pb Age, Hf-O Isotopes, and Inclusions. *J. Geol.* **2016**, *124*, 221–234. [CrossRef]

11. Sutherland, L.; Graham, I.; Yaxley, G.; Armstrong, R.; Giuliani, G.; Hoskin, P.; Nechaev, V.; Woodhead, J. Major zircon megacryst suites of the Indo-Pacific lithospheric margin (ZIP) and their petrogenetic and regional implications. *Mineral. Petrol.* **2016**, *110*, 399–420. [CrossRef]

12. Zeug, M.; Nasdala, L.; Wanthanachaisaeng, B.; Balmer, W.A.; Corfu, F.; Wildner, M. Blue Zircon from Ratanakiri, Cambodia. *J. Gemmol.* **2018**, *36*, 112–132. [CrossRef]

13. Huong, L.T.T.; Vuong, B.T.S.; Thuyet, N.T.M.; Khoi, N.N.; Somruedee, S.; Bhuwado, W.; Hofmeister, W.;

Tobias, H.; Hauzenbeger, C. Geology, gemological properties and preliminary heat treatment of gem-quality zircon from the Central Highlands of Vietnam. *J. Gemmol.* **2016**, *35*, 308–318. [CrossRef]

14. Sutherland, F.L.; Bosshart, G.; Fanning, C.M.; Hoskin, P.W.O.; Coenraads, R.R. Sapphire crystallization, age and origin, Ban Huai Sai, Laos: Age based on zircon inclusions. *J. Asian Earth Sci.* **2002**, *20*, 841–849. [CrossRef]

15. Sutherland, F.L. Alkaline rocks and gemstones, Australia: A review and synthesis. *Aust. J. Earth Sci.* **1996**, *43*, 323–343. [CrossRef]

16. Belousova, E.A.; Griffin, W.L.; O'Reilly, S.Y.; Fisher, N.I. Igneous zircon: Trace element composition as an indicator of source rock type. *Contrib. Mineral. Petrol.* **2002**, *143*, 602–622. [CrossRef]

17. Garnier, V.; Ohnenstetter, D.; Giuliani, G.; Fallick, A.E.; Phan Trong, T.; Quang, V.H.; Van, L.P.; Schwarz, D. Basalt petrology, zircon ages and sapphire genesis from Dak Nong, southern Vietnam. *Mineral. Mag.* **2005**, *69*, 21–38. [CrossRef]

18. Sutherland, F.L.; Meffre, S. Zircon megacryst ages and chemistry, from a placer, Dunedin volcanic area, eastern Otago, New Zealand. *N. Z. J. Geol. Geophys.* **2009**, *52*, 185–194. [CrossRef]

19. Griffin, W.L.; Pearson, N.J.; Belousova, E.; Jackson, S.E.; Van Achterbergh, E.; O'Reilly, S.R.; Shee, S.Y. The Hf isotope composition of cratonic mantle: LAM-MC-ICPMS analysis of zircon megacrysts in kimberlites. *Geochim. Cosmochim. Acta* **2000**, *64*, 133–147. [CrossRef]

20. Visonà, D.; Caironi, V.; Carraro, A.; Dallai, L.; Fioretti, A.M.; Fanning, M. Zircon megacrysts from basalts of the Venetian Volcanic Province (NE Italy): U-Pb ages, oxygen isotopes and REE data. *Lithos* **2007**, *94*, 168–180. [CrossRef]

21. Upton, B.G.J.; Hinton, R.W.; Aspen, P.; Finch, A.; Valley, J.W. Megacrysts and associated xenoliths: Evidence for migration of geochemically enriched melts in the upper mantle beneath Scotland. *J. Petrol.* **1999**, *40*, 935–956. [CrossRef]

22. Piilonen, P.C.; Sutherland, F.L.; Danisik, M.; Poirier, G.; Valley, J.W.; Rowe, R. Zircon xenocryst from Cenozoic Alkaline Basalts of the Ratanakiri Volcanic Province (Cambodia), Southeast Asia—Trace Element Geochemistry, O-Hf Isotopic Composition, U-Pb and (U-Th)/He Geochronology—Revelations into the Underlyi Lithospheric Mantle. *Minerals* **2018**, *8*, 556. [CrossRef]

23. Whitney, D.L.; Evans, B.W. Abbreviations for names of rock-forming minerals. *Am. Mineral.* **2010**, *95*, 185–187. [CrossRef]

24. Osanai, Y.; Nakano, N.; Owada, M.; Nam, T.N.; Miyamoto, T.; Minh, N.T.; Nam, N.V.; Tri, T.V. Collision zone metamorphism in Vietnam and adjacent South-eastern Asia: Proposition for Trans Vietnam Orogenic Belt. *J. Mineral. Petrol. Sci.* **2008**, *103*, 226–241. [CrossRef]

25. Hutchison, C.S. Geological Evolution of Southeast Asia. *Oxf. Monogr. Geol. Geophys.* **1989**, *13*, 368.

26. Barr, S.M.; Macdonald, A.S. Geochemistry and geochronology of late Cenozoic basalts of Southeast Asia. *Bull. Geol. Soc. Am.* **1981**, *92*, 1069–1142. [CrossRef]

27. Rangin, C.; Huchon, P.; Le Pichon, X.; Bellon, H.; Lepvrier, C.; Roques, D.; Hoe, N.D.; Quynh, P.V. Cenozoic deformation of central and south. *Tectonophysics* **1995**, *251*, 179–196. [CrossRef]

28. Hoang, N.; Flower, M.F.J.; Carlson, R.W. Major, trace element, and isotopic compositions of Vietnamese basalts: Interaction of enriched mobile asthenosphere with the continental lithosphere. *Geochim. Cosmochim. Acta* **1996**, *60*, 4329–4351. [CrossRef]

29. Lenz, C.; Nasdala, L.; Talla, D.; Hauzenberger, C.; Seitz, R.; Kolitsch, U. Laser-induced REE[3+] photoluminescence of selected accessory minerals—An "advantageous artefact" in Raman spectroscopy. *Chem. Geol.* **2015**, *415*, 1–16. [CrossRef]

30. Adachi, T.; Osanai, Y.; Nakano, N.; Owada, M. La-ICP-MS U-Pb zircon and FE-EPMA U-Th-Pb monazite dating of pelitic granulites from the Mt. Ukidake area, Sefuri Mountains, northern Kyushu. *J. Geol. Soc. Jpn.* **2012**, *118*, 29–52.

31. Black, L.P.; Kamo, S.L.; Allen, C.M.; Aleinikoff, J.N.; Davis, D.W.; Korsh, R.J.; Foudoulis, C. TEMORA1: A new zircon standard for Phanerozoic U-Pb geochronology. *Chem. Geol.* **2003**, *200*, 155–170. [CrossRef]

32. Paces, J.P.; Miller, J.D.J. U-Pb ages of Duluth Complex and related mafic intrusions, northeastern Minnesota: Geochronological insights to physical, petrogenetic, paleomagnetic, and tectonomagmatic processes associated with the 1.1 Ga midcontinent rift system. *J. Geophys. Res.* **1993**, *98*, 13997–14013. [CrossRef]

33. Griffin, W.L.; Powell, W.J.; Pearson, N.J.; O'Reilly, S.Y. GLITTER: Data reduction software for laser ablation ICP-MS. *Mineral. Assoc. Can. Short Course* **2008**, *40*, 308–311.

34. Ludwig, K.R. User's Mnual for Isoplot 3.70: A geochronological toolkit for Microsoft Excel. *Berkeley Geochronl. Cent. Spec. Publ.* **2008**, *4*, 1–77.

35. Nakano, N.; Osanai, Y.; Adachi, T. Major and trace element zoning of euhedral garnet in high-grade (>900 °C) mafic granulite from the Song Ma Suture zone, northern Vietnam. *J. Mineral. Petrol. Sci.* **2010**, *105*, 268–273. [CrossRef]

36. Elhlou, S.; Belousova, E.; Griffin, W.L.; Pearson, N.J.; O'Reilly, S.Y. Trace element and isotopic composition of GJ-red zircon standard by laser ablation. *Geochim. Cosmochim. Acta Suppl.* **2006**, *70*, A158. [CrossRef]

37. Jackson, S.E.; Longerich, H.P.; Dunning, G.R.; Freyer, B.J. The application of laser-ablation microprobe; inductively coupled plasma-mass spectrometry (LAM-ICP-MS) to in situ trace-element determinations in minerals. *Can. Mineral.* **1992**, *30*, 1049–1064.

38. Andersen, T. Correction of common lead in U-Pb analyses that do not report ^{204}Pb. *Chem. Geol.* **2002**, *192*, 59–76. [CrossRef]

39. Stacey, J.S.; Kramers, J.D. Approximation of terrestrial lead isotope evolution by a two-stage model. *Earth Planet. Sci. Lett.* **1975**, *26*, 207–221. [CrossRef]

40. Lenz, C.; Nasdala, L. A photoluminescence study of REE^{3+} emissions in radiation-damaged zircon. *Am. Mineral.* **2015**, *100*, 1123–1133. [CrossRef]

41. Shnukov, S.E.; Andreev, A.V.; Savenok, S.P. *Admixture Elements in Zircons and Apatites: A Tool for Provenance Studies of Terrigenous Sedimentary Rocks*; European Union of Geosciences (EUG 9): Strasbourg, France, 1997; Volume 65.

42. Sun, S.S.; McDonough, W.F. Chemical and isotopic systematics of oceanic basalts: Implications for mantle composition and processes. *Geol. Soc. Lond.* **1989**, *42*, 313–345. [CrossRef]

43. Snyder, G.A.; Taylor, L.A.; Crozaz, G. Rare earth element selenochemistry of immiscible liquids and zircon at Apollo 14: An iron probe study of evolved rocks on the Moon. *Geochim. Cosmochim. Acta* **1993**, *57*, 1143–1149. [CrossRef]

44. Li, X.H.; Liang, X.R.; Sun, M.; Liu, Y.; Tu, X.L. Geochronology and geochemistry of single-grain zircons: Simultaneous in-situ analysis of U-Pb age and trace elements by LAM-ICP-MS. *Eur. J. Mineral.* **2000**, *12*, 1015–1024. [CrossRef]

45. Borghini, G.; Fumagalli, P.; Rampone, E. The stability of plagioclase in the upper mantle: Subsolidus experiments on fertile and depleted lherzolite. *J. Petrol.* **2010**, *51*, 229–254. [CrossRef]

46. Belousova, E.A.; Griffin, W.L.; Pearson, N.J. Trace element composition and cathodoluminescence properties of southern African kimberlitic zircon. *Mineral. Mag.* **1998**, *62*, 355–366. [CrossRef]

47. Hoskin, P.W.O. Minor and trace element analysis of natural zircon ($ZrSiO_4$) by SIMS and laser ablation ICPMS: A consideration and comparison of two broadly competitive techniques. *J. Trace Microprobe Tech.* **1998**, *16*, 301–326.

48. Pupin, J.P. Granite genesis related to geodynamics from Hf-Y in zircon. *Trans. R. Soc. Edinb. Earth Sci.* **2000**, *91*, 245–256. [CrossRef]

49. Saava, E.V.; Belyatsky, B.V.; Antonov, A.B. Carbonatitic Zircon-Myth or Reality: Mineralogical-Geochemical Analyses. Available online: http://alkaline09.narod.ru/abstracts/Savva_Belyatsky.htm (accessed on 30 December 2018).

50. Campbell, L.S.; Compston, W.; Sircombe, K.N.; Wilkinson, C.C. Zircon from the East Orebody of the Bayan Obo Fe-Nb-REE deposit, China, and SHRIMP ages for carbonatite-related magmatism and REE mineralization events. *Contrib. Mineral. Petrol.* **2014**, *168*, 1041. [CrossRef]

51. Foley, S.F.; Yaxley, G.M.; Rosenthal, A.; Buhre, S.; Kiseeva, E.S.; Rapp, R.P.; Jacob, D.E. The composition of near-solidus melts of peridotite in the presence of CO_2 and H_2O between 40 and 60 kbar. *Lithos* **2009**, *112*, 274–283. [CrossRef]

52. Le Bas, M.J. Carbonatite magmas. *Mineral. Mag.* **1981**, *44*, 133–140. [CrossRef]

53. Baldwin, L.C.; Tomaschek, F.; Ballhaus, C.; Gerdes, A.; Fonseca, R.O.C.; Wirth, R.; Geisler, T.; Nagel, T. Petrogeneis of alkaline basalt-hosted sapphires megacrysts. Petrological and geochemical investigations of in situ sapphire occurrences from the Siebengebirge volcanic field Western Germany. *Contrib. Mineral. Petrol.* **2017**, *172*, 43. [CrossRef]

54. Chi, N.T.; Flower Martin, J.F.; Hung, D.T. Carbonatites in Phong Tho, Lai Chau Province, north–west Vietnam: Their petrogenesis and relationship with Cenozoic potassic alkaline magmatism. In Proceedings of the 33rd International Geological Congress, Oslo, Norway, 6–14 August 2008; pp. 1–45.

55. Thi, T.N.; Wada, H.; Ishikawa, T.; Shimano, T. Geochemistry and petrogenesis of carbonatites from South Nam Xe, Lai Chau area, northwest Vietnam. *Mineral. Petrol.* **2014**, *108*, 371–390.

56. Geisler, T.; Schaltegger, U.; Tomaschek, F. Re-equilibration of zircon in aqueous fluids and melts. *Elements* **2007**, *3*, 43–50. [CrossRef]

57. Cherniak, D.J.; Watson, E.B. Pb diffusion in zircon. *Chem. Geol.* **2000**, *172*, 5–24. [CrossRef]

58. Lewerentz, A. Experimental Zircon Alteration and Baddeleyite Formation in Silica Saturated Systems: Implications for Dating Hydrothermal Events. Master's Thesis, Lund University, Lund, Sweden, 2011.

PERMISSIONS

LIST OF CONTRIBUTORS

Chawalit Chankhantha and Andy H. Shen
Gemmological Institute, China University of Geosciences, Wuhan 430074, China

Habib Ur Rehman
Gemmological Institute, China University of Geosciences, Wuhan 430074, China
Gems & Jewellery Centre of Excellence, University of Engineering & Technology Peshawar, Peshawar 25120, Pakistan

Gerhard Martens
Private Scientific Consultant, D-24558 Henstedt-Ulzburg, Germany

Ying Lai Tsai
Department of Jewelry Technology, Dahan Institute of Technology, Hualian, Taiwan

Pinit Kidkhunthod
Synchrotron Light Research Institute, 111 University Avenue, Muang, Nakhon Ratchasima 30000, Thailand

Terry Moxon
55 Common lane, Auckley, Doncaster DN9 3HX, UK

Galina Palyanova
Sobolev Institute of Geology and Mineralogy, Siberian Branch of Russian Academy of Sciences, 630090 Novosibirsk, Russia
Department of Geology and Geophysics, Novosibirsk State University, Pirogova str., 2, 630090 Novosibirsk, Russia

Maria I. Filina and Natalia N. Kononkova
Vernadsky Institute of Geochemistry and Analytical Chemistry Russian Academy of Sciences (GEOKHI RAS), Kosygin str. 19, 119991 Moscow, Russia

Roman Botcharnikov and Wolfgang Hofmeister
Institut für Geowissenschaften, Johannes Gutenberg Universität Mainz, J.-J.-Becher-Weg 21, 55128 Mainz, Germany

Elena S. Sorokina
Vernadsky Institute of Geochemistry and Analytical Chemistry Russian Academy of Sciences (GEOKHI RAS), Kosygin str. 19, 119991 Moscow, Russia
Institut für Geowissenschaften, Johannes Gutenberg Universität Mainz, J.-J.-Becher-Weg 21, 55128 Mainz, Germany

Stefanos Karampelas
Bahrain Institute for Pearls & Gemstones (DANAT), WTC East Tower, Bahrain

Mikhail A. Rassomakhin
Institute of Mineralogy SU FRC MiG UB RAS, 456317 Miass, Chelyabinsk Region, Russia
Ilmen State Reserve SU FRC MiG UB RAS, 456317 Miass, Chelyabinsk Region, Russia

Anatoly G. Nikolaev
Department of mineralogy and lithology, Institute of Geology and Petroleum Technologies, Kazan Federal University, 420008 Kazan, Russia

Jasper Berndt
Institut für Mineralogie, Westfälische Wilhelms Universität Münster, Corrensstrasse 24, 48149 Münster, Germany

Fatima Mohamed, Hasan Abdulla, Fatema Almahmood, Latifa Flamarzi, Supharart Sangsawong and Abeer Alalawi
Bahrain Institute for Pearls & Gemstones (DANAT), WTC East Tower, Bahrain

Neville J. Curtis and Allan Pring
South Australian Museum, North Terrace, Adelaide, SA 5000, Australia
College of Science and Engineering, Flinders University, Sturt Rd, Bedford Park, SA 5042, Australia

Jason R. Gascooke and Martin R. Johnston
College of Science and Engineering, Flinders University, Sturt Rd, Bedford Park, SA 5042, Australia

Le Thi-Thu Huong, Christoph A. Hauzenberger and Kurt Krenn
NAWI Graz Geocentre, University of Graz, 8010 Graz, Austria

Dorothea S. Macholdt, Ulrike Weis, Brigitte Stoll and Klaus Peter Jochum
Climate Geochemistry Department, Max Planck Institute for Chemistry, 55128 Mainz, Germany

Michael W. Förster
Department of Earth and Planetary Sciences, Macquarie University, Sydney NSW 2109, Australia

Laura M. Otter
Climate Geochemistry Department, Max Planck Institute for Chemistry, 55128 Mainz, Germany
Department of Earth and Planetary Sciences, Macquarie University, Sydney NSW 2109, Australia

Olivier Alard
Department of Earth and Planetary Sciences, Macquarie University, Sydney NSW 2109, Australia
Géosciences Montpellier, UMR 5243, CNRS & Université Montpellier, 34095 Montpellier, France

Kandy K. Wang, Ian T. Graham, Angela Lay and Stephen J. Harris
PANGEA Research Centre, School of Biological, Earth and Environmental Sciences, University of NSW, 2052 Sydney, Australia

Laure Martin
Centre for Microscopy Characterisation and Analysis, The University of Western Australia, 6009 Perth, Australia

Panagiotis Voudouris
Faculty of Geology and Geoenvironment, National and Kapodistrian University of Athens, 157 84 Athens, Greece

Gaston Giuliani
Université Paul Sabatier, GET/IRD et Université de Lorraine, CRPG/CNRS, 15 rue Notre-Dame des Pauvres, BP 20, 54501 Vandoeuvre cedex, France

Anthony Fallick
Isotope Geosciences Unit, S.U.E.R.C., Rankine Avenue, East Kilbride, Glasgow G75 0QF, UK

Alexandre Tarantola, Aurélien Eglinger, Kimberly Trebus, Marie Bitte, Marius Etienne and Chantal Peiffert
GeoRessources, Faculté des Sciences et Technologies, Université de Lorraine, CNRS, F-54506 Vandoeuvre-lès-Nancy, France

Constantinos Mavrogonatos
Department of Geology & Geoenvironment, National and Kapodistrian University of Athens, 15784 Athens, Greece

Christophe Scheffer
GeoRessources, Faculté des Sciences et Technologies, Université de Lorraine, CNRS, F-54506 Vandoeuvre-lès-Nancy, France
Département de Géologie et de Génie Géologique, Université Laval, Québec, QC G1V 0A6, Canada

Benjamin Rondeau
Laboratoire de Planétologie et Géodynamique, Université de Nantes, CNRS UMR 6112, 44322 Nantes, France

Ian Graham
PANGEA Research Centre, School of Biological, Earth and Environmental Sciences, University of New South Wales, Sydney, NSW 2052 Australia

Lee A. Groat
Department of Earth, Ocean and Atmospheric Sciences, University of British Columbia, Vancouver, BC V6T 1Z4, Canada

Dan Marshall
Department of Earth Sciences, Simon Fraser University, Burnaby, BC V5A 1S6, Canada

Anthony E. Fallick
Isotope Geosciences Unit, S.U.E.R.C., Rankine Avenue, East Kilbride, Glasgow G75 0QF, Scotland, UK

Yannick Branquet
Institut des Sciences de la Terre d'Orléans (ISTO), UMR 7327-CNRS/Université d'Orléans/BRGM, 45071 Orléans, France

Vuong Bui Thi Sinh
Graduate School of Integrated Sciences for Global Society, Kyushu University, 744 Motooka, Nishi-ku, Fukuoka 819-0395, Japan

Yasuhito Osanai, Nobuhiko Nakano, Tatsuro Adachi and Ippei Kitano
Division of Earth Sciences, Faculty of Social and Cultural Studies, Kyushu University, 744 Motooka, Nishi-ku, Fukuoka 819-0395, Japan

Christoph Lenz
Institut für Mineralogie und Kristallographie, Universität Wien, 1090 Wien, Austria
Australian Research Council (ARC) Centre of Excellence for Core to Crust Fluid Systems (CCFS) and National Key Centre for Geochemical Evolution and Metallogeny of Continents (GEMOC), Department of Earth and Planetary Sciences, Macquarie University, Sydney, NSW 2109, Australia

Elena Belousova
Australian Research Council (ARC) Centre of Excellence for Core to Crust Fluid Systems (CCFS) and National Key Centre for Geochemical Evolution and Metallogeny of Continents (GEMOC), Department of Earth and Planetary Sciences, Macquarie University, Sydney, NSW 2109, Australia

Index

www.ingramcontent.com/pod-product-compliance
Lightning Source LLC
Chambersburg PA
CBHW080506200326

41458CB00012B/4105